T0191977

Studies in Computational Intelligence 467

Editor-in-Chief

Prof. Janusz Kacprzyk
Systems Research Institute
Polish Academy of Sciences
ul. Newelska 6
01-447 Warsaw
Poland
E-mail: kacprzyk@ibspan.waw.pl

For further volumes:
http://www.springer.com/series/7092

Robert Bembenik, Łukasz Skonieczny,
Henryk Rybiński, Marzena Kryszkiewicz,
and Marek Niezgódka (Eds.)

Intelligent Tools for Building a Scientific Information Platform

Advanced Architectures and Solutions

 Springer

Editors
Dr. Robert Bembenik
Institute of Computer Science
Faculty of Electronics and Information
Technology
Warsaw University of Technology
Warsaw
Poland

Dr. Łukasz Skonieczny
Institute of Computer Science
Faculty of Electronics and Information
Technology
Warsaw University of Technology
Warsaw
Poland

Prof. Henryk Rybiński
Institute of Computer Science
Faculty of Electronics and Information
Technology
Warsaw University of Technology
Warsaw
Poland

Prof. Marzena Kryszkiewicz
Institute of Computer Science
Faculty of Electronics and Information
Technology
Warsaw University of Technology
Warsaw
Poland

Prof. Marek Niezgódka
Interdisciplinary Centre for Mathematical and
Computational Modelling (ICM)
University of Warsaw
Warsaw
Poland

ISSN 1860-949X
ISBN 978-3-642-43515-7
DOI 10.1007/978-3-642-35647-6
Springer Heidelberg New York Dordrecht London

e-ISSN 1860-9503
ISBN 978-3-642-35647-6 (eBook)

Printed on acid-free paper

Springer is part of Springer Science+Business Media (www.springer.com)

Preface

This book is a selection of results obtained within two years of research performed under SYNAT – a nation-wide scientific project aiming at creating an infrastructure for scientific content storage and sharing for academia, education and open knowledge society in Poland. The SYNAT project is financed by the National Centre for Research and Development (NCBiR) in Poland. A network of 16 academic and scientific partners have committed to implement the SYNAT's objectives in the form of a universal open knowledge infrastructure for the information society in Poland. Beyond the system development, a comprehensive portfolio of research problems is addressed by the partners.

The program, initiated in 2010, has been scheduled for the period of three years. In the view of the limited implementation time, the primary goal of the program consists in meeting the challenges of global digital information revolution, especially in the context of scientific information. Novel methods, algorithms and their practical applications are the results of the SYNAT activities. A broad range of new functionalities have been introduced and implemented upon validating their real practical features, such as scalability and interoperability. The result of two year research within the SYNAT project comprises a number of works in the areas of, *inter alia*, artificial intelligence, knowledge discovery and data mining, information retrieval and natural language processing, addressing the problems of implementing intelligent tools for building a scientific information platform.

This book is inspired by the SYNAT 2012 Workshop held in Warsaw, which was a forum for discussions on various solutions based on intelligent tools aiming at significant improvement of the quality of the planned scientific information services. The research issues important for building the platform were subject of the workshop. The book summarizes the workshop's results as well as continues and extends the ideas presented in "Intelligent Tools for Building a Scientific Information Platform" published by Springer in 2012.

The papers included in this volume cover the following topics: Information Retrieval, Repository Systems, Text Processing, Ontology-based Systems, Text Mining, Multimedia Data Processing and Advanced Software Engineering. We will now outline the contents of the chapters.

Chapter I, "Information Retrieval", deals with issues and problems related to automatic acquisition of interesting information from such sources as the internet, text data collections or scanned pages as well as the need to filter the acquired information.

- **Bruno Jacobfeuerborn and Mieczysław Muraszkiewicz** ("Media, Information Overload, and Information Science") argue that although media and ICT strengthen human's propensity to insane intake of information and reinforce bad habits leading to information overconsumption, it is possible to escape from the trap of information overload thanks to administering an information diet based on digital literacy. The paper further argues that new research should be undertaken within information science to address the challenges caused by the internet and the specificity of human perception exposed to the cognitive overload, untrustworthiness, and fuzziness.
- **Tomasz Kogut, Dominik Ryżko, and Karol Gałązka** ("Information Retrieval from Heterogeneous Knowledge Sources based on Multi-Agent Systems") present a multi-agent approach for scientific information retrieval from heterogeneous sources. An experimental platform has been built with the use of Scala language. A case study is shown demonstrating the results with the use of Google and Bing search engines and DBLP database.
- **Marcin Deptuła, Julian Szymański and Henryk Krawczyk** ("Interactive Information Search in Text Data Collections") present new ideas for information retrieving that can be applied to big text repositories as well as describe the infrastructure of a system created to implement and test those ideas. The implemented system differs from today's standard search engine by introducing a process of interactive search driven by data clustering. The authors present the basic algorithms behind their system and measures they used for results evaluation. The achieved results indicate that the proposed method can be useful for improvement of classical approaches based on keywords.
- **Rafał Hazan and Piotr Andruszkiewicz** ("Home Pages Identification and Information Extraction in Researcher Profiling") discuss the problem of creating a structured database describing researchers. To this end, home pages can be used as an information source. As the first step of this task, home pages are searched and identified with the usage of a classifier. Then, the information extraction process is performed to enrich researchers' profiles. The authors proposed an algorithm for extracting phone numbers, fax numbers and e-mails based on generalized sequential patterns. Extracted information is stored in a structured database.
- **Maciej Durzewski, Andrzej Jankowski, Łukasz Szydełko, and Ewa Wiszowata** ("On Digitalizing the Geographical Dictionary of Polish Kingdom Published in 1880") study the problem of digitalizing the Geographical Dictionary of Polish Kingdom published in 1880. They address two kinds of problems: technical problems of converting scanned pages of that dictionary to reliable textual data and theoretical challenges of organizing that data into knowledge. Their solution for organizing data into knowledge is based on defining an appropriate ontology and rules for converting textual data to ontological knowledge. The authors describe methods of extracting simple information like keywords and bootstrapping it into higher level relations and discuss their possible uses.

Chapter II, "Repository Systems", deals with knowledge-base and repository systems allowing for gathering large amounts of information, which are oftentimes augmented with additional functionalities, e.g. semantic data sharing, long-term preservation services, or a built-in modular environment providing an infrastructure of components.

- **Michał Krysiński, Marcin Krystek, Cezary Mazurek, Juliusz Pukacki, Paweł Spychała, Maciej Stroiński, and Jan Węglarz** ("Semantic Data Sharing and Presentation in Integrated Knowledge System") present their concept of creating a convenient semantic environment for ordinary users as well as maintaining rich functionality available for professionals. They present an adaptation of a classic three-layer architecture for the semantic data sharing and presentation purpose. Also, the authors pay special attention to standards and cooperation issues, which makes their system available to other tools implementing semantic data sharing concepts and keeps it open for Linked Open Data environment integration.

- **Piotr Jędrzejczak, Michał Kozak, Cezary Mazurek, Tomasz Parkoła, Szymon Pietrzak, Maciej Stroiński, and Jan Węglarz** ("Long-term Preservation Services as a Key Element of the Digital Libraries Infrastructure") present long-term preservation services developed in Poznan Supercomputing and Networking Center as a part of the SYNAT project. The proposed solution is an extendible framework allowing data preservation, monitoring, migration, conversion and advanced delivery. Its capability to share data manipulation services and used Semantic Web solutions, like the OWL–S standard and the OWLIM–Lite semantic repository, create a potential for broad applicability of the publicly available services and therefore fostering and improving digitization related activities. METS and PREMIS formats accompanied by FITS, DROID and FFMpeg tools are the core elements of the WRDZ services, making them interoperable and innovative contribution to the digital libraries infrastructure in Poland.

- **Jakub Koperwas, Łukasz Skonieczny, Henryk Rybiński, Wacław Struk** ("Development of a University Knowledge Base") summarize deployment and a one-year-exploitation of repository platform for knowledge and scientific resources. The platform was developed under the SYNAT project and designed for the purpose of Warsaw University of Technology. First, the authors present functionalities of the platform with accordance to the SYNAT objectives and the requirements of Warsaw University of Technology. Those functionalities are then compared to other well established systems, like dLibra, Fedora Commons Repository, and DSpace. The architecture of the platform and the aspect of integration with other systems are also discussed. Finally, it is discussed how the platform can be reused for deployment for the purposes of any other university with respect to the specific resources it produces and archives.

- **Tomasz Rosiek, Wojtek Sylwestrzak, Aleksander Nowiński, and Marek Niezgódka** ("Infrastructural Approach to Modern Digital Library and Repository Management Systems") present a modern approach to DLM Systems, where a modular environment provides an infrastructure of components to build DMLS applications upon. A case study of an open SYNAT Software Platform is presented, together with its sample applications, and key benefits of the approach are discussed.

Chapter III, "Text Processing", is concerned primarily with methods and algorithms enabling efficient text processing, storage, terminology extraction, digitization of historical documents, text classification, tagging and recognition of proper names.

- **Piotr Jan Dendek, Mariusz Wojewódzki, and Łukasz Bolikowski** ("Author Disambiguation in the YADDA2 Software Platform") discuss the SYNAT platform powered by the YADDA2 architecture that has been extended with the Author Disambiguation Framework and the Query Framework. The former framework clusters occurrences of contributor names into identities of authors, the latter answers queries about authors and documents written by them. The authors present an outline of disambiguation algorithms, implementation of query framework, integration into the SYNAT platform and performance evaluation of the solution.
- **Mateusz Fedoryszak, Łukasz Bolikowski, Dominika Tkaczyk, and Krzyś Wojciechowski** ("Methodology for Evaluating Citation Parsing and Matching") discuss the topic of bibliographic references between scholarly publications containing valuable information for researchers and developers dealing with digital repositories. Bibliographic references serve as indicators of topical similarity between linked texts and impact of a referenced document as well as improve navigation in user interfaces of digital libraries. Consequently, several approaches to extraction, parsing and resolving references have been proposed. The authors develop a methodology for evaluating parsing and matching algorithms and choosing the most appropriate one for a document collection at hand. They apply the methodology for evaluating reference parsing and matching module of the YADDA2 software platform.
- **Adam Kawa, Łukasz Bolikowski, Artur Czeczko, Piotr Jan Dendek, and Dominika Tkaczyk** ("Data Model for Analysis of Scholarly Documents in the MapReduce Paradigm") are in possession of a large collection of scholarly documents that are stored and processed using the MapReduce paradigm. One of the main challenges is to design a simple, but effective data model that fits various data access patterns and allows performing diverse analyses efficiently. The authors describe the organization of their data and explain how this data is accessed and processed by open-source tools from Apache Hadoop Ecosystem.
- **Małgorzata Marciniak and Agnieszka Mykowiecka** ("Terminology Extraction from Domain Texts in Polish") present a method for extracting terminology from Polish texts which consists of two steps. The first one identifies candidates for terms, and is supported by linguistic knowledge - a shallow grammar used for extracted phrases is given. The second step is based on statistics, consisting in ranking and filtering candidates for domain terms with the help of a C-value method, and phrases extracted from general Polish texts. The presented approach is sensitive to finding terminology also expressed as sub-phrases. The authors applied the method to economics texts, describe the results of the experiment, evaluate and discuss the results.
- **Adam Dudczak, Miłosz Kmieciak, Cezary Mazurek, Maciej Stroiński, Marcin Werla, and Jan Węglarz** ("Improving the Workflow for Creation of Textual Versions of Polish Historical Documents") describe improvements which can be included in digitization workflow to increase the number and enhance the quality of full text representations of historical documents. This kind of documents is well represented in Polish digital libraries and can offer interesting opportunities for

digital humanities researchers. The proposed solution focuses on changing existing approach to OCR and simplifying the manual process of text correction through crowdsourcing.

- **Karol Kurach, Krzysztof Pawłowski, Łukasz Romaszko, Marcin Tatjewski, Andrzej Janusz, and Hung Son Nguyen** ("Multi-label Classification of Biomedical Articles") investigate a special case of classification problem, called multi-label learning, where each instance is associated with a set of target labels. Multi-label classification is one of the most important issues in semantic indexing and text categorization systems. Most of multi-label classification methods are based on combination of binary classifiers, which are trained separately for each label. The authors concentrate on the application of ensemble technique to multi-label classification problem. They present the most recent ensemble methods for both the binary classifier training phase as well as the combination learning phase.

- **Adam Radziszewski** ("A Tiered CRF Tagger for Polish") presents a new approach to morphosyntactic tagging of Polish by bringing together Conditional Random Fields and tiered tagging. The proposal allows taking advantage of a rich set of morphological features, which resort to an external morphological analyser. Evaluation of the tagger shows significant improvement in tagging accuracy on two state-of-the-art taggers, namely PANTERA and WMBT.

- **Michał Marcińczuk, Jan Kocoń, and Maciej Janicki** ("Liner2 — a Customizable Framework for Proper Names Recognition for Polish") present a customizable and open-source framework for proper names recognition called Liner2. The framework consists of several universal methods for sequence chunking which include: dictionary look-up, pattern matching and statistical processing. The statistical processing is performed using Conditional Random Fields and a rich set of features including morphological, lexical and semantic information.

Chapter IV, "Ontology-based Systems", is devoted to systems that utilize ontologies or tools from the ontology field in their design as well as system and lexical ontologies themselves.

- **Krzysztof Goczyła, Aleksander Waloszek, Wojciech Waloszek, and Teresa Zawadzka** ("Theoretical and Architectural Framework for Contextual Modular Knowledge Bases") present an approach aimed at building modularized knowledge bases in a systematic, context-aware way. The authors focus on logical modeling of such knowledge bases, including an underlying SIM metamodel. The architecture of a comprehensive set of tools for knowledge-base systems engineering is presented. The tools enable an engineer to design, create and edit a knowledge base schema according to a novel context approach. It is explained how a knowledge base built according to SIM (Structured-Interpretation Model) paradigm is processed by a prototypical reasoner Conglo-S, which is a custom version of widely known Pellet reasoner extended with support for modules of ontologies called tarsets. The user interface of the system is a plug-in to Protégé ontology editor, which is a standard tool for development of Semantic Web ontologies. Possible applications of the presented framework to development of knowledge bases for culture heritage and scientific information dissemination are also discussed.

- **Janusz Granat, Edward Klimasara, Anna Mościcka, Sylwia Paczuska, Mariusz Pajer, and Andrzej P. Wierzbicki** ("Hermeneutic Cognitive Process and its Computerized Support") discuss the role of tacit knowledge (including emotive and intuitive knowledge) in cognitive processes, recall an evolutionary theory of intuition, academic and organizational knowledge creation processes, describe diverse cognitive processes in a research group, but especially concentrate on individual hermeneutic processes and processes of knowledge exchange while distinguishing their hermeneutic and semantic role. The functionality and main features of the PrOnto system are shortly recalled, and the relation of changes in individualized ontological profile of the user to her/his hermeneutical perspective is postulated. An interpretation of such changes for both hermeneutic and semantic aspects of knowledge exchange is presented. Conclusions and future research problems are outlined.
- **Anna Wróblewska, Grzegorz Protaziuk, Robert Bembenik, and Teresa Podsiadły-Marczykowska** ("Associations Between Texts and Ontology") present a model of lexical layer that establishes relations between terms (words or phrases) and entities of a given ontology. Moreover, the layer contains representations not only of terms but also of their contexts of usage. It also provides associations with commonly used lexical knowledge resources, such as WordNet, Wikipedia and DBPedia.
- **Anna Wróblewska, Teresa Podsiadły-Marczykowska, Robert Bembenik, Henryk Rybiński, and Grzegorz Protaziuk** ("SYNAT System Ontology: Design Patterns Applied to Modeling of Scientific Community, Preliminary Model Evaluation") present an extended version of the SYNAT system ontology, used design patterns, modeling choices and preliminary evaluation of the model. The SYNAT system ontology was designed to define semantic scope of the SYNAT platform. It covers concepts related to scientific community and its activities i.e.: people in science and their activities, scientific and science-related documents, academic and non-academic organizations, scientific events and data resources, geographic notions necessary to characterize facts about science as well as classification of scientific topics. In its current version, the SYNAT system ontology counts 472 classes and 296 properties, its consistency was verified using Pellet and HermiT reasoners.

Chapter V, "Text Mining", deals with the classical problems of mining the textual data like classification, clustering and extraction, applied to scientific information sources.

- **Michał Łukasik, Tomasz Kuśmierczyk, Łukasz Bolikowski, and Hung Son Nguyen** ("Hierarchical, Multi-label Classification of Scholarly Publications: Modifications of ML-KNN Algorithm") present a modification of the well-established ML-KNN algorithm and show its superiority over original algorithm on the example of publications classified with Mathematics Subject Classification System.
- **Tomasz Kuśmierczyk, Michał Łukasik, Łukasz Bolikowski, and Hung Son Nguyen** ("Comparing Hierarchical Mathematical Document Clustering Against the Mathematics Subject Classification Tree") posit that the Mathematical Subject Classification codes hierarchy is highly correlated with content of publications. Following this assumption they try to reconstruct the MSC tree basing on theirs publications corpora and compare the results to the original hierarchy.

- **Sinh Hoa Nguyen, Wojciech Świeboda, and Grzegorz Jaśkiewicz** ("Semantic Evaluation of Search Result Clustering Methods") investigate the problem of quality analysis of the Tolerance Rough Set Model extension which they introduced in previous papers. They propose some semantic evaluation measures to estimate the quality of the proposed search clustering methods. They illustrate the proposed evaluation method on the base of the Medical Subject Headings (MeSH) thesaurus and compare different clustering techniques over the commonly accessed biomedical research articles from PubMed Central (PMC) portal.
- **Piotr Andruszkiewicz and Beata Nachyła** ("Automatic Extraction of Profiles from Web Pages") describe the process of information extraction on the example of researchers' home pages. For this reason they applied SVM, CRF, and MLN models. Performed analysis concerns texts in English language.

Chapter VI, "Multimedia Data Processing", is devoted to data mining approaches to processing multimedia data like music, spoken language, images and video.

- **Andrzej Dziech, Andrzej Glowacz, Jacek Wszołek, Sebastian Ernst, and Michał Pawłowski** ("A Distributed Architecture for Multimedia File Storage, Analysis and Processing") present a prototype of a distributed storage, analysis and processing system for multimedia (video, audio, image) content based on the Apache Hadoop platform. The core of the system is a distributed file system with advanced topology, replication and balancing capabilities. The article evaluates several important aspects of the proposed distributed repository, such as security, performance, extensibility and interoperability.
- **Janusz Cichowski, Piotr Czyżyk, Bożena Kostek, and Andrzej Czyżewski** ("Low-level Music Feature Vectors Embedded as Watermarks") propose a method consisting in embedding low-level music feature vectors as watermarks into a musical signal. The robustness of the watermark implemented is tested against audio file processing, such as re-sampling, filtration, time warping, cropping and lossy compression. The advantages and disadvantages of the proposed approach are discussed.
- **Andrzej Czyżewski** ("Intelligent Control of Spectral Subtraction Algorithm for Noise Removal from Audio") considers 'soft computing' algorithms for audio signal restoration in regard to a practical digital sound library application. The methods presented are designed to reduce empty channel noise, being applicable to the restoration of noisy audio recordings. The author describes a comparator module based on a neural network which approximates the distribution representing a non-linear function of spectral power density estimates. The author demonstrates experimentally that the methods examined may produce meaningful noise reduction results without degrading the original sound fidelity.
- **Danijel Koržinek, Krzysztof Marasek, and Łukasz Brocki** ("Automatic Transcription of Polish Radio and Television Broadcast Audio") describe a Large-Vocabulary Continuous Speech Recognition (LVCSR) system for the transcription of television and radio broadcast audio in Polish. This work is one of the first attempts of speech recognition of broadcast audio in Polish. The system uses a hybrid, connectionist recognizer based on a recurrent neural network architecture. The training is based on an extensive set of manually transcribed and verified recordings of

television and radio shows. This is further boosted by a large collection of textual data available from online sources, mostly up-to-date news articles.

Chapter VII, "Advanced Software Engineering", contains various programming and architectural problems and solutions for performance monitoring, uniform handling of data or dealing with heterogeneous services.

- **Janusz Sosnowski, Piotr Gawkowski, and Krzysztof Cabaj** ("Exploring the Space of System Monitoring") study the capabilities of standard event and performance logs which are available in computer systems. In particular, they concentrate on checking the morphology and information contents of various event logs (system, application levels) as well as the correlation of performance logs with operational profiles. This analysis has been supported with some tools and special scripts.
- **Emil Wcisło, Piotr Habela, and Kazimierz Subieta** ("Handling XML and Relational Data Sources with a Uniform Object-Oriented Query Language") present the data store model used by the SBQL query and programming language and discuss the already developed and future features of the language aimed at uniform handling of relational and XML data.
- **Cezary Mazurek, Marcin Mielnicki, Aleksandra Nowak, Maciej Stroiński, Marcin Werla, and Jan Węglarz** ("Architecture for Aggregation, Processing and Provisioning of Data from Heterogeneous Scientific Information Services") describe the architecture for the scientific information system allowing integration of heterogeneous distributed services necessary to build a knowledge base and innovative applications based on it. They provide overall description of the designed architecture with the analysis of its most important technical aspects, features and possible weak points. They also present initial results from the test deployment of the first prototype of several architecture components, including initial aggregation and processing of several millions of metadata records from several tenths of network services.

This book could not have been completed without the help of many people. We would like to thank the authors for their contribution to the book and their efforts in addressing reviewers' and editorial feedback. We would also like to express our gratitude to the reviewers. We are grateful to Janusz Kacprzyk for encouraging us to prepare this book.

October 2012   Robert Bembenik
Warszawa   Łukasz Skonieczny
                                                                Henryk Rybiński
                                                        Marzena Kryszkiewicz
                                                              Marek Niezgódka

Contents

Chapter III: Text Processing

Chapter IV: Ontology-Based Systems

Chapter V: Text Mining

Chapter VI: Multimedia Data Processing

Chapter VII: Advanced Software Engineering

Chapter I

Information Retrieval

Media, Information Overload, and Information Science

Bruno Jacobfeuerborn[1] and Mieczyslaw Muraszkiewicz[2]

[1] Deutsche Telekom AG, Germany
`Bruno.Jacobfeuerborn@t-mobile.de`
[2] Institute of Computer Science, Warsaw University of Technology,
Warsaw, Poland
`M.Muraszkiewicz@elka.pw.edu.pl`

Abstract. The Information Sword of Damocles hangs over e-literate people, in particular over knowledge workers, researchers, scholars, and students, and hinders their time management schemes. Our lives are replete with information of various kinds, of diverse importance, and of various degrees of relevance to our needs, and at the same time, we have less and less time and will to get acquainted with the incoming information and to follow it up. This paper argues that although media and ICT strengthen human's propensity to insane intake of information and reinforce bad habits leading to information overconsumption, it is possible to escape from the trap of information overload thanks to administering an information diet based on digital literacy. It further argues that new research should be undertaken within information science to address the challenges caused by the internet and the specificity of human perception exposed to the cognitive overload, untrustworthiness, and fuzziness.

Keywords: information science, informational overload, media, digital literacy, information literacy, information ecology, ICT, neuroplasticity.

1   Introduction

The mental world has always been littered with false and trash information because of purposeful misleading endeavours, or just ignorance, stupidity, inadvertence or negligence. Fortunately, it is not the major point that can particularly trouble us these days for techniques have been developed to identify, verify, filter and eliminate most of such information from our daily information intake, and those who want to make use of these techniques can smoothly apply them to the information flow at hand. The problem is rather what to do with the flood of true and potentially relevant and useful information that non-stop overwhelms us and aggressively urges our attention and reaction. Common daily complains of white collar workers, knowledge workers, scientific workers, scholars, and students are of the like: I cannot keep up with all the letters, memos, and reports arriving at my desks, and to cope with a number of emails, voice messages, phone calls, tweets, alarms, notifications, and subscribed business newsletters. Indeed, we live in the world replete with information of various kinds and calibre, of diverse importance, of known and unknown sources, and of various degrees of relevance to our needs, and at the same time, due to the tiredness and ennui

R. Bembenik et al. (Eds.): *Intell. Tools for Building a Scientific Information*, SCI 467, pp. 3–13.
DOI: 10.1007/978-3-642-35647-6_1   © Springer-Verlag Berlin Heidelberg 2013

caused by the information glut we have less and less time and will to get acquainted with the information, to digest it, and to follow it up. The *signum temporis* of contemporary e-literate active people is the Information Sword of Damocles that hangs over their time balances and time management rules and schemes. The anxiety threat feeling is that if it drops on us, it could damage or even ruin both professional and private relationships, deprive us of business and social opportunities, exclude us from communities we belong to and throw us out from "being in the loop". The result is that the fear to miss information starts governing lives of many people whose typical reaction to handle it is to try to devour all incoming information. Not surprisingly the result of such a strategy is what is dubbed an information overload, and metaphorically – information obesity.

The latter note instantly evokes an analogy between absorption of information and consumption of food. Almost throughout the whole history of mankind most people suffered from the persistent dearth of food and periodic famines, and experienced tenacious scarcity of nutrition. As a result, we have developed the habit to eat as much as possible when occasion occurred and to store food for periods of austerity. Metaphorically, such an attitude has become part of human's mental DNA that conditions human behaviour when it comes to food consumption and storing policies. In Maslow's hierarchy of human needs food is placed at the most basic level of the pyramid of needs, meaning that food assurance and consumption belong, along with such essential needs as sex and sleep, to the first priorities of men. Over ages humans have made a long journey as far as food is concerned: from scarcities to abundances, from modest to conspicuous consumption, from frugality to profligacy. Today, in spite of the fact that in developed countries the problems of food supply and starvation are already resolved, this attitude is still present in a sense that people instinctively consume more food than it is necessary to enjoy healthy life. Now, let us theorize: this legacy instinct vis-à-vis food that is a physical object can transform *mutatis mutandis* into the need for accumulating also mental objects such as information. Therefore, we have an imperative to collect and store information, even if it is not necessary at the moment, simply just in case of future use. One can put forward even more convincing explanation: Collecting, processing, communicating and exchanging information on the surrounding environment, i.e. plants, animals, weather changes, and people provided our remote ancestors with a comparative advantage over other species and tribes and was a fundamental part of their survival strategies. This instinct has remained in us till today and produces similar outcomes. Aristotle in his "Metaphysics" was obviously right by claiming that "*All men desire to know*". The problem however comes along when this desire exceeds the necessary and reasonable, and slips out of control. There is but a slim risk of being wrong to attribute the aforementioned propensity for amassing information and vulnerability while facing the information flood also to many knowledge and scientific workers.

This paper argues that although media and information and communications technologies (ICT) strengthen human's propensity to insane intake of information and reinforce bad habits leading to information overconsumption, it is possible to escape from the trap of information overload and obesity by means of devising and applying an information and digital diet strategy. Information science has a role to play towards this end. The structure of the paper is as follows. In the next chapter we shall sketch the role of media and digital literacy; the latter is a condition sine qua non for implementing sensible scenarios of information intake. Then, we shall recall some facts regarding the neuroplasticity of human's brain for they have to be taken into account while developing such scenarios and they will also provide input to the next section of the paper where we present a proposal on how to enhance the conventional scope and coverage of information science. The paper is closed by a note on the need of devising individual information diet strategies for those who are exposed to dense and aggressive information flows.

2   Media

Complaining on mass media and its tyranny has become a trite and boring effort, indeed. Yet, for the sake of completeness of our discourse let us briefly remind, while remembering a positive role of the media and being aware that still there are respected newspapers, weekly magazines, TV stations, and web portals that are benchmarks of professional journalism, the main critique, reproaches and accusations is addressed to the popular media, especially to large TV channels and tabloids. At the expense of the truth and objectivity they skew to create events and "reality" and/or exaggerate the scale, importance and impact of actual events. Everyday we are bludgeoned with a myriad of messages and marketing. They go even further as they attempt to set up agendas for politics, administration, social life, etc. These days, media are often referred to as the fourth power (in addition to an executive, a legislature, and a judiciary) owing to their tremendous influence on what people think and do, and to the ability to direct and control people's attention. Through the selection of topics to dispense, and by means of concocted comments and shows that explicitly or implicitly provide interpretations of information. Media tend to condition the way we think and try to profile our perception. Their pursuit for sensational headlines, breaking news, scandals, product placements, false fears and affirmations, and a tendency to address simple needs and instincts of the audience and to keep readers, viewers and listeners on an invisible tether of emotions and curiosity has mainly one goal: to maximize the audience share and thereby to increase profit or get more subsidies. In this context C. Johnson notes: *"Just as food companies learned that if they want to sell a lot of cheap calories, they should pack them with salt, fat, and sugar—the stuff that people crave—media companies learned that affirmation sells a lot better than information. Who wants to hear the truth when they can hear that they're right?"* [1].

The result is twofold: a relentless torrent of information, including redundant information and meaningless broadcasts, and low quality and reliability of information; all this often packed in eye-catching frames to attract and maintain attention thanks to the compelling form, rather than to the worthy content. Incidentally, one can observe a paradoxical phenomenon, as on the one hand side media are simplifiers of the reality by reducing complex problems to readily digestible headlines with the intent to facilitate and accelerate the communication process between them and the audience, but on the other hand they are complexifiers while trying to turn banality and platitude into valuable information and interesting content by means of scientific tools such as graphs and statistics, or by adding comments of academics, politicians, and/or celebrities. These two trends happen within the process of information industrialisation with all the consequences of this fact, out of which the informational homogenisation of world's diversity is the most worrying. The case of the Huffington Post is emblematic and shows at work the process of aggregation of content from a number of various sources on a wide spectrum of topics starting with politics, to lifestyle and culture, to technology and business, to environment, to entertainment, and local news. Noteworthy, in the internet realm, there are tools for collecting and aggregation of information published by different news providers, for instance news.google.com, available also to individuals, which allow one for personal customisation of a predefined information spectrum. Obviously also in this case we depend on the objectives and policies of aggregation services providers, which typically are not explicitly determined. The industrialisation of media also leads to the gradual replacement of experienced professional journalists and resource persons by a network of lower profile collaborators in order to reduce the costs of information collection, processing, and editing. The U.S. Bureau of Labour Statistics informs that *"Employment of reporters and correspondents is expected to decline moderately by 8 per cent from 2010 to 2020"* [2].

At this point one could ask: How do media know what the audience wants? With the advent of high-circulation newspapers, and later the invention of radio and television, and now the internet, the media have gotten more and more industrialised not only in terms of technology they use but also in terms of developing methods and tools to learn who is the audience, what is the audience segmentation, what are the preferences and expectations, and what is the instant reaction to broadcasted programs. One of the most experienced companies from which one can learn the methods for measuring audience engagement is Nielsen that operates in 100+ countries, which for almost a century has been developing tools to measure and analyse radio, TV, and online stationary and mobile audiences, including tools for measuring cross-platform engagements. Now, owing to mobile and wireless technologies and the internet the measuring of audience engagement and attentiveness, behaviours and preferences of content consumers has become relatively easy. For instance when it comes to social media, in addition to such popular and widely available tools as Google Alerts, Google Analytics, CoTweet, more sophisticated and specialized tools are available at a price or free-of-charge; the interested reader is referred to the site [3].

Indeed, media ubiquity and the way they learn about our proclivities and habits, the efficiency and speed they permeate our public and private lives is the reason why to a large extent we see the world through the lens of the media whose products we consume. Now, the fundamental question is whether and how can we cut through the hype, the inflation of breaking news, the flood of meaningless information and unreliable announcements? How can we release ourselves from the addiction to junk and useless information that pollutes our minds and waists our times? The answer to these questions includes a few stages. The first one is to realize that we should avoid blaming media for the way information is consumed; rather we should address any pretensions to ourselves. Second, we need to elaborate a personal information diet strategy, including clearly defined objectives we want to achieve and tools to get them realised, and last but not least we have to vigorously and persistently adhere to it. Part of the strategy refers to the notion of e-literacy, also referred to as digital literacy. Let us now then take a closer look at this notion.

In a nutshell, *e-literacy* means that an individual has the willingness, knowledge, and skills to use stationary and mobile information and communication technologies, in particular the internet, to support her/his professional and/or other activities. In other words s/he knows how to define queries and search heuristics, how to filter the obtained hits, and then how to synthetize the information acquired. In addition, the skills of how to present and publish the outcomes of the work belong to the e-literacy concept.

Further on, we repackage the term e-literacy into three bunches, namely e-skills, information literacy, and media literacy.

By e-skills we understand a set of capabilities that allow one to smoothly, fluently, confidently and critically make use of the appropriate ICT tools, as well as the resources and facilities constituting the digital universe of the internet. Fluent usage of the digital arsenal, be it ICT devices, operating systems, and/or applications is a necessary condition to exploit opportunities offered by ICT and the internet. At this point we can note that the process of acquiring e-skills is a multi-level task. For most of us just basic skills are sufficient to cope with and benefit from our laptops, smartphones and the internet. However, more advanced use of the digital artefacts requires specialized training and efforts, especially on the part of elder generations who are not digital native. Fortunately, a number of training and learning facilities are widely available, and basic courses of ICT have been introduced to curricula of most schools.

The American Library Association offers the following definition of information literacy *"To be information literate, a person must be able to recognize when information is needed and have the ability to locate, evaluate, and use effectively the needed information"* [4]. For the definition of media literacy let us refer to Wikipedia: *"Media literacy is a repertoire of competences that enable people to analyse, evaluate, and create messages in a wide variety of media modes, genres, and forms"* [5]. It should be noted that this definition includes an act of creation of content, authoring, whereas the notion of information literacy is rather focused on locating, retrieving, and using content only. OFCOM, an independent regulator and competition authority for the UK communications industry, also emphasizes the content development aspect in its definition of media literacy, which reads as follows: *"Media literacy is the ability to use, understand and create media and communications"* [6]. OFCOM carried out 3,171 in-home interviews with adults aged 16 years and elder. Here are some results picked up from the OFCOM report [6].

- *"Communication remains the most popular type of activity to be undertaken at least weekly among internet users (83%). In terms of individual activities carried out online, those most likely to be undertaken at least once a week are email*

(79%), social networking (45%), work/ studies information (45%), banking/ paying bills (33%), playing games online (15%), watching online or download TV programs or films (14%), maintaining a website/blog (10%), and doing an online course to achieve a qualification (5%).

- *Over half of internet users say they have a social networking profile (54%) compared to 44% in 2009. One half of those with a profile (51%) now use it daily compared to 41% in 2009.*
- *Installing software to control or block access to certain websites (20%) and installing security features (18%), are the main activities people say they are interested in but aren't confident doing. Women are twice as likely as men to say they are interested but not confident."*

The picture that emerges from the OFCOM reports (the one we referred to above and the ones of the previous years) shows that the level of media literacy observed on the internet constantly grows yet the pace is not so dramatic as one could expected given the growth of the internet infrastructure and the number of its users.

3   Neuroplasticity

There is no doubt that the phonetic alphabet that is the basis of efficacious writing, and a printing press, which altogether allowed an inexorable dissemination of billions of books towards mass readers, have been significantly moulding human mind and provided a framework that organizes the way how our minds work while reading and interpreting texts. Using a narrative shortcut one might say that they "rewired" our brains in order to be able to follow and cope with linear development of action and reasoning, which is not merely a metaphor but, as neurologists confirm, a physiological effect occurring in the brain neural structure owing to its plasticity and adaptability. It was proved in many neurological studies that thanks to the neuroplasticity neural paths connecting the visual cortex with nearby areas of sense making can intensify their activities when reading, as opposed to meaningless doodles that leave these paths intact, that memorizing and acquiring new habits cause rewiring of our brain circuitry and crisscross the brain in a new way, and that *"the number of synapses in the brain is not fixed—it changes with learning! Moreover, long-term memory persists for as long as the anatomical changes are maintained"* [7].

That said one could instantly ask the following question: does the internet with its hyper-text organization of content and knowledge, and empowered by efficient search engines have the same power as its predecessors to be a mind altering technology? The answer to this question is in the affirmative, i.e. the visible impact of the internet on human mental makeup is actually the case [8]. Non-linear organization of content (links), pervasiveness of multimedia, interactivity, responsiveness (high speed of delivering hits in return of queries), social networking, and distractedness while using the internet make it even more powerful to mould our brains, especially that the universe behind the internet features a seductive emotional power to its many users, especially to the youth.

Nowadays we can see how neurons light up as a result of thoughts, emotions and other brain's activities thanks to the technique called functional Magnetic Resonance Imaging (fMRI). G. Small and G. Vorgan who used this technique while studying the use of the internet noted the following: *"The current explosion of digital technology not only is changing the way we live and communicate but is rapidly and profoundly altering our brains ... stimulates brain cell alterations and neurotransmitter release, gradually strengthening new neural pathways in our brains while weakening the old ones"* [9].

Apart from ancient reservations regarding intellectual technologies expressed by the Egyptian King Thamus in Plato's "Phaedrus" where one can find the striking opinion on the harmfulness of writing and on written texts, it is interesting to note that also contemporary thinkers, already before the advent of the internet and expansion of information technologies, had expressed their concerns about technology in general and information technologies in particular. N. Postman bluntly claimed: *"We always pay a price for technology; the greater the technology, the greater the price. Second, that there are always winners and losers, and that the winners always try to persuade the losers that they are really winners"* [10], and adds *"Technological change is always a Faustian bargain: Technology giveth and technology taketh away, and not always in equal measure"* [11]. W. Ong's critique goes even farther when he expresses his concern as follows: *"Technologies are not mere exterior aids but also interior transformations of consciousness"* [12]. Although both thinkers do not share the concept of technological determinism that assumes that social and economic changes are mainly conditioned and driven by technology progress, the ground of their critique and similar critiques by other thinkers[1] relay on the assertion that technology is not neutral and that its usage does not hinge only on the way people want to use it.

We are sympathetic to this stance for this is also our conviction. We think that each technology inherently contains a potential of autonomy that interacts with the manner we perceive it and intend to use it. Being aware that not everyone could share this stance we nevertheless believe that technology, in particular intellectual technologies, are not neutral agents and therefore shape our mindsets and brains independently of our will.

Let us then come back to the impact of the internet on the way we acquire and evaluate information. After a brief reading of the book [13] in which the author reports testimonies of his colleagues on the way they read information from the web, we repeated this exercise on a sample of our fellow academic and industry colleagues. We obtained the same result as N. Carr. First of all the verb "read" is not relevant to depict what happens. The more relevant word would be "pick up" as they collect a few sentences of the text and decide on what to do aftermath. Usually after a few moments they click on a hyper-link available on the skimmed page and jump to

---

[1] Among the critics of technology one can find such prominent thinkers and scientists as T. Adorno, H. Dreyfus, J. Ellul, F. Fukuyama, J. Habermas, M. Heidegger, B. Joy, J. Illich, H. Marcuse, L. Mumford, T. Roszak. Although the edges of their critique were of different sharpness and aimed at different objectives they didn't reject technology as essential part of human material culture.

another page, or withdraw from the page, or which is statistically the less likely case they start attentively reading the displayed text. But unless the text is exceptionally seductive or is a kind of a "must-reading", they lose attention after reading two or three paragraphs and leave the page seeking something more interesting. They keep skimming the content available on the web. It seems that the reader unconsciously "forces" the internet to dialog with her/him in the way people talk on casual matters by handling a few threads concurrently, but without a deeper engagement in any of them. The interaction with the web somehow resembles a shallow, dispersed and unfocused informal human conversations. Perhaps we unconsciously attempt to anthropomorphize the web and see it as a living partner for dialoguing. We have to concede that our own experience is compatible with the experience of our interviewed colleagues. Incidentally, N. Carr summarized his query as follows *"When we go online, we enter an environment that promotes cursory reading, hurried and distracted thinking and superficial learning"* [13]. He supplements this finding by saying that the net *"fragments content and disrupts our concentration"* and *"seizes our attention to scatter it"*. Indeed, while sitting in front of a computer connected to the internet we are tempted by many opportunities, or to be more exact, many "scatters". Here is a non-exhaustive list of the scatters that distract us from focused reading, writing and/or thinking and to which we are in one way or another, stronger or weaker addicted:

email	mobile apps
instant messaging	online music
Twitter, Facebook, other social networks	internet radio
chats, online forums	videos
blogs	online shopping
news sites, RSS	online games
phones & mobile phones notifications	Internet TV
Skype	e-books
Podcasts	

At this point of the paper we feel almost obliged to quote T. S. Eliot who wrote prophetically in "The Four Quartets", long before the era of the internet and digital media: *"Distracted from distraction by distraction. Filled with fancies and empty of meaning. Tumid apathy with no concentration."* Isn't it the best description of the risk one can encounter while being on the web? Note that the tendency to use the internet according to the formula *"distracted from distraction by distraction"* leads to the situation when media available through the web get gradually transformed into social media for the imperative to be distracted has its origin in a deep psychological need for being continually communicated with other people and human affairs.

4   Information Science and Information Overload

Having briefly examined and justified the conjecture that intellectual technologies that we use by means of computers and the internet change the way we work with information and thereby can rewire our brains' circuitry, now we can express the main claim of this paper. But before that let us draft a definition of information science. This definition reads: Information Science is about developing the understanding and knowledge on how to collect, classify, manipulate, store, retrieve and disseminate any type of information by means of any medium or a combination of them.

We argue that in order to better address emerging challenges caused by the specificity of new digital media, the internet, and human perception exposed to the cognitive overload, fuzziness, untrustworthiness, and unverifiability that are commonplace while interacting with this new digital universe, information science has to include in its methodological repertoire a subjective aspect of information and methods to deal with it. To put it short: information science has to address the human base of information processes, especially regarding complex information systems. But what might it mean in practice? Our answer resorts to the broadly understood concept of information ecology[2] that offers a comprehensive framework to create, process, and understand information and its flow in a multifaceted context of a human ecosystem whose vital elements are not only technological facilities, but also social, cultural, and psychological determinants. In particular, efforts and research founded on cognitive assumptions and cognitive science achievements are needed towards new concepts regarding, *inter alia*, the following topics:

- Methods of knowledge organization, representation, and visualization, which take into account knowledge alterability, dynamics, and perception.
- Big data and deep information analysis (data and text mining) and content comprehension.
- Models of information relevance in volatile, amorphous, and dynamic information environments.
- Measures of information quality, in particular acquired from the web.
- Querying and obtaining answers in a natural language.
- Identification of users needs and their dynamics.
- Behavioural patterns of information users of different categories.
- Adaptability of information systems to users' behavioural patterns.
- Information use/sharing within collaborating communities.
- Social networking.
- Information literacy and media literacy.
- Semantic web.
- New methods of bibliometrics.

Research on these topics, aimed at establishing new ideas and concepts, new algorithms, and new processing tools, is undoubtedly a vital, ambitious and large endeavour going beyond the present frontiers of information science. Fortunately, the importance of setting up a nexus between information science and cognitive sciences is recognised by more and more research centres and researchers themselves dealing

---

[2] In [14] the reader can find an interesting characteristic of information ecology, and in [15] information ecology is discussed in the context of social media and online communities.

with information science, and hopefully diligent studies will be undertaken and carried out, and their results will be put into practice. We are convinced that the partnership of information science with cognitive sciences can be inspirational to both disciplines and take them to a higher level and greater heights in terms of quantitative and qualitative feats. Note that in order to depict the world science strives for objectivity, whereas humans permeate the world subjectively. Perhaps the partnership of information science with cognitive sciences will help science to objectify the subjective world of human beings.

5   Final Remarks

Above we noted that while complaining on and blaming media, we should not forget to complain on ourselves. It seems that the problem of the information universe we live in does not lie in the information overload; it is rather the way we deal with the incoming information and information overconsumption. Thus the famous saying: *"The destiny of nations depends on the manner in which they nourish themselves"* by Jean Brillat-Savarain in his "The Physiology of Taste" published in the year of 1826 can be for the sake of this paper modified as follows: The destiny of nations depends on the manner in which they consume information.

This adage should encourage us to start thinking, devising and setting in motion an information diet strategy, a personal trophic information pyramid and information firewall and filtering routine in order to sensibly govern the incoming information flow and to protect our brains and minds against the flood and toxicity of undesired information. As we argued in this paper increasing and fostering our digital literacy is an essential step towards this direction. For specific recommendations towards this end the reader is referred to practical guidelines included in [1], [16].

Acknowledgments. The authors wish to thank Professor Barbara Sosinska-Kalata of the Institute of Information and Book Studies, University of Warsaw, for inspiring discussions on the evolution of information science.

The National Centre for Research and Development (NCBiR) supported the work reported in this paper under Grant No. SP/I/1/77065/10 devoted to the Strategic Scientific Research and Experimental Development Program: "Interdisciplinary System for Interactive Scientific and Scientific-Technical Information".

References

1. Johnson, C.A.: The Information Diet. A Case for Conscious Consumption. O'Reilly Media (2011)
2. U.S. Bureau of Labour Statistics, http://www.bls.gov/ooh/Media-and-Communication/Reporters-correspondents-and-broadcast-news-analysts.html (retrieved June 12, 2012)
3. http://www.pamorama.net/2010/10/12/100-social-media-monitoring-tools/#.UAv33G-f_uw (retrieved July 12, 2012)

4. American Library Association, `http://www.ala.org/ala/mgrps/divs/acrl/publications/whitepapers/presidential.cfm` (retrieved July 12, 2012)
5. `http://en.wikipedia.org/wiki/Media_literacy` (retrieved July 12, 2012)
6. OFCOM, `http://stakeholders.ofcom.org.uk/market-data-research/media-literacy` (retrieved July 12, 2012)
7. Kandel, E.: In Search of Memory: The Emergence of a New Science of Mind. W. W. Norton & Company (2007)
8. Merzenich, M.: Going Googly (Thoughts on "Is Google Making Us Stupid?") (2010), `http://www.systematic-innovation.com/download/management%20brief%2009.ppt` (retrieved March 12, 2011)
9. Small, G., Vorgan, G.: iBrain: Surviving the Technological Alteration of the Modern Mind. Harper Paperbacks (2009)
10. Postman, N.: Five Things We Need to Know About Technological Change. In: The Speech Given at the Conference "The New Technologies and the Human Person: Communicating the Faith in the New Millennium", Denver, Colorado (March 27, 1998), `http://www.mat.upm.es/%7Ejcm/neil-postman-five-things.html` (retrieved March 12, 2011)
11. Postman, N.: Informing Ourselves to Death. In: The Speech Given at a Meeting of the German Informatics Society, Stuttgart (October 11, 1990), `http://www.mat.upm.es/~jcm/postman-informing.html` (retrieved March 12, 2011)
12. Ong, W.J.: Orality and Literacy, 2nd edn. Routledge (2002)
13. Carr, N.: The Shallows: What the Internet is Doing to Our Brains. W.W. Norton & Company (2010)
14. Malhorta, Y.: Information Ecology and Knowledge Management: Toward Knowledge Ecology for Hyperturbulent Organizational Environment. In: Encyclopedia of Life Support Systems (EOLSS), UNESCO/Eolss, Oxford UK (2002), `http://www.brint.org/KMEcology.pdf` (retrieved March 12, 2011)
15. Finin, T., et al.: The information ecology of social media and online communities. AI Magazine 29(3), 77–92 (2008)
16. Sieberg, D.: The Digital Diet. In: The 4-Step Plan to Break your tech Addiction and Regain Balance in your Life. Three Rivers Press, New York (2011)

Information Retrieval from Heterogeneous Knowledge Sources Based on Multi-agent System*

Tomasz Kogut, Dominik Ryżko, and Karol Gałązka

Institute of Computer Science, Warsaw University of Technology,
Nowowiejska 15/19, 00-655 Warsaw, Poland
T.Kogut@stud.elka.pw.edu.pl, karol.galazka@gmail.com,
d.ryzko@ii.pw.edu.pl

Abstract. This paper presents a multi-agent approach for scientific information retrieval from heterogeneous sources. An experimental platform has been built with the use of Scala language. A case study is shown demonstrating the results with the use of Google and Bing search engines and locally stored DBLP database.

Keywords: multi-agent systems, information retrieval, information brokering.

1 Introduction

One of the main challenges in SYNAT project is the automated acquisition of knowledge from various structured and unstructured sources including the Internet. Despite the overwhelming amount of irrelevant and low quality data, there are several useful resources located on the Web. This includes researchers' homepages and blogs, homepages of research and open source projects, open access journals, university tutorials, software and hardware documentation, conference and workshop information etc. Finding, evaluating and harvesting such information is a complex task but nevertheless it has to be taken up in order to provide the users with a wide range of up to date resources regarding science as well as past and ongoing research activities.

This work demonstrates how the scientific information acquisition can be performed by a multi-agent system in which various agents cooperate in order to provide the users with the information from sources varying by the quality and format of the data stored in them.

2 Previous Work

Several approaches to harvesting information from the Internet have been proposed in the past. The most popular approach nowadays is the use of general purpose search

---

* This work is supported by the National Centre for Research and Development (NCBiR) under Grant No. SP/I/1/77065/10 by the strategic scientific research and experimental development program: "Interdisciplinary System for Interactive Scientific and Scientific-Technical Information".

R. Bembenik et al. (Eds.): *Intell. Tools for Building a Scientific Information*, SCI 467, pp. 15–23.
DOI: 10.1007/ 978-3-642-35647-6_2   © Springer-Verlag Berlin Heidelberg 2013

engines. The improvement in the search quality caused that a vast majority of users say the Internet is a good place to go for getting everyday information [12]. Sites like Google.com, Yahoo.com, Ask.com provide tools for ad-hoc queries based on the keywords and page rankings. This approach, while very helpful on the day to day basis, is not sufficient to search for large amounts of specialized information. General purpose search engines harvest any type of information regardless of their relevance, which reduces efficiency and quality of the process. Another, even more important drawback for scientists is that they constitute only a tiny fraction of the population generating web traffic and really valuable pages constitute only a fraction of the entire web. Page ranks built by general purpose solutions, suited for general public will not satisfy quality demands of a scientist. One can use Google Scholar, Citeseer or other sites to get more science-oriented search solutions. Although this may work for scientific papers and some other types of resources, still countless potentially valuable resources remain difficult to discover.

Another approach to the problem is web harvesting, based on creating crawlers, which search the Internet for pages related to a predefined subject. This part of information retrieval is done for us if we use search engines. However, if we want to have some influence on the process and impose some constraints on the document selection or the depth of the search, we have to perform the process by ourselves. A special case of web harvesting is focused crawling. This method introduced by Chakrabarti et al. [3] uses some labeled examples of relevant documents, which serve as a starting point in the search for new resources.

The task of retrieving scientific information from the web has already been approached. In [14] it is proposed to use meta-search enhanced focused crawling, which allows to overcome some of the problems of the local search algorithms, which can be trapped in some sub-graph of the Internet.

The main motivation for the work envisaged in the PASSIM project is to create a comprehensive solution for retrieval of scientific information from the heterogeneous resources including the web. This complex task will involve incorporating several techniques and approaches. Search engines can be used to find most popular resources with high ranks, while focused crawling can be responsible for harvesting additional knowledge in the relevant subjects. Additional techniques will have to be used to classify and process discovered resources.

Since many users will use the system simultaneously, a distributed architecture will be required. While this has several benefits regarding system performance, additional measures have to taken in order to avoid overlap in the search process [2]. Various parallel techniques for searching the web have already been proposed [1]. In the PASSIM project multi-agent paradigms will be used, which propose intelligent, autonomous and proactive agents to solve tasks in a distributed environment.

3 Basic Concepts

3.1 Personalization

Search personalization is a well known subject. There are many papers explaining its applications dated from early 90's. There is a couple of conventional examples explaining need for personalization. When user types the word "java" search engine doesn't know if she/he means the island or programming language. A similar problem applies for the terms like mouse, jaguar etc., and it can be resolved based on knowledge about user interests.

The application of personalization in a system like PASSIM seems to be obvious. It would be really helpful if the system learns user interests, and use them to give more precise information. However this is not as simple as may seem.

In [4] the researchers from Microsoft pointed out that in contradiction to what was written in most papers, the use of personalization can do more harm than good. They compare a couple of personalization techniques using data from MSN query logs. In short the conclusions were: "Personalization may lack effectiveness on some queries, and there is no need for personalization on such queries." "Different strategies may have variant effects on different queries" "The correctness of personalization is connected with size of search history data. It could be difficult to learn interests of users who have done few searches" "Users often search for documents to satisfy short-term information needs, which may be inconsistent with general user interests"

The nature of the SYNAT project simplifies few things. First of all, users will use this system only for searching data concerning their academic work. Of course, they might participate in many projects with broad range of subjects, which might change in time, but still this leaves out many queries, and makes user profile more consistent. Second, users will give some additional information during registration such as their names, academic background etc. This data will be used to solve problem of cold start and provide helpful feedback even without any search history.

There are two basic concepts of personalized search application. First, we can reformulate query adding some information such as category. Second we can leave query unchanged and modify search results by changing their order of display and deleting irrelevant elements. Usually it is recommended to use only one technique. However in this project it would be insufficient because of its specific architecture.

Let us consider a scenario: Query with additional data goes to brokers. They perform search in many sources. We cannot be sure how and whether they use this information, hence, sorting and filtering of the results is mandatory.

In order to perform personalization there is a need to classify both queries and search results.

3.2 Classification

Classification in this context means giving a set of categories with appropriate ratio from an ontology. This can be done by measuring the similarity between current document and the documents with specified categories. Therefore it is necessary to provide such documents.

Profiles are unified concepts of representing variety of data later used for classification. The concept of creating and combining several users profiles comes from [11]. There are three planned types of profiles: general profile, implicit profile, explicit profile. General profile represents a way to categorize documents regardless of user interests. It is based only on Open Directory Project (ODP) data. Due to huge amount of categories in ODP, the majority of applications use only a couple of first levels from hierarchy. The data from subcategories is combined in representations of their parents. Less categories means shorter computation time. Though, more important is that it serves also to create more accurate classifiers (basing on more training data). In lower levels of ontology there are categories with very laconic descriptions and only few web pages.

Implicit profile is obtained from data collected from users activities such as query history. It does not involve users in doing any additional work. Explicit profile is an underestimated concept. Most works about personalization focus on obtaining best implicit profile. The reason is simple. Users usually do not give any additional feedback. During registration to the platform users will fill in some information about themselves. This data will serve to perform further search. Results such as publications, connections in social services etc. will provide sufficient information to build a specific profile.

4   Agent Framework

4.1   Requirements

The Scala language provides an Actors framework for concurrent computations. The framework can be naturally adopted to the needs of a multi-agent system, since it has characteristics of that kind of system: autonomy, local perspective and decentralization. Actors lacks ability to create a conversation between agents (multiple exchange of ordered messages), and ability to discover other agents, specifying what type of agent we are interested in.

4.2   Establishing a Dialog

To meet the requirement of a conversation a concept of "Dialog" inspired by Session Initiation Protocol (SIP) is introduced. It is defined as an ordered exchange of messages between two agents, initiated by one of them, that is uniquely identified by an ID. A dialog is started by sending the EstablishDialogMessage message with universally unique identifier (UUID), and the reference to the sender. Receiver upon receiving the message creates internal structures that can hold further data that should be persistent between messages exchange. Afterwards dialog is being confirmed by sending an OK message. From now on, two agents are logically connected in a conversation and can interact. What is important is that simultaneously more than one "Dialog" can be established. It gives a possibility to serve multiple users at one time and avoid interference between them. For the Dialog system to work properly, all the agents need to follow strict rules when creating a conversation. Thus the abstract class Agent created on the base of the scala's class Actor handles the process of establishing the Dialog. While processing messages from the queue it tries to match the received message, and handle those that it recognize (e.g. EstablishDialogMessage and OK). All the other messages are passed down to the handleMessage() method that specific agents should implement. In the case of receiving the OK message, the processDialog() method is being called. The purpose of this method is to invoke the task that has been associated and saved within the Dialog structure.

4.3 Yellow Pages Service

Yellow pages are a well known mechanism in the context of multiagent systems. It allows agents to advertise themselves, find each other and query their abilities. Implementing Yellow Pages was crucial for effectiveness and, like in other agent frameworks, it was implemented as an agent with one global instance. Agents that want to be searchable in yellow pages (not all of them need to be and should be) need to include YelloPagesSearchable trait. This causes registration of the agent in yellow pages during the creation of the agent. To query for an agents with desired capabilities, another agent needs to send a message with desired constraints to yellow pages and wait for the reply.

5   Use Case Agents

5.1   Broker Agent

Broker agent main task is to receive a query from the user, pass it to search agents, receive answer from all of them and deliver it to the user. In this use case query is a string of characters that would be entered by user into search engine via the web browser.

Upon receiving query from the user Broker creates a new search task that has assigned UUID. Then contacts and establishes dialog with all known search agents (that he has found from the Yellow Pages). UUID of the task is being associated with UUIDs of the dialogs. Query is being passed to the search agents that know how to perform search on the knowledge source. Broker waits for an answer from all sources. Based on the UUID of the dialog results are mapped and appended to the search task structure. When all results are ready, broker merges them using URL and sends it back to the user for presentation.

5.2   Search Agent

SearchAgent abstract class handles receiving a query. A specific SearchAgent implementations need to provide a search() method that would perform searching for knowledge sources. The provided implementations of search agents can query Bing and Google search engine with use of their public API and the DBLP data stored locally in MongoDB database.

Search agents can be easily added to the system when new knowledge sources need to be added. The search agent upon start registers itself in Yellow Pages and therefore can be queried by the Broker, which should do it periodically to provide information from a wide spectrum of sources.

5.3   User Interface

The User Interface agents were created to connect users with their personal agents. One user can have few user interfaces connected to personal agent at the same time. There are many ways to represent the user interface.

In the current version of the platform there is an agent called RemoteUserInterfaceAgent. Its main task is to behave as a bridge between RemoteClient and Personal Agent. Both RemoteUserInterfaceAgent and RemoteClient are implementing the remote agent pattern and therefore they might exist on two different machines.

5.4 Client

Client is an extension of user interface. It is an abstract concept representing a way of connecting to the system via user interface. It can be written in any language under condition that it will have corresponding user interface.

The Remote Client is written in Scala. The implementation of the remote agent constitutes its main part. The users can query and receive search results through a console.

5.5 Personal Agent

A work flow for a personal agent is as follows:

1. Personal agent receives query from user interface.
2. It initiates the task of query classification by sending requests to the classifiers, and then
3. collects results categories from them.
4. Next, it sends query with categories to brokers.
5. Then it receives search results from brokers and immediately sends them to appropriate user interface.

One user could perform search from couple of the user interfaces simultaneously.

5.6 Classifier

Classifiers are meant to perform the classification tasks. Their clients are the personal agents, as sketched above. In future work they will have autonomy whether to accept job basing on requirements necessary to perform them.

6   Information Sources

Information provided by various knowledge sources can vary. For example comparing to Google which provides URL, title and description, Bing API in addition supports a date field. To avoid loosing of data, a map of attributes is added to the search result. Furthermore those two APIs use different format of data exchange. Google uses JSON whereas Bing can use both JSON and XML. In Bing the XML API was chosen deliberately to show how heterogeneous knowledge sources can be used to provide consistent data. Additionally to mentioned internet sources there has been used a local source of information, the DBLP database. The local MongoDB database operates on a json-similar data structure format. The search agent has to convert the data format used by the external API to the internal system format. This format is recognized by the broker, and thus data in this format can be merged with other data. Some of the map keys can be predefined for Broker to understand them. For example the date attribute mentioned above, can be used for sorting.

7   Example Usage

Querying the system for the "Datamining" returns following resutls:

1 Web result
   Title: Data Mining
   Description: Data mining is a powerful new technology with great potential to help companies focus on the most important information in the data they have collected about ...
   Google: Search Engine
   URL: http://www.laits.utexas.edu/ norman/BUS.FOR/course.mat/Alex/

2 Web result
   Title: Data Mining and Analytic Technologies (Kurt Thearling)
   Description: Information on Data Mining and Analytic technologies (tutorials, papers, etc.)
   DateTime: Thu May 10 07:39:00 CEST 2012
   Microsoft: Bing Engine
   URL: http://www.thearling.com/index.htm
   Date: Thu May 10 07:39:00 CEST 2012

3 Web result
   Title: Data Mining with Microsoft SQL Server 2008 R2
   Description: Use predictive analysis in Microsoft SQL Server 2008 R2 to mine historical data. Provide new insights and form a reliable basis for accurate forecasting.
   DateTime: Sat May 12 11:39:00 CEST 2012
   Microsoft: Bing Engine
   URL: http://www.microsoft.com/sqlserver/en/us/solutions-technologies/business-intelligence/data-mining.aspx
   Date: Sat May 12 11:39:00 CEST 2012

4 Web result
   Title: Data Mining - Web Home (Users) - The Data Mine Wiki
   Description: Feb 2, 2012 ... Launched in April 1994 to provide information about Data Mining (AKA Knowledge Discovery In Databases or KDD). A Twiki site full of guides, ...
   Google: Search Engine
   URL: http://www.the-data-mine.com

...

11 Web result
    Title: Datamining Support for Householding on Beowulf Clusters.
    Description: NO DESCRIPTION
    Dblp: Knowledge base
    Author: ["Daniel Andresen" , "David Bacon" , "Bernie Leung"]
    URL: http://dblp.uni-trier.de/db/conf/pdpta/pdpta2000.html#AndresenBL00

12 Web result
    Title: Datamining et Customer Relationship Management.
    Description: NO DESCRIPTION
    Dblp: Knowledge base

Author: Michel Jambu
URL: http://dblp.uni-trier.de/db/conf/f-egc/egc2001.html#Jambu01

It is worth mentioning that the number of results is smaller than simple sum of results from Bing and DBLP. Duplicates were removed based on the URL, and where possible "date" attribute was added. Because Google results are subset of those provided by Bing there is no benefit of mixing attributes. But if the knowledge sources provide varying sets of attributes, mixing them in the final result will eventually give more comprehensive answer. This eventually can be noticed in the results: Result 1: a Google result lacking date. Result 2: a Bing result that has an date attribute Result 5: a mixed result of Bing and Google with date attributed inherited from Bing. Result 12: one of many DBLP results

8   Conclusions and Future Work

A multi-agent approach to acquisition of scientific information from heterogeneous sources has been shown. The experimental platform has been built which demonstrates the possibility of applying multi-agent paradigms to this task. Presented experiments show that the solution allows seemles integration of various resources. The distributed nature of the systems means it is easy to add additional sources of knowledge as well as degradation of the system is graceful.

A base for information brokering has been established. Built upon the Scala's actors, the agent platform gives a possibility to present user data from various sources that use different data formats. In the next stage of work we will concentrate on the development and improvement of the current features. The inner "search result" term will be expanded to represent custom abstract documents targeting not only web pages but also scientific papers, books, magazines etc. Agent framework will be integrated with SYNAT databases and web-crawlers layers that are supposed to yield information to the user of the system. Further work on Broker Agents will concentrate on the semantics of the query and adding predefined attributes that could help understand what the user is looking for.

At this stage a personal agent platform with simple query classification was established. It allows users to write queries and receive search results. The next step will focus on adding implicit profiles and basic results filtering. Future work will also involve testing more structured and unstructured knowledge sources.

References

1. Bra, P., Post, R.: Searching for arbitrary information in the www. In: The fish-search for mosaic. Second World Wide Web Conference (WWW2) (1999)
2. Baeza-Yates, R., Ribeiro-Neto, B.: Modern Information Retrieval. Addison-Wesley Longman Publishing Co., Inc., Boston (1999)
3. Chakrabarti, S., van den Berg, M., Dom, B.: Focused crawling: A new approach to topic-specific web resource discovery. Computer Networks 31(11-16), 1623–1640 (1999)
4. Dou, J., Song, R.: A large-scale evaluation and analysis of personalized search strategies. In: Dou, J., Song, R. (eds.) Proceedings of the 16th International Conference on World Wide Web, WWW 2007. ACM, New York (2007)

5. Gawrysiak, P., Rybinski, H., Protaziuk, G.: Text-Onto-Miner - a semi automated ontology building system. In: Proceedings of the 17th International Symposium on Intelligent Systems (2008)
6. Gawrysiak, P., Ryżko, D.: Acquisition of scientific information form the Internet - The PAS-SIM Project Concept. In: Proceedings of the International Workshop on Semantic Interoperability (IWSI) (2011) (accepted for publishing)
7. Gomez-Prez, A., Corcho, O.: Ontology Specification Languages for the Semantic Web. IEEE Intelligent Systems 17(1), 54–60 (2002)
8. Hua, J., Bing, H., Ying, L., Dan, Z., Yong Xing, G.: Design and Implementation of University Focused Crawler Based on BP Network Classifier. In: Second International Workshop on Knowledge Discovery and Data Mining (2009)
9. Klusch, M., Sycara, K.P.: Brokering and matchmaking for coordination of agent societies: A survey. In: Coordination of Internet Agents: Models, Technologies, and Applications, pp. 197–224. Springer (2001)
10. Kohlschutter, C., Fankhauser, P., Nejdl, W.: Boilerplate detection using shallow text features. In: Proceedings of the Third ACM International Conference on Web search and Data Mining (2010)
11. Liu, W., Yu, C.: Personalized web search by mapping user queries to categories. In: Proceedings of the Eleventh International Conference on Information and Knowledge Management, CIKM 2002. ACM, New York (2002)
12. Manning, C.D., Raghavan, P., Schuetze, H.: An Introduction to Information Retrieval. Cambridge University Press (2008)
13. McIlraith, S.A., Son, C.T., Zeng, H.: Semantic Web Services. IEEE Intelligent Systems 16(2), 46–53 (2001)
14. Qin, J., Zhou, Y., Chau, M.: Building domain-specific web collections for scientific digital libraries: a meta-search enhanced focused crawling method. In: Proceedings of the 4th ACM/IEEE-CS Joint Conference on Digital Libraries (2004)
15. Suber, P.: Open access overview,
http://www.earlham.edu/~peters/fos/overview.html (2004)
16. Wenxian, W., Xingshu, C., Yongbin, Z., Haizhou, W., Zongkun, D.: A focused crawler based on Naive Bayes Classifier. In: Third International Symposium on Intelligent Information Technology and Security Informatics (2010)

Interactive Information Search in Text Data Collections

Marcin Deptuła, Julian Szymański, and Henryk Krawczyk

Department of Computer Architecture,
Faculty of Electronics, Telecommunications and Informatics,
Gdańsk University of Technology
marcin.deptula@live.com,
{julian.szymanski,henryk.krawczyk}@eti.pg.gda.pl

Abstract. This article presents a new idea for retrieving in text repositories, as well as it describes general infrastructure of a system created to implement and test those ideas. The implemented system differs from today's standard search engine by introducing process of interactive search with users and data clustering. We present the basic algorithms behind our system and measures we used for results evaluation. The achieved results indicates the proposed method can be useful for improvement of classical approaches based on keywords.

Keywords: information retrieval, search engines, Wikipedia.

1   Introduction

The most commonly used search engines today base on a very simple pattern of how they communicate with a user [1]. First of all, the only input that the user can have is text query, which will be processed by the engine. It restricts the user in many ways:

1. He cannot clarify his intention any further. If his query may be interpreted in few different logical contexts, he has no means of specifying which context he is interested in.
2. If the text query was not precise, because for example he or she had used wrong keywords, a user will receive invalid results. Furthermore, the user will receive no feedback on how he could improve his query. This problem is common, when users search for data in domain with which they are not familiar.

Secondly, structure of results is just a flat sorted list of results. Although, it is important, that results are sorted by relevance, research shows that users tend to ignore results that are not at the top positions [2]. This also means that if results contain data from few different domains, they will be flattened into one list, and it will be harder for the user to find information which is important to him.

It is important to point out, that today's commercial search engine research is heading towards more and more sophisticated search algorithms [3], than most popular Google's PageRank [4]. Most of the work was done to improve algorithms search relevance without changing the user experience also the works shows that a lot is done to add social aspect to searching [5], so that the engine can improve results based on user's feedback. One of the most well known is Google's "+1" button, which allows users strengthen a

R. Bembenik et al. (Eds.): *Intell. Tools for Building a Scientific Information*, SCI 467, pp. 25–40.
DOI: 10.1007/ 978-3-642-35647-6_3   © Springer-Verlag Berlin Heidelberg 2013

content that they like, so that for further queries it will higher in the result list. Microsoft took different approach, but in the same direction, and launched collaborative search engine, where one can share his queries as well results and discuss them with other users, which will eventually create thread of connected topics and answers provided [6].

1.1 Alternative Search Methods

System described in this article research issues described in last section, as well leverage the idea of improving results based on user feedback. The basic scenario of how search query is processed has been rebuilt and it can be described by following steps:

1. User sends text query to the system.
2. Result list (flat) is created based on query.
3. Results are grouped into sets of logically connected items.
4. Based on those groups, conceptual directions are computed.
5. Conceptual directions are presented to the user; he can decide to choose one of them, reject one of them or alternatively ask for more results.
6. Based on user's answer, new list is built by filtering and/or adding new elements to the result list.
7. User's answer is used to check if given options were useful to him.
8. Algorithm goes back to step 3.

Described scenario differs from what could be called a "standard searching method" that use keyword based approach [7]. First important differentiator is fact, that this method is interactive and iterative. User can change results interactively, by choosing conceptual directions; furthermore, he can repeat this process until results presented to him are satisfactory. Secondly, results are presented in logically grouped clusters, which increase the readability of results and allow the user to quickly discard elements, which are uninteresting to him and concentrate on his goal. Each group is distinguishable by name, which best describes elements in it. How groups are created is described in detail in section 2.2, but in general, assumption is, that the best names are those created and managed by humans; that is why this system uses category hierarchy from Wikipedia to name and categorize groups of elements.

2   Architecture

The system was created having followed things in mind: modularity, scalability, extensibility and testability. Modularity was achieved, by defining over 10 interfaces (modules), responsible for different part of computational process, such as clustering, processing of user's input, generating data. Modules can be switched easily, so that different algorithms implementing same functionality can be tested. This approach helps testability, because algorithms can be compared on same data sets very easily. Scalability is a very important factor, because potentially, number of users that use system concurrently can be huge. For this reason, system was split into 2 processes:

- Computational process, which is implemented as a standalone service, accessible from network,
- Presentation process, which is implemented as a web service.

Currently external load balancer is needed to schedule user sessions between many presentation processes, although, in the future, it is possible for a web service to balance its work automatically by scheduling users onto different computational processes. System's computational process architecture can be split into two phases, initialization phase and running phase. During initialization, data is imported and prepared for use by the system. This process may be time consuming and is launched only once per data set.

Second phase works on prepared data sets and it is launched for every user session. Running phase is ought to be fast and be able to run concurrently in many instances. Because those phases use different modules and their life cycle is very different, they will be described separately in sections 2.1 and 2.2 respectively.

As stated before, currently system is using Wikipedia articles for data, because of this, specific blocks will be described assuming current system's state, although, source and format of data might change in the future.

2.1 Initialization Phase

Architecture design of initialization phase is presented in Figure 1.

First step of initialization is the process of importing data from arbitrary format to common database format, recognized by the system, called CFF. This process is split into 2 steps:

1. Importing data to common, memory format
2. Exporting data to CFF file.

Importing is handled by SQL dump importer module, which parses *.sql files to build information about imported data. Files needed for import stage are:

1. Category list - used to retrieve list of categories,
2. Article list - used to retrieve list and content of articles,
3. Category links - used to build information about how categories are connected. This is also needed to filter out categories, which are not connected to main root categories,
4. Redirects - used to filter out pseudo articles, that are nothing more than redirect pages,
5. Page props - used to filter out other meta-related articles.

At this stage, all data resides in memory and can be exported, by CFF exporter to data files, recognizable by core modules. It is important to note, that this is the only step that is required to be exchanged, so that whole system can work with different data.

Search engine builder - during this step, data from CFF file is analyzed and matrices that represent different types or relations are built. During this step, few different algorithms can build their own logical representation, which can be later used by any algorithm during running phase. At the moment, in the system two different algorithms are implemented:

1. Category extractor
2. Link extractor

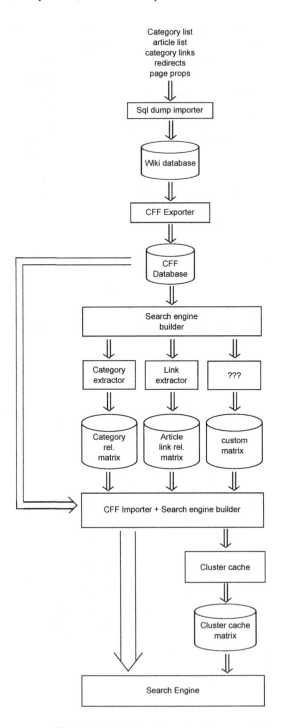

Fig. 1. Initialization phase algorithm

Category extractor builds a graph of connections (associations) between articles and categories they are categorized in, as well between categories. Each existing connection is given weight of 1.0, which is maximum weight. Next, weaker connections are calculated, based on initial graph, according to equation 1.

$$Assoc(a, c) = max\left(\forall_{c_i \in C}\frac{1}{dist(c_i, c) + 1}\right),\tag{1}$$

where: $dist(c_i, c_j)$ is a function of minimum distance in the graph between c_i and c_j. C is a set of categories in which a belongs directly.

Link extractor on the other hand, ignores all data intrinsically given by CFF file and base only on articles texts. Each article is parsed and all links to other articles are found. For each such link, connection is made and written to generated matrix. As a result, matrix of connections between given articles is created.

Next step is to build index of articles, so that they can be searched for efficiently. Currently, the system is using reverse index, generated by Indyx project that creates inverted indexes [8] to selected text repository.

All generated data is then used to create core search engine module, which can be used in running phase. However, for performance reasons, it is possible to pre-generate clusters into a structure called cluster cache, which will speed up clustering process in later phase. If this step is chosen, core system will later use technique called global clustering, instead of local clustering. Difference between those techniques won't be discussed in details in this article.

2.2 Query Phase

Query phase describes processing of the user's session from its beginning - when user sends his first text query, until he resigns from further processing. The processing algorithm is presented in Figure 2. It is composed mainly from two important algorithms: conceptual partitioning and clustering algorithm (DBScan) [9].

Role of the IG (Information Gain) algorithm is to choose the best question (or questions), that can be asked by system; taking into consideration value of information which will be received after user will answer it. Each question is in form of name of a category which represents conceptual direction which is relevant to the user; he can then decide if he is interested in it or not. Currently, the user can choose between up to 10 categories.

DBScan algorithm on the other hand does not influence the results, but it is responsible for how they are presented to the user. It clusters results into logically connected groups, so that they can be presented to the user in this form. Currently, DBScan algorithm is used in the system few times, which is described in details in section 2.2.

In the next sections, each of the algorithm is described more in detail.

Indyx. At the beginning of each user session, his text query is converted into set of articles. What is important during this step, articles are returned as a flat list of results, sorted by their relevance to the input query. The calculated relevance is based on how well an article matches the query based on word similarity. This step, for most general

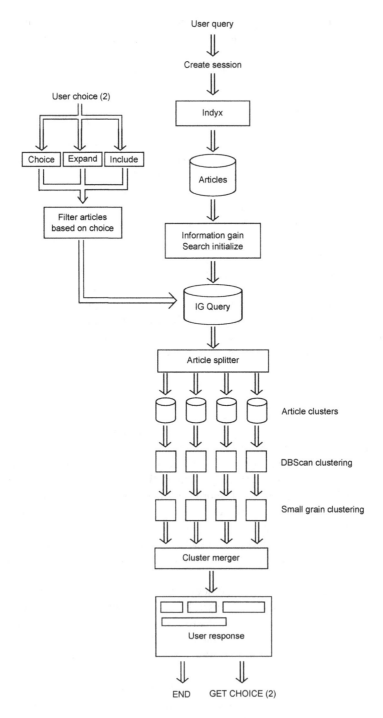

Fig. 2. Query phase algorithm

text search engines, as described in section 1 would be the last step. In the presented system, this is only the first step. Because algorithms described further are compute intensive, list of results is truncated to the best matching 1500 articles. This restriction might be changed in further versions of the system. Fact, that results are truncated will never ruin the quality of user's query, because, using methods described below, user has a method of including new articles, that are relevant to him and which were truncated during this step.

Conceptual Partitioning - Initialization. During this step, new query is initialized. When the query is initialized from list of articles, up to n best matching categories is found. As stated before, categories are chosen based on information value that they give for the system. CategoryUsefullness function has been described at the end of this section as category usability metric. Currently, after testing, parameter n has been set to 10. Pseudo code presented below shows how categories are chosen:

> **SET** $categories \leftarrow AllCategoriesByDocuments(documents)$
> **for all** $category$ in $categories$ **do**
> $r \leftarrow rand(0, 1.0)$
>   **if** $r >= CategoryUsefullness(category)$ **then**
>   **continue**
>   **end if**
> $cc \leftarrow 0$
>   **for all** doc in $documents$ **do**
>   **if** $relation(doc, category) > 0.0$ **then**
> $cc \leftarrow cc + 1$
>   **end if**
>   **end for**
> $$IG_c \leftarrow 1.0 - \left| \frac{(\frac{count(documents)}{2.0} - cc)}{\frac{count(documents)}{2.0}} \right|$$
> $IG_c \leftarrow IG_c * CategoryUsefullness(category)$
> **end for**
> **return** $max_n(categories) ordered by IG_c$

Algorithm, for each category c, which is connected to articles which are in currently analyzed query computes value IG_c. During calculating IG_c, category usability is checked twice. This parameter, which is described in section 2.2, can exclude category from being taken into account, or decrease its value.

Article Splitter. This algorithm is responsible for splitting flat list of articles into lists of given size. This step greatly increases performance of clustering algorithms, especially, when local clustering method is used. Role of article splitter is to split articles in such a way, that one potential cluster will not be split into many different article chunks, because this could lead to a problem, where an important cluster would not be recognized by next steps at all. In the current implementation, simple algorithm that uses only article relevance is used. Chunks are of size 200 articles. Tests shown, that this approach did not have big impact on quality of results.

Clustering. Clustering algorithm is used only to change how results are presented to the final user, without making any impact on how articles are filtered. Currently, the system implements DBScan algorithm, which has been chosen because of few of his advantages. First of all, contrary to many other algorithms, like for example k-means, DBScan does not need to have number of clusters defined a priori. This is an important advantage over many other algorithms.

Another important reason is computational complexity, which for DBScan is $O(n * log(n))$. Using few small optimizations in implementation, this is enough to be able to cluster groups of around 1000 elements in few seconds, which is more than acceptable, considering article splitting method.

Small Grain Clustering. Small grain clustering is a process of clustering articles, which could not be clustered using standard method. This process is implemented using the same algorithm, which is DBScan, but it is parameterized to form much more general clusters.

Cluster Evaluator. Cluster evaluator is used to evaluate relevance of formed clusters, based on user's answers. This allows the system to sort clusters, before presenting them to the user, based on how user responded in the past. Because of this feature, if a conceptual direction starts to form during user's session, clusters which describe this particular concept will be sorted before other ones. Algorithm for currently implemented evaluation is presented below:

> **for all** a in $GetDocuments(cluster)$ **do**
>   **for all** ch in $choices$ **do**
> $R \leftarrow GetRelation(a, ch.Category)$
>   **if** $(ch$ is Yes AND $R > 0)$ OR $(ch$ is No AND $R == 0)$ **then**
> $points \leftarrow points + 1$
>   **end if**
> $maxPoints \leftarrow maxPoints + 1$
>   **end for**
>   **end for**
> **return** $points/maxPoints$

Cluster Merger. Role of a cluster merger is to merge results of clustering for each chunk of articles, which were previously split in the article splitting process. It is important for this algorithm to be able to merge potentially similar clusters. For this reason, currently it is implemented as follows:

- For each cluster in cluster list, choose the best matching category, which will describe the given cluster
- If cluster with given BMC (best matching category) was already found, merge them. Otherwise, create new cluster that is represented by its BMC.

Conceptual Partitioning Algorithms. This set of algorithms exists to respond to the user's choice and modify current result list, before it is processed again. Currently, three different user actions are supported:

- User choice, when the user choices one particular category and whether to include it, or exclude it from the results.
- Expand, when user cannot find articles in which he is interested in the results.
- Include category - when user is particularly interested in one of the categories, he can ask to add all articles related to it to the result list.

User choice can be interpreted as filtrating results based on conceptual choice. When user starts his session, he asks a text query. Answer which is a list of articles might not be satisfactory, because of few factors:

1. Query can be ambiguous. In this situation, articles from different logical domains will be returned to the user.
2. User might want to retrieve information, which is not directly connected to his original query. This might occur, when for example, the user is not able to give precise query using good keywords, because he lacks knowledge in a given topic.

Each time then user chooses a category, he basically chooses a conceptual direction in which he wants to go. Based on this, the system can filter out articles which are not relevant to the topic, as well as add more articles, which were previously not included. Each such choice is used to build answer vector V, which elements will be pairs int form of $category - answer$. At the beginning of the session, this vector is empty and it is expanded after every user's answer. Filtration algorithm, based on user answers, can be illustrated by following pseudo code:

for all doc in $IntialDocuments$ **do**
  $points = ComputeDocumentToAnswerSimilarity(doc, answerHistory)$
  $relPoints = \frac{points}{count(answerHistory)}$
  if $relPoints > 0.0$ **then**
    results $\leftarrow$ **results** $- doc$
  else if $relPoints < 0.0$ **then**
    results $\leftarrow$ **results** $+ doc$
  end if
end for

Currently, in the system, function $ComputeDocumentToAnswerSimilarity$ is implemented like this:

$x \leftarrow 0.0$
for all $choice$ in $answerHistory$ **do**
  $relation = GetRelation(document, choice)$
  if $rel = 0.0$ **then**
    if $choice = No$ **then**
      $y \leftarrow np$
    else
      $y \leftarrow yp$
    end if
  else
    if $choice = Yes$ **then**
      $y \leftarrow rel$

     else
        $y \leftarrow -rel$
     end if
    end if
    $x \leftarrow x + y$
   end for
   return x

Variables np and yp are parameters of the algorithm which describe punishments, when there no relation between document and choice's category. Currently, those values of those variables were chosen based on experiments and they are equal to 0.05 for np and 0.4 for yp. By changing values of those parameters, border between article which should be included or rejected in result set can vary. Values were optimized, so that this border is equal to 0.0. $GetRelation$ function returns relation between a document and a category (being the "choice") from Category relation matrix, which is built by extractor. Process of creating those matrices has been described in section 2.1. $AnswerHistory$ is a list of all answers given by user, which are pairs of category and binary decisions of the user whether given category was interesing to the user.

Category Usability. Base criterion that is measured for each category c in set of all categories C is its usability. This criterion tries to measure, how useful given category's name is for the user. Two parameters are measured, to calculate category usability:

    dk_i - parameter that describes how often, user could not answer (or did not want to) to a question for category c_i.

yn_i - number of positive or negative answers to question for category c_i.

    Based on those 2 parameters, category usability is defined in formula 2:

$$u(dk, yn) = \begin{cases} 1.0 & \text{if } dk_i = 0 \vee yn_i = 0 \\ 0.1 & \text{if } \frac{dk_i}{yn_i + dk_i} < 0.1 \\ \dfrac{dk_i}{yn_i + dk_i} & \text{otherwise} \end{cases} \qquad (2)$$

3   Results

From the user's perspective, the most important factors, when evaluating the system is quality of clusters (groups of articles) that are presented to him, as well as convergence of interactive search process to set of results, which are acceptable and useful. To evaluate quality of computed clusters, for few sample queries, their cohesion was measured, that is, conceptual cohesion of articles that are inside each cluster. This parameter was tested twice, during first test; cohesion was measured subjectively, by analyzing results by a tester. During second test, objective cohesion factors were introduced and measured.

    Convergence of the search process was also measured, based on how subjective cohesion of clusters was changing during the test interactive search session.

For tests, Polish Wikipedia was user. Currently, the system works with wikipedia files, and it cannot use English Wikipedia because of current performance issues. Simple wikipedia is much smaller than any database the system will work in the future, that is why Polish Wikipedia, with over 500000 articles in over 50000 categories was chosen.

3.1 Subjective Cohesion

Subjective cohesion C_c for cluster c is defined as

$$C_c = \frac{|X_c|}{|Y_c|}$$

where:

X_c - set of articles in cluster c which are logically connected to it (based on subjective opinion of a testing person)

Y_c - set of all articles in cluster c

C_c is defined analogously to precision parameter p defined in popular metric - $F-measure$ [10].

Cohesion for a whole query is defined as a sum of cohesion for each cluster, divided by their number. Other measured factors are minimum and maximum cohesion. Results for few test phrases is presented in table 1.

$$C = \frac{\sum\limits_{c \in C} C_c * |Y_c|}{\sum\limits_{c \in C} |Y_c|} \tag{3a}$$

$$C_{min} = \min C_c \tag{3b}$$

$$C_{max} = \max C_c \tag{3c}$$

Table 1. Results for subjective cohesion tests

Phrase	C	C_{min}	C_{max}	no.ofclusters	no.ofarticles	std.dev.C
Widelec	1.000	1.000	1.000	11	124	0.000
Jądro	0.894	0.600	1.000	26	635	0.1071
Niemcy	0.993	0.857	1.000	11	601	0.0410

3.2 Objective Cohesion

Second experiment consisted of calculating objective cohesion (by excluding human factor). To measure that, algorithm based on measurement of minimum distance between articles and their grouping category was created.

To define this this metric, distance between an article and a category had to be defined. If an article a_i belongs to a category c_j, then their relation is equal to 1: $R(a_i, c_j) = 1.0$. If between an article and category there is no relation, $R(a_i, c_j) = 0.0$. If an article belongs to a category indirectly, then R lies between 0.0 and 1.0. Indirect belonging means, that there is such set C_s, that:

- One of a category $c_k \in C_s$ has direct relationship with article a_i.
- One of a category $c_k \in C_s$ has direct relationship with category c_j.
- Each category from set C_s has direct relationship with at least one other category from this set.

Set C_s is called affiliation chain of article and is defined for article - category pair $C_l(a, c)$. Grouping category is defined as c_r and for each subset $A_x \subset A$ it is true that:

$$\forall_{a \in A_x} R(a, c_r) > 0.0$$

Grouping category is a minimal grouping category, if in set C_{ss} which is sum of all affiliation chains between articles from set A_x and c_r there is no other grouping category for set A_x.

Picture 3.2 shows sample set of articles and categories and illustrates defined concepts. Articles are shown as $A1...A5$, categories as $K1...K7$. Edges of the graph are relations of values > 0.0. In given sample, article $A2$ has direct relation to category $K2$, as well as, through affiliation chains, indirect relations with categories $K3, K4, K7$.

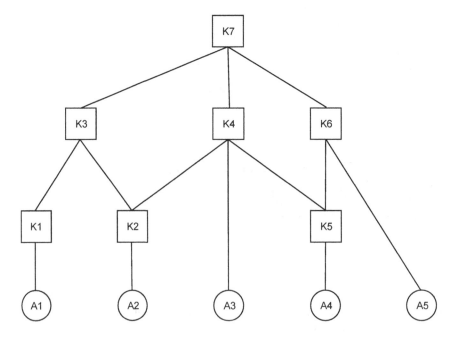

Fig. 3. Sample article and category set

Article $A3$ has one direct relation with category $K4$ and indirect relation with category $K7$. If we define an article subset created from articles $A2, A3$ then, according to the definition, grouping categories for this set will be categories $K4, K7$. Minimal grouping category in this case is the category $K4$, because there is no other affiliation chain, which would be common for articles $A2, A3$ and would contain other grouping category. This is also the reason why category $K7$ is not a minimal grouping category.

Described experiment consisted of defining minimal grouping categories for each cluster (article group) and then calculating weighted relationships between articles and grouping categories. Results are shown in table 2.

Acquired data shows, that for many clusters, distance between articles and minimal grouping category is equal to 1.0. This means, that all articles directly belong to category that represents given cluster. For bigger clusters, which represent more general context, as for example "Science" distance grows. Bigger distance means, that articles should be classified as a separate cluster. Experiments show, that this is the case for clusters of size 60 or bigger.

3.3   Convergence of the Search Process

This experiment measure the quality of convergence of the results in the process of interactive search. Because, IG algorithm is solely responsible for filtration of results after each user response, this algorithm is mostly tested in this experiment.

Measured parameter is subjective, because it requires a tester to decide quality of result clusters.

This experiment has been defined as follows:

– User session is started with a test text query. Also, a group of target articles is chosen by tester.
– Cohesion of articles is calculated, based on result articles and the ones chosen in previous step.
– Best answer is chosen by the user. For each such step, cohesion is calculated.
– Process is stopped, when user decided that the cohesion has reached acceptable value.
– Convergence of cohesion is calculated using formula: $\frac{coh_e - coh_s}{steps}$
  where:
  coh_e - Cohesion at the end of the test
  coh_s - Starting cohesion
  $steps$ - Number of answers given to the system
  Both, quality of cohesion as well as how fast the acceptable limit was reached is taken into account, when calculating the result.

Widelec. Target articles - Sztućce kuchenne, nóż, łyżka, widelec. In Figure 4 shows how cohesion was changing during the interactive search process. At the beggining, number of articles was 124. After first two steps, cohesion value raised about 5 times, mainly because number of articles dropped from 124 to 31. Next answer did not change the result by much. 4th answer made the cohesion worse, because of inappropriate categorization of some articles, however, 5th answer improved results even further.

Table 2. Results for an objective cohesion experiment

Phrase	Articles	Minimum grouping category	Distance
Widelec	-Brudny widelec -A teraz coś ... -Latający cyrk Monty Pythona	Monty Python	1.3333
	-Widelec -Łyżkowidelec -Widelczyk -Chochla -Sztućce -Łyżka	Sztućce	1.0000
	-Widelec rowerowy -Amortyzacja rowerowa -Mostek rowerowy -Kierownica rowerowa -Części rowerowe -Rama rowerowa	Części rowerowe	1.0000
	-Piotr Mocarski -Jacek Janowicz -Krzysztof Szubzda	Polscy artyści kabaretowi	1.0000
	-Kabaret Widelec -Gable -Niezbędnik -(63 innych)	Nauka	7.5693
	-Manewr Worka -Belladonna coup	Rozgrywka w brydżu	1.0000
	-Lech -FIS -WFM -MOJ 130 -Perkun -Podkowa -WSK -SHL -WSK M21W2 S2	Polskie motocykle	1.2222
	-BMW R 17 -BMW R-23 -(14 innych)	Motocykle BMW	1.0000
	-Honda CB 600F Hornet -Honda CBR 125R -Honda FMX 650	Motocykle Honda	1.0000
	-Romet 760 -Romet 210 -Komar	Motorowery Romet	1.0000
	-Lista odcinków serial Latający cyrk Montty Pythona -Lista odcinków serialu animowanego Yin Yang Yo! -Lista odcinków serialu Czarodzieje z Waverly Place	Listy odcinków	1.0000

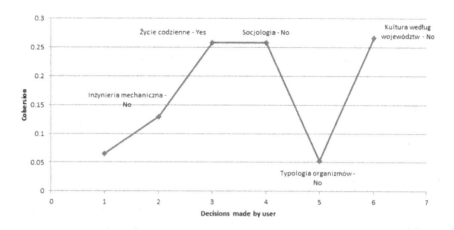

Fig. 4. Results of convergence of cohesion for query "Widelec"

For this test, convergence of cohesion for first 3 steps was equal to 0.0645 and for 5 steps - 0.0404.

4   Further Development

Currently, developed system can be classified as a complete implementation of text search engine. The system can be accessed at `http://bettersearch.eti.pg.gda.pl` for both, Polish and simple English Wikipedia. Sample snapshot from system is presented in Figure 5.

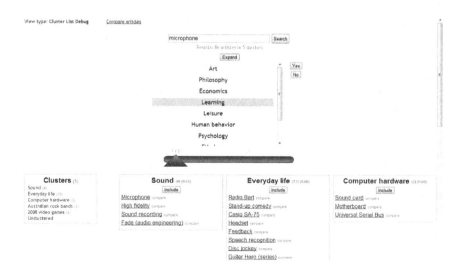

Fig. 5. Sample snapshot from running system

Main development direction right now is to increase performance of the system. This can be achieved by both, optimizing existing source code, as well as introducing algorithmic optimizations. At the moment, limit of the system is around 600000 to 1 million articles in one database. It should, however, be able to work with databases larger than 5 million articles, which would be enough to import and query English Wikipedia.

Second most important branch is to import data from different sources than Wikipedia itself. This would allow importing any text data and work with it using only Wikipedia categories.

Acknowledgments. This work has been supported by the National Centre for Research and Development (NCBiR) under research Grant No. SP/I/1/77065/1 SYNAT: "Establishment of the universal, open, hosting and communication, repository platform for network resources of knowledge to be used by science, education and open knowledge society".

References

1. Salton, G., McGill, M.: Introduction to modern information retrieval (1986)
2. Silverstein, C., Marais, H., Henzinger, M., Moricz, M.: Analysis of a very large web search engine query log. ACM SIGIR Forum. 33, 6–12 (1999)
3. Stonebraker, M., Wong, B., et al.: Deep-web search engine ranking algorithms. PhD thesis, Massachusetts Institute of Technology (2010)
4. Page, L., Brin, S., Motwani, R., Winograd, T.: The pagerank citation ranking: Bringing order to the web (1999)
5. Zhou, D., Bian, J., Zheng, S., Zha, H., Giles, C.: Exploring social annotations for information retrieval. In: Proceeding of the 17th International Conference on World Wide Web, pp. 715–724. ACM (2008)
6. Farnham, S., Lahav, M., Raskino, D., Cheng, L., Ickman, S., Laird-McConnell, T.: So. cl: An interest network for informal learning. In: Sixth International AAAI Conference on Weblogs and Social Media (2012)
7. Sanderson, M., Croft, W.: The history of information retrieval research. Proceedings of the IEEE 100 (2012) 0018–9219
8. Scholer, F., Williams, H., Yiannis, J., Zobel, J.: Compression of inverted indexes for fast query evaluation. In: Proceedings of the 25th Annual International ACM SIGIR Conference on Research and Development in Information Retrieval, pp. 222–229. ACM (2002)
9. Ester, M., Kriegel, H., Sander, J., Xu, X.: A density-based algorithm for discovering clusters in large spatial databases with noise. In: Proceedings of the 2nd International Conference on Knowledge Discovery and Data Mining, vol. 1996, pp. 226–231. AAAI Press (1996)
10. Baeza-Yates, R., Ribeiro-Neto, B., et al.: Modern information retrieval, vol. 82. Addison-Wesley, New York (1999)

Home Pages Identification and Information Extraction in Researcher Profiling

Rafał Hazan and Piotr Andruszkiewicz

Institute of Computer Science,
Warsaw University of Technology,
Warsaw, Poland
R.Hazan@stud.elka.pw.edu.pl, P.Andruszkiewicz@ii.pw.edu.pl

Abstract. In order to create a structured database describing researchers, home pages can be used as an information source. As the first step of this task, home pages are searched and identified with the usage of the classifier. Then, the information extraction process is performed to enrich researchers profiles, e.g., extract phone and e-mail. We proposed the algorithm for extracting phone numbers, fax numbers and e-mails based on generalised sequential patterns. Extracted information is stored in the structured database and can be searched by users.

Keywords: researcher profiling, homepages identification, information extraction, classification, Generalised Sequential Patterns.

1 Introduction

In a modern world, knowledge about currently conducted researches and their results is very important. Analyzing the resources of the knowledge about this domain we focused on the problem of dissemination of information about researchers and research results. Current lack of standards and systems supporting storage of scientific oriented information caused large distribution of repositories, storing data mostly in unstructured formats. Hence, it is really important to have access to a system that collects information about science from different sources and present in a unified form.

In order to create the part of the system for gathering information about researchers, a prototype for extracting information about researchers has been implemented in [1]. Although the work was focused on extracting information about researchers, the experience acquired by this experiment can be easily extended to other object categories. In this paper we will show the algorithms for identifying specific pages and extracting information from the pages, specifically for the objects *researcher*, bearing in mind that a similar approach can be applied for other categories.

In general, in order to enrich knowledge about various objects based on the web resources, we should be able to:

R. Bembenik et al. (Eds.): *Intell. Tools for Building a Scientific Information*, SCI 467, pp. 41–51.
DOI: 10.1007/978-3-642-35647-6_4 © Springer-Verlag Berlin Heidelberg 2013

1. Identify specific pages describing given objects of a certain category;
2. Extract important information from the pages, that is, properties (e.g., position, affiliation of a person, e-mail, telephone number) that characterize the object.

Identification of web pages, the first step of the mentioned process, is described in Section 2 and the second step, information extraction, is discussed in Section 3.

2   Researchers' Home Pages or Introducing Pages Identification

For specific types of objects very specific web resources should be analyzed. For acquiring information about researchers, the home pages of researchers can be used as information sources. In addition, introducing pages (Web pages that introduce researchers) can be utilized as well. Thus, at the beginning of the process of enriching a personal profile we identify a researcher's home page or an introducing page. Then, information published on the page should be extracted and stored in the system. In the first step of home page or introducing page identification, we find home page candidates using Google search engine (we used HTTP client from AsyncHttpClient library [2]), thus we submit the query with the following pattern: first name, if available, last name and 'homepage' keyword.

The results (n first web pages) returned by the search engine are potential candidates to be a researcher's home page or introducing page. Hence, each web page (based on its content) is classified whether it is a home/introducing page or not. For this reason we used the following classifiers to identify a desired web page category:

1. Naïve Bayes classifier (trained over binary, TF, or TF-IDF features).
2. Linear SVM classifier (Liblinear library was used [3]).
3. Classifier based on regular expressions, e.g., when the classifier finds the regular expression: `(?si)(\w+\W+)1,4home\W+page\b`, it assigns the home/introducing page class.
4. Classifier considering the number of pictures on a web page, e.g., for web pages with at least one picture the classifier assigns the home/introducing page class.
5. Web page address classifier, e.g., when the address of a web page contains '.edu', the web page is classified as a home/introducing page.
6. Web page title classifier, e.g., the classifier assigns a web page to the home/introducing page class when a web page title contains at least one of the phrases: 'home page', 'personal page', or first name of the researcher.
7. Web page length classifier: web pages not longer than 2000 words are treated as home/introducing pages.

8. Ensemble classifier - the classifier that consists of a subset of the above classifiers and chooses the final class by weighted voting.

We used naïve Bayes classifier as the simple state of art classifier and the SVM which is known to be one of the best classifiers and often used in this kind of task. We plan to utilize a non-linear SVM classifier in future works.

Naïve Bayes, Linear SVM classifiers and ensemble classifiers that contain at least one of these classifiers need to be trained on prepared data to be used for web pages classification. In order to collect a set of home/introducing pages and non-home pages, we created a list of home pages from DBLP [4], downloaded those pages and manually filtered incorrect addresses. For non-home pages we created the set of HTML documents in the following way: using Google we searched home pages for persons from DBLP, removed all home pages from the results and left only non-home pages. Both sets contained 1100 pages each. The module of the home or introducing page identification can work interactively, for example, the researcher's name is entered by the user or in background mode, then the name(s) of researcher(s) is (are) fetched from the database and then the process of home/introducing page identification is performed separately for each researcher.

As the Scala and Java were chosen to implement the system, the described module was created with the usage of these programming languages.

Having the home/introducing page chosen in the way we presented above, we extract information about the researcher in the automatic way. The process of extraction is described in the next section.

3 Scientific Information Extraction from Text Documents

As stated before, given a home page or introducing page of a researcher, automatic information extraction can be used in order to perform the researcher's profiling. In the process of profiling we find on the web page the values associated with the various properties of a person model, e.g., Warsaw University of Technology for the property: current university of a researcher (affiliation). To this end, we followed a unified approach to researchers' profiling, based on the textual data tagging [5]. In this approach, the problem consists in assigning appropriate tags to extracted parts of the input texts, with each tag representing a specific profile property, such as, e.g., position, affiliation, grade, etc. According to [5], we have defined the following tags: position, affiliation, bsuniv, bsdate, bsdegree, bsmajor, msdegree, msmajor, msuniv, msdate, phddegree, phdmajor, phduniv, phddate, interests, address, email, phone, fax, publication, homepage, and resactivity. Thus, the goal of our profiling task is to find a phrase on a given web page and assign one of the aforementioned tags to it. To this end, for each tag we use various methods, depending on the accuracy of each method for the specific kind of information. The next subsections present the approaches we used to perform researchers' profiling.

3.1 Regular Expressions

In this method a regular expression for a given property should be designed. The value is assigned to a given tag if the regular expression is matched to this value on a web page. For instance, we used

```
""" fax[^\-\(0-9+]{1,15}[\+0-9\-\(\) ] + """.r
```

to find fax numbers. We also used this method for telephone number and e-mail because these are well structured properties, usually preceded by a given term, e.g., fax, e-mail, etc.

E-mails, for security reasons, could be distorted in different ways (one of the examples is `ezrenner [at] uni-tuebingen [dot] de`) or presented as pictures. For distorted e-mails we applied the transformation, which changes different variants of `at` sign to `@` and `dot` to `.` sign. Other solution is to add special regular expressions for distorted e-mails. In order to find e-mails in pictures, all pictures with `e-mail` in the neighborhood (in description, file name, preceding words) can be found and checked by means of OCR.

3.2 Generalized Sequence Patterns in Information Extraction

The next method for automating information extraction proposed within the project is based on Generalized Sequential Patterns (GSP). There are four main steps in this method.

1. Training data transformation,
2. GSP discovery,
3. Weights calculation,
4. Sequence matching.

In order to perform training and testing phase, for this method and the following ones we used data provided by the ArnetMiner team [6]. The set contains of 898 annotated home pages or introducing pages of researches, so it is well suited for performing the quality tests. In the first step, we removed unnecessary annotations and transformed important data, e.g., numbers to special tokens. We will explain the step of data transformation on the example of phone number extraction. The following tasks are performed in this step:

1. Change tags [phone]phone_number[/phone] to keyword GoodCandidate.
2. Remove the rest of tags.
3. Transform the remaining potential phone numbers to keyword Candidate.
4. Remove all tokens except Candidates, GoodCandidates, and neighbors of them.
5. Remove the HTML tags and punctuation marks.
6. Change all sequences of digits to keyword Numbers.

The example of the part of the original web page is shown in Figure 1. The transformed web page is presented in Figure 2.

```
[contactinfo]Contact Information:
   Chang Wen Chen, [phddegree]Ph.D[/phddegree].
   [position]Allen S. Henry Distinguished Professor[/position]
   [position]Director[/position],
   [affiliation]Wireless Center of Excellence[/affiliation]
   [affiliation]Department of Electrical & Computer Engineering
   Florida Institute of Technology[/affiliation]
   [address]150 West University Blvd.
   Melbourne, FL 32901
   U.S.A.[/address]

   Office: [address]305 Olin Engineering Complex[/address]
   Phone: [phone](321) 674-8769[/phone]
   Fax: [fax](321) 674-8192[/fax]
   Email: [email]cchen@fit.edu[/email]][/contactinfo]
<IMAGE src="black-pixel.gif" alt=""/>
Copyright ?2005 Wireless Center of Excellence.
Florida Institute of Technology Prof. Chang Wen Chen
```

Fig. 1. The example of the part of the original web page

```
florida institute of technology
   [numbers]  west university blvd
   melbourne fl  [numbers]
   usa

   office  [numbers]  olin engineering complex
   phone  [candidate]
   fax  [goodcandidate]
   email  cchen @ fitedu

copyright  [numbers]  wireless center
of excellence florida institute of technology
```

Fig. 2. The transformed web page

In the second step, generalized sequential patterns are discovered by means of the GSP algorithm proposed in [7]. Having discovered patterns, we assign weights to the sequences. It is proposed to calculate the weight for a given sequence as the percentage of occurrences of the sequence on the web pages that contain GoodCandidate to the number of all occurrences of the sequence. The list of the example sequences with weights is shown in Table 1.

Table 1. The example sequences

Sequence	Weight
(phone, candidate, fax, goodcandidate, email)	0.98
(phone, fax, goodcandidate)	0.97
(phone, candidate, goodcandidate)	0.97
(goodcandidate, email)	0.60
(candidate, goodcandidate)	0.52

In the last step, the discovered sequences are matched to the web pages that we want to annotate (i.e., the ones that have been positively classified as personal home pages). The web pages are transformed in the same way as the training data, except for the steps 1 and 2, which are omitted this time. Then we try to match sequences to the web page. The values classified as Candidate from the web page could have been matched to both keywords: Candidate and Good-Candidate, but we add a corresponding weight only for candidates matched to GoodCandidates. We accept candidates with averaged weights greater then a given threshold. The presented method has been applied for telephone number, fax number and e-mail tokens. The results obtained by the described method are presented by a prototype application, which was created with Lift Framework [8], and deployed on the servlet Jetty [9]. An example result of Vladimir Vapnik's profile extraction is shown in Figure 3.

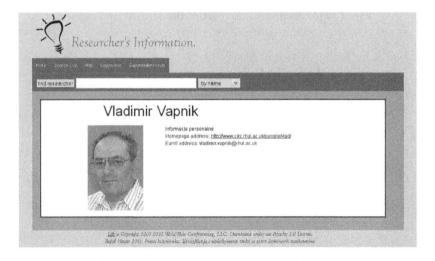

Fig. 3. The example home page identification and profile extraction

In our prototype solution, the extracted profiles are stored in the MongoDB database and can be searched/browsed using the aforementioned application.

4   Experimental Evaluation

In this section we describe experiments we conducted in order to test proposed methods.

4.1   Collecting Candidate Pages in the Process of Home Pages Identification

The first phase of home pages or introducing pages identification was to find candidate pages. We used Google search engine to perform this task. Candidate pages were links returned by the search engine as a result of the following query: firstname lastname "homepage". In order to test this phase, we collected 200 links to home pages and names of researchers from DBLP. Then we asked a query and checked whether there is a home page link in the results. Hence, we are interested in recall because we check whether one of the links matches the link we collected from DBLP.

The number of returned links can be parameterized. Thus, we checked the recall measure for this task with respect to the number of returned links and presented the results in Figure 4. We obtained the best results for 20 returned links, however, we obtained high recall for the cases with at least 10 links and for higher number of links the results were only slightly better. For 15 - 17 number of returned links, the results were slightly worse then for 14. Please, notice that the higher number of returned links, the harder classification task is. Moreover, the time we need to download all web pages returned by the search engine is higher. Hence, we assumed that the optimal number of returned links is 10.

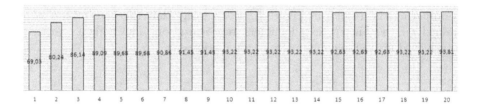

Fig. 4. Recall of the search engine with the respect to the number of results

4.2   Web Pages Classification in the Process of Home Pages Identification

The next phase of home pages or introducing pages identification was to indicate the home page among the returned candidate links. In order to test this part of the process, we created the set of HTML documents containing home pages and non-home pages. To this end we downloaded home pages indicated in DBLP. Then, we manually filtered pages without errors and in English. The set of non-home pages was prepared as a results of queries to the Internet search engine about persons from DBLP database. We filtered out all home pages (not only those for a given person) and part of similar pages. This set should be representative for non-home pages because only pages returned by the search engine are classified in the process of indicating home pages of researchers. We collected 1100 pages for each set.

Naïve Bayes Classifier. During training the naïve Bayes classifier we changed the following parameters to find the best combination.

- Type of the input data, that is, binary data, TF, TF-IDF,
- The maximum number of attributes, i.e., a dictionary size.

First, we checked the influence of words from the stop word list. The accuracy of classification increased and was 90% after the stop word list was applied compared to 88%. Moreover, removing words from the stop word list reduced the size of the dictionary. In addition, we removed all words with count less then 2 and observed that the accuracy of classification was not lower. This step removed additional 4689 words. Usually, the removed words were either proper nouns or HTML code remains.

As a next step, we checked the precision of classification with the naïve Bayes classifier with respect to the type of the input data and the dictionary size. The results of this experiment are shown in the Figure 5.

The training set had 500 pages per class and the test set was created with 600 pages per class. All words with count less then 2 were removed and we preferred words with higher frequency in all documents. The maximum number of words was 29771.

Comparing the results in Figure 5, we may say that the best results were obtained for the binary text document representation. The highest precision was achieved for the maximum size of the dictionary.

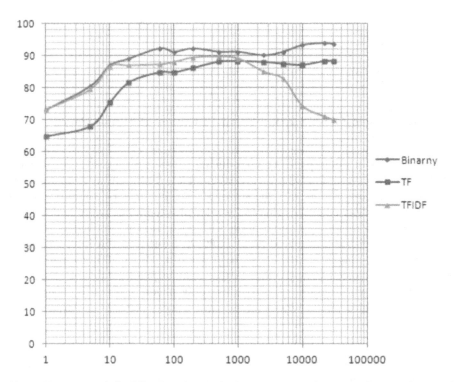

Fig. 5. Precision of the NB classifier with the respect to the type of input data and the size of the dictionary

The classifier with TF-IDF representation yielded high results for the dictionary size up to 1000. For bigger dictionary size the precision dropped. Although TF-IDF representation is the most complicated among these presented in this paper, the naïve Bayes classifier trained with this representation has not achieved the best results.

The bigger dictionary size, the higher precision of the classifier based on TF representation. For the size of dictionary greater than 1000, the classifier with TF representation obtained better results than for TFIDF representation. However, This classifier did not outperform the classifier based on binary representation.

SVM Classifier. We used linear SVM classifier in our tests. Table 2 presents the results of the tests with SVM classifier with respect to the number of attributes. The experiments were conducted in the same way as for naïve Bayes classifier.

Table 2. Precision of the SVM classifier

Number of attributes	Precision
60	93.28
100	94.34
200	93.78
500	93.28
1000	90.1
2500	91.16
5000	89.04
10000	92.22
22000	87.98
29771	89.04

The best precision for SVM classifier is higher by 2% than the best precision for naïve Bayes classifier. However, SVM needs more time in the training phase. The best results for SVM classifier were obtained for 100 attributes, which is much smaller than for naïve Bayes classifier.

Ensembles. In order to improve the results of classification, we created combination of the classifiers. The web page length classifier was designed to eliminate pages which are not home pages. Hence, this classifier has recall in the home pages set equal to 0, but precision in the non-home pages is very high. Web page title and bold phrases classifiers works in the opposite way, thus, have recall equal to 0 in the non-home pages set and high precision in the home pages set. This simple classifiers cannot be used separately and combined with only simple classifiers. Combining SVM classifier with high weight and the simple classifier we obtained results better than single SVM classifier. High weight of SVM classifier corrects errors of simple classifiers and for errors generated by SVM classifier, the simple classifiers improve the final results. The experiments for combined classifiers are shown in Table 3, where the names of the classifiers mean:

- cl-len - web page length classifier,
- cl-title - web page title classifier,
- cl-bold - bold phrases classifier,
- cl-l-t - web page length and title classifiers (weights 1.0),
- cl-svm-l-t-b - SVM, web page length, title, bold phrases classifiers (weights 1.0),
- cl-svm1.5-l-t-b - SVM (weight 1.5), web page length, title, bold phrases classifiers (remaining weights 1.0).

Table 3. The results of the classifiers

Classifier	Overall precision	Precision home pages	Recall home pages	Precision non-home pages	Recall non-home pages
cl-len	70%	0	0	100%	11%
cl-title	89%	100%	33%	0	0
cl-bold	87%	100%	42%	0	0
cl-l-t	86%	92%	31%	72%	10%
cl-svm-l-t-b	93%	91%	94%	92%	92%
cl-svm1.5-l-t-b	96.5%	96%	97%	97%	95%

4.3   Information Extraction

Information extraction was tested on set of about 800 annotated web pages published by ArnetMiner service [6]. 70% of the web pages were used as the training set, the remaining web pages constituted the test set. We compared GSP and regular expression solutions. Table 4 shows the results for fax number, telephone number, and e-mail. The GSP method yielded significantly better results than the method based on regular expressions. However, the GSP method is more time and memory consuming.

Table 4. Accuracy of the classifiers

Token	Measure	Regex	GSP
Fax number	Precision	0.82	0.95
	Recall	0.78	0.95
Telephone number	Precision	0.86	0.95
	Recall	0.46	0.85
e-mail	Precision	0.93	0.98
	Recall	0.79	0.80

5   Conclusions and Future Work

An experimental system described above is a prototype of more general platform which will be implemented in future works. The aim of the platform will be extraction of varied type of information regarding science, but not only personal data from home pages. To give an example, extracted information will include data describing science organizations, relations between co-authors of documents, conferences and other events. Retrieved contents will be converted to instances of appropriate concepts and integrated in a knowledge base. A source of the concepts definitions will be the Ontology of Science [10]. Prepared application enabled us to test number of algorithms and technologies and to select the best solutions for implementation of the target system.

References

1. Hazan, R.: Identyfikacja stron domowych ludzi nauki i wydobywanie z nich informacji. Bachelor's Thesis, Warsaw University of Technology (2012)
2. https://github.com/ning/async-http-client
3. http://www.csie.ntu.edu.tw/~cjlin/libsvm/index.html
4. http://www.informatik.unitrier.de/~ley/db/
5. Yao, L., Tang, J., Li, J.Z.: A unified approach to researcher profiling. In: Web Intelligence, pp. 359–366. IEEE Computer Society (2007)
6. http://arnetminer.org/labdatasets/profiling/
7. Srikant, R., Agrawal, R.: Mining Sequential Patterns: Generalizations and Performance Improvements. In: Apers, P.M.G., Bouzeghoub, M., Gardarin, G. (eds.) EDBT 1996. LNCS, vol. 1057, pp. 1–17. Springer, Heidelberg (1996)
8. http://liftweb.net/
9. http://jetty.codehaus.org/jetty/
10. Synat system ontology, http://wizzar.ii.pw.edu.pl/passim-ontology/

On Digitalizing the Geographical Dictionary of Polish Kingdom Published in 1880

Maciej Durzewski, Andrzej Jankowski, Łukasz Szydełko, and Ewa Wiszowata

Faculty of Mathematics, Informatics and Mechanics,
University of Warsaw, Banacha 2, 02-097 Warsaw, Poland
m.durzewski@mimuw.edu.pl,
{andrzej.adgam,lukiszydelko,ewa.wiszowata}@gmail.com

Abstract. Printed encyclopedic texts provide us with great quality and organized textual data but make us face several problems. Most of the modern data extractors are indexers that create rather simple databases. Such type of database provides keyword search and simple links between subjects (based on co-occurrences). The most advanced extractors are supported by predefined ontologies which help to build relations between concepts. In this paper we study the problem of digitalizing the Geographical Dictionary of Polish Kingdom published in 1880. We address two kinds of problems: technical problems of converting scanned pages of that dictionary to reliable textual data and theoretical challenges of organizing that data into knowledge. Our solution for organizing data into knowledge is based on defining an appropriate ontology and rules for converting textual data to ontological knowledge. We describe methods of extracting simple information like keywords and bootstrapping it into higher level relations and discuss their possible uses.

Keywords: text mining, search engine, semantic, OCR, ontology, geographical, dictionary.

1 Introduction

Nowadays most textual pieces of information are digitalized. There are many accessible, digitalized databases. The most famous ones are Wikipedia and DBpedia, but we also have other abundant sources like online dictionaries, news archives, blogs, discussion groups, forums and so on. Encyclopedias are also useful sources of information. They were usually created for commercial purposes and thus contain reliable information and knowledge. The most popular encyclopedias concern general knowledge, but there are also domain-specific ones like encyclopedias of biology, geography or physics. By encyclopedic texts we mean large sets of terms related to a global topic. In encyclopedias, terms may have links to other ones and different terms may share the same facts, dates, places and so on. Another interesting fact is that most of them are created by a very narrow group of people which causes that terms contain repeatable patterns which are very useful during automated processing.

R. Bembenik et al. (Eds.): *Intell. Tools for Building a Scientific Information*, SCI 467, pp. 53–64.
DOI: 10.1007/978-3-642-35647-6_5   © Springer-Verlag Berlin Heidelberg 2013

Geographical Dictionary of Polish Kingdom (GDPK) [1] published in 1880 is a very useful historical encyclopedia. It contains reliable and detailed information about geography of Polish Kingdom and other Slavic countries. From modern perspective the authors of this opus have done gargantuan job during collecting and organizing very detailed data. This is probably the most precise description of the Polish territory ever made. A scan of this dictionary can be found at http://dir.icm.edu.pl/pl/Slownik_geograficzny.

An encyclopedia is represented as a list of entries, where each entry is represented by the main term and a few sentences which describe that term. All the main terms usually relate to a specific topic in a regular way. In geographical dictionaries we would expect terms related to rivers, mountains, lakes and other elements of environment. GDPK is, however, more urban than geographical, and we also find cities, villages, railway stations, telegraph stations, churches, parishes, rivers (which are usually related to cities) and even population structure. That is, we find there almost everything what is necessary for reconstruction the urban structure of Poland in 1880. This dictionary consists of 16 volumes, each of which has nearly one thousand pages and about eleven thousand entries. Many entries in this dictionary are marked by initials of authors.

As a part of SONCA [2] project which is a part of the SYNAT [3] project, we plan to digitalize GDPK, extract knowledge and share it with other people. In this paper we address two kinds of problems: technical problems of converting scanned pages of that dictionary to reliable textual data and theoretical challenges of organizing that data into knowledge. Current goal which we are going to achieve is to extract data from pictures and organize them into relational model. Our solution for organizing data into knowledge is based on defining an appropriate ontology and rules for converting textual data to ontological knowledge. We describe methods of extracting simple information like keywords and bootstrapping it into higher level relations and discuss their possible uses.

2   An Overview of the Project

The main goal of this project is providing data for semantic search engine. Today word "semantic" is widely used and its meaning is very fuzzy. Under "semantic search" we understand possibilities of searching information in given context and discovering dependencies between concepts. Nowadays people are familiar with search engines using keywords or they are writing full sentences hoping that somebody asked the same question on one of the forums. If we translate this approach to function we will get an intersection of counterdomains of functions (function from keyword to set of documents) in first case and simple function key-value in second case. Although these two approaches are working great in most of cases, they have a few weak points. First of all a user has to know a keyword. This is trivial observation but it has not so trivial consequences. It causes a situation that if there is no common part we will not get any results. We are also unable to use sets of keywords in our queries (giving set instead of one keyword would match any word from the set - not only one). We could easily solve this limitation by introducing complex logic formulas in queries, for instance: ('dog' and ('black' or 'brown') and (not 'barking')) which means that we are searching for a black or brown dog which is not barking. For ordinary use it is very impractical. But here we can use semantic properties of texts.

2.1   Semantic Relations

By semantic we mean connection between given word, the real world and other words. When we talk about semantic we have to understand what kind of relation can exists in texts. We can look at the WordNet [6], it is a very large semantic data base. Words contained therein are connected by well-defined relations (each WordNet is different so I have picked the most common relations): hyponymy, hyperonymy, synonymy and meronymy. If we use such kind of WordNet we can easily define complex queries with even one word only. For example if we are trying to find "dog", search engine will expand "dog" using all hyponyms such like "dachshund", "sheep-dog", "terrier" and many others. Nowadays search engines are able to make such kind of query expansions. Of course this is not the only way for expansion of queries. Solutions based on Concept Similarities are very effective, especially that they are using similar techniques used during building WordNet. We have to mention here the Concept Similarities and its usefulness in queries expansion. One of many methods is based on the Local Context Analysis [8]. Techniques which are presented produce results similar to synonymy (in some cases similar to hyponymy and hyperonymy), so our first results can be compared with results of this method.

2.2   Our Semantic Goals

From high perspective we have two types of goals. Firstly we are going to focus on simple relations between phrases and concepts. These relations can be extracted with simple statistical methods. Another goal is an ontology. Ontology which is able to describe concepts from GDPK and dependencies between them.

Semantic Based on Words. We are going to do almost the same thing as the authors of WordNet [6] did. We are going to extract relations between words based on the text and knowledge provided by experts. There are two big differences between WordNet goals and ours. First of all we do not try to create general method for all kind of words, but we are going to focus on proper names and few popular adjectives in text. Secondly, we are going to find different functions than hyponymy, hyperonymy and others. We are going to find relations from higher abstraction level, for instance: something has been built by someone, someone has been living somewhere. If we had to do it based on ordinary texts it would be extremely difficult. Fortunately, by using this dictionary we have texts which define relations in which we are interested in. Due to this fact we do not have to create sophisticated natural language analysers and we can focus on sentences which contain information in common way. Despite our comfortable situation we are going to achieve situation where all words and phrases are well identified (it means that we have high precision in assigning to each group) and all of them are loaded into our databases for future analysis.

Ontology Relations. So far we are researching two approaches. In the first one we are going to create not very detailed and general ontology and during processing GDPK enrich it. In the second approach we are going to create "ontology" in a bootstrap way, with unnamed relations (how to name them and determine what they mean will be a new problem). Despite that we have to decide which approach will be more universal and more accurate, the global goal remains the same. We want to have data structure (ontology) which contains well-defined concepts and relations between them. The main task is to provide structure which will be useful in our search engine.

2.3   First Approach to Search Engine Enrichment

At the beginning of analytical part of solution we would like to focus on people, cities/villages, events and time points/intervals. The most interesting is not extraction but connection between concepts from different groups. This step allows us to create four-dimensional space which will be a direct enrichment of search engine. The most obvious way how we can add new features is an OLAP cube. Such OLAP cube will have 3-4 dimensions: cities/villages, structure of population, points in time and probably events. Query in such environment can be easily projected on the cube dimensions and results will be more accurate. The important thing is that projecting process will be not direct. With the help of ontology we will be able to use queries from higher level of abstraction. Having information that "many citizens" or "big city" describes city with population over 10K we are able to transform such query to one of the cube dimension which is expressed by raw numbers.

3   Data Processing

By data processing we mean process from image files to data structures based on Cassandra framework. In our solution we use our programs supported with commercial solution. Total process is presented on the following scheme (see Figure 1).

3.1   OCR

In the first step we have to create solution responsible for extraction of text from images. Due to the fact that text is printed we assumed that we will be able to create OCR good enough (what does it mean we explain in next section).

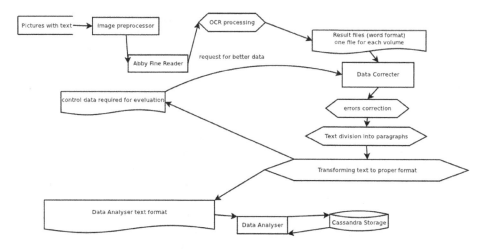

Fig. 1. Data extraction process

First Approach. Our OCR consists of a few very standard components: parts responsible for reading/writing images, part responsible for recognition of columns and lines of text, part responsible for recognition of words and spaces and part (neural network) responsible for recognition of characters. Unfortunately, during the test we realized that number of errors is much bigger that we expected. After research we discovered that the main problem is the situation when two (or sometimes more) characters are joined. Symmetrical problem was the situation where one character ("m" for example) is divided into several parts. Because we do not have additional knowledge about letter statistic (language used in the dictionary is different than modern Polish) we are not able to add this factor on the character recognition level. We have tried to create algorithms to discover these situations but it did not solve our problem. Another problem was the small size of training set. We have only 30 pages of training text. We have huge disproportion in representation of different characters and even normalization has not given much better results. Most errors are located in the names of entries because names are written with bold and unclear fonts. Final results of our solution are shown in table (see Figure 2). We realized that commercial solution should solve our problems (just because it is much better trained). Results generated by Abbyy Fine Reader are much better from characters errors and words errors point of view, but it generates new type of errors - wrong layouts. Because text in GDPK is organized in two columns, the OCR has to determine which entries should belong to which column. Another important feature is indentation. Each entry has indentation before its header. Saving this element is very important for entries detection. Unfortunately, a combination of two columns layout and oblique scans create a situation where many indentations are abandoned and some entries are placed on wrong positions. Then we decided to join two OCRs and we created complex solution.

	our OCR	Abbyy Fine Reader	combined solution
% of character errors	9	1,5	1,5
% of lexemes errors	14	9	9
% of wrongly detected entries	0,2	0,4	0,1
% of missed entries	1,5	3,8	1

Fig. 2. Our OCR results

Final Solution. We transformed our OCR into image preprocessor. It is responsible for taking every page, dividing it into two columns (so far this is encoded but solution is prepared for any number) and dividing each column into parts (each part must contain at most one entry). After creating a new set (it is an ordered set, we know proper position of each entry/part of text) we put it as data into Abbyy Fine Reader. Because of preprocessing we have almost eliminated errors on entries level. Last error was the most typical error - wrongly recognized character. Final results were flawed with 1,5% wrongly recognized characters which resulted in a 9% of wrong lexemes (see Figure 2). By wrong characters we mean every wrongly recognized character - even dots and commas. By wrong lexemes we mean lexemes with at least one character changed. Unfortunately, errors are concentrated in proper names (headers of entries) what causes next two problems. The first problem is that we are not able to correct such lexeme because (in most cases) it exists only once in whole GDPK and we even do not know that there is a mistake. The second problem arises from the first and when we are trying to find for example a village, we can not search with 100% match but we have to assume that there is a error and maybe we have to pick another village. Finally during the search process we have to operate with a set of keywords not one keyword. We have also tried correction of errors not only in proper names. Due to the fact that we do not have corpora from 19th century, the only corpora which we have is GDPK. We proposed a bootstrap process, but more than 40% of all words used in GDPK exist at most two times and we have no other materials to compare. Finally the undetectable errors exist in numerical data (wrong years, numbers of citizens etc.) and so far we have reconciled with them. Such kind of error occurs in a number, date or other lexeme which is unindexable (because without context they mean nothing and they are part of bigger data set). The only possible solution is to process such lexeme in context and by using knowledge database check if given value has rational value.

3.2 Post-OCR Processing

In the next step we have to correct the text as much as possible. Due to the fact that we are not able to base on semantic text, we have to use available syntactic information. First we have to detect headers of entries. We are able to do few simple assumptions about headers of entries. We know that all of them are in lexicographical order and each of them is starting with a capital letter. Simple heuristic based on these two conditions corrected about 50% of total single lines and joined them correctly to proper entries. Other types of errors are errors located on the lexemes level. Most detectable lexemes errors are correctable with predefined rules. Each rule is a translation pattern. If the left side exists in the text it will be replaced with the right part of given formula. These formulas are resilient to greedy apply because predicate pattern of each formula is not right lexeme.

match pattern	proper value
'włość,'	'włośc.'
'włość.'	'włośc.'
'włośo.'	'włośc.'
'przysiółek'	'przysiołek'
'orn,'	'orn.'
'om.'	'orn.'
'mn,'	'mn.'
'poczt,'	'poczt.'
'st,'	'st.'
'dworak,'	'dworsk.'
'dworsk,'	'dworsk.'
'Bozi.'	'Rozl.'
'pry', 'w.'	'pryw.'
'pos,'	'pos.'
'posiadł,'	'posiadł.'
'paraf,'	'paraf.'
'polio.'	'polic.'

These rules were used about 2 thousand times during processing of the second part (800K lexemes) and reduced overall error from 2% to below 1.5% from character error perspective and to 9% from 10% error from lexemes point of view (see Figure 3).

number of pages	926
number of lexemes	813012
number of entities	11250
% of character errors	1,5
% of lexemes errors	9
usages of correct rules	2188

Fig. 3. 2nd volume statistics

4   Raw Data Model

Finally we want to translate textual data into a model which will provide us with a possibility of contextual queries. We decided to create a model where items are organized into relations. Relation can exist not only between two elements but could also be a single argument (we can call it *property*). There can be a relation between three or more elements. Elements in relations can be anything: lexemes, phrases and - what is the most important - entries from other relations. This solution provided us with possibilities for creating relations between relations and in consequence it allows us to create high level relations. We would like to give some examples of such kind of relations:

single argument relations (properties):
ITEM is a "city" (i.e. ITEM belongs to set of cities)
ITEM is a "date" (i.e. ITEM belongs to set of dates)
ITEM is a "person" (i.e. ITEM belongs to set of people)

double argument relations:
(ITEM1, ITEM2) are in relation "adhesion" (i.e. such tuple belongs to set of all double near places)
(ITEM1, ITEM2) are in relation "flows trough" (i.e. such ordered tuple belongs to set of all "flows trough" pairs)

three or more argument relations:
(ITEM1, ITEM2, ITEM3, ITEM4) are in relation "event in time executed by someone" (i.e. ITEM1 is a place, ITEM2 is a date, ITEM3 is a person, ITEM4 is a name of event, and such ordered tuple belongs to set, where all elements describe many events)

A careful reader will notice that the first two type of relations are almost the same as ordinary ontology format, but more argument relations can be emulated by two argument relations. The main advantage over classic ontology is lower complexity and strong organization data around relations. If we look at the data we will get simple relational data model (see Figure 4).

But after a while we realize that some of the tables have open structure and entries could be much bigger than any top bound. We have to decide what kind of operations we will be executing on the data base and how much data we are going to contain.

4.1 Data Model Requirements

For our purposes, we need solution which provides such operations:

```
insert lexeme
get lexeme id
insert phrase
get phrases with given lexeme
check if lexeme exists in entry
check if phrase exists in entry
insert entry
give all entries with given phrases
insert tuple into given set
check if given tuple exists in set
```

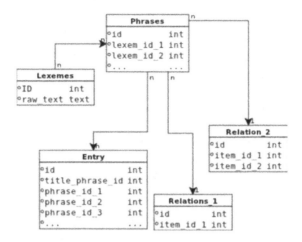

Fig. 4. Simple relational data model

Each of these operations should not have bigger complexity than $O(\log(n))$ (according to data size) due to the fact that each operation is going to be used as a part of iteration algorithm and total number of lexemes is greater than 16M. We realized that *Entries* and *Relations* tables have matrix properties, so we decided that we should use solution which is able to simulate matrix properties. There is one difference from ordinary matrices that we would like to introduce into our indexes instead of numerical. We decided to use Entity Attribute Value solution as the closest to our needs. We have chosen Cassandra framework [5].

4.2   Cassandra Data Model

Our data model assumes that everything is in relation and lexemes are the atomic values in our solution (see Figures 5, 6, 7, 8, 9 and 10).

4.3   Relations

In the first approach we would like to focus on basic - one argument relations. So far we do not have any corporas with texts from the 19th century, so we have to base our extractions on the current knowledge. Due to the fact that authors defined a set of keywords that are used for marking important proper names like villages, parishes, rivers and many others we are able to determine with great certainty what given proper name means. Next important pieces of information are points and intervals in time. Most of them are easily detectable because of clear information that something happened in a given year or, especially in the case of a century, are written in Roman numbers. Finally the most difficult problem on this stage concerns names and surnames. We do not have a complete set of past names and surnames so we cannot do a full match. We are able to

Lexemes	raw_text
lexeme_id	lexeme_value

Entry relation	title_phrase_id	description_id
entry_id	id	id

Single value relation	lexem_1_id	tlexem_2_id	lexem_3_id	lexem_4_id	...
relation_item_id	boolean	boolean	boolean	boolean	boolean

Phrases	lexem_1_id	tlexem_2_id	lexem_3_id	lexem_4_id	...
phrase_id	boolean	boolean	boolean	boolean	boolean

Description relation	lexem_1_id	tlexem_2_id	lexem_3_id	lexem_4_id	...
description_id	boolean	boolean	boolean	boolean	boolean

Fig. 5. Cassandra scheme

lexem_id	raw_text
1	Zabór
2	Pruski
3	1880
4	r.

Fig. 6. Lexemes table

	Lexemes identifiers			
phrase_id	1	2	3	4
1	1	1		
2			1	1

Fig. 7. Phrases table

	Phrases identifiers	
relation_item_id	1	2
1		1

Fig. 8. Single value relation (ie. being a date)

entry_item_id	title_phrase_id	description_id
10	12	13

Fig. 9. Entry relation

	Phrases identifiers	
description_item_id	1	...
101	1	..

Fig. 10. Description relation

use context where given word exists (for instance we can search titles) and we can detect surnames using knowledge about the structure of Polish surnames (suffixes -wski, -cki, and similar).

4.4 Extraction Relations

We based the project on these rules and decided to create three main one-argument relations: places, point time and people. Each of them could have some subclasses, especially in the case of places when we extract a few types. Relations extracting is solved with simple rules. In case of city names we do not have much work because most of them are in headers of entries. Points in time in most cases are explicitly described so we can use very simple rules. Situation with people names is a little more difficult but in many cases such phrases are preceded by proper titles of given person. Extraction of more complex relation poses a real challenge. Currently we experiment with very simple formulas like "date in city description" or "word river in description". By this simple research we want to determine what is a potential for relation extracting.

4.5 Technical Solution

To achieve the objectives described above we created program which is responsible for extracting relations and managing data structures. Our data storage is Cassandra. Program is written in C++ and his Cassandra interface is based on Thrift framework. It is responsible for dumping data to Cassandra, extracting data from Cassandra, extracting relations and processing queries. It is also responsible for many helper functions as counting support indexes and buffering data. One of the most important (from efficiency point of view) is data buffering. If we try to process every query with fresh data extracted from Cassandra then we would be able to process only few thousands (about two) of queries.

5 Semantic Data Model

Final data can be treated like OLAP cube. We are able to search using criteria such as places, time and people. Relations from higher abstraction level provide us with possibilities for searching and making queries not only by keywords like cities or rivers names, but we will also be able to ask about all cities near river. So far we are in the experiments phase. We have to determine which goals are reachable in reasonable time. First deployment of complete system should not take longer than one month. We will be able to fully assess global data state and our possibilities of creating knowledge database. Sample cube described in 2.3 will be four-dimensional: places, citizens, time points, events. Size of each dimension is estimated: number of places is bounded by total number of entries multiplied by constant (this constant describes different places in every entry like communities, parishes etc.); citizens will be grouped into several nationalities, time points are restricted to full years; events in most cases are connected with ownership changes (so not many types). The cube contains references to entries which meet conditions. Because we have about 180K entries, our cube will be sparse matrix. Proper implementation ensures linear size according to number of entries (we do not count data required for containing ontology and relations, because estimates are not precise enough). If we want our engine to be used we have to provide solution with response time less than 1 second per one query. I do not talk about total efficiency in queries per second because it is scalable issue.

6   Conclusions and Futureworks

Our long-term goal is to create methods responsible for automatic discovering and extraction of more complicated relations. Today we do not have full knowledge about how current errors (which remain in data) will generate noise. We realize there is a risk that current error level is too big for decent data processing. We also have to do research on the efficiency model. We have 16M lexemes which translates into similar amount of phrases, but relations between phrases and our goal - relations between relations probably will grow in an exponential way. The next field to research is solution where after program gets a few facts for instance that city was built in some year, it will be able to extract similar information for other cities and dates without any human support. This is necessary in case when the data source is growing constantly.

References

1. Sulimierski, F., Chlebowski, B., Walewski, W.: Słownik Geograficzny Królestwa Polskiego I Innych Krajów Słowiańskich (1880),
   http://dir.icm.edu.pl/pl/Slownik_geograficzny
2. Nguyen, L.A., Nguyen, H.S.: On Designing the SONCA System (2012),
   http://dx.doi.org/10.1007/978-3-642-24809-2_2
3. SYNAT project web page, http://www.synat.pl/
4. Abbyy Fine Reader web page, http://www.finereader.pl/
5. Cassandra project web page, http://cassandra.apache.org/
6. WordNet project web page, http://wordnet.princeton.edu/
7. German project genealogy.net, http://wiki-en.genealogy.net/SlownikGeo
8. Xu, J., Bruce Croft, W.: Improving the Effectiveness of Information Retrieval with Local Context Analysis (2000)

Chapter II

Repository Systems

Semantic Data Sharing and Presentation in Integrated Knowledge System

Michał Krysiński, Marcin Krystek, Cezary Mazurek, Juliusz Pukacki,
Paweł Spychała, Maciej Stroiński, and Jan Węglarz

Poznań Supercomputing and Networking Center,
Noskowskiego 10, 61-704 Poznań, Poland
{mich,mkrystek,mazurek,pukacki,spychala,stroins,weglarz}@man.poznan.pl
http://www.psnc.pl

Abstract. A rapid development of semantic technologies and a growing awareness of a potential of semantic data has a great influence and
effects on an increasing number of semantic data repositories. However,
the abundance of these repositories is hidden behind sophisticated data
schemas. At the same time, the lack of user-friendly tools for presenting data makes this collected knowledge unavailable for average users.
In this paper we present our concept of creating a convenient semantic
environment for ordinary users as well as maintaining rich functionality
available for professionals. Our adaptation of a classic three-layer architecture for the semantic data sharing and presentation purpose reveals
how the great usability and accessibility of data can be achieved. We
also pay special attention to the standards and cooperation issues, which
makes our system available for other tools implementing semantic data
sharing concepts and keeps it open for Linked Open Data environment
integration.

Keywords: semantic data, data management, data sharing, data presentation, Linked Data.

1 Introduction

Integrated Knowledge System (IKS) gathers data concerning cultural heritage
obtained from distributed, heterogeneous sources like digital libraries, digital
museums, scientific and technical information systems. This great amount of
data is analyzed, purged, unified and transformed into a semantic form. A further processing brings an additional value to raw data by introducing knowledge
of meaning of each part of the data. Therefore, raw metadata usually of poor
quality, which is used only to describe cultural heritage objects, is transformed
into the high quality knowledge form. This knowledge can become a subject of
another machine processing due to its highly structural and precisely defined
form or can be visualized and presented to the user in a clear and intuitive way.
Regardless of accepted and implemented scenarios of a further data usage, it is
strongly required to render all data available in the most efficient and accessible

R. Bembenik et al. (Eds.): *Intell. Tools for Building a Scientific Information*, SCI 467, pp. 67–83.
DOI: 10.1007/978-3-642-35647-6_6   © Springer-Verlag Berlin Heidelberg 2013

way. Existing solutions which enable sharing semantic data do not meet expectations and requirements defined by the Integrated Knowledge System. Hence, new advanced semantic data sharing and presentation software components were designed and implemented. It is strongly believed that those components will bring a new quality to the semantic data sharing and presentation process as well as push the whole concept of semantic data sharing to the next level.

The remaining part of this paper is organized as follows: Section 2 presents other projects which deal with the semantic data sharing and presentation problem. In section 3 the semantic data sharing and presentation architecture is proposed and explained. Section 4 presents a motivation underlying data sharing layer creation. Section 5 describes a physical and logical data model used by the data sharing layer as well as its REST[1] API interface. Details concerning how semantic data are visualized and presented to the user is the subject of section 6. Final conclusions and future work plans are given in section 7.

2   State of the Art

While the importance and strength of the semantic technologies are growing, efforts are taken to share and present semantic data in the most effective and useful way. A number of concepts and infrastructures has been proposed and developed within projects funded in recent years. Unfortunately, most of them are domain, purpose or architecture specific. In this section some of well-known projects are presented.

BRICKS "aims to build a Europe-wide distributed Digital Library in the field of Cultural Heritage"[1]. The concept is based on a peer-to-peer network of cooperating organizations. Resources can be searched and presented at any node in the network regardless of where the original resource is stored and maintained. Resource metadata are transformed into RDF format. However, this semantic metadata is not public but summarized and stored in a given metadata schema.

STERNA[2] is a domain specific solution in which all kinds of information related to birds are collected. This project aims to develop a set of best practices on the scope of supporting digital libraries by an integration and semantic enrichment of digital resources.

Apache Stanbol[3] is a set of RESTful independent components which can be combined in order to implement a required usage scenario. The main aim of the Stanbol platform is to support and extend traditional web content management systems with semantic solutions. This involves a dynamic content extraction, natural language processing, content enhancement and reasoning. Persistence services, which store data and knowledge models services used to define and manipulate data models, are also available.

MuseumFinland is a project developed for "...publishing heterogeneous museum collections on the Semantic Web"[5]. MuseumFinland employs a set of domain specific ontologies which were developed solely for this project. These ontologies are used to integrate heterogenous collections of data which were stored

---

[1] REpresentational State Transfer.

in a single repository. Afterwards, processed and enriched data are presented to the end-user via WWW portal.

3 Semantic Data Sharing and Presentation Concept

An architecture of the Integrated Knowledge System (IKS) sharing and presentation functionality was inspired by the classic three layer application model. Required modifications have the desired result in the following functionality layering:

- **Knowledge Base (KB)** is a semantic data repository. It is a lower part of the IKS system, which ensures a durability of data. The KB holds data in the RDF triple format (subject, predicate, object), which is a natural and widely used representation of semantic data. A schema describing a structure of data is expressed as OWL ontology document unlike in the case of relational based systems in which a structure of data is reflected by a structure of a database.
- **Query Processing Module (QPM)** is an implementation of the semantic data sharing and integration layer which is a middle part of the IKS system. It translates detailed semantic data collected in the Knowledge Base into user abstract concepts.
- **Portal** is an implementation of the presentation layer. It takes full advantage of the QPM functionality. In this portal semantic data are presented in the clear and user-friendly way. Moreover, it implements a set of specific scenarios exploiting the potential of semantic data.

The proposed three layer structure makes it possible to separate presentation of resources from data storage and on the other hand - data retrieval and resource creation. This approach guarantees a flexible, rapid and independent development of different components of the system, and, at the same time, preserves the main functionality of the whole application stack.

Both the Knowledge Base and the Query Processing Module are available via REST endpoints. The Knowledge Base implements Sesame repository API [4]. The Query Processing Module defines its own API, which is described in section 5.3. The Portal is a client application for the data sharing layer (QPM) and it introduces a graphical user interface to the system.

4 Data Sharing Layer

The Knowledge Base is a large repository of semantic data, which has been built carefully and with a great effort. The process of constructing the KB is designed as a workflow of actions including data harvesting, normalization, mapping to semantic representation, relation discovery and enrichment. Details concerning KB building procedures can be found in the following paper: "Transforming a Flat Metadata Schema to a Semantic Web Ontology" [6]. One of the main assumptions underlying the concept and design of Integrated Knowledge System is to make all collected data publicly accessible and available for a large number of potential users. Hence, ontology which describes data is based on the commonly used and widely accepted CIDOC Conceptual Reference Model [7]. Further details concerning applied ontology are available in the paper "Applicability of CIDOC CRM in Digital Libraries" [8]. Unfortunately, the schema and the data itself are complex. Moreover, existing standard software libraries (API) and query languages (SPARQL) provide only a low-level support for retrieving and processing data. Therefore, there was a risk that a great abundance of the data remains unavailable for an average user and even in the case of qualified users an access to the data is limited.

A data sharing layer is introduced to the Integrated Knowledge System as a solution in the case of complex data, which addresses the issue of the lack of high level API. The implementation of the core of data sharing layer is called Query Processing Module (QPM). A software module is what mediates between very general, high-level concepts which are typical and natural to operate on by an average user, and atomic and very specific objects collected by the Knowledge Base repository based on ontology used. The main aim of this component is to provide an implementation of common semantic logic which can be shared by different applications, making an application development process faster and simpler.

QPM is implemented as a REST service with a simple API and it provides a semantic view of the collected resources rather then a pure set of triples. It is designed and implemented in accordance with Linked Open Data (LOD) [9] best practices and it meets commonly accepted standards with regard to semantic data structure, format and sharing procedure. A detailed description of the implemented standards is provided in the section 5.2.

4.1   Specificity of Semantic Data Sharing

Sharing semantic data can be achieved in a simple way by publishing schema which describes data and by giving a public access to the repository. This approach is fast and simple but, at the same time, it is very hard to conduct by an average user and highly impractical. As a consequence of such a complex retrieval of important information from the repository, each scenario which involves the usage of data published this way will have to assume a great overhead.

Typical relation based systems created to share text resources use some kind of text indexes to locate requested resources. More advanced scenarios take a meaning of the indexed content into consideration and, in this way, they introduce types or categories of information like for example name, surname, place, date, title, etc. Regardless of the implemented strategy of indexing and searching, a user is provided with an original document which was indexed. Semantic repository does not contain any documents. In contrast to the relation based system, this semantic repository stores fact objects - a minimal amount of data of a specific type, meaning and defined relations to other facts. The initial data content was transformed into unified basic facts. Therefore, there is no single, meaningful document or object that could be presented to the user. It is the role of the Query Processing Module (or more general - data sharing layer) to recreate resources which represent and correspond with general concepts understandable for the user. It should be emphasized that the QPM can recreate a number of various resource types which were not directly mentioned and described in raw source data. In order to be consise with the Integrated Knowledge System, the Knowledge Base is built based on publication records available in digital libraries. Each raw record describes a publication but due to its decomposition, unification and understanding of a semantic meaning of the data, some new concepts are also derived. Hence, raw information from publication records allows a user to create not only publication resources but also related to person, place, work and others.

Each application which employs a semantic data sharing layer receives an access to a great diversity and abundance of resources. Thus, it can be optimized in respect of presenting resources rather than dealing with the Knowledge Base. Furthermore, new resources may become available for all applications at once only by introducing them to the QPM.

4.2   Various Scenarios

A constant development of the Query Processing Module enables semantic data sharing layer to participate in a number of different scenarios involving semantic data retrieval and processing.

The implementation of the QPM is based on the commonly accepted standards concerning sharing semantic data, which were published by the W3C. This especially applies to a data format, structure and content of the information. Compliance with standards renders the usage of generic semantic tools for accessing and processing data possible. Furthermore, fulfilling the Linked Open Data (LOD) [9] requirements and best practices enables the Integrated Knowledge System to become a standard data provider in the LOD environment. This feature places the IKS between such well known semantic services like Geonames [10], Viaf [11] or DBPedia [12].

In the near future, the IKS semantic data sharing layer will enable a dynamic enrichment of retrieved resources with information stored in external semantic repositories. This scenario involves a dynamic schema translation between different data providers. Although this process requires further research, assumed outcomes will entail a new, better quality, diversity and abundance of data which will be presented to the user.

5   Data Sharing Model

Data sharing process requires a definition of what kind of information will be published and how this process will be realized. This particular type of agreement can be described in general as a data model on which both Query Processing Module and its clients will operate to accomplish their functionality.

Data model description is divided into two parts: logical and physical, depending on the level of abstraction it represents.

5.1 Logical Data Model

The semantic data sharing concept assumes that data exchanged between the
Query Processing Module and user application is self-descriptive and that it has
a complete semantic meaning. The main portion of such organized data is called
resource. It is a representation of some abstract concepts like place, person, book,
etc. Available types of resources are derived from the ontology (CIDOC-CRM)
used by the Knowledge Base and real data collected in this repository.

The following types of the resources are currently identified:

- WORK
  http://dl.psnc.pl/schemas/frbroo/F1_Work
  In the FRBR model, a publication, such as a book, is described at 4 different levels: work, expression, manifestation and item. Work is "a distinct
  intellectual or artistic creation"; what the author conceived on the abstract
  level. Work may be a "publication work" (monograph) or a "serial work"
  (series, e.g. a periodical).
- PUBLICATION EXPRESSION
  http://dl.psnc.pl/schemas/frbroo/F24_Publication_Expression
  Expression is "the intellectual or artistic realization of a work", such as an
  edition of a work or a translation, still on the abstract/intellectual level.
- ITEM
  http://dl.psnc.pl/schemas/frbroo/F5_Item
  Item is a physical copy of an expression. A set of identical items is called
  Manifestation.
- PERSON
  http://erlangen-crm.org/current/E21_Person
  Represents a human being. A person description contains, if are available:
  name, surname, dates of birth and death, context - identifiers of things which
  were created by this person or activities in which they were involved in some
  other manner.
- PLACE
  http://erlangen-crm.org/current/E53_Place
  Represents a place. This may be any area which is described, a subject of or
  something which has a significant relationship with other resources like for
  instance a city where some book was published.
- LEGAL BODY
  http://erlangen-crm.org/current/E40_Legal_Body
  Represents an organization/institution, such as a publisher of a book.
- SUBJECT
  http://dl.psnc.pl/schemas/ecrm-extended/E55e_Subject
  This is a publication subject that has not been recognized as a named entity
  such as a person or a place. Many subjects are taken from different lists of
  subject headings such as the Library of Congress Subject Headings or Polish
  KABA.

The resource data content has a graph structure which is built in such a way as
semantic data are represented. Each vertex represents a data object and each

edge represents a relationship between objects. One of the vertexes is always distinguished and is interpreted as a root of the resource. This allows to address this resource as an entity - URI of the object represented by such vertex became an URI of the resource.

5.2 Physical Data Model

Sharing semantic data on the web requires the usage of standards which enables humans and applications to process and understand data in the same way. The biggest initiative whose main purpose is to integrate many sources of semantic data is the Link Open Data (LOD) project [9]. The QPM implementation follows a set of best practices suggested by the LOD regarding a structure of shared data.

Due to those requirements, all data available through the IKS sharing layer are organized in the form of XML document. Depending on a requested QPM functionality, an XML document which is returned may be formatted according to RDF/XML [13] or SPARQL-RESULTS/XML [14] document schemas which are both W3C commonly accepted standards. The RDF/XML schema is used when the returned document holds a description of the resource in terms of logical data model described above. This also applies to the SPARQL query request which returns graph (e.g. construct query). The SPARQL-RESULTS/XML schema is used when a set of triples must be returned. This is the case when SPARQL tuple query was requested (e.g. select query).

Information about the XML response document format is conveyed to a client within a HTTP header. A proper mime type value is used for the Content-Type field.

The remaining LOD requirements are fulfilled by the Knowledge Base or the Integrated Knowledge System network infrastructure. This concerns especially a URI structure and significance as well as a link to the resources remaining outside the Knowledge Base like Geonames [10], Viaf [11] or DBPedia [12] objects.

5.3 Interface REST

The Query Processing Module is implemented as REST service available on the Internet. Despite sophisticated processing realized by the QPM, its API should remain clear and simple. It offers the main functionality of the QPM which is to deliver a valuable description of resources based on common user requirements. Despite the fact that most of implementation details are hidden, a user can still perform sophisticated scenarios with the use of direct SPARQL queries. In such case, the QPM remains a single access point to the Knowledge Base resources.

The following part of this section discusses four QPM API usage scenarios. Each of them presents a different way of gaining data from the Knowledge Base. Every scenario begins with a definition of REST function and contains a description of its functionality and an explanation of parameters meaning.

Resource Detail Description

GET qpm/resource?resourceURI=:uri
        [&pred=:p&limit=:l[&offset=:o]]

This method returns a detailed description of the resource represented by the
:uri. If the *:uri* represents a triple whose type is not defined by a logical data
model, the QPM will attempt to locate a number of triples of the known type
for which there is a graph path between a new and a given triple. In such case,
a list of resources with short descriptions will be returned.

Optional parameters

- **pred** - a part of the resource description may be cropped because of a great
  cardinality of a predicate which connects resource represented by the *:uri*
  with other objects in the repository. The variable *pred* allows deciding which
  predicate variables *limit* and *offset* refer to. A result of such a call will be
  uri of the resource and triples which are in relation *:p* with the resource *:uri*.
  Possible values of the variable *pred* are available in metadata associated with
  the response to the previous call.
- **limit** - a number of objects which should be returned
- **offset** - an offset from the first object

Keywords

GET qpm/resources?key=:v1
        [&key=:v2]...[&limit=:l[&offset=:o]]

This method returns a list of resources along with a short description of each
of them. Selected resources will contain requested key-words in their text content.

Optional parameters

- **limit** - a number of objects which should be returned
- **offset** - an offset from the first object

Semantic Query

GET qpm/semantic?key1=:v1
        [&key2=:v2]...[&limit=:l[&offset=:o]]

The semantic query offers an opportunity to define a template which describes
required resources. A form of templates is still under development, however it
will be based on a simplified form of the SPARQL query. A resource template
creation will be supported by a special graphical tool developed as a part of the
presentation layer.

    Currently, the Query Processing Module implements a simplified model of se-
mantic queries. Each query is represented as a set of pairs (key, value), in which

key denotes a name of the parameter and value denotes a value of this parameter. Names of the parameters and acceptable values are derived from cidoc-crm ontology.

Optional parameters

- **limit** - a number of objects which should be returned
- **offset** - an offset from the first object

Examples

GET qpm/semantic/geo?lat =:lat&long =:long&distance =:dist

This is an example of the query which returns resources based on geographical data. Resources must be somehow connected to the place, for example describe it or be published there. The place is represented as a circle with the middle in *:lat* and *:long* and radius *:dist*.

GET qpm/semantic/subject?subject =:sub

This is an example of the resource subject query. It also involves KABA language subjects. A list of resources associated with a given subject *:sub* will constitute the result of the processing. A value of the variable *subject* assumes the form of uri representing this subject.

SPARQL

GET qpm/sparql?query =:q

This method returns the result of processing query *:q* by the Knowledge Base. In this scenario, instead of processing query and query results, the Query Processing Module restricts its actions to passing a query to the Knowledge Base and query results to a user. The query must be formulated as SPARQL query. The result will be formatted as a RDF/XML [13] or SPARQL-RESLULTS/XML [14] document, depending on the query form.

6   Data Presentation Layer

A primary task of the IKS is to integrate information from multiple data sources, its semantic enrichment and presentation into an accessible form. Ensuring an ergonomic, user-friendly access to this accumulated knowledge and maintaining maximum flexibility and functionality, it all requires a preparation of a set of tools. The first and most important point of access to resources is a portal application. The design of this component is based on existing, proven and widely used solutions - very well-known portals such as Google or Yahoo. In addition to standard searching scenarios, there are also plans for developing context based ones exploiting semantic features of the data storage. It includes geographically oriented searching procedures based on location information associated with resources during data enrichment process, or search procedures based on historical information. There is also a part of the portal available only for registered users, which provides a personal space for a management and storage of private resources.

6.1 Data Conversion

One of main goals of IKS Portal is to present data stored in the knowledge base in a user-friendly manner. This task involves converting data served by Query Processing Module into a simpler format which can easily be used for providing data to user interfaces in various contexts. JSON meets this requirement as its processing is implemented in numerous programming languages and application frameworks. Moreover, as a result of mapping only relevant information is passed to the presentation layer. That is to say, data obtained are simplified and do not directly reflect a structure of the underlying ontology, while maintaining all the advantages of linked data. The way RDF classes are mapped to JSON is configurable via an XML mapping file, which is described in details later in this section.

Mapper is implemented as a stateless session Enterprise JavaBean with a very simple remote interface, which declares only one method of the following signature:

```
String processGraph(String input, int depth)
```

Parameters

- **input** - RDFXML string
- **depth** - a desired level of recursion

Output:

- JSON string

Configuration File - mapping.xml. As it was stated earlier, rather than having the mapping rules hardcoded, Mapper is configurable through an XML file. The following is an excerpt from an exemplary mapping.xml used by Mapper:

```xml
<mapping xmlns:xsi="http://www.w3.org/2001/XMLSchema-
    instance"
        xsi:noNamespaceSchemaLocation="mapping.xsd">
    <namespaces>
        <namespace name="curr"
                value="http://erlangen-crm.org/current
                /"/>
        ...
        <namespace name="rdfs"
                value="http://www.w3.org/2000/01/rdf-
                schema#"/>
    </namespaces>
    <types>
        <type name="person" rdfType="%curr;E21_Person">
```

```
<attribute name="name" singular="true" label
   ="true">
   <element pred="%curr;P1_is_identified_by
      "/>
   <element expected_type="%curr;
      E82_Actor_Appellation"
           pred="%psnc;P3h_has_main_rep"/>
</attribute>
...
<attribute name="creations" singular="false"
   label="false">
   <element pred="%curr;P14i_performed"/>
   <element expected_type="%frbroo;
      F27_Work_Conception"
           pred="%curr;P94_has_created"/>
   <element expected_type="%frbroo;F1_Work
      "/>
</attribute>
</type>
...
</types>
</mapping>
```

The file consists of two main sections, namely *namespaces* and *types*. The former allows for defining shorter representations for recurring fragments of URIs. In the preceding excerpt, URI fragment *http://erlangen-crm.org/current/* is bound to *curr*. As a consequence, each occurance of the mentioned URI fragment in type definitions can be replaced with %curr;. It improves readability and shortens the file. *types* section is where mapping is defined. Each *type* element has the following attributes:

- **name** - specifies a unique name of the mapped type
- **rdfType** - specifies RDF type whose individuals are mapped to this type

Beside these attributes, each *type* contains one or more *attribute* elements whose attributes are as follows:

- **name** - specifies a unique (in the context of *type*) name of the attribute
- **singular** - specifies whether or not the attribute can have more than one value
- **label** - specifies whether the attribute's value should be used as a textual representation of the mapped resource

Finally, every *attribute* contains a sequence of *element* elements which define the way RDF graphs are traversed by the mapper. The following are attributes of *element*:

- **pred**
- **expected_type**
- **regexp**

Mapping Procedure. Firstly, for each context in the RDF graph, a type is checked and matched against the ones defined in the mapping file (*rdfType* attribute of *type*). As a next step, if a match is found, paths of all attributes defined for a given type are processed. The first element of such a path is processed by searching the graph for triples (subject, predicate, object) whose subject is equal to the matched context and predicate is equal to the one defined by the path element. For each triple in the result, an object is taken and processed using the next path element and so on.

Result. The mapping process results in a JSON structure which, apart from data derived from the knowledge base, contains metadata useful for the presentation layer. These metadata properties are available for each mapped resource. Their names are preceded and followed by two underscores in order to singnify their special purpose and to avoid confusion with the actual data. The following is an abbreviated example of such a structure:

```
[{
"__type__":"person",
"__rdftype__":"http://erlangen-crm.org/current/E21_Person
  ",
"__uri__":"http://viaf.org/viaf/61585459",
"__singular__":["death_year","name","birth_year"],
"name":"Rej, Mikolaj",
"birth_year":"1505",
"death_year":"1569",
"creations":
        [{"__type__":"work",
         "__rdftype__":"http://dl.psnc.pl/schemas/frbroo/
           F1_Work",
         "__uri__":"http://dl.psnc.pl/kb/
           F19_Publication_Work...",
         "__singular__":["title", ...],
         "title":"Krotka rozprawa miedzy trzema...",
           ...
        }],
"__label__":"Rej, Mikolaj",
"__complex__":["creations"]
}]
```

- **__type__** - used by presentation layer to choose an appropriate presentation template
- **__rdfType__** - RDF type of the mapped object
- **__uri__** - contains the unique identifier of the mapped object, used for navigation
- **__singular__** - indicates single-valued properties
- **__label__** - contains a textual representation of the mapped resource

- _ _**complex**_ _ - specifies which properties contain objects (as opposed to literals)

6.2   Various Data Views

Basic Tabular View. The basic way of presenting data is the tabular view. The table has a two-column layout with attribute names in the first column and values in the second. If an attribute has multiple values, the second column contains a paginated list. Obviously, resources of different types are likely to be described with the use of different attributes. Therefore, each of the main data types as defined in mapping.xml has its corresponding presentation template which basically defines a set of attributes for the given type as well as it indicates which attributes are to be displayed as lists. It is possible to switch to a graph view by clicking "Graph" button at the bottom of the page.

Fig. 1. Basic tabular view

Map View. The map view allows users to search for information by specifying a geographical area of interest. A user enters a name of a location (such as city, voivodeship, etc.) and radius in kilometers. When this information is submitted, the system searches for resources associated with the specified area and presents them on the map grouped by location, each of which is represented by a marker. After clicking on the marker, the user is presented with an info window containing links to pages displaying details of the resources (tabular view).

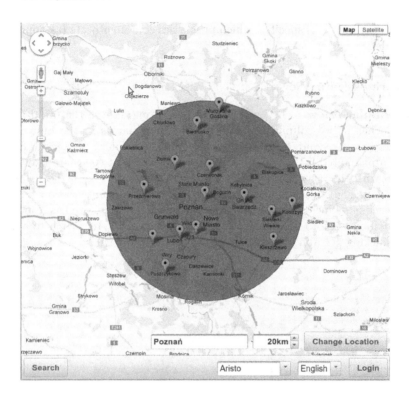

Fig. 2. Map view

Graph View. The main purpose of a graph view is to present how resources relate to one another. Resources such as works, places and so on are visualized as nodes of a directed graph, while relations are represented by edges. Such a graph view is accessible from details page of any resource. At first, it displays only the node representing this resource. More nodes can be added by choosing "Select adjacencies" from context menu available by hovering a node. It opens a dialog containing all nodes adjacent to the selected (hovered) node grouped by relation. A user can then select nodes to be added to/removed from a view and confirms it by cliking "Choose". When a resource is added to the view for the first time, it has to be fetched asynchronously as refreshing the page each time a node is added would be unacceptable from a user experience point of view. This is achieved by making AJAX requests to a REST endpoint, which serves as a simple iterface to Mapper.

GET /portal/rest/node?uri=:v1

This method returns a JSON representation of the resource identified by the URI specified by *uri* query parameter.

SPARQL View. Advanced users can issue SPARQL queries using the SPARQL search form. Depending on a kind of a query, a result can be a list of tuples or

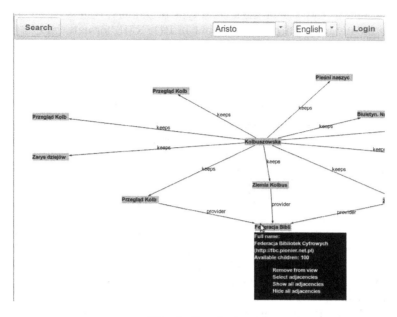

Fig. 3. Graph view

Fig. 4. SPARQL view

an RDF graph. In the case of the second type of a query, results can be further processed by any tool which accepts input in RDFXML format.

7 Summary

This paper presents the semantic data sharing and presentation concept which has been developed and implemented for the Integrated Knowledge System. It also explains functionalities of particular components of the system, and their role in providing semantic data to an end users in a convenient way.

Sharing data concept assumes that the most of a semantic data retrieval is performed by a system component called the Query Processing Module. It is the main access point to the Knowledge Base for a number of applications. The QPM provides common semantic logic required for the resource search purpose, which can be shared by different applications. It is accessible as a web service through the simple REST API. Moreover, the QPM implementation complies with commonly accepted standards concerning semantic data exchange and, therefore, it may become a data provider for the Linked Open Data environment.

A semantic data presentation is implemented by the Internet web portal. The sophisticated graphical user interface enables to search and present semantic data in a user friendly way. The portal creates a convenient semantic environment for the user. It also introduces a unique functionality which renders new scenarios of discovering and processing semantic data possible.

All the concepts and implementations discussed in this paper are developed in order to make more semantic data accessible by an average user. However, the architecture of the system remains open and allows one to create more new and sophisticated tools as well as to implement advanced, new scenarios based on user requirements.

The future work involves a further development and improvement of the portal user interface. New usage scenarios will also be introduced. The Query Processing Module still requires precise performance tests. However, these tests can be performed after a final release of the Knowledge Base repository. The further development of the QPM will focus on implementing a comprehensive support of semantic queries.

References

1. Hecht, R., Haslhofer, B.: Joining the BRICKS Network - A Piece of Cake. In: The International EVA Conference 2005, Moscow (2005)
2. http://www.sterna-net.eu
3. http://incubator.apache.org/stanbol
4. http://www.openrdf.org/doc/sesame2/system/
5. Hyvonen, K., Makela, E., Salminen, M., Valo, A., Viljanen, K., Saarela, S., Junnila, M., Kettula, S.: MuseumFinland - Finnish Museums on the Semantic Web. In: Web Semantics: Science, Services and Agents on the World Wide Web Archive, vol. 3(2-3) (October 2005)

6. Mazurek, C., Sielski, K., Stroiński, M., Walkowska, J., Werla, M., Węglarz, J.: Transforming a Flat Metadata Schema to a Semantic Web Ontology: The Polish Digital Libraries Federation and CIDOC CRM Case Study. In: Bembenik, R., Skonieczny, L., Rybiński, H., Niezgodka, M. (eds.) Intelligent Tools for Building a Scient. Info. Plat. SCI, vol. 390, pp. 153–177. Springer, Heidelberg (2012)
7. http://www.cidoc-crm.org/
8. Mazurek, C., Sielski, K., Walkowska, J., Werla, M.: Applicability of CIDOC CRM in Digital Libraries. In: Proceedings of CIDOC 2011 Knowledge Management and Museums Conference, Sibiu, Romania, September 4-9 (2011)
9. http://linkeddata.org
10. http://www.geonames.org
11. http://viaf.org
12. http://dbpedia.org
13. http://www.w3.org/TR/rdf-syntax-grammar
14. http://www.w3.org/TR/rdf-sparql-XMLres

Long–Term Preservation Services as a Key Element of the Digital Libraries Infrastructure

Piotr Jędrzejczak, Michał Kozak, Cezary Mazurek, Tomasz Parkoła,
Szymon Pietrzak, Maciej Stroiński, and Jan Węglarz

Poznań Supercomputing and Networking Center,
Noskowskiego 12/14, 61–704 Poznań, Poland
`{piotrj,mkozak,mazurek,tparkola,szymonp,stroins,weglarz}@man.poznan.pl`

Abstract. The aim of this paper is to present long–term preservation services developed in Poznań Supercomputing and Networking Center as a part of the SYNAT project[1]. The proposed solution is an extendible framework allowing data preservation, monitoring, migration, conversion and advanced delivery. Its capability to share data manipulation services and used Semantic Web solutions, like the OWL–S standard and the OWLIM–Lite semantic repository, create a potential for broad applicability of the publicly available services and therefore fostering and improving digitisation related activities. METS and PREMIS formats accompanied by FITS, DROID and FFMpeg tools are the core elements of the WRDZ services, making them interoperable and innovative contribution to the digital libraries infrastructure in Poland.

1 Introduction

Polish digital libraries infrastructure, settled in the framework of the PIONIER Polish Optical Internet, has been developed by scientific and cultural heritage institutions since 2002, when the first regional digital library — Wielkopolska Digital Library — was started. The cooperation between various local computing and networking centers and local content providers (e.g. university libraries, museums, archives) resulted in a number of new digital library deployments. Currently, with over 80 digital libraries in Poland and over 1 000 000 digital objects available online, scientists, educators, hobbyists and others have access to various documents connected with the Polish history, culture and science.

The existing infrastructure could not have been built without innovative tools and applications. The first Polish regional digital library was built in 2002 using the dLibra Digital Library Framework (`http://dlibra.psnc.pl/`) developed in Poznań Supercomputing and Networking Center (PSNC), and that framework has since been the basis for the majority of digital libraries in Poland. dLibra–based libraries provide access to approximately 97% of the Polish digital content available online via the Polish Digital Libraries Federation (`http://fbc.pionier.`

---

[1] SYNAT project is financed by the Polish National Center for Research and Development (grant no SP/I/1/77065/10).

R. Bembenik et al. (Eds.): *Intell. Tools for Building a Scientific Information*, SCI 467, pp. 85–95.
DOI: 10.1007/978-3-642-35647-6_7   © Springer-Verlag Berlin Heidelberg 2013

net.pl/), which is the national aggregator and metadata provider for Europeana (http://www.europeana.eu/).

The rapidly growing volume of available digital content and the complexity of the digitisation workflow have introduced new challenges, one of which is the problem of long–term data preservation. Currently available digital content should remain accessible in a year, five years, ten years, etc. despite the fact that software, hardware and data formats change throughout the time. A survey conducted as a part of the SYNAT project showed that almost none of the Polish institutions involved in mass digitisation use the long–term preservation approach and tools for their digital content, proving that there is a real need for a reliable, extensible and efficient software solution. PSNC, working in the scope of the SYNAT project, aims to provide such solution enriched with the mechanism of online accessibility of master files created in the process of digitisation (the source data). The idea is to advance the long–term preservation activities by giving the community means to build services attached to cultural heritage digital archives and digital libraries.

The newly developed prototype tool called WRDZ (*Wielofunkcyjne Repozytorium Danych Źródłowych*, in Polish) consists of a number of network services primarily focused on the preservation of text, images and audiovisual content. WRDZ is an open–source solution for scientific and cultural heritage institutions and provides means to introduce migration according to the OAIS model [5], conversion for the needs of digital libraries, and advanced delivery for research purposes. Moreover, the WRDZ as a whole has been designed in a way that allows its deployed instances to share all their data manipulation services in a P2P–like manner.

Subsequent sections of this article present the WRDZ services with: an introductory overview including a summary of features and characteristics of certain WRDZ services (Section 2), the idea of the source data management mechanism with its advanced storage structure and metadata extraction (Section 3), the migration and conversion mechanism featuring semantic technologies and advanced services orchestration and execution (Section 4), and the idea of the common space of data manipulation services and the service sharing capability (Section 5). The paper ends with the concluding remarks section, where the first deployment of the dArceo package (a production level solution based on the WRDZ services) is described as well.

2   Overview

WRDZ is composed of a set of network services forming an extensible framework for long–term preservation purposes. The structure of a single, fully functional WRDZ instance is presented in Figure 1 and includes the following services:

– **Source Data Manager (SDM)** — responsible for storage and retrieval of source data. In addition, SDM performs automatic metadata extraction and provides versioning mechanisms for both the source data and its metadata. By default, SFTP and local file system can be used as the underlying storage, but the architecture of the SDM allows for easy integration with other data archiving systems via adapter plugins. Moreover, all operations are handled asynchronously, thus allowing the SDM to work with archives where the retrieval of data takes significant time. An internal part of SDM is the OAI–PMH repository, which provides the OAI–PMH interface for stored digital objects' metadata harvesting. It supports both the Dublin Core and METS formats.

– **Data Manipulation Services (DMSs)** — allow the data to be migrated, converted or delivered in a different way. The framework provided by WRDZ allows users to easily add new services and, since DMSs have a clearly defined communication interface, share them with other WRDZ instances via a service registry. In addition to being used on their own, DMSs can be chained together, forming a complex data manipulation flow. There are three distinct groups of DMSs:

  • **Data Migration Services (DMiSs)** migrate the source data to a different format without losing information. They play a significant role in long–term data preservation.

  • **Data Conversion Services (DCSs)** convert the data with some of the information lost in the process (e.g. lossy compression, resolution change, cropping). DCSs are used to create presentation versions of digital objects, as the source data might not be suitable to be presented to the end user in its raw form.

  • **Advanced Data Delivery Services (ADDSs)** deliver the data in an effective, user–dependent way. For example, audiovisual data can be streamed, while large images, such as maps, can be served piece by piece as requested.

– **Service Registry (SR)** — stores information about all registered DMSs. In addition to services added locally, SR can be also configured to harvest other SRs for their publicly available services and synchronise with the central SR. This allows users of a single WRDZ instance to easily locate and access remote services, significantly increasing the number of available data manipulation options.

– **Source Data Monitor (SDMo)** — periodically verifies the data integrity and assesses the risk of data loss in the context of long–term preservation. Data integrity is checked by calculating file checksums and comparing them to the values obtained when the particular file was added. In order to assess the risk of data loss, SDMo analyzes file formats currently used by stored objects and identifies those that might become unreadable in the near future (outdated formats without official support, for example). Corrupted files and file formats with high risk factor are reported to the administrator.

– **System Monitor (SM)** — provides an overview of the WRDZ instance in terms of performance, available resources and usage. Statistics are gathered per user, allowing the administrator to generate reports with a wide range of granularity.

– **Data Migration and Conversion Manager (DMCM)** — allows users to create and execute complex data migration and conversion plans by chaining different data manipulation services. After a plan has been defined, DMCM handles its execution automatically, making even the most complex, multiple–step migrations and conversions easy to perform.

– **Rights Manager** (RM) — allows the administrator to define which digital objects and services a particular user has access to.
– **Notifications Manager** (NM) — provides the system with a unified channel for communication. Allows services to send messages to other services and the administrator in a simple way.

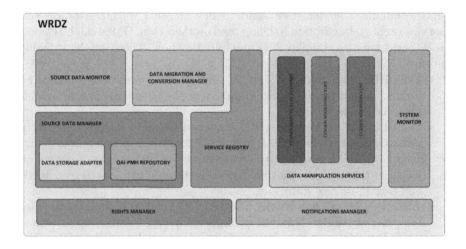

Fig. 1. The structure of a WRDZ instance

Rights and Notifications Managers form the bottom layer of WRDZ and act as supporting services, allowing services from the next layer to communicate and protect their resources while they perform their dedicated tasks. SDMo and DMCM, in addition to being dependent on RM and NM, utilise both the SDM and SR in order to execute their responsibilities. Such a dependency model gives great flexibility in terms of the user needs, as it is possible to set up a WRDZ instance consisting of RM, NM and DMS only that can be used to publish manipulation services, as well as an instance consisting of RM, NM and SDM to implement reliable storage system. Moreover, in terms of scalability, each instance of WRDZ can be deployed on multiple servers. It means that majority of services can be deployed and run on separate machines.

3   Management of the Source Data

The management of the source data lies at the very core of WRDZ and consists of two functions: infinitely archiving the data with its integrity preserved (keeping the bits intact) and ensuring that it remains readable in the future. WRDZ focuses on the latter, relying on existing data storage systems to provide data durability. WRDZ can work with a disk array, Blu–Ray or tape storage accessed via local file system or a network protocol, or a more sophisticated system such as the PLATON–U4 archiving services, which provide data replication in 10 remote localizations in Poland [4].

An integration tier between SDM and a specific data storage system is implemented using the Java EE Connector architecture [3], which, inter alia, provides a framework for processing distributed transactions. This allows to maintain consistency between the administrative metadata of a digital object stored in the relational database of SDM and the data and metadata of this object present in the data storage system.

The structure of any digital object stored in the data storage system managed by SDM is hierarchical, with the root directory of the digital object (named using the object's identifier) divided into version subdirectories, which in turn have the content and metadata subdirectories. Furthermore, the content itself can be hierarchical as well. Every version of the object has the same structure, with every content file having a corresponding subdirectory in the metadata directory with provided and extracted subdirectories, which store the metadata provided with the digital object and the metadata automatically extracted from the file, respectively. The provided and extracted subdirectories are present in the root of the metadata directory as well and contain the corresponding metadata for the whole digital object. A sample structure of a digital object managed by SDM is shown in Figure 2.

It is worth noting that while subsequent versions use the same structure in their respective directories, only the files that have changed are actually stored – WRDZ utilises the administrative metadata of the digital object to store only the references to previous versions for the remaining files, removing redundancy.

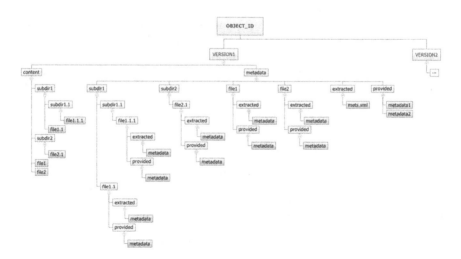

Fig. 2. A sample structure of a digital object stored in a data storage system managed by SDM

Administrative metadata is generated from scratch for every version and stored in the `metadata/extracted/mets.xml` file using the METS schema (`http://www.loc.gov/standards/mets/`). The same file also contains a section in the PREMIS schema (`http://www.loc.gov/standards/premis/`) for changes related to the object (including migrations and conversions), references to optional descriptive metadata provided by users (for the whole object or individual files) and references to technical metadata extracted from and provided for any file of the object.

Descriptive metadata can be supplied in any standard schema (DC[2], DC Terms[3], MARCXML[4], MODS[5], ETD–MS[6], PLMET[7], etc.) or an arbitrary XML file. Technical metadata uses dedicated metadata schemas depending on the file format: TextMD[8] for text files, DocumentMD[9] for documents, MIX[10] for image files, AES–X098B[11] for audio files and VideoMD[12] for video files. Supplying this metadata is optional, since every file that is a part of the digital object's content is processed by the File Information Tool Set (FITS, `http://code.google.com/p/fits/`), which can identify the file format and extract the above mentioned technical metadata automatically. FITS uses the Digital Record Object Identification (DROID) software tool (`http://sourceforge.net/projects/droid/`) to identify the file format. For the purposes of the technical metadata extraction module implemented in WRDZ, FITS was equipped with the FFMpeg tool (`http://ffmpeg.org/`), which can extract technical metadata from audio and video files[13], and the included DROID version was replaced with the latest one. As a result, this modified FITS returns two XML documents for an audio/video file: one in the AES–X098B schema (for the audio track) and another one in the VideoMD schema (for the video track).

4   Migration and Conversion Mechanism

Technical metadata discussed in the previous section is a key component in the migration and conversion mechanism included in WRDZ, and that very mechanism is what allows WRDZ to provide its long–term preservation functionality. Migration process starts in SDMo, which periodically identifies digital objects that have not been migrated (ie. are the most recent "version" of particular

---

[2] `http://dublincore.org/documents/dces/`

[3] `http://dublincore.org/documents/dcmi-terms/`

[4] `http://www.loc.gov/standards/marcxml/`

[5] `http://www.loc.gov/standards/mods/`

[6] `http://www.ndltd.org/standards/metadata/`

[7] `http://confluence.man.poznan.pl/community/display/FBCMETGUIDE/`

[8] `http://www.loc.gov/standards/textMD`

[9] `http://fclaweb.fcla.edu/content/format-specific-metadata/`

[10] `http://www.loc.gov/standards/mix/`

[11] `http://www.aes.org/publications/standards/AES57-2011/`

[12] `http://www.loc.gov/standards/amdvmd/`

[13] FITS fully handles text, document and image files, but has only basic support for audio files by default.

content) and use file formats that might become unreadable in the near future. The Unified Digital Format Registry (UDFR) [1], and its SPARQL endpoint in particular (`http://udfr.org/ontowiki/sparql/`), is used to obtain information about the analysed file format, which allows SDMo to assess the risk associated with that format and notify the administrator if that risk factor is too high.

When the administrator receives the notification, he or she can set up a data migration plan in DMCM. DMCM then uses SR to find the required migration service or services that will allow it to transform any digital object containing files at risk to a new digital object with more contemporary file formats. While a single service might be able to perform the transformation in the best case scenario, WRDZ supports service chaining for situations when no service is capable of transforming the files from the source format directly into the target format and several services need to be chained together. After the appropriate services have been found, DMCM automatically invokes them, thus constructing new digital objects which are saved in SDM. Cross references between the source object and the migrated object are then added to their administrative metadata.

In order to find and use the required services, DMCM needs a computer–interpretable description of them. Following current trends in Semantic Web, the Ontology Web Language for Services (OWL–S) [10] has been chosen as the base solution, with several specific ontologies developed on top of it: *dArceoService* ontology[14] (extending the OWL–S ontology *Service*[15]) with its subordinated *dArceoProfile*[16] (extending the subontology *Profile*[17]) and *dArceoProcess*[18] (extending the subontology *Process*[19]). Those ontologies are mandatory for all services that are to be registered as web services in SR. DMCM supports the Web Application Description Language (WADL) [7], and the *RESTfulGrounding* subontology[20] [6] is required for the grounding part of the service description.

When a service is registered in SR, all transformations between file formats that service can perform are extracted from its process model and saved in the OWLIM–Lite semantic repository (`http://www.ontotext.com/owlim`) using the language of the *dArceoService* ontology. This ontology contains the `File Transformation` class bidirectionally associated with the `Service` class from the *Service* ontology and having two properties called `fileIn` and `fileOut`, which hold the information about the input and output file format of the transformation, respectively. After all the (atomic) transformations performed by the service have been saved, the rule of inference from Figure 3 is applied, which generates complex transformations that include the transformations performed by the added service. Information about such chains of transformations is saved using the `previousTransformation` and `subsequentTransformation` properties. In

---

[14] `http://darceo.psnc.pl/ontologies/dArceoService.owl`
[15] `http://www.daml.org/services/owl-s/1.2/Service.owl`
[16] `http://darceo.psnc.pl/ontologies/dArceoProfile.owl`
[17] `http://www.daml.org/services/owl-s/1.2/Profile.owl`
[18] `http://darceo.psnc.pl/ontologies/dArceoProcess.owl`
[19] `http://www.daml.org/services/owl-s/1.2/Process.owl`
[20] `http://www.fullsemanticweb.com/ontology/RESTfulGrounding/v1.0/`
`RESTfulGrounding.owl`

```
rdf : http://www.w3.org/1999/02/22-rdf-syntax-ns#
process : http://www.daml.org/services/owl-s/1.2/Process.owl#
service : http://www.daml.org/services/owl-s/1.2/Service.owl#
dArceoService : http://darceo.psnc.pl/ontologies/dArceoService.owl#

t1 <dArceoService:fileIn> a
t2 <dArceoService:fileIn> b              [Constraint t2 != t1, b != a]
t1 <dArceoService:fileOut> b
t2 <dArceoService:fileOut> c             [Constraint c != a, c != b]
t1 <dArceoService:performedBy> s1
t2 <dArceoService:performedBy> s2        [Constraint s2 != s1]
--------------------------------------------------------------------
t <rdf:type> <dArceoService:FileTransformation>
p <rdf:type> <process:CompositeProcess>
t <dArceoService:performedBy> s
s <service:describedBy> p
t <dArceoService:fileIn> a
t <dArceoService:fileOut> c
t <dArceoService:previousTransformation> t1
t <dArceoService:subsequentTransformation> t2
```

Fig. 3. The rule of inference that generates complex transformations

order to restrict applications of this rule ad infinitum, a certain limit is imposed on the maximum length of the transformation chain.

The mechanism for conversion and advanced delivery is similar, except the generated chains of services are restricted by the technical parameters of conversion and advanced delivery in the discovering phase. Dedicated subontologies have been developed to allow specifying these parameters: *dArceoText*[21], *dArceoDocument*[22], *dArceoImage*[23], *dArceoAudio*[24] and *dArceoVideo*[25].

5   Common Space of the Data Manipulation Services

As mentioned in previous sections, WRDZ is composed of several services. One of them is dedicated to the data manipulation, including migration, conversion and advanced delivery. In order to gain synergy between institutions involved in the long–term preservation activities, the idea of common data manipulation service space (CDMS) has been introduced. CDMS is a virtual cloud of publicly accessible DMSs from different WRDZ instances created through synchronisation of their SRs. Each SR has a list of other SRs, which it harvests

---

[21] http://darceo.psnc.pl/ontologies/dArceoText.owl
[22] http://darceo.psnc.pl/ontologies/dArceoDocument.owl
[23] http://darceo.psnc.pl/ontologies/dArceoImage.owl
[24] http://darceo.psnc.pl/ontologies/dArceoAudio.owl
[25] http://darceo.psnc.pl/ontologies/dArceoVideo.owl

for available DMSs, storing the obtained information about remote DMSs locally and thus allowing them to be utilised by its own WRDZ internal services (DMCM in particular). On the other hand, that very same SR can be asked by other SRs to provide its own publicly available services, resulting in a P2P network of long–term preservation services being constructed. Moreover, a central service registry can be set up, which would communicate with all WRDZ instances in the network and effectively act as a source of information about all publicly available DMSs. The idea of CDMS is depicted in Figure 4, which presents an example with two WRDZ instances and one central SR. WRDZ A has two publicly available DMSs: DMIS A.2 and DCS A.1. WRDZ B has also two publicly available DMSs: DMIS B.1 and ADDS B.1. Additionally, there are three DMSs available from other instances: ADDS, DMIS and DCS. The DMCM of the WRDZ A can benefit from all the DMSs registered in its SR, which means that the service execution can involve, as presented in the example, DMIS A.1 (not public, but internal for WRDZ A, so can be utilised), DCS (coming from the public space) and ADDS B.1 (coming from WRDZ B). Similarly, WRDZ B and its DMCM can use publicly available remote DMSs, e.g. DCS A.1 (coming from WRDZ A).

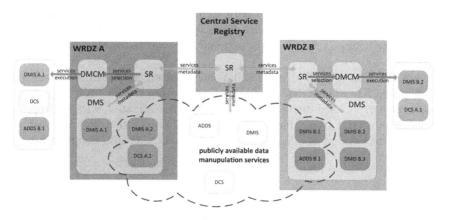

Fig. 4. Data Manipulation Services in the context of multiple WRDZ instances

As the mechanisms to share migration/conversion and advanced data delivery services are already built into the system, it is envisioned that ideally institutions with enough technical potential will serve as service providers, which other (e.g. smaller) institutions can benefit from. It is even possible for an institution to set up only migration, conversion or advanced delivery services and register them in any SR instance. The only requirement is that services need to be described in accordance to the WRDZ semantic model with the usage of the above mentioned OWL–S ontologies.

6   Concluding Remarks

WRDZ consists of a prototype set of services responsible for the realization of the long–term preservation concept, primarily for textual, graphical and audiovisual content. The basic idea behind WRDZ is an OAIS transformation approach to migration, with additional conversion and advanced delivery services attached as an added value to the core preservation services.

These prototype open–source services are the base for the so called dArceo package. The first deployment of the dArceo package, currently in the test phase, exists in scope of the Digital Repository of Scientific Institutes (DRSI). DRSI is a consortium of 16 institutes of the Polish Academy of Sciences with an ambitious plan to digitise their library collection. In the framework of DRSI there are several digitisation centers specialised in digitisation of particular types of documents, multiple libraries utilising these centers for massive digitisation and one portal (http://rcin.org.pl/) for presentation versions of the digital objects being an effect of the digitisation activities. In addition to the online accessibility of the produced resources, DRSI has been cooperating with PSNC in order to develop a software solution dedicated to the management of the digitisation workflow. As an effect PSNC has developed the dLab tool (http://dlab.psnc.pl/), which is capable of handling digitisation tasks composed of certain activities related to particular stages in the digitisation workflow. The archiving stage is done by the dArceo services that create digital objects suitable for long–term preservation.

Future works on WRDZ (dArceo) will focus on closer coupling its services with the Integrated Knowledge System [8] developed within the SYNAT project, and the issue of personalised delivery process will be investigated, with appropriate services introduced if applicable. This solution will also be thoroughly tested and analysed in the context of efficiency, flexibility, scalability and extendibility.

References

1. Abrams, S., Anderson, M., Colvin, L.D., Frisch, P., Gollins, T., Heino, N., Johnston, L., Kunze, J., Loy, M., Powell, T., Reyes, M., Ross, S., Salve, A., Strong, M., Tramp, S.: Unified Digital Format Registry (UDFR). User's Guide (2012)
2. Bearman, D.: Reality and Chimeras in the Preservation of Electronic Records. D–Lib Magazine 5(4) (1999), doi:10.1045/april99-bearman
3. Boersma, E., Brock, A., Chang, B., Chung, E., Gilbode, M., Gish, J., Pedersen, J., Pg, B., Przybylski, P., Tankov, N., Thyagarajan, S., Weijun, T.W.: JSR–000322 — Java EE Connector Architecture 1.6, Sun Microsystems Specification (2009)
4. Brzeźniak, M., Meyer, N., Mikołajczak, R., Jankowski, G., Jankowski, M.: Popular Backup/Archival Service and its Application for the Archival of the Network Traffic in the PIONIER Academic Network. Computational Methods in Science and Technology, Special Issue, 109–118 (2010)
5. Consultative Committee for Space Data Systems, ISO Technical Committee 20: Reference Model for an Open Archival Information System (OAIS). ISO 14721:2003 Standard (2003)

6. Ferreira Filho, O.F., Ferreira, M.A.G.V.: Semantic Web Services: A RESTful Approach. In: White, B., Isaías, P., Nunes, M.B. (eds.) Proceedings of the IADIS International Conference on WWW/Internet, pp. 169–180 (2009)
7. Hadley, M.: Web Application Description Language. W3C Member Submission (2009), http://www.w3.org/Submission/wadl/
8. Krystek, M., Mazurek, C., Pukacki, J., Sielski, K., Stroiński, M., Walkowska, J., Werla, M., Węglarz, J.: Integrated Knowledge System — an Integration Platform of Scientific Information Systems. In: Nowakowski, A. (ed.) Materiały Conference Infobazy 2011, pp. 98–103 (2011) (in Polish)
9. Kuipers, T., van der Hoeven, J.: Insight into Digital Preservation of Research Output in Europe. Parse. Insight Project, Deliverable 3.4 (2009)
10. Martin, D., Burstein, M., Hobbs, J., Lassila, O., McDermott, D., McIlraith, S., Narayanan, S., Paolucci, M., Parsia, B., Payne, T., Sirin, E., Srinivasan, N., Sycara, K.: OWL-S: Semantic Markup for Web Services. W3C Member Submission (2004), http://www.w3.org/Submission/OWL-S/
11. Mazurek, C., Stroiński, M., Werla, M., Węglarz, J.: Digital Libraries Infrastructure in the PIONIER Network. In: Mazurek, C., Stroiński, M., Werla, M., Węglarz, J. (eds.) Materiały Conference Polskie Biblioteki Cyfrowe 2008, pp. 9–14 (2009) (in Polish)

Development of a University Knowledge Base

Jakub Koperwas, Łukasz Skonieczny, Henryk Rybiński, and Wacław Struk

Institute of Computer Science, Warsaw University of Technology,
Nowowiejska 15/19, 00-665 Warszawa, Poland
{J.Koperwas,L.Skonieczny,H.Rybinski,W.Struk}@ii.pw.edu.pl

Abstract. This paper summarizes deployment and a one-year-exploitation of repository platform for knowledge and scientific resources. The platform was developed under SYNAT project and designed for the purpose of Warsaw University of Technology. First we present functionalities of the platform with accordance to the SYNAT objectives and the requirements of Warsaw University of Technology. Those functionalities are then compared to other well established systems like dLibra, Fedora Commons Repository and DSpace. The architecture of the platform and the aspect of integration with other systems are also discussed. Finally we discuss how the platform can be reused for deployment for the purposes of any other university with respect to the specific resources it produces and archives.

Keywords: digital library, knowledge base, scientific resources, repository.

1   Introduction

Despite several years of ongoing efforts to improve the information infrastructure for science in Poland, its state is still highly unsatisfactory. There are several initiatives started at local scientific centers (e.g., Federacja Bibliotek Cyfrowych), as well as two solutions supported by the Polish Ministry of Science and Higher Education (NUKAT, KARO), but they are rather conservative, strongly oriented to share physical objects (books) available in the associated libraries, and not fully make use of today's Internet capabilities. Ironically, the most complete and convenient access to the publications of polish researchers is provided by the external information resources (Web of Knowledge, SCOPUS, DL ACM, IEEE Xplore, etc.), and digital libraries of scientific publishers (Springer, Addison Wesley, Elsevier, etc.). Despite the fact that the formal reporting of projects has long been performed electronically within the system Nauka Polska, neither these projects nor resulting publications are available online. Though undoubtedly the positive results of this system is gathering metadata about doctoral and habilitation dissertations, personal data of Polish scientists and factual data on national scientific and academic centers. It is worth noting, however, that these data are used to a limited extent, primarily due to the following:

- missing source documents, bibliographical resources;
- primitive search tools;

R. Bembenik et al. (Eds.): *Intell. Tools for Building a Scientific Information*, SCI 467, pp. 97–110.
DOI: 10.1007/978-3-642-35647-6_8   © Springer-Verlag Berlin Heidelberg 2013

- the information stored in the system is not visible to the scientific world which is best illustrated by two simple experiment with querying Google Scholar (table 1 and table 2).

Table 1. Number of Polish PhD theses searchable from Google Scholar and Nauka Polska

Query submitted to Google Scholar	Number of documents found in Google Scholar	Number of documents in the Nauka Polska system
"rozprawa doktorska"[1]	4 820	90409
"rozprawa doktorska" Recent (i.e. since 2007)	1240	23953

Table 2. Number of scientific documents concerning Polish and Czech literature in English and native language searchable from Google Scholar

Query submitted to Google Scholar	Number of documents found in Google Scholar	Comments
"Polish literature"	6610	Result mainly from western universities. There are links to digital version.
"Literatura polska"[2]	2760	Mainly monographs not affiliated by universities. No digital versions.
"Czech literature"	3580	Result mainly from western and Czech universities.
"Česká literatura"[3]	2550	Results from the national science portal. Digital versions are sometimes available.

The SYNAT project is meant to address the mentioned deficiencies. The project concept is based on three levels of scientific repositories, whose ultimate goal is to ensure the dissemination of the polish scientific achievements and to improve integration and communication of the scientific community while leveraging existing infrastructure assets and distributed resources. Three levels of scientific repositories are shown in Figure 1.

---

[1] 'PhD thesis' in Polish.
[2] 'Polish literature' in Polish.
[3] 'Czech literature' in Czech.

- The central level (the SYNAT platform and the INFONA portal)
- The domain level (specialized domain-specific repositories)
- The university level (repositories held by the universities).

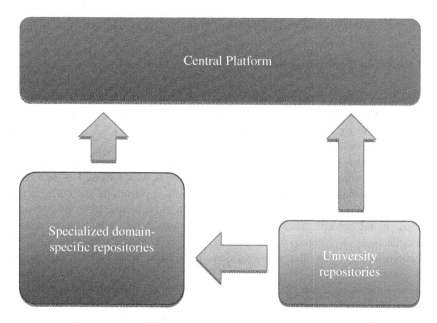

Fig. 1. General SYNAT architecture

In this paper we focus on the university level and postulate that it should be more a knowledge base rather than just a repository. The purposes of the university "Knowledge Base" is to provide:

- an entry point of the entire scientific information (together with full texts) created by universities, with accordance to the principle that the information about publication is entered to the system in the place where it the publication was prepared. This ensures that this information is correct and up-to-date.
- an access point to this information with limiting factors depending on the access type (logged user, access from university, access from the world) and the publishers' copyright policies.
- an easy way to store cleaned, unambiguous and interconnected information of various types, e.g. researchers, institutions, conferences, journals, book series, projects, publications, diplomas etc.
- administrative tools for reporting and evaluating scientific achievements of universities, institutions, units and individual researchers, according to the requirements of Polish authorities.
- interoperability, that is, a both side communication channel with various university services and systems (e.g. homepages, system supporting diploma process, report generators, employee database) and various external services like Google Scholar, publishers' websites, etc.

We have been developing such a "knowledge base" for almost two years now. The project originated from analyzing the needs of Faculty of Electronics and Information Technology, and was first implemented to suit one faculty requirements, but soon it became clear that the system should become an university-wide one. Hence, it is now planned to be deployed at the whole university.

2   Possible Technical Solutions

We considered many existing products which at first glance looked like ready and complete solutions for our needs. The most serious candidates were The Fedora Commons repository system [12], DSpace [6,7], and dLibra [1,2,3,4,5,8], which are briefly described below.

Fedora Commons repository system is an architecture for storing, managing, and accessing digital content. It provides a core repository service exposed as web services as well as a bunch of supporting services and applications including search, OAI-PMH, messaging, administrative clients. It is highly flexible in terms of types of supported digital content. There are numerous examples of Fedora being used for digital collections, e-research, digital libraries, archives, digital preservation, institutional repositories, open access publishing, document management, digital asset management.

DSpace is an open source software that provides tools for management of digital assets. It is commonly used as the basis for an institutional repository, both academic and commercial. It supports a wide variety of data, including books, theses, 3D digital scans of objects, photographs, film, video, research data sets and other forms of content. It is easy to install "out of the box" and is highly customizable to fit the needs of organization.

dLibra is a Polish system for building digital libraries. It has been developed by the Poznan Supercomputing and Networking Center (PSNC). It is probably the most popular software of this type in Poland. The communication and data exchange is based on well-known standards and protocols, such as: RSS, RDF, MARC, Dublin-Core or OAI-PMH. The digital libraries based on dLibra offer to their users expanded possibilities, such as: searching the content of stored elements, searching bibliographic description with use of synonyms dictionary, grouping digital publications and navigating in these structure or precise and expanded access rules to elements.

We have also analyzed a Polish made system Expertus [13], which is a dedicated software solution to manage bibliographies. Its interesting feature is that it supports semi-automated procedure for generating research reports in a form required by the Polish High Education authorities.

Our conclusion was that we needed a system which would combine the functionalities of digital repository (which is the case for Fedora, DSpace and dLibra) with the functionalities of complete institutional bibliographies, information about projects, scientific activities of researchers, and on top of it the possibilities to measure the scientific achievements (at least at the level of Expertus). Additionally, we have planned implementing some more advanced functionalities like "looking for an expert", citation index, and other research evaluation measures.

Another important reason for taking the decision to build a new system resulted from the interoperability requirements. The system was supposed to work tightly with internal faculty systems, (e.g. the system supporting diploma) and we found that such tight, both way connection would be easier and faster to achieve with the new product. We started development of such tailor-made system and successively generalized tight connections making it an highly universal platform possible to deploy within heterogeneous environment.

We also wanted to ensure that the running system is capable to adapt to the ever-changing requirements of authorities regarding reporting methods, evaluation rules, visualization styles and types of stored information.

3   Design and Capabilities of the System

3.1   The Purpose of the System and Its Core Functionalities

The main goals of this project was to:

- create a flexible system for building repository-based knowledge bases for scientific institutions
- build the Warsaw University of Technology knowledge base.

The fundamental assumption for the software was to provide

- a way for storing any typed metadata and digital content
- an easy way for defining custom types of stored metadata
- availability of linking records of different types
- preservation of "historical values" of linked objects in the course of changes
- an efficient full text search capabilities in both metadata and digital objects
- automatically generated, highly ergonomic and customizable GUI
- easy integration with external systems and exposure for external search engines
- extensive access control mechanisms
- capability of deployment in various scenarios

With the application of the implemented system we have built the KB for the University that has following functionalities:

- storing of the University organizational structure
- storing knowledge resources metadata and their correlation, among others:

    o publications(books, articles etc.)
    o diplomas (BSc,MSc,PhD BSc, MSc, PhD)
    o research projects, along with the project documents
    o other scientific documents (reports, reviews etc.)
    o authors and their affiliations

- storing employees data together with their achievements
- evaluation of employees and organizational subunits
- reporting for the internal purposes and for the purpose of the authorities
- import/export in various formats

3.2   System Architecture

The system was built on Java platform, with various popular tools and frameworks. It utilizes popular XML format that naturally fits hierarchical data structure desired for data modeling. The object structures are defined in XML Schema Definition (XSD) documents. As a result of using the JAXB technology, the data are seen in the application logic as regular Java objects.

The system architecture and integration scenarios are shown in Figure 2. It consists of the following 3 tiers:

- web presentation
- business logic
- persistence (repository)

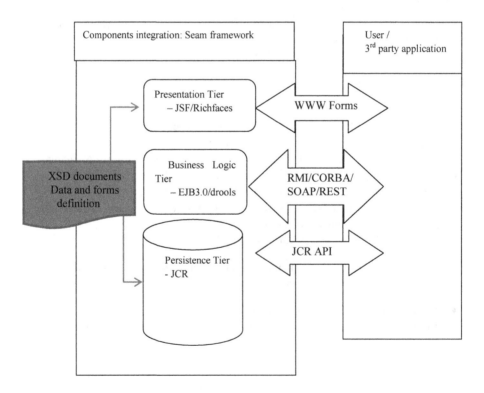

Fig. 2. The system architecture and integration scenarios

The presentation tier provides a highly ergonomic graphic user interface, which is built on the Ajax enabled JSF components form the RichFaces library. It is worth noting that in order to make the system more flexible, a special screen generator has been developed. Based on an XSD specification of various data types/categories the generator is able to generate data entry forms, as well as the results presentation screens. Obviously, the automatically generated screens may be manually customized, mainly for visual effects, as the screen functionality is already provided.

The repository functionality has been implemented with the use of Apache Jackrabbit[4] [9], which is compliant to the JCR standards, namely JSR 170 and JSR 283[5]. JCR provides a standardized API, versioning, content observation, access control layer and other functionalities. At the same time it makes the application provider independent and physical storage type independent. JCR can be configured to persist the data in a storage (a relational database, a file system, or other). Both, metadata and digital objects are indexed with Lucene [10], which gives rise to very powerful and efficient searching capabilities.

Business logic tier is encapsulated in the EJB 3 components. This modern EJB version provides a way of defining services that automatically handle distributed transactions, automatic exposure via SOAP based and REST based web services.

The Java Enterprise Edition 5 server is required to deploy the application, however it is possible to build local transaction version built on Seam, rather than the EJB services. The system has been designed to work either in centralized or distributed way.

3.3   Integration and Interoperability

The system is designed to provide system integrations on many levels. The functionalities of the system are exposed via both SOAP and REST web services. Also storage tier is directly exposed via JCR API or WebDAV protocol and allows distributed deployment as described later in this paper. The system has also the capability of integration with Single Sign On external authentication services, for the purpose of the University the CAS was used. The system at the moment is ready to expose all content to be harvested by Google and Google Scholar search engines.

Additionally, we plan implementing an assessment of citation index. Dedicated algorithms are in the research stage. At the same time for improving access to the documents, DOI resolver is used in order to direct users to the publishers versions.

Our further plans are to utilize ZOTERO [14], and Mendeley [11] as loosely coupled systems. Last but not least the system will be exposed to be harvested by the higher level SYNAT systems (see Figure 1).

---

[4] Jackrabbit is an Apache openApache openopen source implementation of Java Content Repository standard.
[5] These standards define the repository system for storing digital objects organized hierarchically, together with the accompanying metadata.

3.4   The Exploitation Scenarios for the System

The system can be installed in two various ways: centralized or distributed one.

Centralized Installation

Indicate of the central installation it is assumed that system is one coherent application, and all the data are kept in one data store. All organizational units of the university (faculties, departments, institutes, etc.) use the same application. It is worth noting, however, that particular units have specific privileges in terms of updating the common storage. In addition, in the data entry processes each organizational unit, first of all sees specific part of the data(publications, journals etc.), which makes the data entry processes much more friendly. Such approach is highly convenient for the maintenance and consistency reasons. It does not mean that the application must be run on one server, as it would be a single point of failure. To this end, both the application and data store could be clustered and load-balanced. Accepting the centralized solution does not necessarily means the unified presentation for all involved departments.

The system provides robust personalization possibilities for the departments, including custom themes, data presentation formats etc. This way, the consistency and aggregation requirements can be met with keeping the possibility for each organization to preserve its own identity.

Distributed Installation

In the case of the distributed / federated installation it is assumed that no common policy can be agreed between university units so various departments require different business logic or custom data structures, therefore require separate customized installation of the system should be provided. The federated version is also convenient where reliable network connection between the interested parties cannot be established, so that slave nodes can work offline with respect to the master node, and synchronize whenever possible. In such a case the departments hosts their own versions of the system. However still a common knowledge base at the university level is desired. To this end yet another installation has to be provided at the central level, where a collective master knowledge base is built by harvesting data from the distributed slave applications. At this point it is still possible that data are corrected in central knowledge base and the change can be propagated down to the slave database. Such behavior can be configured with custom policies.

The federated version requires additional computer-aided handling of the conflicts and duplicates coming from various slave nodes.

If required, the system can run in a mixed mode, so that some university units work on a central database, whereas other ones have their local installations, synchronizing temporarily with the central installation. Figure 3 illustrates the concept of the distributed deployment.

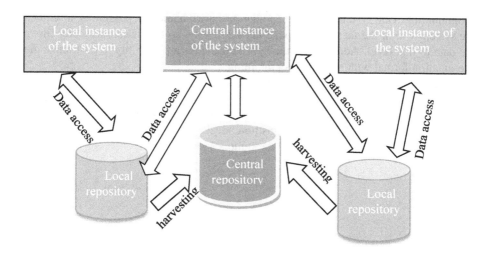

Fig. 3. Illustration of the distributed deployment

3.5   Hierarchical Structure of Knowledge Base

The hierarchy of the knowledge base reflects the hierarchy of the University, therefore the system must be aware of the organizational structure of the units. The structure of organization is kept as one of regular data type in the knowledge base. It has hierarchical structure and can be defined freely as needed i.e. the depth of the organizational tree is unlimited. The repository elements may be assigned to any desired level, therefore some records may be assigned to the global university-level and some may be for example based to an institute level. This in particular affects the University staff, which are stored in the knowledge base along with other assets. The organizational level is utilized in calculating access rights, as well as viewed pool of objects. For example a user assigned to the central University level, may have access to all documents, whereas a user at institute level may access, and/or view only documents assigned to this level. In addition while entering a record one can see first of all the level assigned objects.

The organizational structure may change over time. To preserve historical relationship between objects, the whole referred structure is stored frozen in each record in the knowledge base. For example, if an author changes its affiliation from one institute to another, the papers he contributed before the transfer will be affiliated to the original institute. The same refers to changing, for example, a title of a journal or name of an author. In either case the main object is searchable by old and new values.

The requirement of the university is to maintain a complete and consistent knowledge base, however the subordinated units often require to present their part only on their home pages. Two straightforward solution are possible: keep parallel systems for each faculty, which is error-prone and inefficient with respect to the maintenance, or to integrate faculties with central knowledge base using API (in our case REST or SOAP-based). This approach is very flexible, however it requires additional programming. Our system provides another scenario, namely a possibility to embed the system within the faculty web page in such a way, so that it automatically filters the data to that level. Additionally the system provides a possibility to customize the presentation format with no extra programming.

Let us note that local level user interfaces are sensitive for the local contexts. For example, the autosuggestion lists provide first of all local data, e.g. while looking for a name "Smith" from an institute home page, first of all the suggested values will refer to those Smith's affiliated at the institute level.

These possibilities are presented in the figures 4 and 5.

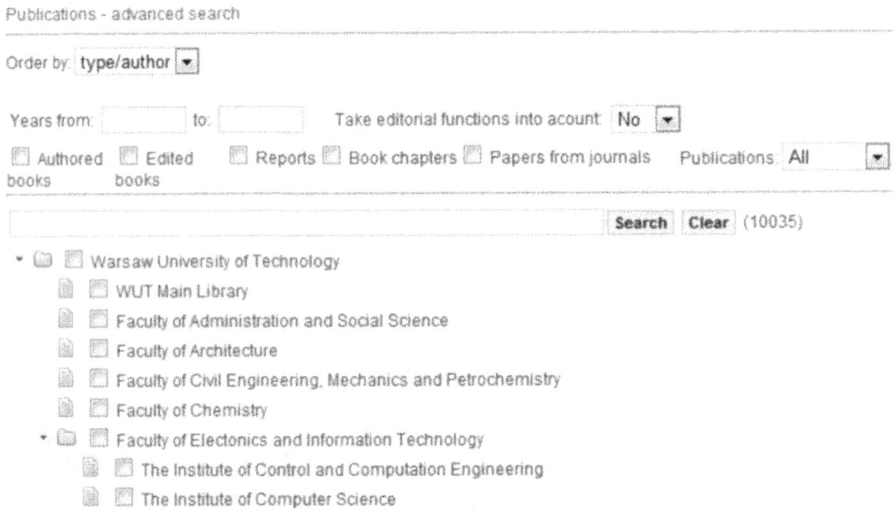

Fig. 4. University level: full organization tree

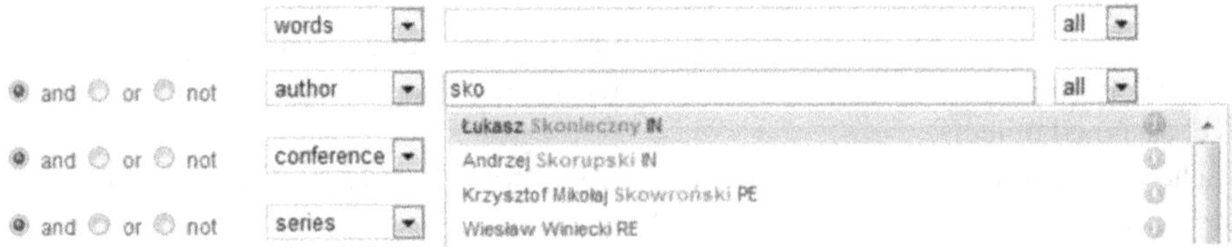

Fig. 5. Faculty-level advanced search – the "help list" depends on the institution level and user affiliation

3.6   System Flexibility

The system architecture is strongly oriented for providing high flexibility in defining various aspects of the system applications. A main idea behind this requirement was to provide easiness in defining custom installations and make possible expanding the system in course of its life, without the need of (re)programming of the software. The following aspects of the system, *inter alia*, can be defined:

- Data structures
- Custom views
- Access rights
- Custom reports

Data Structures and Views Definitions

As mentioned above all the data structure are defined by means of XSD definitions, extended with some extra constructs. A sample record definition is given below:

```
<complexType name="affiliation">
  <annotation>
    <documentation>
    Reprezentuje afiliacje autorów i pracowników
    </documentation>
  </annotation>
  <sequence>
    <element name="id" type="ID"></element>
    <element name="owner" type="ID"></element>
    <element name="affiliationowner" type="ID"></element>
    <element name="namePL" type="string" fw:defaultFormatter='true'
      fw:required='true'>
      <annotation><appinfo>
        <fw:unique required="true"/>
      </annotation>
    </element>
    <element name="nameEN" type="string" fw:required='true'>
      <annotation><appinfo>
        <fw:unique required="true"/>
      </annotation>
    </element>
    <element name="accronymPL" type="string" fw:required='true'>
      <annotation><appinfo>
        <fw:unique required="true"/></appinfo>
      </annotation>
    </element>
    <element name="accronymEN" type="string" fw:required='true'>
      <annotation><appinfo>
        <fw:unique required="true"/></appinfo>
      </annotation>
    </element>
    <element ref="tns:affiliation"></element>
  </sequence>
</complexType>
```

The definitions given that way can be used by especially prepared generator for generating the following pages:

- Search pages
- Record edition page
- Preview page
- Row formatting template for search result page
- Sorting result page definitions

The resulting screens are already fully equipped with all necessary user interface functionalities. Nevertheless additional aspects of user interface can be manually configured so that specific interface needs are satisfied.

Access Rights Control Definition

Another issue tightly coupled with the organization structure is access control and users privileges. Access control system takes into account many various environmental parameters. First of all, basic default privileges of a user result from his/her affiliation to a unit defined within the organizational structure. Additionally the following attributes can be used:

- authentication information about the user
- his/her affiliations (if authenticated)
- the roles assigned to the user (the user may have many roles assigned)
- individual and affiliation ownerships of the objects, which may indicate if the user is authorized to perform actions on the objects (the ownerships are   initialized automatically when the object is created but may be the subject of changing by superuser).
- the selected values of attributes of the objects
- the internet/intranet localization of the user
- the date attributes (including current date)
- the action being performed on the system
- custom permission levels for digital objects

Those aspects are then combined in business rules that compute access rights. The rules can be easily modified, as they are kept in a separate configuration file. Below we provide two examples which illustrate how the rules can be built:

Example 1: A user that has "dataentry" or "superdataentry" role assigned can modify and delete his/her own records. Additionally, if the user enters a publication co-authored with an author from another institute, the other institute will also have access to the record modification (except deletion).

```
rule DataEntryAndSuperDataEntryCanAddCanDeleteAndModfiyItsOwnRecords
when
        check: PermissionCheck(granted ==false,action =="edit" ||
                        action=="delete")
        role: Role(name == 'dataentry' || name=='superdataentry')
        identity:Identity(name:principal.name, affiliation:affiliation)
```

```
            entity:Entity(owner==name,affiliationowner==affiliation)
then
    check.grant();
end

rule DataEntryCanModifyPublicationsOfOwnedAuthors
when
        check: PermissionCheck(granted ==false,action =="edit")
        role: Role(name == 'dataentry' || name=='superdataentry')
        identity:Identity()
        art:Article()
        a:Author(affiliation !=null,
                affiliation.accronymPL==identity.affiliation) from
                art.author
then
        check.grant();
end
```

Example 2: Documents in the system, may have custom permissions as concerning their availability for downloading. Let us define a rule which restricts the access to protected files, unless the user belongs to a given organizational unit, and a protection period time for a document is defined (so that the document before the defined protection date is unavailable and becomes available afterwards). Many other scenarios are possible and will not be discussed later.

```
rule DownloadProtected
when
    check: PermissionCheck(granted ==false,action =="download")
    identity:Identity(af:affiliation, dt:currentDate)
    file:File(permissions:permission)
    perm:Permission(access=="PROTECTED",affiliation==af,dateFrom==null
                    || dateFrom<dt) from permissions
then
        check.grant();
end
```

4   Lessons Learned and Conclusions

The system is being used for more than one year at the moment. At first the system was deployed only in the Institute of Computer Science and managed only scientific publications. After evaluation period the system was extended to support diplomas, employees directory and projects and now operates for the whole Faculty of Electronics and Computer Science. During this year the end user feedback was gathered and used to continuously improve the system.

To operate within FEIT infrastructure the system was integrated with CAS SSO system and the system supporting diploma process, where the camera ready diplomas are automatically pushed to the repository via the REST API. For the purposes of research evaluation standard reports for the University and Ministry authorities have been developed.

Currently, we work on deploying to the Main University Library. It is planned that the system will be deployed in centralized mode as described in section 3.4 and would cover institute and faculty hierarchy of the entire Warsaw University of Technology.

This step requires the system to be prepared for the large number of users. Initial performance have already been made. Those tests show that the system scales linearly with respect to the number of user, however some performance tuning is required to decrease the average response time for user requests.

We are also working on incorporating data and text mining tools, which would benefit from the knowledge base character of the system. As for now, the work has started on adding the functionality of "looking for an expert" and "looking for a team".

References

1. Górny, M., Gruszczyński, P., Mazurek, C., Nikisch, J.A., Stroiński, M., Swędrzyński, I.A.: Zastosowanie oprogramowania dLibra do budowy Wielkopolskiej Biblioteki Cyfrowej. In: Zeszyty Naukowe Wydziału ETI Politechniki Gdańskiej, pp. 109–118 (2003)
2. Gruszczyński, P., Mazurek, C., Osinski, S., Swedrzynski, A., Szuber, I.S.: dLibra Content Maintenance for Digital Libraries. In: Euromedia WEBTEC, Kwiecień, Modena (2002)
3. Mazurek, C., Werla, I.M.: Digital Object Lifecycle in dLibra Digital Library Framework. In: Proceedings of the 9th International Workshop of the DELOS Network of Excellence on Digital Libraries on Digital Repositories, Heraklion, Recte (2005)
4. Parkoła, T.: Podręcznik użytkownika środowiska dLibra (wersja 3.0). PCSS (2007)
5. Skubała, E., Kazan, I.A.: Polskie biblioteki cyfrowe na platformie dLibra–zasób w kontekście tworzenia nowoczesnych kolekcji źródeł informacji dla nauk technicznych. In: Ganińska, H. (ed.) Informacja Dla Nauki a Świat zasobów Cyfrowych: Biblioteka Główna Politechniki Poznańskiej, Poznań (2008) ISBN 83-910677-4-2
6. Smith, M.K., Barton, M., Branschofsky, M., McClellan, G., Walker, J.H., Bass, M., Stuve, D.: DSpace: An Open Source Dynamic Digital Repository. D-Lib Magazine 9(1) (styczeń 2003), http://dspace.mit.edu/handle/1721.1/29465.
7. Tansley, R., Bass, M., Stuve, D., Branschofsky, M., Chudnov, D., McClellan, G., Smith, I.: The DSpace institutional digital repository system: current functionality. In: Proceedings of the 3rd ACM/IEEE-CS Joint Conference on Digital libraries, JCDL 2003, pp. 87–97. IEEE Computer Society, Washington, DC (2003), http://dl.acm.org/citation.cfm?id=827140.827151
8. Werla, M.: Biblioteka cyfrowa jako repozytorium OAI-PMH [on-line]. dLibra". Poznańskie Centrum Superkomputerowo-Sieciowe [Dostęp 15 kwietnia 2009]. Dostępny w World Wide Web, http://dlibra.psnc.pl/community/display/KB/Biblioteka+cyfrowa+jako+repozytorium+OAI-PMH
9. Apache Jackrabbit, http://jackrabbit.apache.org/
10. Apache Lucene, http://lucene.apache.org
11. Mendeley, http://www.mendeley.com/
12. Fedora Commons. Fedora Commons Repository Software (2009), http://www.fedora-commons.org/
13. Splendor Expertus, http://www.splendor.net.pl/
14. Zotero, http://www.zotero.org/

Infrastructural Approach to Modern Digital Library and Repository Management Systems

Tomasz Rosiek[1], Wojtek Sylwestrzak[1],
Aleksander Nowiński[1], and Marek Niezgódka[2]

[1] Centre for Open Science,
[2] Interdisciplinary Centre of Mathematical and Computational Modelling,
University of Warsaw, Poland
T.Rosiek@icm.edu.pl

Abstract. Traditional DLM Systems were usually implemented in the form of an either monolithic or distributed applications. The paper presents a modern approach, where a modular environment provides an infrastructure of components to build DMLS applications upon. A case study of an open Synat Software Platform is presented, together with its sample applications, and the key benefits of the approach are discussed.

Keywords: software platform, software architecture, digital libraries, open repositories, content aggregation, Service Oriented Architecture, cloud computing, Software as a Service.

1 Introduction

During the last 20 years the Digital Library Management Systems evolved from electronic catalogues, OPACs, through ILSes, towards standalone repository-like applications. In the paper we propose a step further - departing from monolithic applications, and developing a content infrastructure software environment, where individual applications share a common set of resources, services, and protocols. The infrastructural approach makes rapid development and deployment of future custom solutions easy, with the new applications being from the start well integrated with the already existing environment. It also allows for the whole system to smoothly evolve, with older applications or services being transparently replaced by their newer implementations. This approach allows to avoid the common scenario of a library being locked into a single vendor system with no possibility to effectively influence its development path. It fits well into the PaaS (Platform as a Service) cloud computing model, thus additionally freeing the solutions developers from the underlying hardware infrastructure burden.

The described approach also allows to create open, future-proof systems effectively and efficiently, because each new application builds upon a set of services, tools and contents already existing in the infrastructure, thus significantly reducing the development cost and time. What is more, the proposed component-based approach promotes collaborative development models, where

R. Bembenik et al. (Eds.): *Intell. Tools for Building a Scientific Information*, SCI 467, pp. 111–128.
DOI: 10.1007/978-3-642-35647-6_9   © Springer-Verlag Berlin Heidelberg 2013

different organisations or developers can cooperate, delivering their own modules being the building blocks of the larger applications. [1]

Finally, the proposed model encourages usage of the contemporary flexible software development paradigms, such as agile programming. These techniques are often difficult to be fully applied in strictly formal environments of public-funded projects. Nowadays in commercial environment most software developers work in close contact with their customers, being able to quickly adapt to dynamically changing functional requirements and technology. At the same time, a typical public-funded project follows old software development paradigms, often originating from the previous century, where the traditional analyse-design-develop cycle is enforced by the way the project is evaluated, managed and reported. Typically, due to long evaluation processes, proposals are written many months before the projects start. It is sometimes possible to negotiate some last minutes changes, but when the project ends in another two or three years, the delivered products are often already outdated. This is further exacerbated by the fact that in such scenarios the formal nature of the contacts with the funding agencies makes introducing rapid changes virtually impossible, thus defeating any agility aspect of the development model. Additionally, applying heavyweight project management processes and methodologies can prevent any effective adaptation to the rapidly changing needs. The infrastructural approach allows to avoid these common pitfalls by decomposing the final products into sets of smaller applications built from services. Many of the services may already exist in the environment as results of previous projects or other activities. This way, it is possible to conduct a larger number of smaller scale developments asynchronously, with their individual results adding to the large DL infrastructure instead of a smaller number of larger projects, with many of them ending in isolated throw-away systems. [1]

1.1 State of the Art / Current Practices

There is a number of solutions available for the digital libraries and repositories. Commercial companies offer systems of different scale and price, from small applications to large integrated systems, or metasearch engines. Open-source products, like DSpace or Fedora, provide good solutions for medium scale institutional repositories. Most of these products deliver reasonable functionality and performance and so far have been sufficient means to create individual digital libraries or institutional repositories. In fact in case of implementation of typical functionalities related to publications there is hardly a need to use any more advanced solutions. However, the main drawbacks of the existing library and repository systems are their diversity and monolithic architectures. Most of the existing systems are just autonomous applications or sets of applications integrating searching, browsing and storing of digital data. Introducing new major features or services may require extensive changes to the existing system and require deep knowledge of the system's structure. Another problem is rather limited ability of existing digital repository systems to interact with each other.

The obvious advantage of digital resource publishing is the ability to access any digital library/content provider from a single Internet station. But a well-known problem remains still how to effectively find the library that hosts the resource the user is searching for. Currently the following options to locate a particular publication are available:

- The user may just know the library which hosts the particular resource.
- The user may use a meta-search engine that aggregates data from a number of digital libraries. An important limitation in this case is the requirement, that the particular library must belong to a specified federation of libraries and must be bound to the specified meta-searching engine.
- The user may make use of a universal search engine like Google or Bing, or may use a publication specific search engine like Google Scholar in hope, that the resource is indexed by these engines. The services described harvests all the data that are available publicly and share these data in a consistent way. However in this approach only data that are available without restrictions (that are in public domain or are openly licensed) will be harvested.

The two latter options provide only limited search functionality based on the available simplest metadata of publications. They do not provide the ability to fully and consistently integrate all the data stored in multiple heterogeneous systems.

Although it is possible to create federated systems and to aggregate multiple libraries is some of the existing tools, in fact these solutions have rather limited functionality and their interoperability capabilities are limited only to systems belonging to the same family. Commercial solutions remain vendor-controlled and do not offer the flexibility and openness necessary to easily extend their functionalities or the resource range.

There are few important data exchange standards intended to be used in repository systems - like OAI-PMH [2], Dublin Core, OAI-ORE, SRU/SRW, or the currently mostly historical Z39.50 information retrieval protocol for digital libraries [3]. Unfortunately they allow just for integration of repositories on the data level - to aggregate or, in the best case replicate data between repositories. We still miss the tools allowing to consistently integrate separate services included in different repositories - like using a common OCR engine in multiple repository applications or sharing of indexes. Currently most of the repository software allows only for some data-level integration using the few widely acknowledged standards. This involves mostly OAI-PMH - the protocol for metadata retrieval and Dublin Core - data format that allows to describe different kinds of publications. The OAI-PMH protocol allows to harvest publication's metadata from a repository in order to index this data or to perform any other processing. The protocol is not bound to any specific metadata description format, however the main format supported by OAI-PMH servers is Dublin Core. OAI-PMH has limited abilities of declaring the range of data to be harvested. It is possible to fetch only specified dataset and to declare conditions based on the metadata timestamps.

The OAI-PMH protocol is strictly related to the Dublin Core format that allows to describe the metadata of different resources like articles, but also multimedia like videos and audio tracks.

The main advantages of the Dublin Core protocol are its simplicity and wide usage - virtually every repository system is able to export its contents in the Dublin Core format. What is the main weakness of this format is that it is very generic and does not allow to describe publications distinctly. Attributes of publications are defined in a way allowing many possible interpretations. Furthermore, basic Dublin Core describes only the most essential properties of publications (like title or author) without distinction of different types of authors or for example keywords. These problems led the users to create multiple mutually incompatible extensions of the format that allow to precisely store additional information but result in loss of compatibility.

In these terms Dublin Core remains quite good format allowing to feed aggregating services and meta-searchers but hardly fits as a main format for storing and replication of repositores.

One very interesting communication protocol that enables interoperability of repository systems is OpenSearch. It is a RESTful protocol based on the Dublin Core format and RSS protocol that provides machine interface allowing to search into publication repositories. Using OpenSearch protocol it is possible to perform search in the publication index - as a result the metadata list of found documents is returned. The response format is a Dublin Core wrapped within the RSS. The protocol seems to be a good solution to expose the data in order to create meta-searchers. Still it inherits many problems of the Dublin Core format like e.g. ambiguity.

It is also worth to mention the trend of creating solutions that share knowledge resources in a form of semantic data - using Linked Data paradigm. This approach is based on sharing the data in a form of RDF triplets [4]. RDF data from multiple repositories may be merged and used to make complex searches and reasonings. Linked Data paradigm is not only limited to publication repositories but allows to merge data from different domains. For example we are able to combine publication data with person databases and with datasets describing geographical locations or chemical molecules. [5]

1.2 Known Issues

Existing solutions mainly allow to fulfill basic requirements of common users, like searching for contents or redirecting the user to a repository providing a particular resource. The problem arises when we need to create a new service that allows to look into data from a particular dataset from a new point of view. This includes new ways of data exploring requiring semantic indexes, domain indexes (geographic index, multimedia indexes) and tools that allow to enhance and postprocess existing data (OCR, automatic classification). Introducing such tools in existing services is possible in two ways - extending existing software (which needs in-depth knowledge of the architecture of existing system), or creating a new system that makes use of the new tools and aggregates data from existing systems using standard protocols. Neither scenario makes it possible to efficiently introduce the same postprocessing and enrichment component into multiple different systems at once. If the data are published in Dublin Core format we have to be aware of the nuances of the format usage specific for the particular provider. Having harvested the data, we may then perform the desired analytic or indexing operations and store their results in a way specific to the service. In addition we have to create our own user interface or machine interface that would allow to access the results of processing. However in this approach we have to find a way to integrate the new interface with all existing applications. The cost of implementing and integrating the new components is rather high, and integration of existing post processing services to alternate services requires repetitive work and significantly increases the systems complexity. The main cause is the fact that the only available way of interoperability is sharing of the data not sharing of the services.

The problem of integration cost is especially visible during research projects that deal with repository data. Usually there is a need to create prototype of an ad-hoc tool that analyses the data stored in the repository. This type of tool usually has short lifetime, and the effort needed to create is should not be significant. We found it desirable to develop an approach that significantly reduces the effort needed to create tools that work with data stored in a particular network of repositories and instead allows to focus on research problems themselves. In addition none of existing solutions allows to consistently store results of performed post processing.

The described problems may seem not so important in case of small amounts of data and prototype solutions, but when there is a need to switch from one source repository to another or to use prototype services in production environment, the problems become significant.

2   Proposed Solution

2.1   Service Oriented Architecture

What we propose as a solution for heterogeneous network of digital data repositories is loosely-coupled, service oriented architecture. This approach is well known and used in many enterprise systems by multiple organizations.

The main benefits of the proposed architecture are:

- the ability to easily combine heterogeneous systems,
- the ability to scale repositories in order to efficiently support increasing load,
- the ability to merge production-level systems and prototype tools in one consistent software ecosystem.

The proposed approach assumes that each of the repository systems consists of a set of generic low-level services providing common standardised functionalities like fulltext-indexing or metadata storing, and a set of business-level services that are aware of business rules of the particular system, its content licensing policies and supported data formats.

A single service ecosystem should be able to work with multiple incompatible data formats or multiple versions of the same data format. This requirement is essential as we expect that one ecosystem should be able to combine often very different repositories. The other requirement is an ability to dynamically switch relations between services in the ecosystem in order to be able to upgrade service implementations of particular components without the need to shutdown the systems that depend on them.

2.2   Lightweight Service Integration Platform

The challenges of scalable heterogeneous distributed systems are well known for a long time, and there is a number of service integration solutions available on the market already. Most of them are intended to be used as a base of enterprise systems. The list of the most important SOA platforms and solutions include:

- Java EE - popular architecture that is based on the Java language. Java EE provides many standard mechanisms providing such aspects of enterprise systems like relational database connectivity, distributed transactions, remote service access, asynchronous messaging and much more. The problem with JEE is the fact that all technologies provided by the framework are tightly coupled. In addition, partial upgrades of JEE based systems are complicated which makes frequent releases hard. [6]
- OSGi - the framework and dynamic service integration platform that implements complete component integration model. OSGi allows to create software modules that can be loaded and unloaded dynamically in a specified environment. OSGi is based on the Java language and its scope is reduced to a single virtual machine. Raw OSGi standard does not address interoperability between remote services. OSGi offers many sophisticated features, but the overhead required to create and maintain dynamic components is relatively high. [7]
- Enterprise Service Bus approach - the model of interoperability in which the system consists of a set of loosely coupled, independent services. Services do not interact with each other directly but send requests and responses to an interoperability-layer called enterprise service bus. The bus acts as a broker and a buffer in the communication between services. Introducing additional layer allows to separate services and simplifies implementing such mechanisms as load balancing, buffering and introduction of new services. ESB systems do not restrict users to use particular technologies or protocols. Typical solutions used with EJB systems include SOAP, message queueing, RMI, REST and many more. Frameworks like Java EE allow to create ESB modules in a simple way. Actually, there is no single standard of creating ESB systems. There are lots of commercial products like - webMethods ESB, IBM WebSphere ESB and few open source ones like Apache Synapse.

We decided to create our own lightweight ESB platform based on open source components including Spring Framework and Spring Integration Framework.

What was important to our approach was loose-coupling of the services and the ability to flexibly distribute the service environment over complex infrastructure consisting of multiple nodes managed by different operators (service or content providers). The platform is expected to be resistant to temporary losses of connectivity between particular parts of the infrastructure. On the other hand we did not need some features often available in products used in commercial systems like strict transactivity and real time processing. Also we treat some of security constraints in another way than enterprise systems.

2.3   Architecture of the Platform

Our solution - the Synat Software Platform allows to instantiate and host collection of services that may be implemented in different technologies and be ran in a distributed hardware environment (figure 1). In this architecture each system consists of:

- the set of universal and application specific services
- the infrastructure that integrates all components of the system (service containers and service registries)
- end user application or applications that use the services to implement the desired functionality

The Platform allows us to create an ecosystem of services which consists of multiple nodes, where each node is a container for multiple services and allows to separate physical location of the service and its architecture from its place in organizational structure. Each service is executed in a service container that is a specific application that may be started directly from the operating system and provides some essential aspects of the service lifecycle like configuration management, resource management, ability to export service interfaces over the network and so on. Service containers are quite lightweight, allowing even to enable an architecture with separate containers for each service.

An important component of the Platform is the Service Registry. The registry is ran as an application in the operating system and its role is maintaining information about active services in the ecosystem. The information about the services is available along with their service properties such as the exposed interfaces, supported protocols and additional features like support for authentication. The registry will also act as a watchdog, verifying if the services are up and are available. When a client needs to connect to a service, it first performs a query to the registry to fetch all information required to make the connection. Importantly, the service registry enables flexible management of the services. It is possible to define service namespaces using aliases, which allows us to dynamically reconnect particular applications from one service to another. In addition, the proposed architecture allows for easy implementation of load balancing and failover mechanisms. Since it is the service registry that is responsible for decision about where a client connects, it is very easy and natural to define clustering policies as service aliases pointing to multiple instances of the same service.

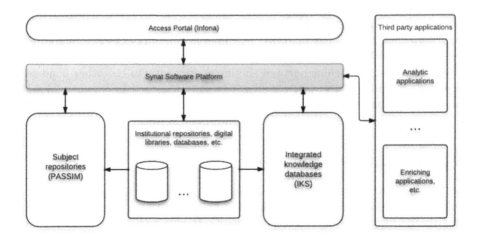

Fig. 1. The Synat Software Platform serves as an interconnect and provides core services to various components of the systems business logic

A service client may be an application that uses services of the Platform ecosystem. Any application may act as a service client, also the service itself may be a client of other services. Each service client has to use proper client connectivity library that is responsible for establishing connections to the registry and other containers. Since there is no single remote invocation protocol defined by the platform, the client should also provide proper libraries enabling support for protocols used by desired services. The current implementation of the Platform and the available connector library provide a set of standard remote invocation protocols - like HTTPInvoke, Hessian and - in limited scope - RMI and REST. Architecture of typical Synat Platform based application is shown on the figure 2.

Our approach assumes that distributed architecture of the system might spread over multiple autonomous organizations. It is therefore necessary to address several security concerns. We have identified the following security aspects:

– service-to-service authorization and authentication
– client-to-registry, service-to-registry authentication
– end-user authorization, authentication and identity propagation

The major security concern is related to assuring that all participants of the conversation are trusted. In order to verify that, it was necessary to establish public-private key infrastructure that can be used to sign connections between particular components. The second problem is the ability to limit access to services only to specified clients. The authorization aspects are considered separately on two levels: coarse-grained service-to-service authorization and fine-grained application-specific authorization. Coarse-grained authorization allows to restrict access to low level services to particular service clients (not separate users) and may be

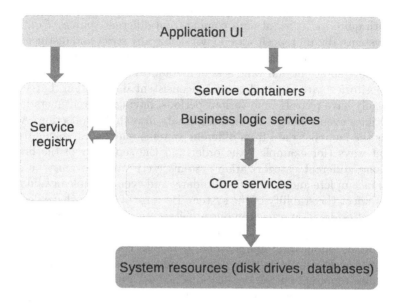

Fig. 2. Interdepedencies of logical components of the Synat Software Platform

used to allow access to separate datasets or tools to particular institutions. Coarse-grained authorization does not give us information about logged-in user and the context of the operation. Fine-grained authorization, in turn, reflects application specific security policy which may be related to sophisticated business logic. This part of security layer should not be standardized in the Synat Platform and should be defined by individual applications instead.

2.4   Document Oriented Processing

There is one essential difference between typical enterprise and repository systems. Business applications usually use data models consisting of lots of fine-grained domain objects that are strictly related to each other. E.g. in case of banking applications a model has to consider relations between customers, accounts, transfers, credit cards, employers and so on. Each of the specified objects has strictly defined relations with many others. Bank operation cannot exist when there is no account that the operation is related to. Enterprise systems expect that operations on the data will be transactional and data structure will be consistent. The approach to repository systems is obviously different. The primary domain object which we have to take in account is a single generalized publication. Each publication has an extensive set of attributes like titles, keywords, abstracts, authors, references, dates and so on. This set can be very rich and complex (multiple sets of translated keywords, detailed, parsed references, non-textual, compound objects, relations to other publications etc.), but on the other hand it may be basic information just about the title and author of the work. The problem is, that an integrated repository system frequently has to deal with quite inconsistent data coming from different sources. Contrary to banking systems, digital repositories do not define any strict and distinct universal set of document metadata. Although many newer, digital-born publication collections have sufficient and valid sets of metadata, often we have to deal with legacy repositories with incomplete and inconsistent information. Large collections typically have records from various periods, having different metadata sets and sometimes even different metadata formats in a single collection. Another problem is the fact that data from different systems may define some attributes in different ways (for example - the order and the separator of the firstname and surname, different categorization systems etc.). We also must be able to work with incomplete and erroneous metadata, and even multiple invalid records should not affect the stability of the system. Because of that we have decided to use document oriented approach in our architecture. Each publication in Synat Platform is represented by a record consisting of one or more files reflecting the metadata of the publication and zero or more files containing contents of the publication. The Platform remains flexible and does not define the amount of metadata files representing the publications nor does it expect anything about the format of the data. The format of records is specific to particular application implemented using Synat Platform. All records with contents of the repository are kept in the Storage Service. The service allows to store large amounts of records with or without contents. What is important - the Storage Service itself does not provide any ways of advanced searching and structure validation of its contents. The main benefit of this approach is the ability to support documents in different formats coming from different repositories in a single service. It also allows to handle inconsistent and invalid data in order to keep track of them and be able to fix them manually or with appropriate automated processes. This approach also implies that relations between objects in our dataset are loosely-coupled and it is not required that the target object of particular relation is available in the repository. This allows the platform to handle partial sets of data, deal with missing objects or distribute the data among various organisations and systems. It also makes partitioning of the Storage Service easy, using shard paradigm. The approach is different to the solutions typically used in enterprise applications, where the data are stored in relational databases with extensive sets of indexes and constraints.

Another issue that differs the Platform from the RDBMS oriented approach is the way how searching and browsing of the data is done. The document oriented approach has a significant drawback: relations are typically unidirectional, i.e. within the document there is information, that it is somehow related to another. But there is no easy way to find the reverse relation pointing back to the document, unless it has been already discovered. This usually requires a set of services, which keep track of relations and it is the accepted tradeoff for the benefits. The Platform introduces the concept of a generalized index service. A generalized index is any service which:

1. Is updated by adding information from documents (iteratively) - indexing
2. Can respond to a query (specific for the index) and responds with one or more identifiers of the documents provided in step (1)

This behaviour is of course also typical for the full-text search engine: documents are indexed, and then query is made to obtain search results. But this pattern applies as well to other services: similarity index (a query is an identifier of one of the documents in the repository), semantic search and other. A very specific type of the index is the relational index, which extracts relations from the elements and creates a graph of the relations, which may be then queried in various ways. The same pattern may be applied to a dozen of the possible services, which will offer the required data browsing and searching capabilities.

A very important feature of the document-oriented system is an easy adoption of variety of metadata formats and versions. The typical problem of many library systems is that imported data is cut-down to the largest common denominator, which in many cases may be as small as Dublin Core basic metadata set. Each update requires then a significant amount of work, and an effort to convert all previously collected data into an updated version. The Synat Platform allows multiple versions of the metadata for each record. Both the original, source metadata, as well as converted and updated (possibly by automatic enhancement procedures) versions are kept. This allows an easy operation and adoption of the new features of the models. The cost of maintaining multiple versions of the same record is negligible, while the profits during various processing procedures are significant. It also allows gradual improvement of the conversion quality, as well as advanced deduplication and merging procedures.

3   Generic Reusable Services

Idea of the universal repository platform is to define and implement a set of generic reusable services that, by proper configuration, may be used in different repository applications.

The common core components should implement all domain-independent aspects of repository applications in such a universal way that designers of the new systems can focus on implementing the business logic. To this end, the following functional aspects have been identified as being common in most repository systems:

- data storage
- fulltext indexing
- user management
- importing of the data
- batch processing of the data
- storing additional contents created by users (comments, ratings, notes, etc.)

All core services are intended to retain their generic nature and should not be bound to any particular data model nor particular set of business rules. The services should enable coexistence of multiple domain models based on the same core set implementation. Basing on already mentioned list of common aspects of functionalities that occur in each repository product we defined and implemented the following set of services that will belong to the Software Platform:

Metadata and content storage - the service that is responsible for storing the metadata of publications and other objects in the system

Indexes - a class of services which allow to easily find publications and other documents stored in the system. The family of index services includes fulltext searching index, index that traces relations between objects, indexes that are specific for particular sets of data (images, source data etc.) We also expect inclusion of semantic indexes.

Content enrichment services - services responsible for postprocessing and enriching of the data available in the store. This class of services includes automatic categorization systems, OCR-engines, conversion engines etc.

Annotation catalog - the service responsible for storing lightweight contents created by users in scope of particular objects in the system. Typical annotations may include comments, notes, user messages, publication ratings or bug reports.

User catalog - the service or the interface to the external service service the data about the user credentials. The user catalog allows to verify user rights.

Audit service - the service that allows to log all important events that occurred in the service.

Underneath we briefly describe each of these service types.

3.1   Metadata and Content Storage

The service is responsible for storing all kinds of documents containing both text and binary data. The main entity stored in the the storage service is a document which owns globally unique identifier. Each document consists of one or more content parts holding text or binary data - each part has its own name. This allows to store multiple contents within one document. In typical cases the document representing the publication contains the following content parts:

– one or more metadata parts - where each metadata part contains metadata of the publication in particular format. It is possible to store metadata of the one document in different formats and different versions. It is also possible to store both source and enriched metadata versions of the document.
– zero or more content parts - containing contents of the publication. Contents of the publication may be splitted into multiple parts (like separate chapters of the book). Also contents of the publication may be stored in multiple parts in different formats (PDF, plaintext, sequence of TIFF files) or different quality (thumbnail, low resolution, high resolution)

The main assumption of the storage is that the service is not aware of the format of data stored within. It is possible to store different kinds of documents having different formats in the same instance of the store. It is a responsibility of the application using the store to perform proper operations on proper kinds of objects.

The main drawback of this approach is that searching abilities of the storage service are very simple. It is only possible to fetch the document having specified identifier or to iterate over documents fulfilling some criteria based on tags. The other issue is that the service doesnt perform any ways of data consistency validation and is not aware of relations between particular objects. This means that all aspects of indexing, searching and maintaining consistency of the data is the responsibility of the application. Since indexing and searching of the data is one of the most essential functionalities of typical repository systems we defined a class of services called Indexes.

3.2   Indexes

In the Synat Platform the index is defined as an instance of the service which allows to perform the following operations: putting and removing document from the index and performing query on the index. The following indexes are being developed during the Synat project:

- Fulltext Index - the service which allows to index and search for documents consisting of multiple textual fields of specified types. The service may include functionalities of hit highlighting, faceted search and including new custom data types.
- Relational Index - the service which allows to store relations between documents in the store. We identified the following relations between documents - aggregation (article belongs to the journal), referencing-referenced, continuation of/continued by (between journals).
- Similarity Index - the service that allows to search for documents similar to existing one.
- Citation Index - the index that analyzes references of articles and allows to restore relations between publications in one repository. Unlike Relational Service, the Citation Index doesnt need exact reference to the referenced document in the metadata. Instead it tries to parse human readable form of citation.
- Personality Index - the service that allows to store information about particular persons and items contributed by them. The Personality index allows to merge identity informations about one person from miscellaneous sources, and to distinct between multiple persons having the same name. The Personality Index may be fed with information coming from the publication metadata (this data are usually inconsistent and incomplete), the data coming from external reference datasources (like national bibliography directories) and from the user catalog of the application using the Index.

The Index service is related to Storage Service which acts as a datasource, and with Process Manager which initiates process of creating and updating indexes.

3.3 Annotation Storage

The service is responsible for storing lightweight data created by application users. Typical digital library application allows users to enrich and label the data available with custom contents. This may include notes, ratings, comments, discussions etc.

We decided that this set of functionalities makes necessary to define new type of service called Annotation Storage. The Annotation Storage is similar to the Metadata Storage and allows to store documents that in addition to having some unique identifier are related to particular user (author of the annotation) and may be related to particular object in the system (publication, another annotation). Contrary to the Metadata Storage, the Annotation Storage expects that each annotation has only one content and this content has a text form. The service is very generic and does not make any strict assumptions about the format of the annotation - and in fact the format is one of textual properties of annotations. In addition, it is possible to define set of attributes of each annotation. The attributes may be later used during searching. We expect that most types of annotations stored in the Storage will be defined as XML documents. The service provides restricted set of search operations on annotations like searching for all annotations of specified type created by specified user and searching for all annotations having particular values of some attributes.

In fact Metadata Storage, Annotation Storage, User Catalog, Profile Catalog and Audit Service are the only one services that need strict backup features. The data stored in other services may be safely restored from previous ones.

3.4 Process Manager

The service is responsible for performing all indexing and enrichment operations on the data in the repository. The Process Manager allows to define and run data processing flow which consists of simple atomic nodes. It allows to dynamically define processes in runtime, start and control their flow. The service enables parallel execution of particular workflow nodes and allows to separate particular nodes with asynchronous queue.

The Process Manager is one of the repository services (along with UI) where the business logic of the system is located. Systems based on the Synat Software Platform mainly consist of the set of properly configured services, frontend UI applications and set of processing nodes and process definitions.

Processes can be started manually, using scheduler and initiated by an asynchronous event raised by the one of the other core services (Store, Annotation Service). Typically the set of processed defined in a repository application consists of:

- general indexing process - which iterate through all data in the Metadata Storage, enriches it and puts into general purpose indexes (like fulltext and relational). The process is responsible for initial indexing of the data after the first acquisition.

- another version of general indexing process that is executed as a result of event raised on the Store. The process in responsible for incremental indexing of the new data that is stored in the Metadata Storage
- data acquisition processes - family of processes that are responsible for importing data from external systems and converting it into system specific format. Such processes may be called manually or triggered by some kind of external event (asynchronous queue or HTTP server/filesystem polling)
- specific indexing processes - some of the indexes doesnt support incremental update, updating of the other ones is time and resource consuming. It is possible to separate some specific indexing tasks that update less critical indexes (like similarity, semantic analysis) and configure the application that some specific indexes will be updated periodically (e.g. every 72 hours). The same pattern may apply to some research and experimental tools that may be executed parallelly with general indexing process.
- long running enrichment processes - in the same way as specific indexing processes, some data enrichment processes are not critical to proper work of the system, so it is possible to separate them to single process that may be run manually or periodically. In addition resource consuming enrichment services may use separate hardware resources. This pattern allow to enhance existing systems with tools like OCR.

3.5   User Catalog and Profile Catalog

User Catalog and Profile Catalog are responsible for storing data about users of the system. The main difference between these services is in their responsibility. While the User Catalog is responsible for storing and maintaining authentication data of users like groups, roles and credentials, the Profile Catalog is another storage service responsible for holding all application specific data related to applications. Profile Catalog allow to define and store binary or text data related to particular users. Each user can have multiple profiles created by multiple applications.

Possible contents stored in Profile Catalog are user personal details, application settings, picture of the user and so on. It is possible to share the same instance of Profile Catalog among multiple applications. In such case each application may create its own set of profiles.

User Catalog - like other services belonging to the Synat Platform has its well defined, open interface that allows to create custom implementations. This enables creation of wrappers over existing authentication sources like LDAP.

3.6   Audit Service

The service is responsible for storing the trace about all actions performed by users (like entering particular pages, performing searches, clicking on the search results). The same service may be also used in order to trace all errors that occurred within the system - like processing problems, access violations and exceptions.

The service allows to store tuples containing timestamp of the event, the module in which the event occurred, type of the event and set of additional event attributes. The service is intended to be fast and lightweight. Stored logs may be periodically processed and analyzed by proper processes - for example it is possible to find the most popular publications.

4   Interoperability

The Synat Platform is primarily based on Java and Java-related technologies like Spring Framework. This means that the reference implementations rely on Java specific protocols and technologies like HTTPInvoke, RMI [8], etc. However the usage of the platform is not limited to these protocols. As described earlier, the platform defines three kinds of components: services (within containers), registries and clients. The only assumption related to communication is that any exchanges between registries and other components are based on the REST standard. It has been chosen because of simplicity of its adaptation on different software platforms and technologies. No assumptions are imposed on protocols used to communication between services and service clients. The protocol is negotiable and multiple instances of the same service may be available through different protocols for the same client, while the selection of proper protocol is performed transparently. The only requirement is to provide proper protocol support library to the client.

Services are normally executed within their service container. Along with the Platform, a generic Java based generic container is being developed. The container allows to instantiate and run services that are based on the JVM standard.

The problem that appears is how to run and instantiate software tools based on other technologies. We proposed two approaches that might be used depending on particular needs. The first approach assumes that it is necessary to build proper technology-specific service container that implements the service container contract. Since the contract of the container, the task might not be complex in case there is large amount of components in the particular technology. The second approach is to use existing implementation of Java based service protocol and create inter-technology bridge which may be based on queues or by calling another process in the operating system. We expect that bridge approach may be also used for interaction between the Synat Platform and other complex data processing solutions like cloud systems that rely on mapreduce paradigm. In such case, the whole map-reduce based infrastructure is available in the Synat Platform ecosystem as a monolithic service available through a well defined interface. This approach will be used in reference applications to implement user-traffic and statistics analysing tools.

5   Scalability and High-Availability

One of the main benefits that may be acquired when applying distributed service-oriented approach is the natural ability to implement scalability and high-availability features of the system. The Platform separates physchical location of services from its organizational structure and allows to dynamically bind services to logical names. This feature enables transparent switching the client between multiple instances of the same service. The feature of dynamic service aliases - where one alias points to multiple services at once - allows to introduce load-balancing and failover policies to the system.

Load balancing of the system can be implemented with multiple instances of the same service (e.g. index) containing the same data. All instances are available using one service alias which is used by service clients. The service registry is responsible for resolving alias and providing proper physical service location to clients. The registry may apply proper load balancing algorithms that take care about equal load of all nodes (e.g. like round-robin), it may also enable sticky sessions mechanism. Introduction of a new node of the service to the system is relatively simple, as it only requires registering the node with the service registry. While the approach perfectly solves problems of the load generated by end users it might not be enough to process and analyze large amounts of data. We expect that for some services like automatic data analysis we would need to make use of cloud computing and mapreduce paradigm. As described earlier we expect that whole cloud subsystem will be available as a single service having well defined protocol.

Another important aspect of the Platform is its ability to support high-availability and failover of the services. This feature is most essential for services that store relevant data. In case of such services their particular implementations have ability of data replication. A master-slave model is adopted, where normally all operations are performed on the master node and then replicated to slave nodes. In case of failure, client requests are transparently switched to on-line slave nodes by service registries. The Platform implements also service directory replication between service registries in order to avoid the registries being single points of failure. A network of service registries can be created within the Platform ecosystem in a way resembling the typical architecture of DNS services.

6   Conclusions

The Synat project introduces a modern approach to digital libraries and in general to research digital content management systems. Instead of developing yet another monolithic application, it focuses on building an open software platform that will serve as a base for development of research content services and application. Besides the software environment itself, the project will deliver a number of applications using the platform services, particularly an integrated content provisioning system (Infona portal) enriched with social functions.

The loosely-coupled, component-based nature of the platform, coupled with its service oriented architecture allows not only for extensibility but also for future natural dynamic refactoring of individual modules, thus extending the lifetime of the resulting applications beyond what would be possible in a traditional model.

Furthermore, it is planned to release the code of the platform under an open license so that both third party and independent developers can participate in development of the platform itself, along additional services and applications. The platform is designed so that it can support both PaaS and SaaS models of cloud computing services, thus allowing to build sustainable, business aware service and content providing solutions.

Acknowledgments. This work is partially supported by the National Centre for Research and Development (NCBiR) under Grant No. SP/I/1/77065/10 by the Strategic scientific research and experimental development program: Synat - Interdisciplinary System for Interactive Scientific and Scientific-Technical Information.

References

1. Ioannidis, Y., Maier, D., Abiteboul, S., Buneman, P., Davidson, S., Fox, E., Halevy, A., Knoblock, C., Rabitti, F., Schek, H., Weikum, G.: Digital library information-technology infrastructures. International Journal on Digital Libraries 5, 266–274 (2005)
2. Lagoze, C., Van de Sompel, H.: The open archives initiative: building a low-barrier interoperability framework. In: Proceedings of the 1st ACM/IEEE-CS Joint Conference on Digital Libraries, JCDL 2001, pp. 54–62. ACM, New York (2001)
3. Lynch, C.: The z39. 50 information retrieval protocol: An overview and status report. ACM SIGCOMM Computer Communication Review 21, 58–70 (1991)
4. Klyne, G., Carroll, J.J.: Resource description framework (rdf): Concepts and abstract syntax. Structure 10, 1–20 (2004)
5. Bizer, C., Heath, T., Berners-Lee, T.: Linked data-the story so far. International Journal on Semantic Web and Information Systems (IJSWIS) 5, 1–22 (2009)
6. Desertot, M., Donsez, D., Lalanda, P.: A dynamic service-oriented implementation for java ee servers. In: IEEE International Conference on Services Computing, SCC 2006, pp. 159–166. IEEE (2006)
7. Hall, R., Cervantes, H.: An osgi implementation and experience report. In: First IEEE Consumer Communications and Networking Conference, CCNC 2004, pp. 394–399. IEEE (2004)
8. Waldo, J.: Remote procedure calls and java remote method invocation. IEEE Concurrency 6, 5–7 (1998)

Chapter III

Text Processing

Author Disambiguation in the YADDA2 Software Platform

Piotr Jan Dendek, Mariusz Wojewódzki, and Łukasz Bolikowski

Interdisciplinary Centre for Mathematical and Computational Modelling,
University of Warsaw, Warsaw, Poland
{p.dendek,m.wojewodzki,l.bolikowski}@icm.edu.pl

Abstract. SYNAT platform powered by the YADDA2 architecture has been extended with the Author Disambiguation Framework and the Query Framework. The former framework clusters occurrences of contributor names into identities of authors, the latter answers queries about authors and documents written by them. This paper presents an outline of the disambiguation algorithms, implementation of the query framework, integration into the platform and performance evaluation of the solution.

Keywords: author disambiguation, record deduplication, software architecture, YADDA2 software platform.

1 Introduction

1.1 Record Deduplication

A common challenge among databases is a record deduplication, which is the term describing the situation when a real-world object is described by many separate records. "Record deduplication" itself is known in different communities as "record linkage" [11], "data cleaning" [28], "data scrubbing" [3], "mirror detection"[41], "instance matching" [2], etc. This issue may occur as a result of multiple formats used for representing an attribute, for example in the case of an address or a person name. The problem is particularly acute when information is gathered over a long period of time. Even when there is only one standard of representing a record, some misspelling or diacritics handling issues may occur. Another concern is merging data from multiple heterogeneous sources, when normalization levels and record definitions may not match. Last but not least, there are cases of automatic data acquisition (e.g. from the Internet), which is an instance of combining information from many origins. Those concerns are especially problematic when coping with big datasets.

The challenge of the accurate record linkage had been addressed many times over the last five decades. Elmagarmid *et al.* [10] et al. coherently enumerate all mainstream approaches to the record linkage. Authors focus on specific aspects of the problem, which are:

R. Bembenik et al. (Eds.): *Intell. Tools for Building a Scientific Information*, SCI 467, pp. 131–143.
DOI: 10.1007/978-3-642-35647-6_10   © Springer-Verlag Berlin Heidelberg 2013

1. Data preparation, covering a parsing and a standardization procedures. [31,33,1,5,19]
2. Attribute matching techniques, as approximate string matching, token based and phonetic based. [29,16,32,22]
3. Duplicate detection, covering supervised and unsupervised machine learning techniques as well as hand-crafted ones. [36,6,38,39]
4. Problem decomposition approaches as blocking, k-nearest neighbour, clustering. [20,21,14]

1.2   Author Disambiguation

Author disambiguation is an instance of the record linkage problem, where instances to match are authors, typically represented by first names, a surname, an affiliation and metadata of co-authored articles. It is clear that none of the mentioned attributes can single-handedly solve the entire problem. There are attributes that determine identity with a high degree of certainty, but they are frequently not present, e.g. an e-mail address appears only in 10% of articles [30].

In author disambiguation all typical object deduplication obstacles arise, beginning from many standards for writing a name ("J.Smith", "John Smith", "J.Smith Jr.", "Smith, J."), misspellings ("J.Smiht", "J.Smth"), an attribute value change over time ("Eleonore Smith", "Eleonore Smith-Black"), diacritic handling ("José Gonçalves", "Jose Goncalves","Jos? Gon?alves"), transliterations (e.g. translating "Angela Johnson" to Japanese equivalent "アンジェラ·ジョンソン" and back to English result is "anjira jyonson") [15] and extraction artifacts ("Smith Machine"). As Torvik and Smalheiser [37] investigated, about 1,3% authors whose e-mail addresses match have different surnames, most likely due to inconsistencies enumerated here.

The other attributes also need further consideration, e.g. some errors may occur in an e-mail address, like "jsmith@@institution.org" or "jsmith"institution. org", hence requiring a rectification step.

Many researchers explored specifically the subject of the author disambiguation. Han et al.[12] compared Naive Bayes and Support Vector Machines (SVM) classifiers for this task, whereas in [13], they examined efficiency of k-Way Spectral Clustering. Concurrently, Dai and Storkey [8] applied hierarchical Dirichlet process and nonparametric latent Dirichlet allocation models, whereas Levin and Heuser [17] included in their solution enhancements derived from the genetic programming.

Typically, authors tried to conduct pairwise comparisons on a set of records with the same value of a major feature (e.g. surname) to determine whether two candidate author items are the same. In contrast to performing analysis in respect of all given features, Qian et al. [30] proposed to perform initial clustering with a limited number of features to obtain High Precision Clusters in the first step and then merge clusters into High Recall Clusters in the second step. They also proposed to introduce a human judgement clustering in the final step. When utilizing user feedback, it is crucial to distinguish between experts and regular person, especially preventing acts of vandalism, considered as sending false information.

Culotta *et al.* [7] proposed to pre-assess each contributor block to determine the likelihood of duplicates. For example, if all authors are affiliated to a few institutions or e-mail addresses then the cluster of candidate items is more likely to have duplicates than the one that contains contributions associated with a high number of e-mails and institutions. Authors claim that the usage of so called first-order features over sets of records may eventually reduce error rate that outperforms a regular binary classification by up to 60%. Tan *et al.* [35] decided to extend an available set of information by employing Internet search engines and adopting as a feature home pages containing given article.

The rest of this paper is organized as follows. Section 2 describes both the SYNAT project and the YADDA2 architecture. Section 3 presents the Author Disambiguation Framework (ADF) developed for purposes of YADDA2 [4] and the results of further examination [9]. Section 4 shows adaptation of the ADF to SYNAT platform with an emphasis on its presentation layer – the Query Framework (QF). Section 5 contains evaluation of the ADF and the QF. Finally, Section 6 concludes the paper and proposes further improvements.

2 SYNAT and YADDA2

SYNAT project aims to build an "Interdisciplinary System for Interactive Scientific and Scientific Technical Information." It is a strategic project commissioned by the Polish National Centre for Research and Development. The system is based on the YADDA2 architecture [34], presented in Figure 1, developed at ICM UW.

YADDA2 is an open, loosely-coupled, service-oriented and modular framework that facilitates development of digital repository applications. The framework contains a number of reusable modules that provide, among others: storage, relational and full-text indexing, process management, authorization/authentication and asynchronous communication. The above are so-called base services, providing general functionalities which are independent of the type of content being processed. On top of them, there is a number of more specialized compotents that implement a business logic layer.

The Author Disambiguation Framework, as well as the Query Framework described later in this paper, are good examples of such specialized components.

3 YADDA2 Author Disambiguation Framework

3.1 Vocabulary

In our previous papers [4,9] we have established the vocabulary for ADF description, which, after further adjustments, can be described as follows. A **contributor** entity reflects the fact that a **person** was a co-author of a **document**. When data about a contributor is extracted from document medatada, it may

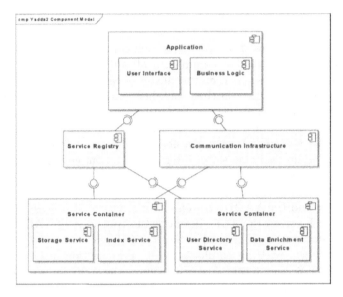

Fig. 1. Component model of the YADDA2 framework

be treated as a one-to-one binding between a person and a document. For each document, the number of corresponding contributor instances is equal to the number of co-authors. Occasionally, we may also reuse the concept of a document to represent a different tangible outcome that can be attributed to a person or persons, for example, a log of user actions in an Internet service. Having a set of contributors, the goal is to cluster them into groups containing publications written by one person.

To do so in the efficient way, we perform a coarse-grained grouping into **blocks** of contributors according to a **hash function**. Typically, a hash function yields a result which is a function of surname, like a diacritics removal and lower-casing. Depending on authors' surnames, a more sophisticated hash function, e.g. Soundex or Double Metaphone phonetic transformations, may be chosen. This division step corresponds to the "map" phase of the MapReduce paradigm.

Consequently, the "reduce" stage is performed, in which **crude features** are calculated. A crude feature is an integer representing a number of common values, e.g. identical key words. Afterwards, we obtain a **feature** by scaling a crude feature into the [0,1] range, which is multiplied by a feature weight yielding an **atomic affinity**. A sum of atomic affinities is called a **total affinity** and constitutes the input data for a clustering algorithm.

As mentioned above, this approach is customizable with respect of a hash function, a set of features with associated weights and a clustering function. Finally, this solution is fully applicable in a MapReduce workflow.

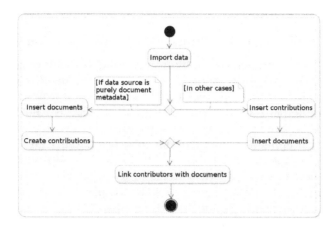

Fig. 2. The process of data import to the Author Disambiguation Framework. It may either extract data about contributors or utilize pre-generated ones.

3.2 Author Disambiguation Framework Flow

In [4] we presented the Author Disambiguation Framework flow, which is briefly summarised in this section. The ADF flow consists of five steps:

1. Data import. Data is transferred to an ADF database.
2. Blocking. All contributors are split into relatively small, computationally less expensive subsets.
3. Affinity calculation. Pairwise comparison against a set of crude features accompanied with their weights is performed in each block, eventually generating a total affinity.
4. Clustering. Contributors which are recognized as similar are inserted into the same group.
5. Result persistence. A connection between a contributor and a cluster is stored either in a database or in text files.

The framework may be initialized by passing two kinds of input data as presented in Figure 2:

1. a document collection from which contributors are extracted,
2. a collection that contains information about contributors with information associated to them and optionally contributors' documents.

Calculations on blocks may be processed in parallel taking advantage of a multicore computer architecture.

4   Implementation

4.1   Author Disambiguation Framework Implementation

The ADF has been implemented purely in Java using two databases: bigdata® [24] and Sesame [27]. In both of them the basic entity is a ordered triple, containing information about subject, predicate and object, $t_i = \langle s, p, o \rangle$, where the first two are Uniform Resource Identifier (URI) objects and the last one can either be an URI or a literal (non-URI) object. ADF methods are implemented using Resource Description Framework Storage And Inference Layer (RDF SAIL) which is a standard set of interfaces defining an API for RDF repositories. As the result, it can connect to standard triple stores such as the ones mentioned above. bigdata® is capable of fitting about 12.7 billion triples in its hard drive journal file, whereas Sesame may accumulate 70 millions of triples in a memory[25]. Due to the size of a database and its localization (a hard drive file vs. memory) the second mentioned triple store outperforms the other one in terms of a communication time by about tree orders of magnitude.

Taking into account these facts, the ADF uses bigdata® to collect all imported data, whereas for each block, the ADF creates a cache in Sesame memory store where all data needed for calculating affinity are transferred. This particular approach proved to be the most fruitful, synergistic strategy in terms of the performance. Eventually, resulting person objects may be written back to a bigdata® or to CSV files.

4.2   Query Framework Implementation

Data Structures. Query Framework (QF), similarly to ADF, is written entirely in Java, but it takes advantage of the Neo4j database[26]. Neo4j, as the example of a NoSQL database has been the subject of detailed comparison [40] with the traditional, relational approach.

The reason for choosing Neo4j was its flexibility in a model construction as well as its high efficiency. A few specialized data structures have been applied to obtain a better performance. For instance, to increase a search performance, identifiers and attributes of stored objects are indexed with full-text indexes embedded into Neo4j.

The structure of data is built as follows. The top element is a **root**, which due to Neo4j restrictions always exists in a database, even if not inserted explicitly. A root is bound by a **root relation** with a **person**, which, as a reflection of a real-world author, should point by an **identity relation** to all **contributor** instances corresponding to that person. A **contributor** stores information about publications or activities associated with a **person** object. In case when a data source is solely documents metadata, contributor-document is a one-to-one relation. On the contrary, when a data source is derived from any other origin, contributor-document is a one-to-many relation (one-to-zero also applies).

The database structure is presented in Figure 3.

During import, the QF takes metadata (containing information about contributors and optionally documents) and information about persons from the

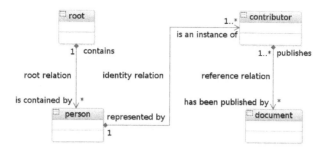

Fig. 3. Classes appearing in the Query Framework's Neo4j database. The top instance is singular root object, which always exists in the Neo4j store. Person objects are created as a reflection of the Author Disambiguation Framework results. Contributors are instances of persons. Documents are publications written by given contributor and transitively by a person.

ADF. Finally, the QF constructs its internal structures described in the following section.

Model Creation. Currently, data are imported from files generated in the process of disambiguation. In order to add to the service the information about a person:

1. The system checks if the person already exists to omit duplicates, which may occur whenever a next import extends only a few blocks of previously digested data.
2. Person is created and linked to the root.
3. Contributors and documents are created and connected in the top-down manner.

A contributor can exist without any document, as a result of employing another data source which does not provide details about publications and activities.

The above description is reflected in the the activity diagram in Figure 4.

Model Usage. Queries directed to the service may be routed through a website, or can be called directly by the appropriate method of the QF API. After loading the QF database, a user can construct queries as follows:

- "Find one person object" – used to construct queries that are aimed at finding one person with all contributors that are in identity relation with it. Input parameter is a person identity.
- "Find one person object and related publications" – used to construct queries that are designed to find one person and associated contributors together with dependent documents. Input parameters are: person identity, order and attribute criteria.

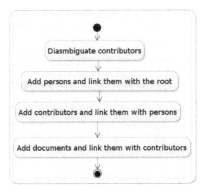

Fig. 4. The Query Framework preparation. The first step is the disambiguation procedure conducted by the Author Disambiguation Framework. Next, results of the ADF are inserted to the QF and connected with the Neo4j root. Then, the regular metadata are imported.

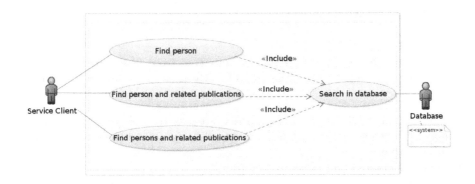

Fig. 5. Actions available in the Query Framework

- ”Find many person objects” – used to construct queries that are aimed at finding persons in a certain order with previously specified constraints. Input parameters are: order and attribute criteria.

Activities mentioned above are also reflected in Figure 5.

5   Evaluation

We conducted performance tests on a 4 cores machine (Intel®Core™ i5 CPU U520@1.07GHz) with 4GB RAM under control of Ubuntu Linux OS, kernel version 2.6.32-41-server. We focused on the BazTech bibliographic database[23] containing document metadata with 257784 contributors.

5.1   Author Disambiguation Framework Performance

Let $BlockSize$ denote a set of contributors sizes and the function $BlockOcc(x)$ returns a number of occurences of a block of size x The disambiguation time for a block of size x can be closely approximated by the formula:

$$T_{disambiguation}(x) = 0,0608099 \cdot x + 0,0005811 \cdot x^2 \qquad (1)$$

where the part $0,0005811 \cdot x^2$ is mainly the effect of performing pairwise comparison, but also employing a single-link hierarchical clustering, which has $\mathcal{O}(N^2)$ computation complexity. It has to be mentioned that a clustering procedure is at least one order of magnitude faster [18] then similarity computation, so choosing another clustering method would not have a crucial impact on an overall disambiguation time, which is given by the following equation:

$$T_{overall}(BlockSize) = \sum_{x \in BlockSize} BlockOcc(x) \cdot T_{disambiguation}(x) \qquad (2)$$

5.2   Query Framework Performance

To evaluate the QF we carried out two types of tests. The first one was a data import of the BazTech metadata to the Neo4j database, which took about 2 hours. The second one covered three queries described in the Section 4.2. The "Find one person object" query repeated 10000 times finished in the total time of 3 seconds, whereas the query "Find one person object and related publications" executed 10000 times took about 6 seconds. Nevertheless, in the evaluation process we were targeting a different person in each query, to avoid implicit use of database cache mechanisms. The time increased as a natural effect of retrieving more information, located further from the root. The last query we examined, "Find 10000 person objects", was executed in 4 seconds. Described results are presented in the Table 1.

Table 1. Results of the Query Framework performance tests against provided query types

Query goal	Number of repetitions	Overall time
Find one person object	10000	3s
Find one person object and related publications	10000	6s
Find 10000 person objects	1	4s

6   Conclusions

In this article we have presented the Author Disambiguation Framework developed in the YADDA2 architecture and adopted for the SYNAT platform by means of the Query Framework. We have described the details of disambiguation as well as the method of adaptation and a presentation layer. We proposed the time complexity of the disambiguation process accompanied with the Query Framework performance tests.

Future work will include applying more sophisticated disambiguation techniques, followed by the use of more efficient computing architectures. Ideally, each framework should rely on the same database and data structures. Furthermore, we plan to extend the disambiguation procedure to cover activities other then publishing. Another promising direction is the enhancement of disambiguation results with a user feedback to fully utilize SYNAT platform capabilities.

Acknowledgment. This work is supported by the National Centre for Research and Development (NCBiR) under Grant No. SP/I/1/77065/10 by the Strategic scientific research and experimental development program: "Interdisciplinary System for Interactive Scientific and Scientific-Technical Information".

References

1. Agichtein, E., Ganti, V.: Mining reference tables for automatic text segmentation. In: Proceedings of the 2004 ACM SIGKDD International Conference on Knowledge Discovery and Data Mining, KDD 2004, p. 20. ACM Press, New York (2004), http://portal.acm.org/citation.cfm?doid=1014052.1014058
2. Aono, M., Seddiqui, M.H.: Scalability in ontology instance matching of large semantic knowledge base. In: AIKED 2010 Proceedings of the 9th WSEAS International Conference on Artificial Intelligence, Knowledge Engineering and Data Bases, pp. 378–383 (2010)
3. Berman, J.J.: Concept-Match Medical Data Scrubbing. Archives of Pathology & Laboratory Medicine 127(6), 680–686 (2003)
4. Bolikowski, Ł., Dendek, P.J.: Towards a Flexible Author Name Disambiguation Framework. In: Sojka, P., Bouche, T. (eds.) Towards a Digital Mathematics Library, pp. 27–37. Masaryk University Press (2011)
5. Borkar, V., Deshmukh, K., Sarawagi, S.: Automatic segmentation of text into structured records. In: Proceedings of the 2001 ACM SIGMOD International Conference on Management of Data, SIGMOD 2001, pp. 175–186. ACM Press, New York (2001), http://portal.acm.org/citation.cfm?doid=375663.375682
6. Cohen, W.W., Richman, J.: Learning to match and cluster large high-dimensional data sets for data integration. In: Proceedings of the Eighth ACM SIGKDD International Conference on Knowledge Discovery and Data Mining, KDD 2002, p. 475. ACM Press, New York (2002), http://portal.acm.org/citation.cfm?doid=775047.775116
7. Culotta, A., Kanani, P., Hall, R., Wick, M., McCallum, A.: Author Disambiguation using Error-driven Machine Learning with a Ranking Loss Function. In: Sixth International Workshop on Information Integration on the Web (2007)
8. Dai, A.M., Storkey, A.J.: Author Disambiguation: A Nonparametric Topic and Co-authorship Model. In: NIPS Workshop on Applications for Topic Models Text and Beyond, pp. 1–4 (2009)
9. Dendek, P.J., Bolikowski, Ł.: Evaluation of Features for Author Name Disambiguation Using Linear Support Vector Machines. In: Proceedings of the 10th IAPR International Workshop on Document Analysis Systems, pp. 440–444 (2012)

10. Elmagarmid, A., Ipeirotis, P., Verykios, V.: Duplicate Record Detection: A Survey. IEEE Transactions on Knowledge and Data Engineering 19(1), 1–16 (2007), http://ieeexplore.ieee.org/lpdocs/epic03/wrapper.htm?arnumber=4016511
11. Fellegi, I.P., Sunter, A.B.: A Theory for Record Linkage. Journal of the American Statistical Association 64, 1183–1210 (1969)
12. Han, H., Giles, L., Zha, H., Li, C., Tsioutsiouliklis, K.: Two supervised learning approaches for name disambiguation in author citations. In: Proceedings of the 2004 Joint ACM/IEEE Conference on Digital Libraries - JCDL 2004, p. 296 (2004), http://portal.acm.org/citation.cfm?doid=996350.996419
13. Han, H., Zha, H., Giles, C.L.: Name disambiguation in author citations using a K-way spectral clustering method. In: JCDL 2005: Proceedings of the 5th ACM/IEEE-CS Joint Conference on Digital Libraries, pp. 334–343. ACM Press, New York (2005)
14. Hernández, M.A., Stolfo, S.J.: Real-world Data is Dirty: Data Cleansing and The Merge/Purge Problem. Data Mining and Knowledge Discovery 2(1), 9–37 (1998)
15. Knight, K., Graehl, J.: Machine Transliteration. Computational Linguistics 24(4), 599–612 (1998)
16. Kukich, K.: Technique for automatically correcting words in text. ACM Computing Surveys 24(4), 377–439 (1992), http://portal.acm.org/citation.cfm?doid=146370.146380
17. Levin, F.H., Heuser, C.A.: Using Genetic Programming to Evaluate the Impact of Social Network Analysis in Author Name Disambiguation. In: Laender, A.H.F., Lakshmanan, L.V.S. (eds.) Proceedings of the 4th Alberto Mendelzon International Workshop on Foundations of Data Management Buenos Aires Argentina, Citeseer, May 17-20., vol. 619 (2010), http://citeseerx.ist.psu.edu/viewdoc/download?doi=10.1.1.173.5987&rep=rep1&type=pdf, http://ceur-ws.org/Vol-619/paper2.pdf
18. Manning, C.D., Raghavan P., Schütze, H.: Introduction to Information Retrieval (2008), http://nlp.stanford.edu/IR-book/html/htmledition
19. McCallum, A., Freitag, D.: Maximum entropy Markov models for information extraction and segmentation. In: Proceedings of the Seventeenth International Conference on Machine Learning (2000), http://courses.ischool.berkeley.edu/i290-dm/s11/SECURE/gidofalvi.pdf
20. McCallum, A., Nigam, K., Ungar, L.H.: Efficient clustering of high-dimensional data sets with application to reference matching. In: Proceedings of the 6th ACM SIGKDD International Conference on Knowledge Discovery and Data Mining, KDD 2000, pp. 169–178. ACM Press, New York (2000), http://portal.acm.org/citation.cfm?doid=347090.347123, http://doi.acm.org/10.1145/347090.347123, http://dl.acm.org/citation.cfm?id=347123
21. Monge, A., Elkan, C.: An Efficient Domain-Independent Algorithm for Detecting Approximately Duplicate Database Records. In: Proc. Second ACM SIGMOD Workshop Research Issues in Data Mining and Knowledge Discovery, pp. 23–29 (1997)
22. Navarro, G.: A guided tour to approximate string matching. ACM Computing Surveys 33(1), 31–88 (2001), http://portal.acm.org/citation.cfm?doid=375360.375365
23. Polish Technical Journal Contents, http://baztech.icm.edu.pl/
24. Bigdata Database Webpage, http://www.systap.com/bigdata.htm

25. Large Triple Stores Description, http://www.w3.org/wiki/LargeTripleStores
26. Neo4j: The World's Leading Graph Database,
    http://www.w3.org/wiki/LargeTripleStores
27. Semame Database Webpage, http://www.openrdf.org/
28. Park, K., Becker, E., Vinjumur, J.K., Le, Z., Makedon, F.: Human behavioral de-
    tection and data cleaning in assisted living environment using wireless sensor net-
    works. In: Proceedings of the 2nd International Conference on PErvsive Technolo-
    gies Related to Assistive Environments - PETRA 2009, pp. 1–8. ACM Press, New
    York (2009), http://portal.acm.org/citation.cfm?doid=1579114.1579121
29. Philips, L.: The double metaphone search algorithm. C/C++ Users Journal 18(6),
    38–43 (2000)
30. Qian, Y., Hu, Y., Cui, J., Zheng, Q., Nie, Z.: Combining Machine Learning and
    Human Judgment in Author Disambiguation Framework. In: Proceedings of the
    20th ACM International Conference on Information and Knowledge Management,
    pp. 1241–1246. ACM Press (2011),
    http://research.microsoft.com/pubs/154452/CIKM_CameraReady.pdf
31. Raman, V.: Potter's wheel: An interactive data cleaning system. In: International
    Conference on Very Large Data (2001),
    http://www.vldb.org/conf/2001/P381.pdf
32. Ristad, E., Yianilos, P.: Learning string-edit distance. IEEE Transactions on Pat-
    tern Analysis and Machine Intelligence 20(5), 522–532 (1998),
    http://ieeexplore.ieee.org/lpdocs/epic03/wrapper.htm?arnumber=682181
33. Sutton, C., Rohanimanesh, K., McCallum, A.: Dynamic conditional random fields.
    In: Twenty-first International Conference on Machine Learning, ICML 2004, p. 99.
    ACM Press, New York (2004),
    http://portal.acm.org/citation.cfm?doid=1015330.1015422
34. Sylwestrzak, W., Rosiek, T., Bolikowski, L.: YADDA2 Assemble Your Own Digital
    Library Application from Lego Bricks. In: Proceedings of the 2012 ACM/IEEE on
    Joint Conference on Digital Libraries (2012)
35. Tan, Y.F., Kan, M.Y., Lee, D.: Search engine driven author disambiguation. In:
    Proceedings of the 6th ACM/IEEE-CS Joint Conference on Digital Libraries -
    JCDL 2006, p. 314. ACM Press, New York (2006),
    http://portal.acm.org/citation.cfm?doid=1141753.1141826
36. Tejada, S., Knoblock, C.A., Minton, S.: Learning object identification rules for
    information integration. Information Systems 26(8), 607–633 (2001),
    http://www.sciencedirect.com/science/article/pii/S0306437901000424,
    http://linkinghub.elsevier.com/retrieve/pii/S0306437901000424
37. Torvik, V.I., Smalheiser, N.R.: Author name disambiguation in MEDLINE. ACM
    Transactions on Knowledge Discovery from Data 3(3), 1–29 (2009),
    http://portal.acm.org/citation.cfm?doid=1552303.1552304
38. Verykios, V.S., Moustakides, G.V.: A generalized cost optimal decision model for
    record matching. In: Proceedings of the 2004 International Workshop on Informa-
    tion Quality in Informational Systems, IQIS 2004, p. 20. ACM Press, New York
    (2004),
    http://portal.acm.org/citation.cfm?doid=1012453.1012457
39. Verykios, V., Moustakides, G., Elfeky, M.: A Bayesian decision model for cost
    optimal record matching. The VLDB Journal The International Journal on Very
    Large Data Bases 12(1), 28–40 (2003),
    http://www.springerlink.com/Index/10.1007/s00778-002-0072-y

40. Vicknair, C., Macias, M., Zhao, Z., Nan, X., Chen, Y., Wilkins, D.: A comparison of a graph database and a relational database. In: Proceedings of the 48th Annual Southeast Regional Conference on - ACM SE 2010, p. 1. ACM Press, New York (2010), http://portal.acm.org/citation.cfm?doid=1900008.1900067
41. Widom, J.: Research problems in data warehousing. In: Proceedings of the Fourth International Conference on Information and Knowledge Management, CIKM 1995, pp. 25–30. ACM Press, New York (1995), http://portal.acm.org/citation.cfm?doid=221270.221319

Methodology for Evaluating Citation Parsing and Matching

Mateusz Fedoryszak, Łukasz Bolikowski,
Dominika Tkaczyk, and Krzyś Wojciechowski

Interdisciplinary Centre for Mathematical and Computational Modelling,
Warsaw University, Warsaw, Poland
{m.fedoryszak,l.bolikowski,d.tkaczyk,k.wojciechowski}@icm.edu.pl

Abstract. Bibliographic references between scholarly publications contain valuable information for researchers and developers involved with digital repositories. They are indicators of topical similarity between linked texts, impact of the referenced document, and improve navigation in user interfaces of digital libraries. Consequently, several approaches to extraction, parsing and resolving said references have been proposed to date. In this paper we develop a methodology for evaluating parsing and matching algorithms and choosing the most appropriate one for a document collection at hand. We apply the methodology for evaluating reference parsing and matching module of the YADDA2 software platform.

Keywords: citation parsing, citation matching, evaluation, test set, YADDA2 software platform.

1 Introduction and Related Work

This paper discusses methods of evaluating algorithms for matching scholarly citations. Citation matching attempts to cluster bibliographic references to the same document, and possibly link them with the referenced document (provided that it is present in a collection). This can be seen as an instance of a broader problem of record linkage in databases [5,4]. Citation matching is a fundamental step in creating a digital library of scholarly publications. Links between documents conveying the fact that document A references document B (represented e.g. by the `references` term in the Dublin Core standard) are needed for a number of reasons:

- more user-friendly interfaces – similarly to hypertext links, citation links allow a user to navigate between documents [11].
- scientometrics – number of citations received by documents is an established measure of impact of individual researchers, journals, institutes and counties. For example, Impact Factor [7] and Hirsch Index [10] depend critically on availability of a citation graph.
- link-based classification – bibliographic references provide an excellent context for classifying documents or determining author identities [3,14].

R. Bembenik et al. (Eds.): *Intell. Tools for Building a Scientific Information*, SCI 467, pp. 145–154.
DOI: 10.1007/978-3-642-35647-6_11   © Springer-Verlag Berlin Heidelberg 2013

Historically, citation matching was performed manually [6]. Hitchcock et al. [11] demonstrated a proof-of-concept system that performed autonomous linking within Cognitive Science Open Journal. Citeseer was one of the first large-scale systems [8] that autonomously indexed scholarly citations. Pasula et al. [16] proposed a probabilistic model for citation matching.

Citation matching is sometimes divided into two phases: segmentation and entity resolution. Segmentation, or citation parsing, aims to deconstruct a bibliographic reference into functional pieces such as author names, title, year of publication, etc. Entity resolution clusters records representing the same document. However, several authors depart from such a division, most notably Wellner et al. [20], Poon and Domingos [17], Liao and Zhang [13], or Goutorbe [9].

Introduction of autonomous citation matching solutions created a need for their evaluation. Lawrence et al. [12] created a training set by hand and evaluated performance of their citation matching approaches by calculating the percentage of fully correct groups of citations. Their set is often referred to as "CiteSeer set." McCallum et al. [15] demonstrated how to employ machine learning techniques in automating construction of digital libraries. They created a number of training data sets[1], including one for evaluating citation matching. This set is often referred to as "Cora set." Most other authors train and evaluate their algorithms on one or both of these sets.

Our paper proposes a different take on evaluating performance of citation matching algorithms. In particular, we propose a method of *autonomous generation* of training and test sets and a *wider range of metrics* for evaluation of citation matching solutions.

2   Methodology

In this section we shall present a method of evaluating a citation matcher. We shall demonstrate how we create a test set and then propose a set of metrics used to measure matcher correctness.

We define citation matching problem in a slightly different way than usual: we assume we have a set of citations and a set of documents' metadata in a database. We want to assign to each citation a database record or information that the store does not contain the appropriate document.

2.1   Test Set Preparation

To generate the test set we have used the metadata of 1400 randomly selected publications from the Spanish Digital Mathematics Library (DML-E) aggregated in the European Digital Mathematics Library (EuDML, [18]). The library contains mathematical publications written in either English or Spanish, of which the oldest date back to late 50's of the previous century. The metadata was available as XML files in NLM format, each describing one publication.

---

[1] See: `http://people.cs.umass.edu/~mccallum/data.html`

We have divided our document set into 3 subsets containing 1000, 200 and 200 documents respectively. We have used metadata of documents in 1st and 2nd subset to generate citations and put those from subsets 1st and 3rd into the database (see Fig. 1).

We wanted to generate many citation strings using various popular bibliographical styles. The easiest way to achieve that was to use BIBTEX. We only needed to create a database file with the metadata of all the documents we wanted to create citations for and supply bibliographical styles[2]. BIBTEX generated files containing LATEX bibliography. From these files we extracted citation strings and converted them to plain text by removing all the LATEXcommands. Each style constitutes one test set, metrics described in the following section are defined per a set.

Finally, we asked the matcher under evaluation to match citations to the database records.

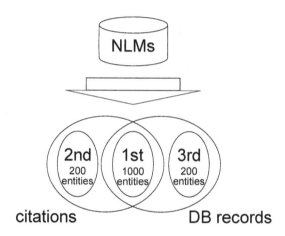

Fig. 1. Test set generation

2.2 Metrics

We have designed a set of metrics which we used in our experiments. They can be divided into several groups according to the aspect of the matching they deal with. In the following paragraphs we shall present these groups and define the metrics they contain.

Grouping Correctness Metric. As we have mentioned, citation matching task is often defined as finding among the set of citations those that refer to the same paper. Such a formulation of the problem appears, among others, in the classic paper by Lawrence et al. [12]. Let there S be the set of correct citation

---

[2] We have used `abbrv`, `acm`, `alpha`, `apalike`, `ieeetr`, `jpc`, `pccp`, `plain`, `ppcf` and `revcompchem`.

groups and R the set of groups returned by a matcher. *Grouping correctness* is then defined as

$$G = \frac{|S \cap R|}{|S|}$$

To make our results comparable with this approach, we shall define a similar metric in our evaluation framework. We can assume we have two types of group elements: citations and database records.

Reference group set S, representing the correct clustering that we hope to reconstruct, will consist of (see Fig. 2):

- 1000 2-element groups each containing a citation and an appropriate record,
- 200 1-element groups each containing a citation only,
- 200 1-element groups each containing a database record only.

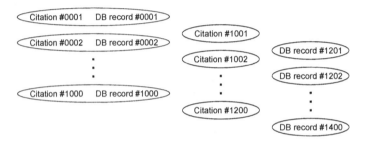

Fig. 2. Groups in the test set

As for result set, it will be defined as follows. Let there P be the set of (citation, database record) pairs returned by a matcher. Two elements x and y are in the same group if $(x, y) \in P \lor (y, x) \in P$. Result set R is a set of such groups.

Now we can introduce grouping correctness metric into our framework in an analogous way.

References-Based Metrics. We can also look at matching task as finding (citation, database record) pairs. Let there C be the set of such pairs that exist in test set and P be the answers of an algorithm such that their database records are not empty. We can now define the following metrics:

- reference precision $P_r = \frac{|C \cap P|}{|P|}$
- reference recall $R_r = \frac{|C \cap P|}{|C|}$
- reference F-measure $F_r = 2\frac{P_r P_r}{P_r + P_r}$

Nonexistent Record Metrics. There is a number of citations that do not have a corresponding record in the database. The metrics defined in this section are to cover them. M is the set of citations that do not have a database record and N is the set of citations for which an algorithm did not match any record. We define:

- nonexistent precision $P_\perp = \frac{|M \cap N|}{|N|}$
- nonexistent recall $R_\perp = \frac{|M \cap N|}{|M|}$
- nonexistent F-measure $F_\perp = 2\frac{P_\perp P_\perp}{P_\perp + P_\perp}$

Miscellaneous Metrics. Let E be the set of all correct (citation, database record) pairs where citation is not empty and P the set of pairs returned by a matcher We define *accuracy* as

$$Acc = \frac{|E \cap P|}{|E|}$$

3   Evaluated Bibliographic Reference Matcher

In order to demonstrate the methodology described in Section 2, we will apply it to evaluate our in-house citation matcher implemented in the YADDA2 platform. In this section we shall briefly describe our matcher, while the next section (Section 4) shall be devoted to presentation of its results.

Bibliographic reference matcher often has to deal with a collection of documents containing bibliographic references. References can exist in different forms, from raw text strings to hierarchical structures with tagged metadata information. Our requirements for bibliographic reference matcher include:

- identifying all pairs *reference — referenced document* in the collection,
- finding all documents referenced by a new document added to the collection,
- finding all documents referencing a new document added to the collection.

The whole implementation of bibliographic references matcher we have evaluated consists of three parts:

- bibliographic references parser used to extract valuable metadata information from bibliographic reference strings,
- metadata store that indexes and allows to search the metadata information of both the documents and the references from the collection,
- the bibliographic reference matcher being able to identify documents referenced by a given document and documents referencing a given document, based on comparing various metadata information of documents and references.

Figure 3 shows the process of matching references in the collection. First all reference strings are parsed and reference metadata is extracted. In the second step the metadata of both the documents and the references included in the collection is added to the metadata store and indexed. After this the matcher is ready to match references with documents by comparing various metadata fragments found in the metadata store.

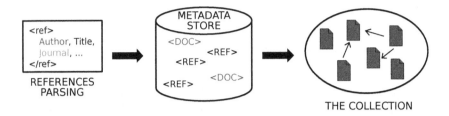

Fig. 3. The process of references matching

3.1   Bibliographic Reference Parser

One should not assume that the references in the collection are in the form of parsed structures with tagged metadata information. In many cases the matcher has to deal with references in the form of raw text strings. As a result the matcher requires a method for extracting metadata information from reference strings.

The goal of parsing the bibliographic reference strings is to identify fragments of the strings containing meaningful pieces of metadata information. The information our parser extracts include: *author, title, journal, volume, issue, pages, publisher, location* and *year*. Extracted metadata information fragments are indexed in the next step, which allows us to match them with the metadata of the documents in the collection.

The implementation of bibliographic reference parser is based on a Hidden Markov Model. First the reference string is tokenized into substrings containing only letters and digits or a single character of another type. In our model HMM sequence is composed of reference's tokens, labels of tokens are treated as unknown states and vectors of features computed for every token are visible observations. The Viterbi algorithm is used to determine the most probable sequence of token labels based on initial, transition and emission probability obtained from a training set. More details of the parser implementation can be found in [19].

The citation sample we have used to train the parser contained 100 references, each in 10 different bibliographic styles (described by BibTeX styles `abbrv`, `acm`, `alpha`, `apalike`, `ieeetr`, `jpc`, `pccp`, `plain`, `ppcf` and `revcompchem`), 1,000 references in total. The citations were generated in a similar way we did it in test set building. The metadata information extracted from the bibliographic references has been used in further matching steps.

3.2   Metadata Store

The metadata store indexes the metadata information of both the documents from the collection and bibliographic references contained by them. The reference matcher uses the metadata store to search for documents and references based on various metadata information.

The implementation of the metadata store can be based on any software able to index and search data. Our first implementation used Apache Solr search platform [1], for reference matching evaluation we used PostgreSQL database [2].

3.3 Bibliographic Reference Matching

Bibliographic reference matcher is based on comparing various metadata information of documents and references. The matcher allows to:

- find the document referenced by a given bibliographic reference,
- identify all the documents referencing a given document, that is the documents containing references that reference a given document.

In both cases the matching process consists of several matching steps executed in a certain order. The result of each step is a set of matched objects. If the matcher tries to find the document referenced by a given bibliographic reference, the first matched object is returned and the whole process exits. If the matcher attempts to identify all the references referencing a given document, all steps are executed and the results are combined into one set of matched objects.

Each matching step consists of two phases:

1. selecting candidates from the metadata store according to a specific criterion,
2. evaluating the candidates by comparing corresponding metadata information.

In our evaluation process we have used two matching steps.

During the first step the candidates have been selected based on the following metadata information: authors' surnames, year of publication and hash of journal name. Then the candidates have been evaluated by comparing: authors' full name, journal name, volume, issue and year of publication.

During the second step the candidates have been selected based on only authors' surnames and year of publication. Then the candidates have been evaluated by comparing: authors' full name, journal name, volume, issue, year of publication and title.

Comparing various metadata information is not a trivial task due to different formats, abbreviations, typos, etc. In our implementation we use different methods of comparing for different metadata information. For example author full name, volume, issue and year of publication are considered equal if the corresponding strings are identical. Journal names are considered equal if one string is a subsequence of the other. In the case of title we make use of Levenshtein distance: two titles are equal if one of them is a subsequence of the other or if both are long and Levenshtein distance between them is less than a small fixed number.

4   Results

We have followed described evaluation path for our in-house citation matcher. Numerical values achieved are presented in Table 1.

Table 1. Matcher evaluation. Each column shows numerical values of metrics defined in Section 2.2. Each row represents a single test set generated using one bibliographic style.

Style	Acc	P_r	R_r	F_r	$P_\perp$	$R_\perp$	$F_\perp$	G
alpha	0.70	0.98	0.64	0.78	0.36	0.98	0.52	0.74
abbrv	0.87	0.98	0.85	0.91	0.57	0.98	0.72	0.88
ieeetr	0.87	0.99	0.85	0.91	0.57	0.98	0.72	0.89
plain	0.41	0.99	0.29	0.45	0.22	0.99	0.36	0.49
apalike	0.71	1.00	0.65	0.79	0.36	1.00	0.53	0.75
acm	0.87	0.99	0.84	0.91	0.57	0.98	0.72	0.88
jpc	0.94	0.99	0.93	0.96	0.77	0.98	0.86	0.95
pccp	0.94	0.99	0.93	0.96	0.76	0.98	0.85	0.95
ppcf	0.71	1.00	0.66	0.79	0.37	1.00	0.54	0.76
revcompchem	0.93	0.98	0.92	0.95	0.73	0.98	0.83	0.94
Average	0.79	0.99	0.75	0.84	0.53	0.98	0.67	0.82

Accuracy, being the most basic metric, can be treated as a single-valued benchmark of the overall performance. It tells us that in general our matcher does fairly well. From references-based metrics we conclude that if our algorithm matches a citation to the database record it almost always does that correctly, but there are many more entities that should have been linked (i.e. high precision P_r, relatively low recall R_r). Similar information can be drawn from nonexistent record metrics: almost all citations that have no database record are correctly classified as such (high recall $R_\perp$), but some citations identified as not having a corresponding database record in fact do have one (low precision $P_\perp$).

5   Conclusions

We have presented a methodology for evaluating a reference matcher. We have shown how a test set can be automatically built using existing publication metadata and BibTeX, removing the need for a laborious construction by hand. We have also proposed some metrics which should be used to generate numerical values reflecting algorithm performance. Finally, we have described and evaluated our in-house matcher using presented methodology.

However, one can point out some weaknesses of proposed test set creation method: generated citations are very consistent in terms of formatting and contain no punctuation errors. Moreover, we use the same metadata for citation generation and matching. That means that we do not deal with some matching issues, e.g. different ways of abbreviating journal names. This issues can be at least partially addressed by introducing some arbitrary errors. To simulate them we could substitute a number of random characters in a citation string for different ones.

Nevertheless this small flaw should not make us forget about obvious advantages of a proposed method, among them the huge scalability as we are able to create unlimited number of citations using arbitrarily many bibliographic styles.

Acknowledgements. This work is supported by the National Centre for Research and Development (NCBiR) under Grant No. SP/I/1/77065/10 by the Strategic scientific research and experimental development program: "Interdisciplinary System for Interactive Scientific and Scientific-Technical Information."

References

1. Apache Solr, `http://lucene.apache.org/solr/`
2. PostgreSQL, `http://www.postgresql.org/`
3. Bolelli, L., Ertekin, S., Giles, C.L.: Clustering Scientific Literature Using Sparse Citation Graph Analysis. In: Fürnkranz, J., Scheffer, T., Spiliopoulou, M. (eds.) PKDD 2006. LNCS (LNAI), vol. 4213, pp. 30–41. Springer, Heidelberg (2006)
4. Christen, P.: A survey of indexing techniques for scalable record linkage and deduplication. IEEE Transactions on Knowledge and Data Engineering (2011)
5. Elmagarmid, A., Ipeirotis, P., Verykios, V.: Duplicate Record Detection: A Survey. IEEE Transactions on Knowledge and Data Engineering 19(1), 1–16 (2007)
6. Garfield, E.: Citation Indexing: Its Theory and Application in Science, Technology, and Humanities. John Wiley & Sons, New York (1979)
7. Garfield, E.: The history and meaning of the journal impact factor. Journal of the American Medical Association 295(1), 90–93 (2006)
8. Giles, C., Bollacker, K., Lawrence, S.: CiteSeer: An automatic citation indexing system. In: Proceedings of the Third ACM Conference on Digital Libraries, pp. 89–98. ACM (1998)
9. Goutorbe, C.: Document Interlinking in a Digital Math Library. In: Towards a Digital Mathematics Library, pp. 85–94 (2009)
10. Hirsch, J.E.: An index to quantify an individual's scientific research output. Proceedings of the National Academy of Sciences of the United States of America 102(46) (2005)
11. Hitchcock, S.M., Carr, L.A., Harris, S.W., Hey, J.M.N., Hall, W.: Citation Linking: Improving Access to Online Journals. Proceedings of Digital Libraries 97, 115–122 (1997)
12. Lawrence, S., Giles, C.L., Bollacker, K.D.: Autonomous citation matching. In: Etzioni, O., Müller, J.P., Bradshaw, J.M. (eds.) Proceedings of the Third Annual Conference on Autonomous Agents AGENTS 1999, vol. 1, pp. 392–393. ACM Press (1999)
13. Liao, Z., Zhang, Z.: A Generalized Joint Inference Approach for Citation Matching. In: Wobcke, W., Zhang, M. (eds.) AI 2008. LNCS (LNAI), vol. 5360, pp. 601–607. Springer, Heidelberg (2008)
14. Macskassy, S.A., Provost, F.: Classification in Networked Data: A Toolkit and a Univariate Case Study. Journal of Machine Learning Research 8, 935–983 (2007)
15. McCallum, A., Nigam, K., Rennie, J.: Automating the construction of internet portals with machine learning. Information Retrieval, 127–163 (2000)
16. Pasula, H., Marthi, B., Milch, B., Russell, S., Shpitser, I.: Identity uncertainty and citation matching. In: Proceedings of NIPS 2002. MIT Press (2002)

17. Poon, H., Domingos, P.: Joint Inference in Information Extraction. In: Artificial Intelligence, vol. 22, pp. 913–918. AAAI Press (2007)
18. Sylwestrzak, W., Borbinha, J., Bouche, T., Nowiski, A., Sojka, P.: EuDML Towards the European Digital Mathematics Library. In: Towards a Digital Mathematics Library, pp. 11–26 (2010), http://www.eudml.eu/
19. Tkaczyk, D., Bolikowski, L., Czeczko, A., Rusek, K.: A modular metadata extraction system for born-digital articles. In: 10th IAPR International Workshop on Document Analysis Systems, pp. 11–16 (2012)
20. Wellner, B., McCallum, A., Peng, F., Hay, M.: An integrated, conditional model of information extraction and coreference with application to citation matching. In: Proc. UAI, pp. 593–601 (2004)

Data Model for Analysis of Scholarly Documents in the MapReduce Paradigm

Adam Kawa, Łukasz Bolikowski, Artur Czeczko,
Piotr Jan Dendek, and Dominika Tkaczyk

Interdisciplinary Centre for Mathematical and Computational Modelling,
University of Warsaw, Warsaw, Poland
{a.kawa,l.bolikowski,a.czeczko,p.dendek,d.tkaczyk}@icm.edu.pl

Abstract. At CeON ICM UW we are in possession of a large collection of scholarly documents that we store and process using MapReduce paradigm. One of the main challenges is to design a simple, but effective data model that fits various data access patterns and allows us to perform diverse analysis efficiently. In this paper, we will describe the organization of our data and explain how this data is accessed and processed by open-source tools from Apache Hadoop Ecosystem.

Keywords: Data model, Apache Hadoop Ecosystem, MapReduce.

1 Introduction

At CeON ICM UW [6] we are in possession of vast collections of scholarly documents to store and analyze. Currently, there are about 10 million of full texts (PDF or plain text) and 17 million of document metadata records that together occupy several terabytes of disk space. The data grows at the rate of approximately one hundred thousand of document metadata records and PDF files (i.e. about 50 GB) a month.

XML-based BWMeta format [30] is used to describe the document metadata records. It contains information like title, subtitle, abstract, keywords, references, contributors and their affiliations, publishing venue and so on.

We are using this data as input for algorithms to analyze and discover various relationships between documents, contributors, publishers, references and other entities. Some of the algorithms are relatively simple such as searching documents with given title or finding scientific teams, but some of them are quite complex and require implementation of state-of-the art machine learning and network analysis methods e.g. author name disambiguation, classification code assignment or finding most influential papers in given domains.

Our current approach suffers from performance/scalability issues since the data was usually located on several separate machines, but processed by a single machine. Although we can easily add more machines to store more data, it will not solve our performance problems since the computation is not fully distributed. Therefore, we made a decision to move our computations from a single-machine to a multi-machine configuration.

R. Bembenik et al. (Eds.): *Intell. Tools for Building a Scientific Information*, SCI 467, pp. 155–169.
DOI: 10.1007/978-3-642-35647-6_12   © Springer-Verlag Berlin Heidelberg 2013

2   Problem Definition

The research problem can be stated as designing a scalable and efficient storage scheme for RDF triples. However, rather than designing one-fit-all storage scheme, we are interested in a solution that is better adapted to our data and access pattern. As an additional restriction, the solution should be implementable utilizing open-source tools from Apache Hadoop Ecosystem.

We investigated various frameworks for scalable, distributed and efficient storage and processing of a large amounts of data. Apache Hadoop [1] and related projects like Apache HBase [2], Apache Hive [3] and Apache Pig [4] attracted us the most since they are commonly used, open-source solutions that provide reliable and cost-effective way to persist and process big data.

We focused on designing a simple, but still handy data model that copes with storing terabytes of detailed information about scholarly documents. Various requirements regarding data access were imposed on the model, mainly in terms of flexibility (possibility to add, update and delete records, and enhance their content by implicit information discovered by our algorithms), latency requirements (from batch offline processing to random, realtime read and write requests) and client interface (accessed by programmers and analysts with diverse language preferences and expertise).

We observed that our data can be represented in a flexible way as a collection of triples, each representing a statement of the form *subject-predicate-object*, which denotes that a resource (*subject*) has an attribute (*predicate*) with some value (*object*). In other words, such a collection of triples describes a directed, labeled graph, where nodes represent subjects and objects, and edges represent predicates which connect subject nodes to object nodes.

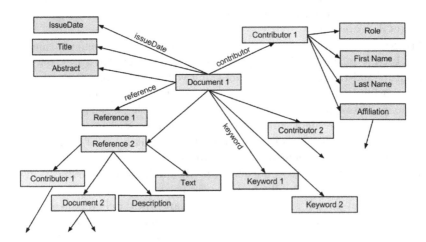

Fig. 1. An example RDF graph data for scholarly documents (some predicate names are omitted for clarity)

The concept of triples is used by the Resource Description Framework (RDF) [17] to describe information about Web resources (documents, files, images, services etc.) and model various relationships between them (see Fig. 1 for an example). The RDF data model is extremely flexible since it allows anyone to make any statements about any resource using the *subject-predicate-object*. In a similar way, we can describe any relationship that explicitly exists in our data, or may be discovered later by our machine learning algorithms.

3   Related Work

During recent years, significant efforts have been made to develop scalable, high-performance and low-cost distributed systems for storing and querying large collections of triples (called triplestores) [24], [37]. Some of them were focused on taking advantage of open-source projects such as Apache Hadoop, Apache HBase and Apache Pig to attain this goal [34], [35], [31].

Rohloff and Schantz [34] introduce a concept of scalable triple-store built on Hadoop and HDFS[12] and being processed with MapReduce paradigm (see [25] for a description of this paradigm). The paper and resulting system (called SHARD) has aroused significant interest as it leverages Hadoop to scale sub-graph pattern matching (which is quite difficult task) using technique that is 2–3 times more efficient than the naive way of using Hadoop for this task [23]. The system persists RDF triples in "flat" text files in HDFS (each line stores all triples associated with a different subject). The disadvantage of this approach is decreased efficiency for queries that require the inspection of only a small number of triples [34] and impossibility of random modification of the triples due to the nature of HDFS.

One-table-per-property approach that uses a concept of vertical-partitioning of RDF data in column-oriented stores was presented in [24]. Here, separate two-column tables are constructed for each property (predicate), where the first column contains the subjects that define that property and the second column contains the object related to those subjects. This approach has several advantages such as support for multi-valued attributes, reduced IO costs and a potential use of linear merge joins, however it suffers from performance drawbacks on queries that are not bound by a predicate value [37].

Weiss et al. [37] address above-mentioned performance problems by enhancing the idea of vertical partitioning and creating a six indices structure for storing RDF triples. The six indices (*pso, pos, spo, sop, ops* and *osp*) cover all the six possible ordering of three elements in a triple. This format allows for quick lookups and partial scans of data at the price of an increase of storage space and complication of the update operation. Since data is stored with multiple indices, all first-step pairwise joins are fast (linear) merge-joins.

Papailiou et al. [33] adopt the idea of [37], but reduce the number of indices to three. Here, all triples are persisted in HBase using three "flat-wide" tables (*sp_o, po_s* and *os_p*). The authors argue that the six-index approach can have better performance only for certain queries that contain filters on variables, while for all other queries, three indices suffice for optimal performance. The paper also presents several different join strategies which take the query selectivity and the inherent features of the MapReduce and HBase into account to minimize the processing time.

Performance of six different HBase storage schemas on a subset of queries from SP^{2}Bench [21] is evaluated and discussed in [31]. The hybrid storage schema (consisting of three "flat-wide" tables each indexed by subjects, predicates and objects plus set of two tables for every unique predicate, each indexed by subjects and objects) achieved the best performance results.

The default way of querying RDF data stored in Hadoop is execution of MapReduce algorithms. Since developing such algorithms is still considered challenging, especially for non-technical analysts, some researchers investigated alternative methods for querying RDF data on a Hadoop cluster exploiting high-level languages. One of the approaches is to utilize Apache Pig framework. The paper [35] introduces PigSPARQL, a system that translates SPARQL [18] (a query language for RDF) queries to PigLatin programs and executes them on Hadoop cluster. Here, RDF triples are stored in HDFS files and according to the vertical partitioning idea, triples with the same predicate share the same folder and each predicate has its own folder.

A very impressive performance results of querying RDF data are presented in [29]. The paper describes optimization techniques like RDF graph partitioning scheme to exploit the spatial locality inherent in graph pattern matching. In this approach, a higher replication factor is set for the data on the border of any particular partition. The authors claim their optimizations result in a 1000 fold improvement in efficiency based on experiments done using Lehigh University Benchmark (LUBM) [20]. However, it might be argued [26] that the idea of a higher replication factor would be inappropriate for graph applications that modify the RDF graph data, because it would be very hard to maintain consistency among the replicas of the boundary vertices as they change.

On the basis of papers mentioned above, we can state that the greatest challenge for processing big collections of triples is a large number of join operations. While joins are expensive, they cannot be avoided in practice (at least for more complex queries). Some systems aim at reducing the number of joins by either storing related data co-located in the same row (in HBase) [36], [33], [31] or line of text (in HDFS) [34], while other systems optimize the join executions by using specialized join techniques like multi join, merge join or skewed joins [37], [35], [33]. Optimizations like property tables [38], [39] and materialized paths are also proposed [24].

4   Data Storage

Similarly to [36], [33], and [31], we have selected Apache HBase as a main tool for our storage layer. There are several reasons for that:

- **Flexible data model.** Simple, yet flexible data model provided by HBase gives us a control over data layout and format. Namely, we can dynamically add new columns (e.g. containing detailed information about triples such as certainty of relationship, data source, or inferring algorithm) or delete existing ones. There is a possibility to store multiple versions of data in a particular cell distinguished by a timestamp. Additionally, HBase does not require a fixed definition of data types during table creation.

- **Random read and write.** HBase provides random and realtime read-write access to the data allowing us to easily add, update and delete triples. HBase seems to be more suitable for semi-structured RDF data than HDFS where files cannot be modified randomly and the whole file must be read sequentially to find subset of required records.
- **Many clients available.** HBase can be accessed through interactive clients like native Java API, REST or Apache Thrift [5] as well as through batch clients like Java MapReduce, Pig and Hive. The integration with batch clients, especially Java MapReduce and Pig, seems to be crucial for us since it gives the possibility to run distributed computation asynchronously in the background, scanning and processing large amounts of data in parallel.

  As mentioned earlier, our data is analyzed by many researchers (including non-technical ones) with various language preferences. Availability of many clients will allow anyone to choose preferable way and language to access the data (e.g. Java, Python, HiveQL or PigLatin).
- **Automatically sorted records.** HBase stores the data sorted lexicographically by a row key. When storing huge amounts of data, this feature becomes really important since data can be looked-up and scanned quickly. If joins are required (what is often the case), they can be possibly done using simple and fast (linear) merge join [24], [37].

5   Storage Schema

Following recommendations in [28], we have organized our data in "tall-narrow" layout (many rows, few columns). The main problem with "flat-wide" (few rows, many columns) layout is worse load balancing since a single row is never split across HBase regions. Moreover if a row has an unlimited number of columns, it may outgrow the maximum region size and work against the region split facility [28]. Such a situation seems to be possible in our case as we deal with tens of billions of triples. Moreover storing really wide rows with millions of columns (e.g. a given publisher may be in relation with tens of millions of documents and authors, or a predicate *type* can be related to hundreds of millions of entities like documents, authors, references, keywords etc.) also reduces performance due to non optimal load balancing (as mentioned above) and row-level locking mechanism [32].

Our HBase storage schema layout is presented in Fig. 2. We adopted the ideas of [24], [37], [31] with some changes to make it more suitable for our use cases. The schema consists of two "tall-narrow" tables *pso* and *pos*, each indexed by *predicate-subject-object* and *predicate-object-subject* respectively. For a given triple, its subject, predicate and object values are concatenated and

pso table   pos table

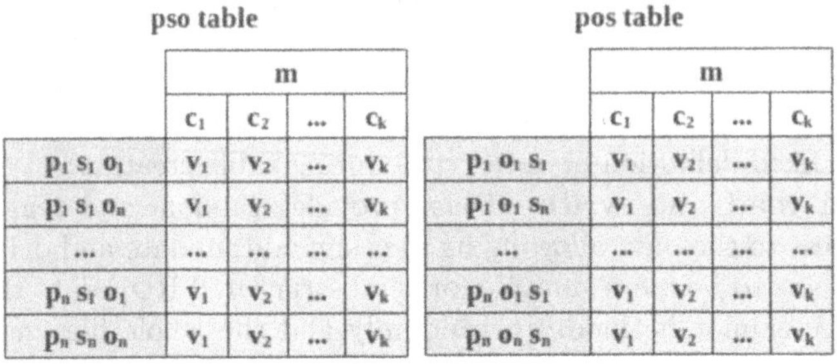

	m						m			
	c_1	c_2	...	c_k			c_1	c_2	...	c_k
$p_1\ s_1\ o_1$	v_1	v_2	...	v_k		$p_1\ o_1\ s_1$	v_1	v_2	...	v_k
$p_1\ s_1\ o_n$	v_1	v_2	...	v_k		$p_1\ o_1\ s_n$	v_1	v_2	...	v_k
...	...	...	...	...		...	...	...	...	...
$p_n\ s_1\ o_1$	v_1	v_2	...	v_k		$p_n\ o_1\ s_1$	v_1	v_2	...	v_k
$p_n\ s_n\ o_n$	v_1	v_2	...	v_k		$p_n\ o_n\ s_n$	v_1	v_2	...	v_k

Fig. 2. Predicate Indexed Layout - HBase Storage Schema

stored entirely in a row key and one row is added to each of two tables. Each table consists of a one column family called m (metadata). Columns dynamically added to this column family contain additional information about triples (this way we can potentially describe statements about statements).

The proposed scheme requires that data is stored twice (HBase has no native support for secondary indexes [28]). Tables *pso* and *pos* can be used to efficiently retrieve triples with known predicate, predicate-subject and predicate-object values. However, retrieval of triples based on subject or object values (where predicate is unknown) may be less efficient since it requires a scan of multiple parts of an appropriate table (*pso* and *pos*, respectively). To remedy this problem, we could create four additional tables with remaining indices (*spo*, *sop*, *ops*, *osp*). Although such a solution, may provide significant performance improvement for many interesting queries [37], currently the majority of our queries is predicate-bound (similarly to queries in commonly used benchmarks e.g. LUMB, SP^{2}Bench). Moreover our set of predicates is relatively small and most of the queries require only few predicates at the same time (so that only several partial scans are needed).

Basically, we favour starting with a simpler approach (what results in lower redundancy, lower storage requirements and slightly easier update operations) and potentially add new tables with required indices if they appear to be really necessary in the future.

The main advantages of this layout can be outlined as:

- **Support of multi-valued properties.** Multi-valued properties (such as a document with multiple titles in different languages) and many-to-many relationships (such as the document and authorship relationship where a document can have multiple authors and an author can write multiple document) are easy to handle in this approach [24]. Actually, there is no difference between single-valued and multi-valued properties. If a subject has more than one object value for a particular property, then each distinct value is stored in a successive row in the table [24].
- **Support of reified statements (statements about statements).** As mentioned, new columns qualifiers can be dynamically added to column

family m and contain additional information about triples such as certainty, data source or inferring algorithm.

- **First-step (predicate-bound) pairwise joins as fast merge-joins.** For a given predicate, all triples are sorted by subject (*pso* table) and object (*pos* table). As a result, every pairwise join (with specified predicate) performed during the first step of query processing is a fast, linear-time merge join [37].
- **Reduced redundancy.** Although data is stored twice, the redundancy is still significantly smaller than in storage schemas presented in [36], [33] as well as in 5 out of 6 schemas from [31].

We can potentially optimize the consumed storage space by directly adopting the idea of vertically partitioned layout [24], [31] and splitting *pso* and *pos* tables into multiple separate tables. This way, for every unique predicate, two tables are created (each respectively indexed by subjects and objects). The storage savings are gained by moving a predicate name to the name of the table and completely eliminating the storage of a predicate as a part of row keys [31].

Despite advantages mentioned above, this HBase schema layout has several significant disadvantages:

- **Increased number of joins.** As triples are stored in "tall-narrow" tables, a larger number of join operations is required to process data. In comparison, "flat-wide" tables approaches entirely remove the need for joins to answer queries about the same subject (or object) however, they still require joins to answer RDF path queries.

  This performance drawback can be alleviated using specialized join techniques such as multi join (when multiple sets are joined by the join key), merge join (when sets are sorted by the join key), replicated join (when one set is very large, while other sets are small enough to fit into memory) and skewed join (when a large number of records for some values of the join key is expected).

- **Increased storage for reified statements.** In HBase, each cell is stored in a "fully qualified" way (together with its row key, column family, column qualifier and timestamp) on disk. Adding cells representing reified statements causes that a row key (consisting of *predicate, subject, object*) is repeated and stored multiple times on disk, thus increasing the storage space. This is another reason, why a special care should be taken not to make a row too wide.

  This storage overhead can be slightly reduced by using standard compressions mechanisms. Moreover, a key prefix compression feature, that will be available in HBase 0.94, may additionally benefit to minimization of disk space used.

- **Complexity of update operation.** Although the update operation requires modification in lower number of HBase tables than in the approaches presented in [36], [33], [31], this operation still can not be considered as trivial.

HBase does not provide native support for cross-row atomicity (e.g. in the form of transactions), so that the consistency of *pso* and *pos* tables cannot be guaranteed. This can be partially overcome by using recreating one table from another, if any inconsistency is discovered. One MapReduce, highly efficient bulk import job [8] could be implemented for this task.

6   Optimizations

This data model allows us to improve the queries performance when more use cases are defined and a deeper knowledge about our data access pattern is gained. Following optimizations can be proposed:

- **Adding new indices.** As mentioned, introducing new indices will improve performance for queries which are not bound by the predicate [37].
- **Property tables.** Property tables was proposed by researchers developing the Jena Semantic Web toolkit, Jena2 [38], [39]. By definition, a property table contains clusters of properties that tend to be defined together. One example are type, title, issue date defined as properties for scholarly documents. Such properties can be stored together in the same row for a quick access.

  In fact, this idea is already exploited by us for the storage of reified statements (statements about statements). We take a step forward with this approach and dynamically add new columns qualifiers to a new column family *p* (property) that contains useful properties related to a particular subject in a triple. Rows with the predicate *type* can be simply selected to hold these additional column qualifiers (Fig. 3). Note that some properties may be appropriate to only one or two types, while being totally inappropriate for others. This increases the sparsity of the table, however, HBase deals with this issue very efficiently as NULLs are not stored on disk.

pos table

	m			p			
	c_1	..	c_n	email	fullName	issueDate	title
...	...	...	...	...	...	...	...
type_auth _s_1	v_1	..	v_k	email$_1$	fullName$_1$		
type_auth _s_2	v_1	..	v_k	email$_2$	fullName$_2$		
...	...	...	...	...	...	...	...
type_doc_ s_1	v_1	..	v_k			issueData$_1$	title$_1$
type_doc_ s_2	v_1	..	v_k			issueData$_2$	title$_2$
...	...	...	...	...	...	...	...

Fig. 3. Graphical presentation of table *pos* exploiting the idea of a property table

While property tables can significantly improve performance by reducing the number of self-joins, they introduce several problems:

- **Increased complexity.** Property clustering (based on explicit or implicit knowledge about data and its access pattern) must be carefully done to create rows that are not too wide, while still being wide enough to answer most queries directly [24].
- **Increased redundancy.** While a particular *predicate, subject, object* is always stored in row keys, it might be also additionally stored in a column qualifier or a cell.
- **Increased storage.** The storage space will insrease in the same way as with reified statements, what was described in the previous chapter.

- **Materialized path expressions.** The concept of materialized path expressions was presented in [24] and discussed in [37]. Path expressions are expressions that match specific paths through a RDF graph (see Fig. 4). Querying path expressions is a common operation on RDF data, but can be quite expensive due to the fact that it requires subject-object joins. Basically, a path of length n requires $n - 1$ subject-object joins. Since our data schema contains *pso* and *pos* indices, the first of joins in a path is a linear merge-join, while the rest $n - 2$ are sort-merge joins, i.e. each one requires one sorting operation [37].

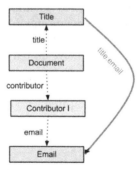

Fig. 4. Graphical presentation of an exemplary expression path query to find title and emails of contributors for each given document

The idea of precalculation and materialization of the most commonly used path expressions in advance may significantly improve performance of queries. For example, we analyze various statistics about contributor cooperation e.g. finding hubs (i.e. people that collaborate with many other people), finding international scientific teams (i.e. people of different nationalities who collaborate with each other very often), calculating the Erdos number [9] for each person and so on. Generally speaking, the above-mentioned algorithms take pairs of collaborating people as input data. Precalculation and materialization of triples in form of *predicate-subject-object*, where *predicate* is *contributorPair*, *subject* is a document and *object* is a pair of people contributing to this document, would improve performance of these algorithms (see Fig. 5).

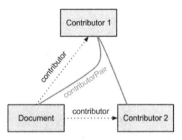

Fig. 5. Graphical presentation of an expression path query to find all pairs of people who contributed to each given document

Basically, we can think about materialized paths as an output of our algorithms (some of them very simple) that we want to store together with the input data in the same HBase tables. For these triples, a special prefix (e.g. *mp*) can be added to row keys to distinguish them from input triples.

All of described optimizations contribute to the improvement of the performance at the price of increased storage requirements and increased complexity of update operations. In general, when new data is added or changed, properties tables and materialized paths must be recalculated. We do realize that optimizations above do not solve the problem of large number of joins in a general fashion. They are also not fully automated; however, in many cases the necessary calculations can be computed by regularly scheduled MapReduce jobs.

7   Processing

This section describes how our data residing in HBase tables is accessed by open-source tools from Apache Hadoop Ecosystem. We take advantage of multiple existing clients to meet our various demands like latency requirements (batch offline processing and random, realtime access) and programming preferences (object-oriented, scripting, and declarative languages).

7.1   Pig

Pig is an Apache open source project that provides an engine for executing data flows in parallel on Hadoop. It includes a high-level language (called PigLatin) for expressing data analysis programs. PigLatin supports many operators for the traditional data operations (such as join, union, sort, filter) as well as custom UDFs (user-defined functions) for reading, processing and writing data [27]. PigLatin programs are automatically translated into a serie of MapReduce jobs.

Joining Data. Processing our data with Pig is very convenient since it supports multiple specialized join implementation (such as multi join, merge join, merge-sparse join, replicated join, skewed join) [19]. Having data sorted both by subject and object (for a given predicate), we are able to perform first step subject-subject joins and subject-object joins using simple and fast merge join operation.

Currently, Pig does not allow programmer to use any UDF in the foreach statement between the load of the sorted input and the merge join statement [14]. Unfortunately, since our subject, object, and predicate are concatenated in a row key, we need to use one UDF (called STRSPLIT) to split a row key into three separate parts (i.e. into a tuple with three fields containing subject, object and predicate which needs to be unnested by FLATTEN operation). As a workaround, we can extend our data model to store additional three column qualifiers for storing subject, predicate and object values, so that execution of STRSPLIT function can be avoided. Alternatively, we can load sorted input data from HBase, split and flatten each row key into three separate parts and then store them in HDFS. After loading this recently stored data from HDFS, a merge join operation can be executed. The first approach requires more storage space and increases data redundancy, however, is much faster in comparison to the second one. So far, have we decided to use the first approach until this issue is resolved by us or Apache Pig community (PIG-2673 [15]).

Extending Pig Functionality. The list of functions provided by Pig can be easily extended by user by developing his own UDFs [27]. Currently, the UDFs can be written in Java or Python.

The integration with Python is even deeper since PigLatin can be embedded in Python scripts. Thus, Python's control flow constructs like "if" and "for" (which are not natively supported by Pig) may be used to either repeat processing a certain number of times, or to branch based on the results of an operator. Since many machine learning algorithms require repeating a calculation until a certain error value is within an acceptable bound, this feature seems to be very useful [27].

7.2 MapReduce

Using HBase as a storage layer gives also us a possibility to process data by implementing Java MapReduce jobs that read data from and write data to HBase tables. We can implement such jobs if we want to obtain an increased performance or exploit legacy Java code. According to PigMix2 benchmark[16], PigLatin programs are slower by a factor of 1.37–1.76 when compared with native MapReduce programs regarding operations like "order by", "outer join", and "group by" (where the key accounts for a large portion of the record).

The greatest current limitation of running Java MapReduce jobs over HBase tables is lack of support for using multiple tables and scanners as input to the mapper in MapReduce jobs (HBASE-3996 [10]). However, most of our algorithms require input records from multiple parts of two HBase tables. We alleviate this problem by preparing input data for MapReduce jobs in Pig (using LOAD, UNION and JOIN operations) and then running MapReduce jobs directly from Pig scripts with the mapreduce command [27]. This way we can avoid the burden of implementing complex (like join) or even not supported yet (like multiple tables and scanners as input) operations in Java MapReduce, but still have a possibility to incorporate processing using the MapReduce paradigm in our big data analysis.

7.3   Interactive Clients

While offline, complex analysis are performed by batch processing clients, there are some use cases where we need interactive access to relatively small subset of our data. One of such examples is a web-based client that may search for entities with attributes matching a certain query and navigate amongst interconnected entities by sending client API calls on demand (such as get and scan).

HBase is a good fit for such use cases, as it is designed to perform random, realtime read and write requests to big data. Interactive clients include, e.g. native Java API, REST or Apache Thrift.

7.4   Other Clients: Hive

Hive is a data warehouse system for Hadoop that supports easy data summarization, ad-hoc queries, and analysis of large datasets. It provides a SQL-like language called HiveQL to query the data, thus makes Hadoop accessible to analysts who already know SQL. Hive is integrated with HBase, so that HiveQL statements can access HBase tables for both read (SELECT) and write (INSERT)[11]. Similarly to Pig and MapReduce, all these reads and writes are bulk operations.

The main advantage of using Hive over HBase is possibility to use JOIN and UNION statements. This way more complex analysis on data from HBase can be performed similarly as in SQL. As many of our researches already know SQL, this approach can be widely used for a rapid development of ad-hoc queries.

8   Configuration

Our Hadoop cluster consists of a four "fat" slave nodes and a virtual machine on separate physical machine in the role of HDFS NameNode HDFS Secondary-NameNode, Hadoop JobTracker, HBase Master and ZooKeeper node. Although it seems like a lot of processes is running on a single machine, none of them is particularly resource intensive.

Each worker node has four AMD Opteron 6174 processors (48 cores in total), 192 GB of RAM, four 600 GB disks which work in RAID 5 array with an access to 7TB LUN of NetApp disk storage over FC. Each slave node runs 25 map and 15 reduce tasks maximally, each consuming 2GB of RAM by default (may be increased for memory-consuming applications). We are using Cloudera's Distribution including Apache Hadoop (CDH)[7], mainly CDH3u3 which includes

Hadoop 0.20.2, HBase 0.90.4 and Oozie 2.3.2 by default. We decided to update to Hive (from 0.7.1 to 0.9.0) and Pig (from 0.8.1. to 0.10.0) to benefit from newly implemented features. We compress data using LZO compression [13].

9   Conclusions

We have described how our data can be stored and processed using open-source tools from Apache Hadoop Ecosystem. The main goal was to provide a simple data model that can be conveniently accessed using various client interfaces. Our initial work proves that Apache Hadoop and its related projects are good framework for persisting and processing of millions of scholarly documents that together occupy several terabytes of data. We see benefits of using Apache Hadoop Ecosystem tools for this task when compared with our previous, not fully distributed approach, because now the data is stored in a reliable way (replicated in HDFS) and processed faster thanks to the parallelism offered by MapReduce paradigm.

10   Plans for Future

Deep examination and specification of our future algorithms may contribute to selection of the most suitable indices for storing and querying triples or precalculation of the most common materialized paths.

We intend to investigate other HBase storage schema layouts, mainly composed of "flat-wide" tables, where the number of columns can be limited to a certain number (or its approximate value) in some way. Such an approach would allow us to reduce number of rows (and result in faster seeks) and joins (as data is collocated in the same rows) and shorten row key size (to save storage space, as the row key is stored with each KeyValue pair) [22], while the rows will still have predictable size and do not work against the Hadoop region split facility. Such a functionality looks to be possible to implement using HBase coprocessor functionality.

References

1. Apache Hadoop - framework for the distributed processing of large data sets across clusters of computers using simple programming models,
   http://hadoop.apache.org
2. Apache HBase - scalable, distributed database that supports structured data storage for large tables, http://hbase.apache.org
3. Apache Hive - data warehouse infrastructure that provides data summarization and ad hoc querying, http://hive.apache.org
4. Apache Pig - high-level data-flow language and execution framework for parallel computation, http://pig.apache.org
5. Apache Thrift - software framework, for scalable cross-language services development, http://thrift.apache.org

6. Centre for Open Science in Interdisciplinary Centre for Mathematical and Computational Modelling (CeON ICM), University of Warsaw, http://ceon.pl/en
7. Cloudera's Distribution including Apache Hadoop (CDC), http://www.cloudera.com/hadoop
8. Efficient method for loading large amounts of data into HBase table, http://hbase.apache.org/book.html#arch.bulk.load
9. Erdos Number, http://en.wikipedia.org/wiki/Erd%C5%91s_number
10. HBASE-3996 Jira Issue, https://issues.apache.org/jira/browse/HBASE-3996
11. HBase module for an integration with Hive, https://cwiki.apache.org/confluence/display/Hive/HBaseIntegration
12. HDFS - distributed file system that provides high-throughput access to application data, http://hadoop.apache.org/hdfs
13. Lempel-Ziv-Oberhumer (LZO) - lossless data compression algorithm that is focused on decompression speed, http://wiki.apache.org/hadoop/UsingLzoCompression
14. Merge Join in Apache Pig, http://pig.apache.org/docs/r0.10.0/perf.html#merge-joins
15. PIG-2673 Jira Issue, https://issues.apache.org/jira/browse/PIG-2673
16. PigMix - set of queries used test Apache Pig performance from release to release, https://cwiki.apache.org/confluence/display/PIG/PigMix
17. Resource Description Framework (RDF) - standard model for data interchange on the Web, http://www.w3.org/RDF
18. SPARQL - query language for RDF, http://www.w3.org/TR/rdf-sparql-query
19. Specialized join implementations in Apache Pig, http://pig.apache.org/docs/r0.10.0/perf.html#specialized-joins
20. The Lehigh University Benchmark (LUBM) that facilitates the evaluation of Semantic Web repositories in a standard and systematic way, http://swat.cse.lehigh.edu/projects/lubm
21. The SP^{2}Bench SPARQL Performance Benchmark, http://dbis.informatik.uni-freiburg.de/forschung/projekte/SP2B
22. Lessons learned from OpenTSDB (2012), http://www.cloudera.com/resource/hbasecon-2012-lessons-learned-from-opentsdb
23. Abadi, D.J.: Hadoop's tremendous inefficiency on graph data management (and how to avoid it) (2011), http://dbmsmusings.blogspot.com/2011/07/hadoops-tremendous-inefficiency-on.html
24. Abadi, D.J., Marcus, A., Madden, S.R., Hollenbach, K.: Scalable semantic web data management using vertical partitioning. In: VLDB, pp. 411–422 (2007), http://dl.acm.org/citation.cfm?id=1325900
25. Dean, J., Ghemawat, S.: MapReduce: Simplified Data Processing on Large Clusters. Communications of the ACM 51(1), 1–13 (2004)
26. Demirbas, M.: Scalable SPARQL Querying of Large RDF Graphs (2011), http://muratbuffalo.blogspot.com/2011/12/scalable-sparql-querying-of-large-rdf.html
27. Gates, A.: Programming Pig. O'Reilly Media (2011)
28. George, L.: HBase: The Definitive Guide. O'Reilly Media (2011)
29. Huang, J., Abadi, D.J., Ren, K.: Scalable SPARQL Querying of Large RDF Graphs. VLDB Endowment 4 (2011)
30. Jurkiewicz, J., Nowiński, A.: Detailed Presentation versus Ease of Search — Towards the Universal Format of Bibliographic Metadata. Case Study of Dealing with Different Metadata Kinds during Import to Virtual Library of Science. In: García-Barriocanal, E., Cebeci, Z., Okur, M.C., Öztürk, A. (eds.) MTSR 2011. CCIS, vol. 240, pp. 186–193. Springer, Heidelberg (2011)

31. Khadilkar, V., Kantarcioglu, M., Thuraisingham, B., Castagna, P.: Jena-HBase: A Distributed, Scalable and Efficient RDF Triple Store. Tech. rep. (2012), http://www.utdallas.edu/ vvk072000/Research/Jena-HBase-Ext/tech-report.pdf
32. Lipcon, T.: Is there a limit to the number of columns in an HBase row? (2011), http://www.quora.com/Is-there-a-limit-to-the-number-of-columns-in-an-HBase-row
33. Papailiou, N., Konstantinou, I., Tsoumakos, D., Koziris, N.: H2RDF: Adaptive Query Processing on RDF Data in the Cloud. In: Proceedings of the 21st International Conference on World Wide Web (WWW Demo Track), pp. 397–400 (2012)
34. Rohloff, K., Schantz, R.: Clause-Iteration with MapReduce to Scalably Query Data Graphs in the SHARD Graph-Store. In: Proceedings of the Fourth International Workshop on Data-intensive Distributed Computing, pp. 35–44. ACM (2011), http://www.dist-systems.bbn.com/people/krohloff/papers/2011/Rohloff_Schantz_DIDC_2011.pdf
35. Schätzle, A., Przyjaciel-Zablocki, M., Lausen, G.: PigSPARQL: Mapping SPARQL to Pig Latin. In: 3rd International Workshop on Semantic Web Information Management (SWIM 2011), in conjunction with the 2011 ACM International Conference on Management of Data (SIGMOD 2011), Athens, Greece (2011), http://www.informatik.uni-freiburg.de/~schaetzl/papers/PigSPARQL_SWIM2011.pdf
36. Sun, J., Jin, Q.: Scalable RDF Store based on HBase. In: 3rd International Conference on Advanced Computer Theory and Engineering (ICACTE), pp. 633–636 (2010)
37. Weiss, C., Karras, P., Bernstein, A.: Hexastore: sextuple indexing for semantic web data management. In: Proceedings of the VLDB Endowment, pp. 1008–1019 (2008), http://dl.acm.org/citation.cfm?id=1453965
38. Wilkinson, K.: Jena Property Table Implementation. Tech. rep. (2006)
39. Wilkinson, K., Sayers, C., Kuno, H., Reynolds, D.: Efficient RDF Storage and Retrieval in Jena2. Tech. rep. (2003)

Terminology Extraction
from Domain Texts in Polish

Małgorzata Marciniak and Agnieszka Mykowiecka

Institute of Computer Science, Polish Academy of Sciences,
Jana Kazimierza 5, 01-248 Warsaw, Poland
{mm,agn}@ipipan.waw.pl

Abstract. The paper presents a method of extracting terminology from
Polish texts which consists of two steps. The first one identifies candidates
for terms, and is supported by linguistic knowledge—a shallow grammar
used for extracted phrases is given. The second step is based on statistics,
consisting in ranking and filtering candidates for domain terms with the
help of a C-value method, and phrases extracted from general Polish
texts. The presented approach is sensitive to finding terminology also
expressed as subphrases. We applied the method to economics texts,
and describe the results of the experiment. The paper closes with an
evaluation and a discussion of the results.

Keywords: terminology extraction, shallow text processing, domain
corpora.

1 Introduction

Finding information in large text collections efficiently is tightly connected with
the problem of formulating appropriate questions. Without knowing the specific
terminology used in a particular domain it can be hard to localize the documents
which contain the information needed. It is particularly true if words which are
used in a query are ambiguous or if the specific subject we are looking for is rarely
described. Availability of appropriate vocabularies can improve both the process
of precise question formulation and proper document indexing, but multiplicity
of different domains and their quick changes in time make the task of manual
preparation of such resources practically infeasible. The solution to this problem
can lie in automation of the dictionary creation process. In our paper we present
the complete procedure of terminology extraction from specialized Polish texts
and results we obtained for the chosen domain—economics.

Terminology extraction is a process of identifying domain specific terms from
texts and usually consists of two steps. The first one identifies candidates for
terms and is usually supported by linguistic knowledge. The second step, based
on statistics, consists in ranking and filtering candidates for domain terms. An
overview of existing approaches to automatic terminology extraction is included
in [9]. The first step of the process is crucial in the sense that the result can
contain only those phrases which are defined at this stage. In practice, from an

R. Bembenik et al. (Eds.): *Intell. Tools for Building a Scientific Information*, SCI 467, pp. 171–185.
DOI: 10.1007/978-3-642-35647-6_13   © Springer-Verlag Berlin Heidelberg 2013

algorithmic point of view, the existing solutions differ at this stage only in the degree in which they use linguistic knowledge and in choice of the type of phrases they operate on. Minimalistic solutions totally neglect linguistic information [16] while others use POS tagging and simple shallow syntactic grammars [4]. Much more diversity is observed at the second processing stage when the initially identified phrases are filtered and ranked.

Although research on automatic terminology extraction has been carried on for many years now, very little was done in this area for Polish. The inflectional character of the language, together with the lack of specialized tools and resources, make processing Polish domain specific texts challenging. Till recently only the related task of collocation recognition has been explored, e.g [3]. In this paper we present an attempt to use slightly modified methods of automatic term extraction presented in [4] to unstructured Polish texts. In comparison to the first experiment which was performed on Polish clinical data [8], the texts which were analyzed in this experiment are better edited and are more similar in both sturucture and vocabulary to general newspaper texts.

In the terminology extraction process presented in this paper we adopt a corpus based approach. Our rough definition of a term is as follows: "a term is a nominal phrase which is used in the domain texts frequently enough to make it plausible that it represents something important which can be asked for by Internet readers and it does not occur equally frequently in texts on different subjects". We do not try to interpret these terms in relation to any external formal domain related knowledge, as we concentrate on collecting language constructs which can be identified within real texts. The normalization of these terms which may result in representing complex phrases as instances of relations between different objects, should constitute the subsequent processing level.

Our approach to terminology extraction consists of two standard steps enumerated above. At the first step we identify both single-word and multi-word nominal phrases using a simple shallow grammar operating on morphologically annotated text. Then, we use a slightly modified version of a popular approach of [4] which also allows us to recognize phrases occurring only inside larger nominal phrases. In this method, all phrases (external and internal) are assigned a C-value which is computed on the basis of the number of their occurrences within the text, the context in which they occur and their length. The procedure was performed separately on the selected set of specialized texts and on the texts which are representative for general usage of Polish (the balanced one million word subcorpus of NKJP [13]). As domain specific phrases, we chose those which are relatively more frequent in the first set than in the second one.

The paper is organized as follows. First we present a characteristic of the type of terminology we are interested in. In the section 3 problems connected with processing specialized texts are presented, together with the accepted solutions. Section 3 is devoted to the description of the types of phrases which are considered as candidates for terms while the next part of the paper presents the method of establishing a ranked list of all identified phrases. Section 6 contains the description of the obtained results.

2   Domain Term Characteristics

The first step towards terminology extraction is to define the exact goal, i.e. to define the type of phrases which can constitute domain related terms. But even before doing that, one has to decide what is to be considered as a domain term and what is not. The decision is far from being straightforward, as it can be easily seen while inspecting various already existing terminological resources. In standard dictionaries we can find general, not very precise definitions of a term and terminology. Terminology is defined for example as "a system of words used to name things in a particular discipline" or "the technical or special terms used in a business, art, science, or special subject" (http://www.merriam-webster.com/dictionary/), while a term is "a word or expression that is used in a particular variety of English or in realtion to a particular subject" [15], "a name, expression, or word used for some particular thing, especially in a specialized field of knowledge" (http://www.collinsdictionary.com/dictionary/), or "word or expression that has a precise meaning in some uses or is peculiar to a science, art, profession, or subject" (http://www.merriam-webster.com/dictionary/term). In manually created terminology resources, the decision what to include in the terminology lexicon depends on the authors' judgment supported by a set of adopted rules. In computational approaches to terminology extractions, terms are usually nominal phrases which have the specified structure and which fulfill some defined selection criteria. The final evaluation of obtained results is again done by domain specialists either directly or indirectly by comparing them to already existing resources. In the result, what is considered a domain term and what is not, depends on human judgment of the importance of a particular phrase for the chosen domain and on its exact definition, as well as the accepted description level.

Using already defined terminology resources is not easy as frequently they do not contain the exact terms which are needed in the specific task. One of the problems is the fact that while in specialized dictionaries there are a lot of very detailed terms, some general expressions can still be omitted. For example, one of the phrases which is typically connected with medical care is in Polish *badanie USG*, frequently shortened to *USG*. The Polish equivalent of the MESH thesaurus (http://www.ncbi.nlm.nih.gov/mesh) —http://slownik.mesh.pl— containing 24359 general and 546931 compound descriptors, lists 154 terms in which the word *badanie* 'examination' occurs but there is no term *badanie USG*. Only one subtype of this kind of an examination is mentioned: *Badanie USG prenatalne* 'Ultrasonic Prenatal Diagnosis'. General information is placed under the heading *USG*. Among those mentioned above 154 terms containing *badanie*, a generally used phrase *badanie diagnostyczne* 'diagnostic examination' is not listed at all.

What is common to all domain vocabularies is that the vast majority (if not all) of terms are noun phrases. Thus, we also decided to concentrate on these kinds of phrases, although there are approaches, like [14], in which verbal phrases are also taken into account.

The internal structure of terminological noun phrases can vary, but the number of construction types are limited. In Polish, domain terms most frequently have one of the following syntactic structures:

- a single noun or an acronym, e.g. *grawitacja* 'gravity', *PKB* 'GDP';
- a noun followed (or, more rarely, preceded) by an adjective, e.g. *stosunki$_n$ gospodarcze$_{adj}$* 'economic relations' or *nowe$_{adj}$ technologie$_n$* 'new technologies';
- a sequence of a noun and another noun in genitive: *prawo$_{n,nom}$ ciążenia$_{n,gen}$* '(Reilly's) law of (retail) gravitation';
- a combination of the last two structures: *europejski$_{adj}$ rynek$_{n,nom}$ usług$_{n,gen}$ finansowych$_{adj}$* 'European market for financial services';
- a noun phrase modified by a prepositional phrase, e.g. *wierzytelność podatnika wobec skarbu państwa* 'taxpayer liability to the State (Treasury)'.

Other features of Polish nominal phrases which complicate the recognition process are: relatively free word order, genitive phrase nesting: the sequences of genitive modifiers can have more than two elements, internal prepositional phrases, and coordination. In section 4 of the paper we present a shallow grammar defining noun phrase types which we consider to cover most Polish nominal domain related terms.

3 Linguistic Analysis of Domain Specific Texts

The process of automatic domain terminology extraction starts with gathering an appropriate corpora containing domain texts. The extraction procedure begins with text segmentation into tokens: words, numbers and punctuation marks. The next step is morphological analysis and disambiguation, which results in tagging all word forms with information about their base form, part of speech, and morphological feature values. For this purpose we can use publicly available Polish taggers: TaKIPI tagger [10] or Pantera [1]. They cooperate with different versions of the morphological analyser Morfeusz [17] (former SIAT and current SGJP respectively).

The quality of results obtained from a terminology extraction system strongly depends on the quality of the domain corpus, especially on the quality of its tagging. Texts which deal with specific domains usually contain a lot of words which are not present in general dictionaries, there are also specific token types and domain related acronyms or abbreviations. Processing them with general language tools trained on newspaper or literature data usually leads to incorrect descriptions of many tokens, so to obtain reliable results we have two choices: to train a tagger (a program which disambiguates morphological descriptions) for domain texts which requires a lot of correctly tagged data, or to correct data obtained with an available tagger. As for the first solution, the appropriate training data is currently unavailable, we had to choose the latter option.

In Morfeusz, an abbreviation is assigned a *brev* POS (part of speech) with characteristic if the full stop is necessary after it (attribute *pun* and *npun*).

We decided to extended the description of an abbreviation with information about the type of word or phrase it abbreviates. The type of an abbreviated phrase/word is necessary if we want to construct a grammatical phrase containing the abbreviation. So, we add an attribute *btype* that has following values: *nphr, nw, prepphr, adjw, etc.* that informs if the abbreviated is a word or a phrase (ending *w* or *phr* respectively) and gives a grammatical type of an abbreviated phrase or POS of the abbreviated word. For example *nr* (*numer* 'number') has attribute btype *nw*, while *SA* (*Spółka Akcyjna* 'joint-stock company') is of *nphr* type.

Taggers can only annotate tokens for which descriptions are available to them. To deal with out-of-dictionary words, both taggers can cooperate with the *Guesser* module [11] which suggests tags for words which are not analyzed by Morfeusz. As in the experiment described in [6] only 20% of suggested descriptions were fully correct (this means that: base form, POS and its complete morphological characterization are correctly assigned), we decided not to use this module. In our case, unknown tokens are assigned the *ign* tag. The Pantera tagger allows us to define a separate dictionary in which we can describe unknown tokens.[1] Strings defined in that dictionary cannot contain spaces nor punctuation marks such as periods. Thus the common abbreviation *m.in.* (*między innymi* 'among others') cannot be introduced into this dictionary. However it is possible to introduce the full declination of the foreign word *outsoursing* 'outsourcing' that is used with Polish endings: *outsoursingiem*$_{inst}$ or *outsoursingu*$_{gen}$.

Another method of improving tagger results, is to define rules in Spejd [12]. The advantage of this method is the possibility of taking into account contexts. For example in the following string *Dz.U.* that abbreviates the phrase *Dziennik Ustaw* 'Journal of Law' the string *U* is interpreted as the preposition 'at', but in this context it is the abbreviation of the word *ustawa* 'law'. The places where the description of *U* ought to be changed is recognized by the Spejd rule (1)[2] which is named *DZU*. The rule indicates the modified element after the *Match* string: the orthographic form of the string equal to *U*. The contexts are described after the keywords *Left* and *Right*. In this case only the left context is defined. It consists of two tokens: the first has the orographic form *Dz* and the second is the full stop (without spaces between tokens—indicated by *ns*). The new description of *U* is given after the *Eval* keyword, e.g.: the new base form *ustawa* 'law' and the new tag that describes the abbreviation of a noun word.

(1)   Rule "DZU"
       Left: [orth~"Dz"] ns [orth~"."] ns;
       Match: [orth~"U"];
       Eval: word(Modif-morf, "base:ustawa#ctag:brev:pun:nw#");

Spejd rules are particularly helpful in correcting some regular tagging errors in often occurring phrases. For example, in the frequent phrase *osobami zarządza-*

---

[1] TaKIPI tagger does not allow to use an external dictionary. Some methods of correcting TaKIPI results are described in [6].

[2] Spejd also provides different methods of correcting word annotations.

jącymi lub nadzorującymi emitenta 'persons managing or supervising the issuer' the Pantera tagger assigned the gender of participle *nadzorującymi* 'supervising' to impersonal masculine instead of feminine. This can be corrected by the rule (2), that changes the description of the morphological features of the matched element, including the gender. The same error occurs in phrases where instead of the word *emitenta* 'issuer$_{gen}$' there is another phrase like *(między) osobami zarządzającymi lub nadzorującymi a Spółką* '(between) persons managing or supervising and the company', so the rule (2) is applied independently to a right context.

(2)   Rule "nadzorującymi"
      Left: [orth˜"osobami"] [orth˜"zarządzającymi"] [orth˜"lub"];
      Match: [orth˜"nadzorującymi"];
      Eval: word(Modif-morf, "ctag:pact:pl:inst:m3:imperf:aff#");

4   Phrase Selection

For recognizing the selected types of nominal phrases, we defined a cascade of shallow grammars consisting of six small sets of rules being regular expressions in which morphological information is used.

The first set of rules describe the basic types of phrase elements. Head elements of nominal phrases can have tags denoting nouns, gerunds or nominal abbreviations (*subst, ger, brev:xx:nw, brev:xx:nphr*). The first two types are recognized as inflecting elements, so they should agree with adjectival modifiers in number, case and gender, while the other two do not inflect, so they can be combined with modifiers of different forms. Adjectival modifies are also divided into two classes. The first class describes those elements which show inflection, i.e. adjectives, past participles and a special kind of complex adjectives which are built up from a special adjectival form ending with '-o', a hyphen, and an ordinary adjective (e.g. *społeczno-ekonomiczny* 'socioeconomic'). The second, non-inflecting group consists of adjectival abbreviations, e.g. *ang. (brev:pun:adjw)* 'English'.

The second set of rules describe adverbial modifications of adjectives, while the third one describes adjectival phrases which can consist of up to five adjectives optionally separated by commas. The last adjective in a sequence can be preceded by a conjunction *i* 'and'. At the next level, nominal phrases which consist of a nominal element and an optional pre or post adjectival modification are formed. A nominal phrase has gender, number and case assigned on the basis of the characteristic of its main element. In the case when the modified element is of a non-inflecting type, the inflectional description of the phrase is assigned on the basis of the adjectival features.

The next level of complex phrase formulation consists in building sequences of nominal genitive modifiers. The possibility of placing adjective modifications after the genitive one is also described. The last grammar level accounts for nominal modification in nominative case (apposition) and for modification by prepositional phrases. The examples of the nominal phrases recognized by the rules are the following:

- cenę$_{n,acc}$ ropy$_{n,gen}$ 'price of oil';
- współczesna$_{n,nom}$ struktura$_{n,nom}$ systemu$_{n,gen}$ transportowego$_{n,gen}$ 'contemporary structure of the transportation system';
- poziom$_{n,nom}$ wykorzystania$_{n,gen}$ standardowych$_{adj,gen}$ jednostek$_{n,gen}$ rozliczeniowych$_{adj,gen}$ 'degree of utilization of standard units of account';
- cena$_{n,nom}$ na$_{prep}$ nowy$_{adj,acc}$ produkt$_{n,acc}$ 'price on a new product';
- cena$_{n,nom}$ równowagi$_{n,gen}$ kształtowana$_{ppas,nom}$ przez$_{prep}$ relację$_{n,acc}$ podaży$_{n,gen}$ 'price of equilibrium shaped relative to supply'.

Applying these general rules to our data resulted in a set of phrases which included an easily distinguishable subset of non-domain terms. These were phrases beginning with modifiers describing that a concept represented by a subsequent subphrase is occurring, desired or expected, for example, *(w) trakcie$_n$ sesji* 'during the session'. To eliminate such phrases we defined a set of words which were to be ignored during phrase construction and modified the set of rules accordingly. The excluded words belong to the following classes:

- general time or duration specification, e.g. *czas 'time'*, *miesiąc 'month'*;
- names of months, weekdays;
- introductory/intension specific words, e.g. *kierunek 'direction'*, *cel 'goal'*;
- general adjectives which can modify nearly every phrase, e.g. *inny 'other'*, *sam 'alone'*, *niektóry 'some'*, *który 'that'*, *każdy 'every'*, *taki 'such'*.

The set of phrases obtained using this modified grammar constituted a starting point for terminology selection procedure. As Polish is an inflectional language, phrases which are identified within the text are of different forms (e.g. cenę$_{n,acc}$ ropy$_{n,gen}$, cena$_{n,nom}$ ropy$_{n,gen}$) so the usual processing stages like counting phrase frequencies and preparing a list of phrase types became difficult. To overcome this problem we produce an artificial base form of every identified phrase occurrence taking base forms assigned by the tagger, i.e. cena$_{n,nom}$ ropa$_{n,nom}$.

To allow for the recognition of terms which are nested inside other more complex terms, we add information about internal phrase structure, i.e. we mark limits of substrings matched by rules applied at the subsequent levels of the grammar cascade. The annotation style is minimalistic, i.e. only the end of the phrase and its type is marked by the '>' sign with the type name. A phrase can not be divided on a $>_a$ (adjective) marker. For the selected examples, the grammar output looks as follows:

- cena $>_n$ ropa $>_n$
  price $>_n$ oil $>_n$
- współczesny $>_a$ struktura $>_n>_t$ system $>_n$ transportowy $>_a>_t>_{ng}$
  contemporary $>_a$ structure $>_n>_t$ system $>_n$ transport $>_a>_t>_{ng}$
- poziom $>_n$ wykorzystać $>_n$ standardowy $>_a$ jednostka $>_n>_t$ rozliczeniowy $>_a>_n>_{ng}$
  degree $>_n$ utilization $>_n$ standard $>_a$ unit $>_n>_t$ accounting $>_a>_n>_{ng}$
- likwidacja $>_n$ zagraniczny $>_a$ konto $>_n>_t$
  cancellation $>_n$ foreign $>_a$ account $>_n>_t$.

On the basis of the structural information we can identify nominal subphrases. For example, in the second phrase enumerated above there are four such subphrases (given here in a properly lematized form): *współczesna struktura, struktura systemu, współczesna struktura systemu*, and *struktura systemu transportowego* 'contemporary structure', 'structure of the system', 'contemporary structure of the system', 'structure of the transportation system'. In the last example only one of two substrings is a proper subphrase: *zagraniczne konto* 'foreign account'.

5   Term Identification

The set of phrases constitute input data for the term selection algorithm. This stage aims to identify subterms which occur inside other terms—internal terms—and to eliminate phrases which come from general language and should not be placed within the domain dictionary. To eliminate phrases from general language, we compare frequencies of selected phrases in a domain corpus and in a corpus of general language.

For the purpose of ranking terms we adopted one of the most popular solutions to this problem proposed by [4], [2]. In this approach, all phrases (external and internal) are assigned a C-value which is computed on the basis of the number of their occurrences within the text and their length. Internal phrases are not just every substring of the identified phrases, but only those sequences of phrase elements which would be accepted by our grammar as correct nominal phrases (the exact identification process was characterized in the previous section). As we also wanted to take into account phrases of the length 1, for one word phrases we replace the logarithm of the length with the constant 0.1.[3] This slightly modified definition of the C-value is given below (p – is a phrase under consideration, LP is a set of phrases containing p):

$$
C - value(p) = \begin{cases} lc(p) * freq(p) - \frac{1}{\|LP\|} \sum freq(lp), & if\ \|LP\| > 0,\ lp \in LP \\ lc(p) * freq(p), & if\ \|LP\| = 0 \end{cases}
$$

where $lc(p) = log_2(length(p))$ if $length(p) > 1$ and 0.1 otherwise.

The general idea of this coefficient is to promote phrases which occur in different contexts as it is more likely that they constitute separate terms than in the case where most of their occurrences have the same context. For example, *system bankowy* 'banking system' has occurred in the analyzed texts 5 times of which 5 were inside a wider nominal phrase of 5 different types, for *bezpieczeństwo publiczne* 'public security' these numbers were respectively 4-1-1. In both these cases we have clear evidence that these phrases constitute separate terms. Similarly, the phrase *waluta narodowy* 'national currency' which never occurred in

---

[3] The value 0.1 was chosen arbitrary from the interval 0–1 to balance lower frequencies of phrases of length 2 in comparison to single words. If many very long phrases are recognized within a particular data, the appropriate coefficient should also be modified as terms consisting of very many words practically do not occur.

isolation, but occurred five times in four different contexts, should be considered as a phrase. The phrase *wysoki przedział* 'high interval' would be much lower on the term ranking list as it occurred 3 times but only in one type of context.

The choice of the C-value coefficient was supported by evaluation done in [5] and [9] which showed its high efficiency for term identification task. Apart from the relatively high usage of this coefficient, there are nevertheless some problems in interpreting the notion of the LP set from the definition given above. If we consider a following set of phrases: *kapitał zakładowy*, 'share capital' *pozyskanie kapitału*, 'acquisition of capital' and *pozyskanie kapitału zakładowego* 'acquisition of share capital' it is not straightforward how to count contexts in which the basic phrase 'capital' occurs. According to the original definition of C-value method three different contexts would be counted. In our approach, to cover possible change of word order, we decided to count right and left contexts separately so we count only two.

6   Experiment Description

The experiment aimed at automatic extraction of domain specific terms was conducted on the economics articles taken from the Polish Wikipedia. Only textual content of these articles was taken into account. Texts were collected within the Nekst project (*An adaptive system to support problem-solving on the basis of document collections in the Internet* POIG.01.01.02-14-013/10), by Łukasz Kobyliński in 2011 in order to test word sense disambiguation methods. The data contains 1219 articles that have economics related headings and articles linked to them. The data contains about 450,000 tokens.

The plain texts were processed by Pantera tager working with Morfeusz SGJP. We defined, through Pantera, an additional domain dictionary containing 741 entries of word-forms (not recognized by the Morfeusz analyzer) and their descriptions. Additionally 156 Spejd rules were created to correct Pantera decisions, and to extend descriptions of abbreviations. Spejd rules corrected or extended about 5500 token descriptions.

As the reference set we have chosen the balanced subcorpus of NKJP [13]— *nkjp-e*. It was originally tagged using Pantera and then manually corrected, so for this set we did not prepared any additional dictionaries nor correction rules. The entire set consists of about 1,200,000 tokens.

The results of applying the grammar described in section 4 to the economics texts (*wiki-econo*) and to the general corpus texts (*nkjp-e*) are presented in Table 1 in which the distribution of phrase lengths and frequencies are given.

The list of phrases obtained after processing economics texts had to be cleaned up from two kinds of expressions. First, some phrases occurring within these texts are coming form general language and should not be treated as economic. The examples of these phrases could be: *pierwsza próba* 'the first trial', *sposób liczenia* 'the way of counting', and *czynnik zewnęrzny* 'external factor'. The second group of phrases are those which resulted from extracting internal nominal phrases and in practice never occur alone, for example *konwersja części* 'conversion of

Table 1. Distribution of phrases lengths and frequencies

phrase length	data set wiki-econo	nkjp-e	common nb	%	phrase freq	data set wiki-econo	nkjp-e
$\sum$	104847	232099	13359	12.74	$\sum$	104847	232100
1	8214	35249	6270	76.38	=1	85747	197713
2	39607	96088	6051	15.27	2-10	16623	29599
3	27826	53680	832	2.99	11-50	1885	3602
4	15980	27518	149	0.93	51-100	280	631
5	7836	12200	42	0.53	101-1000	305	545
6-9	5292	7257	15	0.28	1000-	7	10
>=10	92	107	0	0			
max	13	14	-	-	max	1565	2414

a part' being a part of *konwersja części długu* 'conversion of a part of debt' or *dokument wystawiony* 'documment issued' extracted from *dokument wystawiony przez podmiot* 'document issued by a given subject'.

The first problem is addressed by comparing the resulting data set to the list of phrases obtained for the general texts, while the second one is (partially) solved by ordering the phrases according to their C-value and eliminating those which are low on this list.

To compare the terminology extracted from economics texts with phrases extracted from the general corpus of Polish we analyzed terms identified in both corpora. Table 3 shows how many terms are recognized in both corpora and how many of them have greater C-value in each data set. 4% multi-word terms recognized in economics texts are also recognized in *nkjp-e* data—the longest common phrases have 6 words. 2.2% of multi-word terms have higher C-value in economic than in *nkjp-e* data. For economics texts C-value is higher for example for the phrase *papiery wartościowe dopuszczone do publicznego obrotu* 'securities admitted to public trading' while for *nkjp-e* subcorpus such a phrase is *minister właściwy do spraw finansów publicznych* 'minister responsible for public finances'. This example illustrates the observation that some of multi-word terms with higher C-value in *nkjp-e* data are in fact related to the economic domain. There are quite a lot of phrases with greater C-value for *nkjp-e* subcorpus and relevant to the economic domain, below a few more examples of such phrases are given:

- *walne zgromadzenie akcjonariuszy* 'general meeting of shareholders';
- *Narodowy Bank Polski* 'Polish National Bank';
- *narodowy fundusz inwestycyjny* 'national investment fund';
- *rada nadzorcza* 'supervisory board';
- *skarb państwa* 'state treasury';
- *urząd skarbowy* 'treasury office'.

In our opinion the situation described above results from the popularity of economic topics in newspaper articles and Parliamentary speeches which are included in the *nkjp-e* subcorpus. Because of that, it would be desirable to inspect

Table 2. The most frequent phrases

wiki-econo			nkjp-e		
	phrase	occur.		phrase	occur.
1	cena 'price'	1565	1	pan 'mister'	2414
2	spółka 'company'	1364	2	człowiek 'human'	1789
3	rynek 'market'	1300	3	sprawa 'case'	1500
4	koszt 'cost'	1277	4	praca 'work'	1373
5	podatek 'tax'	1214	5	osoba 'person'	1196
...			...		
92	papier wartościowy 'stock'	281	386	pan poseł 'member of Parliament'	134
130	działalność gospodarcza 'economic activity'	220	556	unia europejska 'European Union'	100
192	osoba fizyczna 'natural person'	156	612	projekt ustawy 'project of an Act '	91
209	podatek dochodowy 'income tax'	148	641	minister właściwy 'appropriate minister'	87
244	fundusz inwestycyjny 'investment fund'	128	723	rada ministrów 'Council of Ministers'	79
...			...		
431	kodeks spółki handlowej 'code of commercial companies'	70	841	minister właściwy do spraw 'minister responsible for the task'	70
538	spółka z ograniczoną odpowiedzialnością 'limited liability company'	38	1600	Jan Paweł II 'John Paul II'	37
539	koszt uzyskania przychodów 'cost of revenues'	38	1739	II wojna światowa 'II World War'	33
540	prowadzenie działalności gospodarczej 'running a business'	38	1902	sojusz lewicy demokratycznej 'Democratic Left Alliance '	30
563	ustawa o rachunkowości 'Accounting Act'	33	2162	wejść w życie 'come into force'	26

Table 3. Comparison with general corpus

Terms	common	C-value greater in econom.	C-value greater in *nkjp-e*
1-word	5535	1563	3972
2-words	3526	1963	1563
3-6-words	360	224	136
Total	9421	3750	5671

manually phrases recommended for removing from the domain terminology on the basis of comparison with phrases created from *nkjp-e* subcorpus. However, as manual inspection of so many phrases is hard to perform, we decided to stick to automatic approach and eliminate from the result all phrases which have greater

Table 4. C-value distribution

C-value	initial nb of phrase types	nb after selection
=0	53513	51888
<1	4597	3357
<2	25239	24512
<5	18987	18551
<10	1454	1324
<100	1043	943
>=100	14	14
Total	104847	100589

Table 5. Phrases considered as terms

	1st annotator			2nd annotator			final annotation		
	domain	general	wrong	domain	general	wrong	domain	general	wrong
top	412	72	16	413	72	15	409	75	16
middle	322	135	43	290	141	69	278	263	59
end	246	170	84	209	181	110	206	187	107

C-value counted in the context of general texts than that counted for the economic data. After this step the term list consisted of 100589 nominal phrases with the distribution of C-value given in Table 4.

To evaluate the quality of the results we performed a manual verification of three groups of 500 randomly chosen phrases. The first set was drawn from the top terms, which have C-value greater or equal 5, while two others represent terms which have C-value below or equal 2 and above 1 (a set named *middle*) or equal 0 (*end*). These lists were checked by two annotators. They had to qualify each term either as a domain specific, general or wrong (i.e. sequences which are not terms at all or have wrong syntactic structure). The results of this verification are given in Table 5. In this table we can observe that the task of judging what is and what is not a domain terminology is highly subjective—the number of phrases judged as general differ a lot. Two examples of such differently judged phrases are *konkurs ograniczony* 'limited competition' and *matematyka stosowana* 'applied mathematics'. But in spite of these differences, the obtained results show that the applied method of automatic term extraction can give reliable results. More than 80% of terms from the *top* list are judged as domain related by both annotators, while this percentage lowers significantly towards the end of the ranked list. 40% of domain related terms in the *end* group of phrases is probably the result of the rather small size of a data set.

To confront automatic extraction with manual dictionary creation, we compared the results of our experiment with the economic terms dictionary constructed by Agata Savary within the already mentioned Nekst project. It contains terminology manually collected from different economics dictionaries and consists of about 10,000 multi-word terms. All possible grammatical forms of terms were created with help of *Topostaw* tool [7].

To perform the comparison, from our list of terms that have C-value grater than in NKJP data, we removed terms that have C-value less then 1. Because the manually constructed dictionary consists only of multi-word phrases, we deleted also one word terms. To perform the comparison of such prepared list of terms with the manually created dictionary we had to identify all their occurrences in texts. Then, we compared our list of terms (their occurrences) with that manually collected and declined. The results are given in Table 6. As we can see about 90% phrases recognized in manually created dictionary was represented in our texts as full phrases while 10% appeared only as subphrases.

Table 6. Comparison with manual dictionary

Phrases	our method	in manual. dict.
full phrases	41031	2142
subphrases	2943	245
Total	43974	2387

The results of this comparison show that on our list there are many terms which should be taken into account in economics dictionary. For example the manually created dictionary contains 25 different phrases describing *aktywa* 'assets' (we considered phrases where 'asset' is the head element), 7 of them were recognized in our data: *aktywa finansowe* 'financial assets', *krótkotrwałe aktywa finansowe* 'short-term financial assets', *aktywa obrotowe* 'current assets' and *aktywa trwałe* 'fixed assets', *aktywa netto* 'net assets', *oficjalne aktywa rzeczowe* 'official tangible assets', *aktywa rezerwowe* 'reserve assets' . Another 25 phrases describing assets (not present in the dictionary) were recognized by our method. 20 phrases are correct domain terms like: *aktywa firmy* 'assets of the company', *zagraniczne aktywa* 'foreign assets', *aktywa niefinansowe* 'non-financial assets', *aktywa trwałej wartości* 'assets of lasting value', *płynne aktywa* 'liquid assets'. 2 phrases are not classified as the domain terms: *wszystkie aktywa* 'all assets' and *pozostałe aktywa* 'other assets' while 3 phrases are incorrect.

7   Conclusions

In this paper we presented results of automatic terminology extraction from domain unstructured texts. Such tools can be valuable for processing texts from domains for which no electronic terminological or ontological resources exist. Although good results presented in this paper were obtained on the basis of relatively clean data—grammar rules operate on the results of a general tagger which were corrected by a set of dedicated rules, the method can also be used on uncorrected data. In that case the results would be worse but they still may be of a practical value. The performed comparison with the manually created terminological lexicon showed that automatic terminology extraction can be also a valuable method for enriching already existing dictionaries, especially in domains which quickly change in time.

Although using the adopted method, a lot of relevant phrases can be identified, the selection procedure should be further improved. In further work we plan to enhance a definition of potential term structure and to define more sophisticated rules of candidates ranking which would be more suitable for Polish phrases.

Acknowledgments. This work was supported by SYNAT project financed by the Polish National Center for Research and Development (SP/I/1/77065/10).

References

1. Acedański, S.: A Morphosyntactic Brill Tagger for Inflectional Languages. In: Loftsson, H., Rögnvaldsson, E., Helgadóttir, S. (eds.) IceTAL 2010. LNCS, vol. 6233, pp. 3–14. Springer, Heidelberg (2010)
2. Barrón-Cedeño, A., Sierra, G., Drouin, P., Ananiadou, S.: An Improved Automatic Term Recognition Method for Spanish. In: Gelbukh, A. (ed.) CICLing 2009. LNCS, vol. 5449, pp. 125–136. Springer, Heidelberg (2009)
3. Broda, B., Derwojedowa, M., Piasecki, M.: Recognition of structured collocations in an inflective language. System Science (4) (2008)
4. Frantzi, K., Ananiadou, S., Mima, H.: Automatic recognition of multi-word terms: the C-value/NC-value method. Int. Journal on Digital Libraries 3, 115–130 (2000)
5. Korkontzelos, I., Klapaftis, I.P., Manandhar, S.: Reviewing and Evaluating Automatic Term Recognition Techniques. In: Nordström, B., Ranta, A. (eds.) GoTAL 2008. LNCS (LNAI), vol. 5221, pp. 248–259. Springer, Heidelberg (2008)
6. Marciniak, M., Mykowiecka, A.: Towards morphologically annotated corpus of hospital discharge reports in Polish. In: Proc. of the BioNLP, ACL/HLT 2011 Workshop, Portland, Oregon (2011)
7. Marciniak, M., Savary, A., Sikora, P., Woliński, M.: Toposław – A Lexicographic Framework for Multi-word Units. In: Vetulani, Z. (ed.) LTC 2009. LNCS, vol. 6562, pp. 139–150. Springer, Heidelberg (2011)
8. Mykowiecka, A., Marciniak, M.: Terminology extraction from medical texts in Polish. In: Ananiadou, S., Pyysalo, S., Rebholz-Schuhmann, D., Rinaldi, F., Salakoski, T. (eds.) Proceedings of the 5th International Symposium on Semantic Mining in Biomedicine, SMBM 2012 (2012)
9. Pazienza, M.T., Marco Pennacchiotti, M., Zanzotto, F.M.: Terminology Extraction: An Analysis of Linguistic and Statistical Approaches. In: Sirmakessis, S. (ed.) Knowledge Mining. STUDFUZZ, vol. 185, pp. 255–279. Springer, Heidelberg (2005)
10. Piasecki, M.: Polish tagger TaKIPI: Rule based construction and optimisation. Task Quarterly 11(1-2), 151–167 (2007)
11. Piasecki, M., Radziszewski, A.: Polish Morphological Guesser Based on a Statistical A Tergo Index. In: Proceedings of the International Multiconference on Computer Science and Information Technology — 2nd International Symposium Advances in Artificial Intelligence and Applications (AAIA 2007), pp. 247–256 (2007)
12. Przepiórkowski, A.: Powierzchniowe przetwarzanie języka polskiego. Akademicka Oficyna Wydawnicza EXIT, Warsaw (2008)
13. Przepiórkowski, A., Bañko, M., Górski, R.L., Lewandowska-Tomaszczyk, B. (eds.): Narodowy Korpus Języka Polskiego. Wydawnictwo Naukowe PWN, Warsaw (2012)
14. Savova, G.K., Harris, M., Johnson, T., Pakhomov, S.V., Chute, C.G.: A data-driven approach for extracting "the most specific term" for ontology development. In: Proc. of AMIA (2003)

15. Sinclair, J. (ed.): Collins Cobuid English Language Dictionary. Collins Publ. (1990)
16. Wermter, J., Hahn, U.: Massive Biomedical Term Discovery. In: Hoffmann, A., Motoda, H., Scheffer, T. (eds.) DS 2005. LNCS (LNAI), vol. 3735, pp. 281–293. Springer, Heidelberg (2005)
17. Woliński, M.: Morfeusz — a Practical Tool for the Morphological Analysis of Polish. In: Kłopotek, M., Wierzchoń, S., Trojanowski, K. (eds.) Intelligent Information Processing and Web Mining, IIS: IIPWM 2006 Proceedings, pp. 503–512. Springer, Heidelberg (2006)

Improving the Workflow for Creation of Textual Versions of Polish Historical Documents*

Adam Dudczak, Miłosz Kmieciak, Cezary Mazurek, Maciej Stroiński, Marcin Werla, and Jan Węglarz

Poznan Supercomputing and Networking Center,
ul Z. Noskowskiego 12/14, 61-704 Poznan, Poland
{maneo,milosz,mazurek,stroins,mwerla,weglarz}@man.poznan.pl
http://www.man.poznan.pl

Abstract. This paper describes improvements which can be included in digitisation workflow to increase the number and enhance the quality of full text representations of historical documents. This kind of documents is well represented in Polish digital libraries and can offer interesting opportunities for digital humanities researchers. Proposed solution focuses on changing existing approach to OCR and simplifying the manual process of text correction through crowdsourcing. Tools required to implement these enhancements are available in Virtual Transcription Laboratory (VTL) prototype, developed in the framework of SYNAT (http://www.synat.pl) project. In the last chapter paper describes results of the experiment with the custom OCR recognition profiles which proves that proposed approach is a viable alternative to existing OCR practices.

Keywords: OCR, digital libraries, digitsation.

1 Introduction

Historical documents available in Polish digital libraries can be used as a base for various digital humanities research including linguistics, social sciences, philology and history itself. These materials are accessible through metadata-based search offered by services like Polish Digital Libraries Federation[1][1]. Metadata associated with these objects usually offers only selective (and sometimes very subjective) view on object's content. This may be a serious obstacle for researchers interested in specific aspects of a given document as these aspects may not be included in general description. One possible approach to increase the visibility and usefulness of historical documents is to facilitate creation of its machine readable textual representations. These textual versions of scanned books, newspapers, etc. can be used as a base for full text search engine which can offer better retrieval capabilities than simple metadata search.

---

* Presented results have been developed as a part of PSNC activities within the scope of the SYNAT project (http://www.synat.pl) funded by the Polish National Center for Research and Development (grant no SP/I/1/77065/10).

[1] http://fbc.pionier.net.pl

R. Bembenik et al. (Eds.): *Intell. Tools for Building a Scientific Information*, SCI 467, pp. 187–198.
DOI: 10.1007/978-3-642-35647-6_14 © Springer-Verlag Berlin Heidelberg 2013

This paper outlines flaws of existing workflows and tools used by Polish digital librarians in process of digitisation of historical documents. The following chapters describes details of proposed improvements and their implementation in the Virtual Transcription Laboratory (VTL) prototype. Paper also presents results of evaluation of the custom Optical Character Recognition (OCR) service which shows that developed set of tools may help to increase the number and enhance the quality of textual versions of the historical documents available in Polish digital libraries.

2   Description of Existing Digitisation Workflows for Textual Materials

Digitisation workflow can be divided into various steps, starting from document scanning, then its graphical post-processing (e.g. fixing orientation, dewarping of pages, etc.), creation of textual version via OCR, composing document version for presentation on the web. This process is well established and works well for modern textual resources. In case of historical documents [2] number of factors are causing huge decrease in quality of textual representation. In [2] author points to possible reasons for this, list includes inconsistent line and character spacing, use of non-standard fonts, text layouts and finally bad condition of digitised material. Some of mentioned problems can be handled through improvements in OCR engines, but in some cases only manual correction makes sense.

In order to get more insight on digitisation practices related to historical textual materials in Polish digital libraries, a survey was conducted. It was open from 28th of September till 7th of October 2010. Request to fill in the survey was sent directly to institutions responsible for creation of Polish digital libraries registered in the Digital Libraries Federation, information about the survey was also published in several professional portals.

The survey was divided into three parts: general, related to digitised resources and to usage of OCR in digitisation of historical documents. The last part of the survey was the most expanded one. It contained questions related to OCR tools, file formats in use and practices related to evaluation and correction of OCR results.

As an outcome we have received responses from 26 institutions responsible for the creation of the biggest Polish digital libraries including Digital Library of Wielkopolska[3], Małopolska Digital Library[4], National Digital Library "Polona"[5] and Digital Library of Wrocław University[6]. In summary these digital libraries held more than 70% of all objects available via the Polish Digital Libraries Federation Portal at that time.

---

[2] Historical in this context usually means dated before year 1850.
[3] http://www.wbc.poznan.pl
[4] http://mbc.malopolska.pl
[5] http://polona.pl
[6] http://www.bibliotekacyfrowa.pl

Polish digital libraries offer access to various kinds of documents, in 2010 most of them were described as textual resources (around 95%[7]). According to the results of the survey the percentage of resources that had been a subject to OCR was 41,6% (appr. 130 000).

Documents (mainly newspapers and books) were scanned in resolutions between 300 PPI and 600 PPI. Choice of color depth depended on type of given document and its features, starting from 1 BPP for 19th century newspapers to 24 BPP for unique manuscripts. Results of scanning were usually stored in a TIFF file format and this was also the dominant input format for OCR tools.

In order to clarify which features of the OCR are useful from practical point of view, several questions related to resources were included in the survey. It is commonly agreed that quality of the OCR output depends on support for particular language in given tool. To clarify which languages are relevant questions related to language of resources were included. All respondents were working with documents in Polish, but also documents in German, English, French and Latin were quite common. It is worth to mention that 70% of respondents had experience with documents written in more than one language e.g. Polish, German and Latin. Another important factor causing decrease in quality is support for given type of fonts. Apart from traditional antiqua-based documents institutions were also working with resources printed using gothic (8 institutions) and cyrillic (8 institutions) scripts.

The third part of the survey was dedicated to features of OCR systems used by librarians. Results of the survey showed that in most cases librarians are using ABBYY FineReader[8] or Document Express[9]. Document Express has a built-in OCR but it have to be mentioned that it is used mainly as a tool for creation of DjVu files. Respondents reported that results of OCR produced with FineReader were imported into Document Express in order to produce final DjVu output. That is why FineReader was considered as the most important OCR engine in our later works described in this paper.

ABBYY FineReader allows for customization of OCR process e.g. to train recognition engine for particular type of font. Only 3 institutions had experience with this feature. Nine out of 26 institutions claimed that they obtained good results without additional training. The rest of surveyed respondents would like to improve the quality of their OCR output via training of the OCR engine, but they suffered lack of time or technical knowledge.

Apart from training capabilities survey included several other questions related to e.g. use of scripting APIs, layout analysis and integration of OCR software into the digitisation infrastructure of the institution. Full text of the survey and more elaborate description of its results can be found in [3].

The last important issue raised in the survey was associated with the evaluation of results and manual correction of the textual versions of documents. Most of (80%) the respondents claims that they skim the random sample of the

---

[7] This is still true in 2012.

[8] http://abby.com

[9] http://www.caminova.com/

OCR outcome. If the OCR quality is very poor, librarians may decide to publish image-only version of the document. It have to be mentioned that OCR results are seen rather as a search aid than high quality textual representation of the document. This might be also important justification of the fact that no one was performing manual correction of the OCR results. In the survey respondents usually answered that lack of correction was caused by high cost of such an activity and lack of dedicated human resources.

3   Improvements in Digitisation of Textual Materials

On top of results of the survey two possible areas in which existing workflow can be improved were identified. First area is dedicated to enhancements in quality of the OCR output, in particular to creation of custom recognition profiles. The second area focuses on decreasing the cost of manual correction of both newly digitised and existing historical documents.

3.1   Enhancing Existing OCR Tools

As was already mentioned, FineReader allows to perform custom training in order to recognise specific glyphs from given document. Results of this training can be saved in a separate file and used for other documents. These documents needs to have exactly the same resolution and font sizes in order to make this work. Users can perform training on their own or ask ABBYY to do so (training of the whole engine for custom materials is offered as a paid services).

It is obvious that FineReader training process was not designed for portability and reuse. That is the main reason why it is hard to share effort and use the results of OCR training between institutions interested in digitisation of similar document types. Another drawback related to FineReader training process is the fact that format of training data is proprietary and in fact it does not allow to use the same training data with other OCR engines or combine several "recognition profiles" together.

In order to address drawbacks of existing solution an OCR service was developed. At the moment Tesseract version 3.0.1 [4] is used as a OCR engine in this service, but the service architecture does not exclude usage of other OCR engines (i.e. OCRopus[10] or Gamera[11]). Service is available via simple REST interface and it is supplemented with webapp called Cutouts[12] (for preparation of training material) and dedicated tools which enables the creation of custom recognition profiles. Overall architecture of this solution is depicted on Figure 1.

Training process consists of several steps: preparation of training material in Cutouts, then processing of training material in Page Generator and finally creation of the new recognition profile and its upload to OCR service.

---

[10] http://code.google.com/p/ocropus/
[11] http://gamera.informatik.hsnr.de/
[12] http://wlt.synat.pcss.pl/cutouts

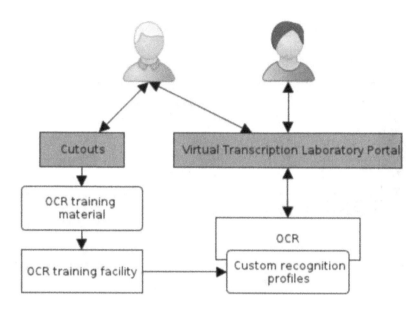

Fig. 1. Custom OCR recognition profiles creation and usage

In first place sample images should be uploaded to Cutouts and than prepro-cessed using OCR. As a result, for each image a corresponding .box file is created. This file contains results of initial recognition for all glyphs in the image, each line contains unicode of a given glyph and its boundaries. Results of this initial recognition is than processed further by users in Cutouts web interface. Via this interface, the user needs to decide whether given selection is a real letter. In case of old documents it is quite common to encounter both false positive and false negative errors of glyph detection. If selection contains valid character it might be necessary to adjust the boundaries of the selection to get the whole shape of the given glyph.

Figure 2 presents second step of Cutouts processing. In this step user can adjust the binarization level of the image, mark special features of the glyph (e.g. mark it as a gothic script) and correct its initial unicode representation. This view offers also a possibility to remove overlapping parts of other glyphs. When processing is done user can proceed to the next element in the scanned image.

The following step is done using Page Generator toolkit which converts Cutouts output into Tesseract training images. When this is ready the simple bash script is launched in order to perform training which results in a new recognition profile. This profile can be uploaded to OCR service and than used in Virtual Transcrip-tion Laboratory which offers web-based user interface for OCR service.

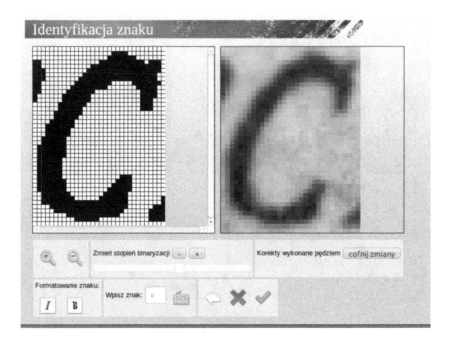

Fig. 2. Preparation of the training material for OCR service in Cutouts

Cutouts is a web application, designed to allow multiple users to work on preparation of the training material simultaneously. This should accelerate creation of training data for newly scanned documents.

In some cases the training material might be already available. Chapter 4 presents results of the experiment conducted on top of ground truth textual data released by PSNC as one of the results of IMPACT (IMProving ACCess to Text, http://digitisation.eu) project. Results of experiment proves that creation of custom recognition profiles might be viable alternative for historical documents.

3.2   Simplifying Manual Correction of OCR Results

If OCR fails, the only way to improve quality of results is to perform manual correction of the text but it is a very laborious task and as survey shown, no one in Polish digital libraries is doing it. On the other hand, examples of successful crowdsourcing projects [5] proves that correction can be easily handled by group of volunteers interested in work with given type of documents.

There are several projects which are employing the power of the crowd to create textual representation of historical documents. The list is of the projects includes Transcribe Bentham[13], Trove Newspapers[14], Distributed Proofreaders[15],

---

[13] http://www.ucl.ac.uk/transcribe-bentham/
[14] http://trove.nla.gov.au/newspaper
[15] http://www.pgdp.net/c/

WikiSource[16], Virtual Manuscript Room[17] and IBM Concert [6]. These projects were created to achieve different goals (i.e. transcribe medieval manuscripts, correct dirty OCR of XIXth century novels) and were aimed at different communities. Nevertheless they have a lot of in common, offering very similar user interfaces (preview of the original document and some sort of editing area) and tracking history of changes made in text. Usually they also allow to export results of work in machine readable form (XML, plain text or PDF). In some cases they feature very strict quality control of the output (e.g. WikiSource, Distributed Proofreaders), but this is not a rule. For example in Trove Newspapers project users' dedication is the only promise that textual content is trustworthy.

Review of existing systems led to identification of two interesting features which might simplify and enhance process of manual correction of transcription: direct access to OCR tools from the web-based transcription editor and representation of information related to position of the text in the scanned image.

Among reviewed platforms only IBM Concert makes direct use of OCR service. It uses information provided in rich output from IBM Adaptive OCR or ABBYY Finereader to guide users through correction process. Results of correction are passed back to OCR as training data. This is quite unique approach, because in both Trove Newspapers and Distributed Proofreaders (which are the biggest among described platforms) materials need to be OCRed before submission to the service. This can be an obstacle for some individuals and institutions who cannot afford to get access to specialized OCR engines dedicated to recognise non standard fonts like gothic script.

Second important thing which is missing and might be useful addition for both users and service itself is preserving information about position of the text. This might be important for researchers interested in analysis of the text structure and very useful for users looking for occurrences of a given word.

This kind of information might be also useful for creation of more advanced recognition algorithms. Position of the text is usually provided as a part of results retrieved from OCR and might be useful even when results of recognition are very bad e.g. in case of manuscripts.

Described analysis led to creation of prototype of Virtual Transcription Laboratory (VTL, `http://wlt.synat.pcss.pl`), a crowdsourcing platform that uses OCR service described in chapter 3.1. Users can upload scans of their documents (in future also direct import from digital library will be possible), process images using one of the available recognition profiles and collaborate on correction of the textual versions of documents. If results of recognition are not satisfying project owner may decide to import only information about position of text lines and use it in a VTL's transcription editor as a help in manual keying.

Transcription editor (see Figure 3) can use information about text position from OCR service but allows for manual creation and adjustment of bounding boxes of text lines. Apart from this it of course allows for manual keying and correction of existing transcription.

---

[16] `http://wikisource.org/wiki/Main_Page`
[17] `http://vmr.bham.ac.uk/`

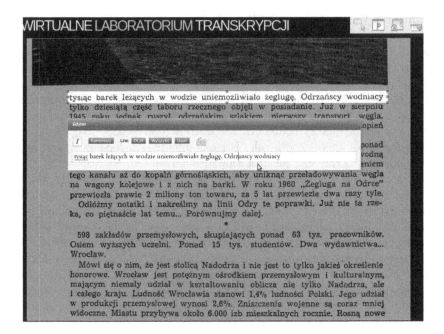

Fig. 3. Transcription editor of Virtual Transcription Laboratory

System keeps track of all changes made in the text of transcription, so the project owner can remove unwanted modifications at any time. Results of work can be exported in hOCR [7] format and included into a full text search index of a digital library which holds scanned version of given document.

Having online access to efficient OCR service with multiple recognition profiles and versatile web-based transcription editor should simplify process of creation and correction of full text versions of both existing and newly scanned materials.

4   Evaluation of Custom Recognition Profiles

Prototype was evaluated on real life data of 475 page scans coming from 9 Polish historical documents created between 16-18th century. Scans were split into two groups of pages (datasets), antiqua font based pages and fraktur font based pages (example page was depicted in Figure 4). Each image has a corresponding ground truth which includes information on full text content of the image (with expected characters accuracy of 99.95%) as well as coordinates of paragraph regions. In addition, the ground truth includes glyphs coordinates and font type indication, which are used as training data for a custom OCR profile.

Among gathered data, most of the selected documents have a rather sophisticated layout that includes signature marks, catchwords, headings and marginalia. Moreover, datasets contain initials - ornamented letters at the beginning of a chapter or a paragraph. Finally, some documents may contain handwritten notes

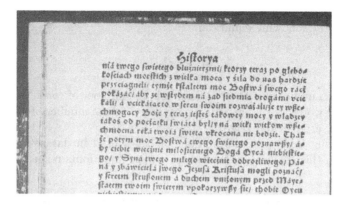

Fig. 4. Example page from evaluated dataset, printed in fraktur font

or even damaged pages, that brings additional noise to the recognition process. All this influence the overall recognition accuracy and hence need to be taken into account when evaluating the OCR engines.

For antiqua dataset, the customised recognition profile was trained to distinguish between regular and italic glyphs. In case of fraktur dataset, however, this distinction was not provided, as italics are represented by antiqua glyphs and therefore were excluded from the customised profile for fraktur glyphs. Please note, that this may decrease the general performance of customised profile in case of common pages with both fraktur and antiqua typefaces found in the same sentence or line.

Characters set found in the gathered ground truth exceeds characters used in today's print (or even modern latin alphabet), as contains e.g. non standard ligatures, long-s letter and acute accents. Since not all of special characters are currently included in the Unicode standard, the ground truth was encoded using Unicode with MUFI extension. Ground truth files contain also replacement character for glyphs that were not recognised by a human, but these were excluded from the data in this experiment.

The original data had to be transformed into the form of Cutouts output, in order to subsequently process it with Page Generator toolkit. In case of pages from the fraktur dataset, cut glyphs were filtered out by the font type, excluding non fraktur fonts. In the second dataset of antiqua fonts, italic glyphs were marked in order to provide additional information to the OCR training process. Layout was simplified to contain only text main paragraphs (e.g. marginalia were skipped) in both cases. In the result of page generation denoised versions of original images were obtained and used for creation of custom OCR profiles.

The evaluation scenario is based on the hold-out test and train approach, where dataset is randomly split into two disjoint subsets. First one is used only for the training purpose in order to create customised OCR engine and is referred to as training set. Later on, a customised OCR engine is applied to the second subset, a testing set. This results in full text content for each page, of which quality is verified with the ground truth. The overall OCR accuracy is then expressed by the following equation:

$$R = \frac{\sum_1^n c_i}{\sum_1^n a_i} \tag{1}$$

where n is the number of evaluated files, c_i is the number of correctly recognised characters (or words) for a particular page and a_i is the number of all characters (or words) for a particular page.

The following table presents number of pages used in datasets for training and testing in this evaluation, for fraktur and antiqua fonts respectively.

Table 1. Number of pages used in fraktur and antiqua datasets for training and testing

Dataset	Number of training pages	Number of test pages
Fraktur	59	228
Antiqua	38	148

Two variants of testing sets were used. The first one *cleaned* contains denoised versions of original images, whereas the second one *real* is made of the original images without any additional preprocessing.

Table 2. Results of the evaluation for fraktur and antiqua datasets. Accuracy value at the level of characters and words is presented for both versions of testing set: cleaned and real.

Dataset	Characters		Words	
	Cleaned	Real	Cleaned	Real
Fraktur	72.43%	54.27%	39.24%	30.77%
Antiqua	76.06%	49.23%	43.10%	29.77%

Results of the evaluation presented in Table 2 show that for cleaned versions of images Tesseract results in OCR accuracy of more than 70% in case of both antiqua and fraktur datasets at the level of characters. However, at the level of words, the overall accuracy equals about 40%. In case of real data, that is pages without any preprocessing, accuracy equals about 50% and 30% at the level of characters and words.

Difference in quality of results between cleaned and real dataset shows how important image preprocessing and layout analysis are. Tesseract 3.0.1 features a basic layout analysis but its image preprocessing capabilities are limited and were designed for modern documents in first place. This drawback of Tesseract can be most likely addressed by using additional, external preprocessing services which will be a subject of further works.

Significant difference between word and character accuracy suggests that there there was a lot of single character errors and they where well distributed among words. It should be easy to correct this kind of mistakes using a dictionary lookup, this issue will be investigated in the future.

Lack of appropriate dataset of Polish documents created before 18th century caused that similar experiments were not conducted in the past. In [8] authors describe results of evaluation of OCR on a Burney Collection[18] which contains English newspapers from 17-18th century. Authors report character accuracy at 75.6% with the word accuracy at 65%. It is impossible to make a direct comparison between these to experiments but results from [8] gives a good indication of difficulty of the problem.

It would be very interesting to perform similar evaluation as described in this chapter with materials from Burney collection. Unfortunately the collection is not freely available for institutions from outside of United Kingdom.

5 Summary and Further Works

This paper outlines possible improvements in the process of digitisation of historical documents. Among possible solutions, two seems to be the most promising: usage of custom OCR recognition profiles and employing the power of the crowd for manual correction.

Creation of textual representations for historical documents is a challenge for modern out-of-the-box OCR engines. That is why creation of custom, reusable OCR recognition profiles dedicated to this kind of materials is a viable alternative to expensive, proprietary OCR engines dedicated to a narrow range of historical documents. Results of the experiments described in Chapter 4 confirms the value of described approach. Nevertheless creation of high quality textual version of historical materials definitely require additional manual processing and correction.

Described prototype allows to crowdsource majority of operations required to create full text version of historical documents. Virtual Transcription Laboratory prototype offers access to two crowdsourcing platforms, one simplifies creation of training data for OCR service (Cutouts) and second (VTL itself) offers access to versatile transcription editor which allows to perform manual correction of historical documents. Apart from this VTL offers web-based access to OCR service which was developed especially for historical documents. Interested user does not have to install any additional OCR software on their computers in order to create a textual version of given document.

Further work will include evaluation of the VTL as a crowdsourcing platform, including its usability and quality produced by real users. Possibilities related to additional tools accelerating process of text correction and OCR training will also be investigated.

---

[18] http://www.bl.uk/reshelp/findhelprestype/news/newspdigproj/burney/index.html

References

1. Lewandowska, A., Werla, M.: PIONIER Network Digital Libraries Federation – Interoperability of Advanced Network Services Implemented on a Country Scale. Computational Methods in Science and Technology, 119–124 (2010)
2. Klijn, E.: The current state-of-art in newspaper digitization: A market perspective. D-Lib Magazine 14(2) (2008)
3. Dudczak, A., Mazurek, C., Stroiński, M., Werla, M., Węglarz, J.: Analiza dostępnych rozwiązań w zakresie automatycznego tworzenia transkrypcji różnego rodzaju materiałów tekstowych w szczególności manuskryptów, starodruków i książek współczesnych. Technical report, Poznań, Supercomputing and Networking Center, Poznań, SYNAT project (2010)
4. Smith, R.: An Overview of the Tesseract OCR Engine. In: Ninth International Conference on Document Analysis and Recognition, ICDAR 2007, vol. 2, pp. 629–633 (2007)
5. Holley, R.: Many Hands Make Light Work: Public Collaborative OCR Text Correction in Australian Historic Newspapers. National Library of Australia Staff Papers, 1–28 (March 2009)
6. Neudecker, C., Tzadok, A.: User Collaboration for Improving Access to Historical Texts. Liber Quarterly 20(1), 119–128 (2010)
7. Breuel, T.: The hOCR Microformat for OCR Workflow and Results. In: Ninth International Conference on Document Analysis and Recognition, ICDAR 2007, vol. 2, pp. 1063–1067 (2007)
8. Tanner, S., Munoz, T., Ros, P.H.: Measuring Mass Text Digitization Quality and Usefulness: Lessons Learned from Assessing the OCR Accuracy of the British Library's 19th Century Online Newspaper Archive. DLib Magazine 15(78) (2009)

Multi-label Classification of Biomedical Articles

Karol Kurach, Krzysztof Pawłowski, Łukasz Romaszko, Marcin Tatjewski,
Andrzej Janusz, and Hung Son Nguyen*

Faculty of Mathematics, Informatics and Mechanics, The University of Warsaw,
Banacha 2, 02-097, Warsaw Poland
{kkurach,kpawlowski236,lukasz.romaszko,tatjew,andrzejanusz}@gmail.com,
son@mimuw.edu.pl

Abstract. In this paper we investigate a special case of classification
problem, called multi-label learning, where each instance (or object) is
associated with a set of target labels (or simple decisions). Multi-label
classification is one of the most important issues in semantic indexing
and text categorization systems. Most of multi-label classification methods are based on combination of binary classifiers, which are trained
separately for each label. In this paper we concentrate on the application of ensemble technique to multi-label classification problem. We
present the most recent ensemble methods for both the binary classifier
training phase as well as the combination learning phase. The proposed
methods have been implemented within the SONCA system which is a
part of SYNAT project. We present some experiment results performed
on PubMed Central biomedical articles database.

Keywords: Data Mining, Topical Classification, Multi-label Classification, Explicit Semantic Analysis, PubMed, MeSH.

1 Introduction

This paper presents some recent multi-label classification methods which are the
main components of a semantic indexing system for textual data. In multi-label
learning approach, each object is associated with a set of labels (binary decisions). Thus the standard definition of a concept learning can be treated as a
special case of a multi-label classification problem. Most of multi-label classification methods are based on a two-layer learning methodology. The lower layer
consists of simple binary classifiers, which are trained separately for each label.
The higher layer consists of one or more classifiers that combine the results of
binary classifiers from lower layer to generate a set of the most adequate labels.

Multi-label learning has proven to be a successful method for solving problems
in different domains, ranging from scene, image and video annotation [1] to multiple applications in bioinformatics, e.g. functional genomics [2] [3]. Multi-label

---

* This research was supported by the National Centre for Research and Development
(NCBiR) under grant SP/I/1/77065/10 by the strategic scientific research and experimental development program: "Interdisciplinary System for Interactive Scientific
and Scientific-Technical Information".

R. Bembenik et al. (Eds.): *Intell. Tools for Building a Scientific Information*, SCI 467, pp. 199–214.
DOI: 10.1007/978-3-642-35647-6_15   © Springer-Verlag Berlin Heidelberg 2013

classification is also one of the main issues in semantic indexing and semantic search systems, where the problem can be understood as assigning to each document a subset of concepts from a given thesaurus or ontology. This topic is the main interest of an ongoing project, called SONCA (Search based on ONtologies and Compound Analytics), realized at the Faculty of Mathematics, Informatics and Mechanics of the University of Warsaw. It is a part of the SYNAT project: "Interdisciplinary System for Interactive Scientific and Scientific-Technical Information" (http://www.synat.pl/). SONCA is a hybrid database framework application, wherein scientific articles are stored and processed in various forms. SONCA is expected to provide interfaces for intelligent algorithms identifying relations among various types of objects. It extends the typical functionality of scientific search engines by more accurate identification of relevant documents and more advanced synthesis of information. To achieve this, concurrent processing of documents needs to be coupled with an ability to produce collections of new objects using queries specific to analytic database technologies.

Multi-label learning problem over the freely available biomedical articles repository called PubMed Central [4] has been investigated in SONCA system. This was also the objective of the JRS'2012 Data Mining Competition, which has been successfully organized by our team and partially supported by the SYNAT project. The aim of this contest was to investigate methods and techniques, that allow accurate predictions of topics from Medical Subject Headings (MeSH) given by human experts.

The most popular multi-label classifier is called ML-KNN [5]. This is, in fact, an extension of the standard KNN and Naive-Bayes classifiers. However, this popular method is not accurate enough for the main task of JRS'2012 Data Mining Competition. The best results of the contest, including the result of the top three teams, were achieved by ensemble methods.

Thus the objective of this paper is to present the most recent ensemble multi-label classification methods. We analyse the best solutions of JRS'2012 Data Mining Competition and propose some new ensemble methods for multi-label classification. We present the accuracy of the proposed methods on the same data set as it was used during the contest.

2   Semantic Indexing of Biomedical Research Papers

One of the possible applications of the multi-label classification methods and the SONCA engine is in the domain of biomedicine. Biomedical research papers stored in a PubMed Central (PMC) repository [4,6] provide a unique opportunity to evaluate usefulness of our automatic indexing algorithms. Apart from the open access to a large set of articles from many international journals and conferences (Open Access Subset), they provide semantic indexes in a form of MeSH *heading/subheading* pairs [7,8]. Those indexes were assigned by human experts and can be used as a reference for our tagging algorithms.

In our experiments with articles from PubMed Central Open Access Subset we divided the indexing task into three separate phases. During the first step we were focusing on predicting the main MeSH headings associated with a given set of papers. In the second step we want to predict subheadings which can be assigned to particular papers. In the last step, we are planning to construct the heading/subheading pairs based on information about the tags assigned in the previous step and some domain knowledge. This knowledge can be given in a form of constraints enforced by the MeSH ontology or as general rules and guidelines for human indexers. Such an approach is in accordance with the evaluation method used in [9].

As a result of our work on automatic semantic indexing we adapted Explicit Semantic Analysis method (ESA) [10] to work with such sources of domain knowledge as DBpedia [11] or the MeSH ontology [12,13]. We have also designed a supervised learning algorithm to extend functionality of ESA and incorporate user feedback into the tagging process [14]. After experiments, we measured accuracy of our supervised tagging algorithm on a corpus consisting of nearly 40000 articles and we showed that our improvement with regard to the original ESA algorithm exceeded 107% in terms of the *recall* and F_1-*score* of predicted headings.

In order to investigate possible methods for predicting MeSH subheadings (the second step in our plan), we organized a data mining contest associated with Joint Rough Sets Symposium (JRS'2012, http://sist.swjtu.edu.cn/JRS2012/). The task for the participants of the JRS'2012 Data Mining Competition [13] was to devise algorithms capable of accurately predicting MeSH subheadings (topics) assigned by the experts to articles from PMC, based on the association strengths of the automatically generated tags corresponding to MeSH headings. Each document could be labelled with several subheadings and this number was not fixed, so the challenge could be regarded as a multi-label classification of textual data. Although this problem is slightly different than predicting the exact MeSH heading/subheading pairs, we believe that the ability to efficiently predict subheadings could substantially aid experts in manual tagging. Additionally it would allow us to effectively utilize domain knowledge and the tagging guidelines in order to perform fully automatic indexing as a subsequent step.

The data set for the contest was provided in a two-dimensional tabular form as two files, namely a training data set and a test set. Each of those tables contained 10, 000 rows corresponding to articles from PCM (20, 000 total). Each row of the data files represented a single document and, in the consecutive columns, it contained integers ranging from 0 to 1000, expressing association strengths to the corresponding MeSH terms. The total number of attributes in the data was 25, 640. Additionally, available was a file containing labels, whose consecutive rows corresponded to entries in the training data set. Each row of that file was a list of subheading identifiers assigned by domain experts. The labels can be regarded as a generalized classification of a journal article. This information was not available for the test set and the task for participants was to predict it using models constructed on the training data. A detailed description of the competition data can be found in [13].

The quality of label predictions submitted by participants of the competition for a single test instance was measured using F_1-$score$, which is defined as a harmonic average of $precision$ and $recall$. Let $TrueTopics_i$ denote labels assigned by experts to i-th test document and let $PredTopics_i$ be a set of predicted labels. The precision of a prediction for the i-th object is defined as:

$$Precision_i = \frac{|TrueTopics_i \cap PredTopics_i|}{|PredTopics_i|}, \qquad (1)$$

whereas the recall of this prediction is:

$$Recall_i = \frac{|TrueTopics_i \cap PredTopics_i|}{|TrueTopics_i|}. \qquad (2)$$

The quality measure used in the competition was an average F_1-$score$ over all test documents, which can be defined as:

$$F_1\text{-}score_i = 2 \cdot \frac{Precision_i \cdot Recall_i}{Precision_i + Recall_i}, \qquad (3)$$

$$AvgF_1\text{-}score = \frac{\sum\limits_{i=1}^{N} F_1\text{-}score_i}{N} \qquad (4)$$

In the above formula N is the total number of test documents. This quality measure is also used in all post-competition research presented in the following sections of this paper.

The competition itself turned out to be a huge success. There was a total of 396 teams with 533 members registered to the challenge. Among them, there ware 126 active teams who submitted at least one solution to the competition's leaderboard [15]. The total number of submissions was 5964. The competition attracted participants from 50 different countries across six continents. Such a big interest in our contest confirms the importance of this topic.

The participants were able to significantly improve over the baseline result, which was obtained by assigning five majority classes to all objects in the test set. The improvement in terms of the F_1-$score$ exceeded 125%. The best team (ULjubljana [16], see Section 4) obtained a result of 0.53579, while the baseline was 0.23721. The most noticeable fact, discovered after analysis of the competition results, was that nearly all of the top-performing solutions were in fact ensembles of multiple learning models.

Blending, or in other words ensemble construction, is a machine learning technique in which multiple base predictors are combined into a single classification model [17,18]. Ensembles are known to be able to produce better results than any of the individual base models [19]. This approach has been successfully used in many data mining competitions, including the Netflix Prize [20].

To demonstrate its usefulness we made an experiment in which we merged solutions of the five best teams in our competition. The new classification of the test documents was constructed by a simple majority voting algorithm. The number of labels to be assigned for each document was decided based on an average length of predictions made by each of the considered teams. Quality of this ensemble was measured using the same evaluation method as in the case of the real contest. As expected, the obtained result, which was 0.536 for the preliminary set and 0.53976 for the final test set, was slightly better than the score of the winners of JRS'2012 Data Mining Competition. This result motivates our further research on the topic of ensemble construction for the multi-label classification task.

3   Overview of Multi-label Classification Methods

In this section we present a brief overview of single-model approaches that we tested during JRS'2012 Data Mining Competition [13]. Results achieved by those models can be regarded as baselines and used for evaluation of more complex classifiers. They may also constitute a base for constructing classifier ensembles [17,21,22]. We describe a simple heuristic which can serve that purpose and we show how effective it can be, even in a case when only a few models are merged.

3.1   A Multi-label k-NN Approach

The most intuitive approach to multi-label classification is based on the classical k-nearest neighbors (k-NN) algorithm [23,24]. For the purpose of the contest we designed and implemented a multi-label version of k-NN which uses the following steps to predict labels for a given document.

Firstly, it employs the weighted k-nearest neighbors algorithm (parameter $k = 90$, higher weights for more similar instances) [23,25]. Cross validation showed that if k was between 20 and 120, the results were almost the same (relative difference below 1%). Given an instance for which the algorithm tries to predict labels (called *current* from now on), solution iterates over all instances (called *compared*) to find neighbors. Similarity is evaluated by finding common non-zero attributes of the *current* and *compared* instances. It is worth emphasizing that several different similarity functions were tested. The best occurred to be a function that sums up attribute values of the *compared* instance (only the common attributes). The Algorithm 1 describes computation of a similarity degree.

Subsequently, the comparing function was improved by a genetic algorithm [26]. At the beginning, a few arrays containing random weights for each attribute were generated. Attributes influence was used in the comparing function. Usefulness of the arrays was estimated by cross validation. Those with the highest score were mutated and the new ones were used in the next iteration. After all iterations, the best array was used to create the output.

Labels of the test cases were chosen on the basis of their frequency assigned to neighbor instances. The most significant improvement of score was obtained by taking into consideration attribute influence on particular labels. Before running kNN, solution checked whether the presence of an attribute (in a whole training

Input: Attibute values of two data instances: current *Curr* and to compare
      compared.
Output: *Measure* - similarity measure
1 *Measure* ← 0;
2 **for** *attribute in Attributes* **do**
3 | **if** *(Curr[attribute]* > 0 *and compared[attribute]* > 0 **then**
4 | | *Measure* ← *Measure* + (*Curr[attribute]*)$^{0.8}$;
5 | **end**
6 **end**
7 **return** *Measure*;
      Algorithm 1: Computation of a similarity degree

part of the data) implies particular label. That one should get much higher rank. For example: if particular attribute implies a label L in 80% cases, a rank of L is multiplied by 2.0.

One of the major difficulties was how to decide how many labels should be assigned to the instance (in the training dataset, for the most cases this number ranged between 1 and 8). Neither the average cardinality of neighbor instances' label sets, nor the *current* instance total tag weights (which might imply a higher number of related labels) correlated with the expected output length. Finally, the length of the output was decided on a rank of highly evaluated label. The cut-off point was determined by a ratio between the rank of the currently processed label and the maximally evaluated one.

A speed of the presented algorithm is very high: when the comparing function is determined, the average prediction time for a single query on a training dataset with 10000 samples is 0.12 seconds (standard 2Ghz processor, code written in Java). The solution also gives reasonable results when applying the default comparing function. The score obtained by this algorithm in JRS'2012 Data Mining Competition was 0.492.

3.2   SVM Models for a Multi-label Classification

A very common approach to the multi-label classification problem is based on Support Vector Machine (SVM) model [27,28]. We applied SVM to perform a binary classification separately for each of the 83 labels. The main tool used for producing this solution was the SVM^{perf} software developed by Thorsten Joachims - the creator of the SVM^{light} library [29,30].

The linear kernel was assessed as performing best on JRS'2012 Data Mining Competition data. It is important to understand which parameters are vital for tuning a linear SVM:

▷ **c** - Trade-off between training error and margin. Low c prevents model overfit. High c may lead to underfit.
▷ **t** - Threshold informing what is the minimal prediction score that marks an occurrence of a label.

▷ **j** - Cost-factor, by which training errors on positive examples outweigh errors on negative examples.

When applying the SVM model, one should not fix on considering exclusively 0.0 cutoff point for interpreting predictions. Especially in situations where positive and negative examples are not balanced, it is worth testing how the result changes when the cutoff point is shifted in both, negative and positive, axis directions. In our solution, the t (threshold) parameter is responsible for moving the cutoff point. The parameter j also has a significant impact on the issue of unbalanced classes, yet value of j differently impacts the classification hyperplane computed by SVM. While t only moves the hyperplane in parallel, changes on j may result in obtaining a completely different hyperplane.

Our best solution during the JRS'2012 Data Mining Competition was obtained using the following procedure:

1. Random partitioning of the training data into data for learning models (80% of training data) and evaluating models (20% of training data).
2. Performing a grid search [31] in order to tune parameters c and t of SVM learning and classification. In this step, c and t were tuned globally for all 83 classifiers.
3. Final adjustment of the t parameter, separately for each of the 83 label classes, in order to maximize average F_1-score on the evaluation data.

The result of our SVM model obtained a score of 0.515 in the JRS'2012 Data Mining Competition final evaluation. It was observed that the third step of the procedure above gave only a minimal final score enhancement, which was probably due to the overfitting which could have been omitted by using 10-fold cross-validation instead of partitioning the training set into fixed learning and evaluation subsets.

The following steps should be also tested as means of further improving the SVM model:

− Apart from c and t, also j parameter should be optimized. It would be interesting to observe to what extent t and j are substitutes.
− All three parameters, not only t, should be fixed separately for each of 83 labels.
− Parameter learning should be performed using 10-fold cross-validation.

3.3 A Simple Ensemble Heuristic

We tested several heuristics for joining results of individual models in order to create an ensemble for the multi-label classification. Below we present the results of two distinct heuristic approaches, obtained from joining k-NN and SVM models. The individual results of those base models, in terms of average F_1-score, were 0.492 and 0.515, respectively.

1. Simple heuristic based on the sorted output labels.

   The individual results are different - results achieved during cross validation showed that SVM is much better than kNN, therefore a simple intersection or sum of labels of both solutions could not improve the result. The reasonable way was to use kNN as an auxiliary model. Lower ranked solution (kNN) is applied to decide whether to include a label in the output, in the case when the SVM classifier is unsure. These labels are returned only if kNN returned them too. Algorithm 2 outlines this heuristic.

   Input: Labels returned by SVM, sorted by score, desc: $Labels_{SVM}$; Labels not returned by SVM, sorted by score, desc: $UnretLabels_{SVM}$; Labels returned by kNN $Labels_{kNN}$;
   Output: $Labels_{Final}$
   1 $L \leftarrow UnretLabels_{SVM}[0:2]$ //three best unreturned labels
   $Labels_{Final} \leftarrow Labels_{SVM} \cup (Labels_{kNN} \cap L)$;
   2 **return** $Labels_{Final}$;   **Algorithm 2**: HEURISTIC 1

   Obtained score is 0.518.

2. Algorithm using generated probabilities.

   The main idea of this solution is that having labels' ranks computed by kNN nad SVM, we can generate new label scores. In this approach, a value associated with each label is a multiplication of probability returned by SVM and a normalized kNN rank. A cutoff point is chosen based on values of the new scores and the length of the SVM output sequence, as presented in Algorithm 3.
   The achieved score 0.525 is much higher than scores of kNN and SVM solutions.

   Input: Labels returned by SVM: $Labels_{SVM}$; Labels in pairs with their scores obtained from multiplying SVM score and kNN score for each label, sorted by score, descending: $Scores$;
   Output: $Labels_{Final}$
   1 **for** $< label, score >$ in $Scores$ **do**
   2     $length \leftarrow len(Labels_{Final})$;
   3   **if** *(length > 0 and score < 0.1) or length = 9* **then**
   4 | break;
   5   **end**
   6   **if** $length \geq len(Labels_{SVM})$ *and* $score < 0.3 \cdot prevScore$ **then**
   7 | break;
   8   **end**
   9     $Labels_{Final} \leftarrow Labels_{Final} \cup \{label\}$;
   10    $prevScore \leftarrow score$;
   11 **end**
   Algorithm 3: HEURISTIC 2

4   The Winner's Approach to JRS'2012 Data Mining Competition

This section describes a solution of the winners of JRS'2012 Data Mining Competition[1] [16]. Team ULjubljana consisted of Jure Žbontar, Miha Zidar, Blaž Zupan, Gregor Majcen, Marinka Žitnik, Matic Potočnik from Faculty of Computer and Information Science Ljubljana, Slovenia. They provided a detailed description of their approach which allowed us to re-implement their method. In this way, we were able not only to confirm its top quality, but we also gained a better insight into how it works. During the competition, ULjubljana's winning solution achieved a final score of 0.53579.

4.1   Base Learners

The Team ULjubljana's winning solution consists of three main stages.

1. 14 base learners are used to generate predictions for each of 83 labels.
2. For each label, 14 base predictions are combined using the stacking(4.2) method into one meta-prediction.
3. A thresholding(4.2) technique, which takes 83 meta-predictions for every row, is applied to choose the final set of labels for the row.

Every base learner takes 25640 attributes and generates predictions for each of 83 labels. A prediction is a real number between 0 and 1, where value close to 1 indicate high likelihood of having this label. There are 5 types of base learners that are used with different parameters making total of 14 distinct base learners.

For evaluation of base learner's results, thresholding (4.2) technique is used with threshold value equal to 0.25. However, the stacking algorithm 4.2 uses raw, continuous predictions of the base learners as an input.

The only parameter to each base learner type except Random Forest is λ. It controls the regularization. All logistic regression and neural network algorithms are optimized with L-BFGS [32]. We describe each base learner type below.

Logistic Regression (lr): L2 regularized logistic regression is run once for each one of 83 labels independently. It uses sigmoid cost function. Each instance takes 25640 attributes as an input and outputs a single real value between 0 and 1.

Four instances of this base learner are used, each different only in the value of regularization parameter λ.

Logistic Regression with Neural Network (lr_nn): The drawback of previous learner was the assumption of independence of attributes, whereas real-world values are not completely independent. To overcome this obstacle, stacking

---

[1] All the results are available on-line at
   http://tunedit.org/challenge/JRS12Contest

algorithm is used. Neural network is trained on 83 inputs that are fed with pre-dictions generated by linear regression instances. There are 100 hidden layers and 83 outputs. Each output is a real value between 0 and 1 that corresponds to the likelihood that given label shall be included in the answer for current row.

Logistic regression with neural network achieves substantially better results than any other base learner algorithm – perhaps due to the fact that it is an ensemble already. Two instances of this base learner are used.

F-score Logistic Regression (flr): This application of the logistic regression attempts to predict all the 83 labels in a single run, as opposed to constructing 83 independent binary models. The advantage of this approach is that the F-measure can be optimized directly. In order to do so, the F-measure derivative is approximated.

F-score logistic regression achieves results similar to the plain logistic regres-sion. Four instances of this base learner are used in the final blend.

Log F-score Logistic Regression (flr_log): Log F-score is a modification of the algorithm from the previous section. The cost function is changed by applying the logarithm operation.

The log F-score achieves slightly better results than the F-score logistic re-gression, despite a very similar structure. Three instances of this base learner are used in the ensemble.

Random Forest (rf): Random forest [33] is the final class of base learners used in the winning solution. Due to large number of original attributes, feature selection is used. Random forest is trained only on attributes having at least 50 non-zero entries in the training set. Such feature selection reduces the number of attributes from 25640 to just over 7000. For each label a separate random forest is trained.

Random forest achieves a disappointing results of 0.46062 evaluated individ-ually and does not improve the final score on the test set. A single instance of this base learner is used in the blend, as it contains no regularization parameter.

4.2   A Stacking Technique

Stacking is a type of ensemble learning, in which predictions from base learners are used as input for one meta-learner. Algorithm 4 shows a pseudo code for this method.

In the winner's solution, neural network with 14 input units, 20 hidden units and one output unit is used as the meta-learner algorithm. Ensemble model is trained for each of 83 labels using five cross-validation folds.

Assigning Final Labels by Thresholding. The output from ensemble method described above are 83 real numbers for every row. In order to predict final an-swer for given row, we need to choose subset of labels. Simple method which

Input:
Let a = number of attributes (25640 in this task)
Let k = number of algorithms used in blending (14 in this task)
Let m = number of rows in training set (10000 in this task)

Data set $D = \{(x_1, y_1), \ldots, (x_m, y_m)\}$, $x_i \in \mathbb{R}^a, y_i \in \{0, 1\}$
Base learners: $\{L_1, L_2, \ldots, L_k\}$
L_i = function(x, y) $(x \in \mathbb{R}^a, y \in \{0, 1\})$ which returns model (function) $\mathbb{R}^a \to \mathbb{R}$
Meta learner: $Meta$: function(x, y) $(x \in \mathbb{R}^k, y \in \mathbb{R})$ which returns model $\mathbb{R}^k \to \mathbb{R}$

Output: Ensemble model (function $\mathbb{R}^k \to \mathbb{R}$)

```
1  for f ← 1 to folds do
2  │   // Every train set consist of m * (1 − 1/folds) elements.
   │   D^f_train ← {D_i ∈ D | i ≢ f (mod folds)};
3  │   for i ← 1 to k do
4  │   │   baseModel^f_i ← L_i(D^f_train.x, D^f_train.y);
5  │   end
6  end
7  meta_input ← ∅;
8  for i ← 1 to m do
9  │   for j ← 1 to k do
10 │   │   // baseModel^i mod folds was NOT trained on i-th row (D.x_i)
11 │   │   // so it can be used to predict the result of i-th row. This predicted
12 │   │   // value will be one of the inputs to the blending algorithm (Meta).
13 │   │   pred_j = baseModel^i mod folds_j (D.x_i);
14 │   end
15 │   meta_input ← meta_input.append((pred_1, ..., pred_k), D.y_i);
16 end
17 ensembleModel ← Meta(meta_input.pred, meta_input.y);
18 return ensembleModel;
```

Algorithm 4: STACKING FOR ONE LABEL

takes all labels with score greater than 0.25 would be enough to score 0.53378 and take 2nd place in the competition. To improve this result, Team ULjubljana adjusts thresholds for every label separately using greedy algorithm. In the first step, each of the 83 thresholds is initialized to 0.25. Then, for each label, threshold is chosen to maximize the average F-score on training set, assuming that all other thresholds are fixed.

4.3 Discussion of the Results

Of 14 base learners that are tested, Logistic Regression with Neural Network (4.1) achieves the best result. We present performance of the base learners in Figure 1. It shows the strength of ensemble technique. It is worth noting that Logistic Regression with Neural Network algorithm uses the same number of base algorithms as the number of labels it predicts - showing that good result can

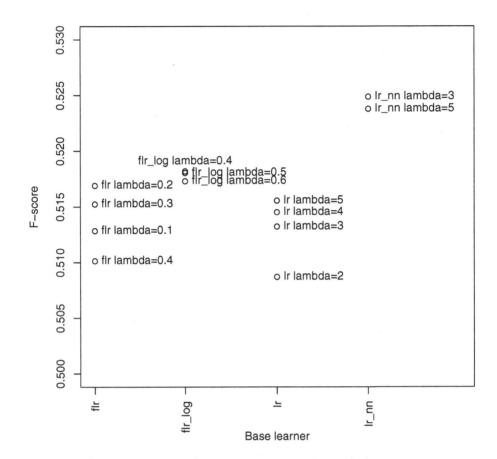

Fig. 1. A performance comparison of Team ULjubljana's base learners. Each label represents a result of one (algorithm, lambda) pair (algorithm names are defined section 4.1). The result of the random forest model was 0.46062. It is omitted in order to ensure clarity of the graph.

be achieved without increasing the diversity and providing the general approach that can be used to solve multi-label classification problems.

Logistic Regression, F-Score Logistic Regression and Log F-Score Logistic Regression achieve quite similar results. Even though their result is significantly worse than that of best algorithm their presence can be justified by diversity. Log F-score Logistic Regression gets a good improvement over F-score Logistic Regression that shows that a simple operation such as taking log of cost function can be surprisingly effective.

Random Forest achieves poor results. That may be caused by feature selection that was necessary to make it run in time - probably at the cost of accuracy.

4.4 Cross-Validation

In order to check the reproducibility of the results we perform 10-fold cross-validation on 20,000-row dataset that is a concatenation of training set and test set that were used in the contest. We evaluate base learners and the meta learner and compare their cross-validation scores with the scores obtained on train/test set used in the contest. The results are presented in table 1.

Table 1. Scores of classifiers evaluated on the contest data using 10-fold cross-validation

learner	lambda	contest	cv
1 lr	2.0	0.5087	0.5099
2 lr	3.0	0.5132	0.5163
3 lr	4.0	0.5145	0.5197
4 lr	5.0	0.5155	0.5214
5 lr_nn	3.0	0.5249	0.5273
6 lr_nn	5.0	0.5238	0.5300
7 flr	0.1	0.5128	0.5210
8 flr	0.2	0.5169	0.5239
9 flr	0.3	0.5152	0.5233
10 flr	0.4	0.5101	0.5211
11 flr_log	0.4	0.5182	0.5265
12 flr_log	0.5	0.5180	0.5269
13 flr_log	0.6	0.5173	0.5264
14 blend		0.5337	0.5390

Correlation coefficient between those two score sequences is 0.9123 thus it appears that the described methods generalize well for other train/test datasets. Performance of classifiers trained under cross-validation is noticeably better (around 0.0064 higher score on average). We conject that it is caused by larger training set sizes – CV classifiers are trained with 18,000 rows while the classifiers trained under contest conditions had only 10,000 rows available.

5 Plans for the Future

For the purpose of this paper we have analyzed, re-implemented or improved several solutions submitted for JRS'2012 Data Mining Competition. Presented methods show their usefulness for semantic indexing of biomedical research papers as they provide means to efficiently predict MeSH subheadings associated with the given set of texts. The discussed algorithms can also be considered in a more general context of a multi-label classification. The paper overviews the state-of-art in this topic.

Our work done with the chosen solutions from JRS'2012 Data Mining Competition clearly presents the higher performance of ensemble approaches over single-model approaches to multi-label classification. It also gives an intuition about the level of a possible improvement that could be achieved by using ensembles of base learners. We plan to further enhance our top-performing single-models and develop improved blending methods in order to draw a more clear boundary between capabilities of base learners and their mixes.

We are also working on improving the methods for indexing textual data with terms from arbitrary ontologies, such as DBpedia or MeSH [11,14]. Our current version of the tagging algorithm is able to significantly outperform the method which was used for generating data sets in JRS'2012 Data Mining Competition. In the nearest future we would like to investigate the impact of the improved input data quality on performance of the selected multi-label classification algorithms.

Acknowledgements. We would like to thank Team ULjubljana for providing code for their winning method. It was invaluable for us to verify the result, deeply understand the solution and re-write it in R.

References

1. Zhou, Z., Zhang, M.: Multi-instance multi-label learning with application to scene classification. In: Advances in Neural Information Processing Systems, vol. 19, p. 1609 (2007)
2. Barutcuoglu, Z., Schapire, R.E., Troyanskaya, O.G.: Hierarchical multi-label prediction of gene function. Bioinformatics 22(7), 830–836 (2006)
3. Zhou, Z., Zhang, M., Huang, S., Li, Y.: Multi-instance multi-label learning. Artificial Intelligence 176(1), 2291–2320 (2012)
4. Roberts, R.J.: PubMed Central: The GenBank of the published literature. Proceedings of the National Academy of Sciences of the United States of America 98(2), 381–382 (2001)
5. Zhang, M.L., Zhou, Z.H.: Ml-knn: A lazy learning approach to multi-label learning. Pattern Recognition 40(7), 2038–2048 (2007)
6. Beck, J., Sequeira, E.: PubMed Central (PMC): An archive for literature from life sciences journals. In: McEntyre, J., Ostell, J. (eds.) The NCBI Handbook. National Center for Biotechnology Information, Bethesda (2003)
7. United States National Library of Medicine: Introduction to MeSH - 2011 (2011), http://www.nlm.nih.gov/mesh/introduction.html
8. Canese, K., Jentsch,J., Myers,C.: PubMed: The Bibliographic Database (2002) (updated August 13, 2003)
9. Névéol, A., Shooshan, S.E., Humphrey, S.M., Mork, J.G., Aronson, A.R.: A recent advance in the automatic indexing of the biomedical literature. J. of Biomedical Informatics 42(5), 814–823 (2009)
10. Gabrilovich, E., Markovitch, S.: Computing semantic relatedness using wikipedia-based explicit semantic analysis. In: Proceedings of the Twentieth International Joint Conference for Artificial Intelligence, Hyderabad, India, pp. 1606–1611 (2007)
11. Szczuka, M., Janusz, A., Herba, K.: Clustering of Rough Set Related Documents with Use of Knowledge from DBpedia. In: Yao, J., Ramanna, S., Wang, G., Suraj, Z. (eds.) RSKT 2011. LNCS, vol. 6954, pp. 394–403. Springer, Heidelberg (2011)

12. Ślęzak, D., Janusz, A., Świeboda, W., Nguyen, H.S., Bazan, J.G., Skowron, A.:
    Semantic Analytics of PubMed Content. In: Holzinger, A., Simonic, K.-M. (eds.)
    USAB 2011. LNCS, vol. 7058, pp. 63–74. Springer, Heidelberg (2011)
13. Janusz, A., Nguyen, H.S., Ślęzak, D., Stawicki, S., Krasuski, A.: JRS'2012 Data
    Mining Competition: Topical Classification of Biomedical Research Papers. In:
    Yao, J.T., Yang, Y., Słowiński, R., Greco, S., Li, H., Mitra, S., Polkowski, L. (eds.)
    RSCTC 2012. LNCS(LNAI), vol. 7413, pp. 422–431. Springer, Heidelberg (2012)
14. Janusz, A., Świeboda, W., Krasuski, A., Nguyen, H.S.: Interactive Document In-
    dexing Method Based on Explicit Semantic Analysis. In: Yao, J.T., Yang, Y.,
    Słowiński, R., Greco, S., Li, H., Mitra, S., Polkowski, L. (eds.) RSCTC 2012.
    LNCS(LNAI), vol. 7413, pp. 156–165. Springer, Heidelberg (2012)
15. Wojnarski, M., Stawicki, S., Wojnarowski, P.: TunedIT.org: System for Au-
    tomated Evaluation of Algorithms in Repeatable Experiments. In: Szczuka,
    M., Kryszkiewicz, M., Ramanna, S., Jensen, R., Hu, Q. (eds.) RSCTC 2010.
    LNCS(LNAI), vol. 6086, pp. 20–29. Springer, Heidelberg (2010)
16. Zbontar, J., Zitnik, M., Zidar, M., Majcen, G., Potocnik, M., Zupan, B.: Team
    ULjubljana's Solution to the JRS 2012 Data Mining Competition. In: Yao, J.T.,
    Yang, Y., Słowiński, R., Greco, S., Li, H., Mitra, S., Polkowski, L. (eds.) RSCTC
    2012. LNCS(LNAI), vol. 7413, pp. 471–478. Springer, Heidelberg (2012)
17. Caruana, R., Munson, A., Niculescu-Mizil, A.: Getting the most out of ensem-
    ble selection. In: Proceedings of the 6th IEEE International Conference on Data
    Mining, pp. 828–833 (2006)
18. Caruana, R., Niculescu-Mizil, A., Crew, G., Ksikes, A.: Ensemble selection from
    libraries of models. In: Proceedings of the 21st International Conference on Machine
    Learning, pp. 137–144. ACM Press (2004)
19. Janusz, A.: Combining Multiple Classification or Regression Models Using Genetic
    Algorithms. In: Szczuka, M., Kryszkiewicz, M., Ramanna, S., Jensen, R., Hu, Q.
    (eds.) RSCTC 2010. LNCS(LNAI), vol. 6086, pp. 130–137. Springer, Heidelberg
    (2010)
20. Bennett, J., Lanning, S.: The netflix prize. In: KDD Cup and Workshop in con-
    junction with KDD (2007)
21. Dietterich, T.G.: An experimental comparison of three methods for constructing
    ensembles of decision trees: Bagging, boosting, and randomization. Machine Learn-
    ing 40(2), 139–157 (2000)
22. Bauer, E., Kohavi, R.: An empirical comparison of voting classification algorithms:
    Bagging, boosting, and variants. Machine Learning 36(1-2), 105–139 (1999)
23. Patrick, E., Fischer III, F.: A generalized k-nearest neighbor rule. Information and
    Control 16(2), 128–152 (1970)
24. Mitchell, T.M.: Machine Learning. McGraw Hill series in computer science.
    McGraw-Hill (1997)
25. Coomans, D., Massart, D.: Alternative k-nearest neighbour rules in supervised
    pattern recognition: Part 1. k-nearest neighbour classification by using alternative
    voting rules. Analytica Chimica Acta 136, 15–27 (1982)
26. Michalewicz, Z.: Genetic Algorithms + Data Structures = Evolution Programs.
    Springer (1996)
27. Joachims, T.: A support vector method for multivariate performance measures. In:
    Proceedings of the International Conference on Machine Learning, ICML (2005)
28. Vapnik, V.N.: The nature of statistical learning theory. Springer-Verlag New York,
    Inc., New York (1995)

29. Joachims, T.: Training linear svms in linear time. In: Proceedings of the ACM Conference on Knowledge Discovery and Data Mining, KDD (2006)
30. Yu, C.N.J., Joachims, T.: Sparse kernel svms via cutting-plane training. In: Proceedings of the European Conference on Machine Learning (ECML), Machine Learning Journal, Special ECML Issue (2009)
31. Chang, C.C., Lin, C.J., Hsu, C.W.: A practical guide to support vector classification
32. Byrd, R.H., Lu, P., Nocedal, J., Zhu, C.Y.: A limited memory algorithm for bound constrained optimization. SIAM Journal on Scientific Computing 16(6), 1190–1208 (1995)
33. Breiman, L.: Random forests. Machine Learning 45(1), 5–32 (2001)

A Tiered CRF Tagger for Polish

Adam Radziszewski

Institute of Informatics, Wrocław University of Technology,
Wrocław, Poland

Abstract. In this paper we present a new approach to morphosyntactic tagging of Polish by bringing together Conditional Random Fields
and tiered tagging. Our proposal also allows to take advantage of a rich
set of morphological features, which resort to an external morphological
analyser. The proposed algorithm is implemented as a tagger for Polish.
Evaluation of the tagger shows significant improvement in tagging accuracy on two state-of-the-art taggers, namely PANTERA and WMBT.

Keywords: morphosyntactic tagging, conditional random fields, Polish,
tiered tagging.

1 Introduction

Conditional Random Fields (CRF) is a relatively new mathematical model that
may be employed to solve sequence labelling problems (Lafferty et al. 2001). The
model has been successfully applied to various Natural Language
Processing (NLP) tasks, including Part-of-Speech tagging of English. The list of
successful applications also includes processing of Polish, e.g., concept-tagging of
spoken language corpora (Lehnen et al. 2009), named entity recognition
(Marcińczuk and Janicki 2012) and NP chunking (Radziszewski and Pawlaczek
2012).

There have been attempts at morphosyntactic tagging of Polish using CRF.
Kuta (2010) reports on using the CRF++ toolkit (Kudo 2005) on Polish data.
However, the results are given only for a set-up, where the morphosyntactic
tags are limited to grammatical class (roughly speaking, Part-of-Speech). In
this paper we argue that it is very difficult to use a single CRF model to
perform morphosyntactic tagging of Polish using the full tagset. We propose
a simple solution to overcome this difficulty: a tiered CRF tagger. The proposed algorithm is largely inspired by the WMBT memory-based tagger for
Polish (Radziszewski and Śniatowski 2011), including its later enhancements
(Radziszewski 2012). We perform evaluation of the algorithm and show that it
performs significantly better than WMBT, but also PANTERA, another state-
of-the-art Polish morphosyntactic tagger (Acedański 2010).

Progess made in morphosyntactic tagging is important, since the error rate
at this level contributes to lower quality of next processing stages, e.g. parsing
(Hajič et al. 2001). This was also shown for Polish: a CRF-based NP chunker
exhibits 1.7 times higher error rate on automatically tagged data than when
using reference manual tagging (Radziszewski and Pawlaczek 2012).

R. Bembenik et al. (Eds.): *Intell. Tools for Building a Scientific Information*, SCI 467, pp. 215–230.
DOI: 10.1007/978-3-642-35647-6_16   © Springer-Verlag Berlin Heidelberg 2013

In the next section of this paper we introduce CRF and discuss the difficulties in using the model for morphosyntactic tagging of Polish. What follows is a short introduction to the realms of tagging Polish: available corpora, their tagsets, available taggers, common practices and assumptions. In the next section we propose our method of tagging Polish that combines CRF with tiered tagging. The method consists of training and testing algorithms, as well as a feature set designed for Polish. Afterwards we present evaluation, which shows that the tagger implementing our method outperforms two available Polish taggers. The paper is summarised with conclusions.

2   Conditional Random Fields for Tagging

CRFs are a family of statistical models similar to Hidden Markov Models (HMM). The fundamental difference is as follows: HMM is a *generative* model, that is, models the joint probability distribution of observations and tags, while CRF is a *conditional* model, since it models the conditional probability distribution of tags given observation sequence, i.e. $P(s_1 \ldots s_K | w_1 \ldots w_K)$. This allows to avoid undesired assumptions. Most importantly, the probability of transition between tags is not forced to depend on the current observation only (Sutton and McCallum 2011). This enables reasoning based on wide contexts, which seems especially important in the case of NLP tasks, where one faces long-distance syntactic dependencies. What is more, CRF work well with many features, possibly mutually-dependent (Lafferty et al. 2001). This property is particularly welcome when dealing with large and structured tagsets: one may naturally introduce features corresponding to values of various grammatical categories inferred from tags.

  Linear-chain CRF is the most popular class of CRFs suitable for tagging that may be descibed using the equation (1). The first sum corresponds to subsequent tokens in text (observations and their tags), the second one deals with features f_j and their weights λ_j. The model assumes that features are characteristic functions. The value of 1 denotes that the predicate represented by a feature is satisfied for i-th position of the labelled token sequence (Lafferty et al. 2001). An example feature function could test if the token at i is a noun, having a word form of to and the preceding token is a preposition. CRF training consists in estimation of weight values. A high weight value means that strong evidence has been found to support the relation between observations and tags as expressed by the feature. Z is a normalising function as defined in (2); T^K denotes a set of possible tag sequences of length K assuming tagset T.

$$P(s_1 \ldots s_K | w_1 \ldots w_K) = \frac{1}{Z(w_1 \ldots w_K)} \exp \sum_{i=1}^{K} \sum_{j=1}^{J} \lambda_j f_j(i, s_i, s_{i-1}, (w_1 \ldots w_K))$$

$$(1)$$

$$Z(w_1 \ldots w_K) = \sum_{(s'_1 \ldots s'_K) \in T^K} \exp \sum_{i=1}^{K} \sum_{j=1}^{J} \lambda_j f_j(i, s'_i, s'_{i-1}, (w_1 \ldots w_K)) \quad (2)$$

Tagging with a trained CRF consists in finding a tag sequence that maximises the conditional probability. Z is independent from the sought sequence, hence the problem may be reduced to maximising the sum of $\lambda_j f_j(i, s_i, s_{i-1}, (w_1 \ldots w_K))$ expressions. This may be achieved using dynamic programming methods (Wallach 2004). Training of a CRF tagger requires finding weight values that maximise the probability (1), which may be implemented using maximum likelihood or other estimation methods (Lafferty et al. 2001).

The assumption that features are characteristic functions brings about practical difficulties. If a feature function is to test word forms and/or tags for particular values, those values must be closed over with the function. For each such value configuration, a separate function must be provided in advance. In practice, *feature templates* are used to solve this problem. Feature templates are essentially functions with parameters, whose closures are used as feature functions. There are two types of feature templates commonly used: *unigram templates* (3a), defining functions of the tag and word form occupying the current position and *bigram templates* (3b), also dependent on the tag at the previous position. During training, all the encountered tags and word forms are gathered, and then, the templates are expanded to all their possible instances (Kudo 2005).

$$f_{t_A,v_A}(i, s_i, s_{i-1}, (w_1 \ldots w_K)) = \begin{cases} 1, & s_i = t_A \wedge w_i = v_A \\ 0, & \text{else} \end{cases} \tag{3a}$$

$$f_{t_A,t_B,v_A}(i, s_i, s_{i-1}, (w_1 \ldots w_K)) = \begin{cases} 1, & s_i = t_A \wedge s_{i-1} = t_B \wedge w_i = v_A \\ 0, & \text{else} \end{cases} \tag{3b}$$

The above formulation assumes that the features operate on word forms and tags. One may similarly define feature templates referring to any transformation of word forms or tags, e.g. returning suffixes of word forms or extracting attribute values from positional tags.

Lafferty et al. (2001) apply CRF to Part-of-Speech tagging of English, reporting 94.5% accuracy using a basic feature set and 95.7% when also using simple transformations of word forms.

Unfortunately, a straightforward application of this model to full morphosyntactic tagging of Polish is computationally unfeasible. This is due to the fact that the tagsets used for Polish contain over 1000 different tags appearing in actual texts (Przepiórkowski 2005), while the time complexity of linear-chain CRF training is quadratic in the size of the label set (that is, the number of different tags appearing in the training data). Strictly speaking, the complexity is $O(T^2 \cdot J \cdot D^2)$, where T is the number of different tags, J is the number of features defined and D is the number of tokens in the training data (Cohn 2007).

These considerations were confirmed during our initial experiments: attempts at training the CRF++ tagger (Kudo 2005) with the morphosyntactic annotation from the National Corpus of Polish (Przepiórkowski et al. 2010) had to be quickly terminated due to the exhaustion of RAM on our server (24 GB) before the first iteration had been completed. This is also the most likely reason why

Kuta (2010) gives 'n/a' instead of actual CRF++ performance figure in tables presenting accuracy values for full morphosyntactic tagging of Polish.

3   Tagging Polish

3.1   Corpora and Tagsets

There are two large Polish corpora containing a part with manual morpho-syntactic annotation, each with its own tagset:

1. the IPI PAN Corpus (Przepiórkowski 2004, Przepiórkowski and Woliński 2003) contains a 884 273-token manually annotated part (we will refer to this part as IPIC),
2. the National Corpus of Polish (version 1.0; Przepiórkowski et al. 2010) contains a 1 215 513-token manually annotated part (henceforth NCP).

NCP is the largest publicly available Polish corpus suitable for tagger evaluation. Its annotation was performed according to high quality standards, involving the 2+1 model (Przepiórkowski and Murzynowski 2009). On the other hand, about 30% of the IPIC corpus is not publicly available, while the free part was compiled in the 1960s. NCP, on the contrary, represents contemporary Polish (Przepiórkowski et al. 2010). For these reasons we decided to limit our experiments to NCP.

The NCP tagset is a conservative modification of the IPIC tagset (Przepiórkowski 2009). The basic assumption is that the *grammatical classes* (generalisation of the usual part-of-speech notion) are distinguished primarily on the grounds of inflection (Przepiórkowski and Woliński 2003). In consequence, over 30 grammatical classes are defined. Each class is assigned a set of *attributes* (grammatical categories) whose values must be provided. For instance, nouns are specified for number, gender and case, infinitives are specified for aspect. For example, the `subst:sg:nom:m2` tag denotes an animate masculine (`m2`) noun (substantive) in a nominative singular form. A small subset of attributes are deemed *optional* and their values may be omitted.

The size of tagset is frequently quoted as a major source of difficulty of morpho-syntactic tagging (Vidová-Hladká 2000, Hajič et al. 2001, Piasecki and Godlewski 2006). On the other hand, if a principled tagset design may facilitate manual annotation (Przepiórkowski and Woliński 2003), the same may be hoped for automatic tagging.

3.2   Taggers

We are aware of three publicly available, ready-made taggers designed for Polish: TaKIPI, PANTERA and WMBT.

TaKIPI (Piasecki and Godlewski 2006) is based on a small set of hand-written rules and a substantial number of decision tree classifiers that acquire tagging rules automatically. The rule acquisition process is partially driven by hand-written expressions that constitute building blocks for the rules. TaKIPI employs *tiered tagging* (Tufiş 1999), that is parts of the tags, corresponding to subsets of attributes, are dealt with sequentially. TaKIPI defines three tiers, corresponding to grammatical class, number and gender (together), and then case. Unfortunately, the tagger is tied to the IPIC tagset and cannot be tested on NCP.

PANTERA (Acedański 2010) is also based on automatic rule induction. Unlike TaKIPI, almost no language-dependent information is required (besides the training corpus itself). The rule induction is driven by a modified version of Brill's transformation-based learning algorithm (Brill 1992). There are two major modifications that allow to obtain better results for inflectional languages. The main enhancements is that of tiered tagging (two tiers are used). The other one is generalisation of rule patterns to operate on attribute level instead of whole tags.

WMBT (Radziszewski and Śniatowski 2011) is a memory-based tiered tagger. The tagger uses as many tiers as there are attributes in the tagset (plus one for the grammatical class). The algorithm iterates over tiers; tagging of one tier involves classification of subsequent tokens with a k-Nearest Neighbour classifier. The classification process benefits from a rich feature set, including values of particular attribute (grammatical class, number, gender and case for tokens surrounding the token being tagged), but also tests for morphological agreement on number, gender and case.

Later, WMBT was enhanced to better deal with words uknown to the morphological analyser (Radziszewski 2012). Two modifications have been proposed:

- training separate classifiers for known and unknown words,
- morphological reanalysis of training data.

3.3   Reanalysis of Training Data

Morphological reanalysis of training data is a procedure that allows for a better usage of the available resources, i.e., the available reference corpus and available morphological analyser. The procedure has been proposed as a means for improving WMBT accuracy (Radziszewski 2012).

Taggers that divide the tagging process into morphological analysis and disambiguation usually assume (TaKIPI, PANTERA and WMBT do) that the training data, besides reference tagging, contains results of morphological analysis. That is to say, the reference corpus assigns each token a set of possible tags, out of which one is *highlighted as contextually appropriate*. This information is used for training. Best results are to be expected if this "reference morphological analysis" matches the behaviour of the analyser used during tagging. This concerns the same forms being unrecognised, but also the same sets of tags attached to recognised forms.

A quick work-around could be to use exactly the same version of the analyser as used for corpus annotation. This solution, however, has two serious shortcomings. First, this forces attachment to a particular version of the analyser, which might be outdated, deprecated or even not available — assuming that the information on the exact analyser version employed for annotation is available at all. What is more, one may be willing to extend the analyser dictionary in hope of improving tagging accuracy, which should be encouraged.

Reanalysis of training data is a simple alternative that allows to update the morphological annotation of the reference corpus with the data taken from the version of the morphological analyser that will be used when tagging. The procedure may be sketched as follows. As NCP and IPIC also assign lemmas to tokens, here we will use the term *morphosyntactic interpretation* to denote a tuple consisting of a tag and a lemma.

1. The training data is turned to plain text.
2. Plain text is fed through the morphological analyser and sentence splitter (for best results it should be exactly the same configuration as used when tagging).
3. The analyser output is synchronised with the original training data in the following manner:
   - Tokenisation is taken from the original training data; should any segmentation change occur, the tokens subjected to it are taken from the original data intact (for simplicity).
   - The remaining tokens (vast majority) are compared; if the *highlighted* interpretation also appears in the reanalysed token, the reanalysed token is taken and the correct interpretation is marked as *highlighted*.
   - If the *highlighted* interpretation is not present there, it means that the tagger would not be able to recover it, hence it is an unknown word. This token is marked as such: we retain only the *highlighted* interpretation and add a non-highlighted interpretation consisting of the "unknown" tag (lemma is set as token's orthographic form, lower-cased). This is to let the tagger see it the same way as when tagging plain text.

The above procedure employs a trick to mark the forms recognised as 'unknown words'. Note that forms that were unknown to the analyser used during manual annotation are marked in a very similar fashion in NCP. An example is presented in Fig. 1. The boxed tags and lemmas correspond to the interpretations highlighted as contextually appropriate in the corpus. Note the first interpretation of ofiarowuje, which is marked as 'unknown word' — it consists of the unknown word tag (ign) and artificial lemma None. The above procedure uses word forms as lemmas for unknown words to increase the chance of getting the lemmatisation right, although in this paper we are only concerned with tagging accuracy, while lemmatisation is not assessed.

The purpose of this trick is to give the tagger a valuable information that may be exploited during training: if a particular word form is not present in the actually used morphological dictionary, this form will crop up as an unknown word when tagging with the trained model. Even if, according to the reference

Mars

subst:sg:nom:m1	Mars	
subst:sg:nom:m2	Mars	
subst:sg:nom:m2	mars	
subst:sg:nom:m3	mars	
subst:sg:acc:m3	mars	

ofiarowuje

ign	None
fin:sg:ter:imperf	ofiarowywać

Fig. 1. NCP annotation of two tokens, the second being an unkown word

corpus, this word form is assigned a set of possible interpretations, this is practically an unknown word and the training algorithm is likely to benefit from this
information. Conversely, if a word form is unknown according to the reference
corpus, but the new version of the analyser is able to recognise it correctly, the
above procedure will update the reference corpus to account for this.

Reanalysis of training data also handles a third scenario: when a word form
is known according to both the reference corpus and the morphological analyser employed, but the sets of possible tags attached in both sources do differ.
Such situations are likely causes for tagging errors, as they let the same sentence
be represented differently when training and when using the trained model to
disambiguate. This problem is a practical one. For instance, the reference morphological analysis in NCP version 1.0 is apparently not fully consistent. For
instance, different instances of the word form TAK (*yes, so, thus*) are assigned
different sets of possible tags. There are 12 different sets appearing throughout
the whole corpus. Figure 2 presents three of them. Such inconsistencies are likely
to negatively impact the quality of the trained model.

```
tak
    adv              tak
    qub              tak
    subst:pl:gen:f taka
tak
    adv:pos          tak
    qub              tak
    subst:pl:gen:f taka
tak
    adv:pos          tak
    qub              tak
    subst:pl:gen:f taka
    conj             tak
```

Fig. 2. Three variants of reference morphological analysis of the word form tak in NCP
1.0

4   Proposed Algorithms for Tagger Training and Performance

The proposed method is based on the improved WMBT tagger. Our main modification is to substitute k-nearest neighbour classification of each sentence token
with employing a trained CRF to predict the tagging of the whole sentence at
a time. Also, as we perform no classification at token level, the distinction into
known and unknown words is used only for pre-processing (appending the list
of typical interpretations to unknown words).

4.1   Training

The training procedure as sketched below assumes that morphological analysis
has already been applied. The input is therefore tokens assigned sets of possible morphosyntactic interpretations. We operate according to *tiers* (layers): the
first one is responsible for selecting the correct value of the grammatical class,
the subsequent correspond to attributes as defined in the tagset (grammatical
categories, e.g. number, case). The action performed at each tier is to generate
training examples and reproduce the behaviour of normal tagger performance.
The latter is executed as partial disambiguation after generating the training
examples: the set of interpretations attached to a token at the moment is limited to those that have a particular value of a given tagset attribute (or the
grammatical class in case of the first tier). This particular value is taken from
the tag highlighted as contextually appropriate. This way we simulate tagger
performance under the assumption that a correct decision was made.

The algorithm is parametrised with a set of *features*. For clarity, we assume
here that each of these functions transforms the context of a token into a symbolic
value, not necessarily binary-valued. The transformation of these features into
characteristic functions according to templates defined is the job of the whole
procedure described here simply as 'train CRF with training data for tier a and
save'. Note that this is also how the actual implementation works: the CRF++
toolkit, which we use, assumes training data is a list of feature vectors (and
sentence delimiters) and during training feature templates are used to expand
the features into characteristic functions (Kudo 2005).

The training procedure is given as Algorithm 1. The procedure starts with
collecting tags that are typical for unkown words. This list is then used to aid
tagging of unkown words: each token marked in the training data as unknown
word (thanks to the reanalysis stage, Sec. 3.3) is extended with artificial interpretations, each based on a tag from the frequency list. The interpretations
added consist of a tag from the list and a lemma equal to the word form (lower-
cased). This step is referred to as 'add tags from the frequency list to *token*'.
The next stage consists in generating training examples. Each example consists
of a sequence of feature values and the class label, being the correct value of
the attribute (taken from the reference tagging). Next, partial disambiguation
is performed: the interpretations are removed where the attribute corresponding

---

Algorithm 1. Training of the WCRFT algorithm

---

gather frequency list of tags assigned to unknown words in training *corpus*
remove tags appearing less than U times from the list
for *sentence* $\in$ *corpus* **do**
    for *token* $\in$ *sentence* **do**
        if *token* is an unknown word **then**
            add tags from the frequency list to *token*
        end if
    end for
    for $a \in [class, attr_1, \dots, attr_k]$ **do**
        for *token* $\in$ *sentence* **do**
            $f\_values \leftarrow [f(token, sentence)$ **for** $f \in features(a)]$
            $decision \leftarrow$ correct value of a for *token*
            store training example $(f\_values, decision)$ for tier a
            remove interpretations from *token* with incorrect value of a
        end for
        store sentence delimiter for tier a
    end for
end for
for $a \in [class, attr_1, \dots, attr_k]$ **do**
    train CRF with training data for tier a and save
end for

---

to the current tier is assigned value other than the one inferred from the manual tagging.

4.2 Performance

Tagging is done according to Algorithm 2. An array of trained CRF models is used to disambiguate subsequent attributes. As CRF model classifies a whole sentence at a time, first a sentence representation is gathered, being essentially a list of feature vectors (a vector for each sentence token). Note that, as in WMBT, if some classifier decision would require choosing an attribute value that is not present in the list of possible interpretations attached to a token (including those added to unknown words), this decision is not made, leaving the ambiguity pending. The ambiguity may be resolved later when dealing with subsequent attributes. If any ambiguity remains at the end, an arbitrary decision is made.

5 Implementation and Feature Set

Our implementation is a direct modification of the WMBT tagger. This way we could re-use parts of the existing code, also benefitting from the same tools:

- WCCL[1] (Radziszewski et al. 2011), a toolkit and formalism for morpho-syntactic feature generation,

---

[1] Available at `http://nlp.pwr.wroc.pl/redmine/projects/joskipi/wiki/`

Algorithm 2. Disambiguation of a single sentence with the WCRFT algorithm

```
for token ∈ sentence do
    if token is an unknown word then
        add tags from the frequency list to token
    end if
end for
for a ∈ [class, attr₁, . . . , attrₖ] do
    sent_repr ← []
    for token ∈ sentence do
        f_values ← [f(token, sentence) for f ∈ features(a)]
        append f_values to sent_repr
    end for
    label_list ← classify sent_repr with CRF trained for tier a
    for (token, label) ∈ zip(sentence, label_list) do
        if label ∈ possible values of a for token then
            remove interpretations from token where value(a) ≠ label
        end if
    end for
end for
for token ∈ sentence do
    force "tagset-first" tag if multiple left
end for
```

- Corpus2[2] library (Radziszewski and Śniatowski 2011) for efficient corpus I/O, tagset and tag manipulations.

Both toolkits are distributed with Python wrappers, enabling rapid development of language processing applications.

As the underlying classifier we chose the CRF++ toolkit (Kudo 2005), which also comes with Python wrappers. Thanks to all the above, the implementation was quite straightforward.

The proposed algorithm has been implemented as a configurable tagger for positional tagsets, named WCRFT[3].

A WCRFT configuration specifies tagset to use, features defined using WCCL expressions and CRF++ feature templates for each attribute. As mentioned above, features are understood here as functions transforming the context of a token into symbolic values, not necessarily binary. These 'high-level' features are later expanded into characteristic functions for CRF via *feature templates*. First we will describe the features and then we will turn to employed feature templates.

---

[2] Available at http://nlp.pwr.wroc.pl/redmine/projects/corpus2/wiki/

[3] *Wrocław CRF Tagger* to follow the naming scheme of its predecessor, WMBT (*Wrocław Memory-Based Tagger*). The code has been released under GNU LGPL 3.0 and may be obtained from
http://nlp.pwr.wroc.pl/redmine/projects/wcrft/wiki

Both the feature set and the templates are taken exactly the same as those used for NP chunking (Radziszewski and Pawlaczek 2012), which in turn are very similar to the original WMBT definitions. Exactly the same feature set is used for all tiers. The features are as follows:

1. word form of the current token,
2. possible values of the grammatical class of the token,
3. possible values of grammatical number,
4. possible values of gender,
5. possible values of grammatical case,
6. a predicate checking if there holds a grammatical agreement of the current and the next token with respect to number, gender and case,
7. a similar predicate that checks the agreement of the previous, current and the next tokens (-1, 0, 1),
8. if the current token's orthographic form starts with an upper-case letter,
9. if it starts with lower-case letter.

Below we list the feature templates employed. We use the notation (p, f) to denote the f-th feature (according to the above enumeration) evaluated at the p-th position, relatively to the current token. Each template generates a number of characteristic functions closed over the domain of the f-th feature as well as the domain of the attribute a corresponding to the current tier. For instance, $(0, 1)$ refers to the first feature from the above list at the position 0, that is the word form of the current token. The expanded template generates tests of the form: 'if the word form of the current token equals w and the value of the current tier attribute is v', where for each word form w appearing in the training data and each value v of the current attribute a separate characteristic function is generated. Similarly, $(-1, 5)$ refers to the value of the grammatical case (feature no. 5) of the token preceding the one currently tagged (hence $p = -1$). The notation $(p_1, f_1)/(p_2, f_2)$ is a template that produces all the conjunctions of the two tests encountered in the data, one for the template (p_1, f_1) and the other one for $(p2, f_2)$.

1. Word forms: $(p, 0)$ for $p \in \{-2, -1, 0, 1, 2\}$
2. Word form bigrams: $(-1, 0)/(0, 0)$ and $(0, 0)/(1, 0)$
3. Grammatical class: $(p, 1)$ for $p \in \{-2, -1, 0, 1, 2\}$
4. Class bigrams: $(-2, 1)/(-1, 1)$, $(-1, 1)/(0, 1)$, $(0, 1)/(1, 1)$, $(1, 1)/(2, 1)$
5. Class trigrams: $(p - 1, 1)/(p, 1)/(p + 1, 1)$ for $p \in \{-2, 0, 1\}$
6. Case: $(p, 2)$ for $p \in \{-2, -1, 0, 1, 2\}$
7. Gender: $(p, 3)$ for $p \in \{-2, -1, 0, 1, 2\}$
8. Number: $(p, 4)$ for $p \in \{-2, -1, 0, 1, 2\}$
9. Agreement: $(-1, 5)$, $(0, 5)$, $(-1, 6)$, $(0, 6)$, $(1, 6)$,
10. Orthographic: $(0, 7)/(0, 8)$

Besides the above templates we used one that employed bigrams of attribute values: 'if the current token's word form is w, the value of the attribute for the current token is v_1 and the value of the attribute for the previous token is v_2'. In

the CRF++ terminology such templates are called *bigram templates*, as opposed
to unigram templates listed above (hence they test only the value of the attribute
related to the current token). Note that we use here the term *attribute* to refer
to the class label of the sequence classification tasks, which in the case of the
tagger presented here is the grammatical class or a tagset attribute, depending
on the tier. We do so to avoid confusion between class labels in general and
grammatical classes of the tokens in particular.

6   Evaluation

Radziszewski and Acedański (2012) argue that morphosyntactic taggers should
be evaluated as whole systems, that is:

1. The testing material should be turned to plain text.
2. A tagger should perform tokenisation and all the necessary steps (in case of
   the taggers evaluated here, these include morphological analysis).
3. Tagger output should be compared to the reference annotation and tokenisation of the test material.
4. Differences in tokenisation should be penalised, that is, each token from the
   reference corpus that is not explicitly present in the tagger output should be
   treated as tagger's error. This is to promote effort at getting the tokenisation right instead of tweaking with evaluation procedure to match particular
   taggers' behaviour.

This testing procedure corresponds to a statistic called *accuracy lower bound*.
The statistic is based on the assumption that we count the number of reference
tokens that are present in tagger output intact (they undergo no segmentations
changes) **and** are correctly tagged (the tagger assigned the same tag as in the
reference corpus). Accuracy lower bound is the number of such tokens divided
by the total number of tokens in the reference corpus, cf. (4).

$$Acc_{lower} = \frac{|\{i : tag(i) = ref(match(i)), i \in match\}|}{N} \tag{4}$$

The above formalisation introduces the mapping $match : N \to N$. The mapping assigns tokens output by the tagger to tokens in the reference corpus. The
mapping is only defined for those tokens from the reference corpus that have a
lexically corresponding token in tagger output, therefore $i \in match$ denotes that
the i-th token of the reference corpus undergoes no segmentation changes when
re-tagged (as observed when trying to match tagger output to the reference corpus). $tag(i)$ is used to indicate the i-th tagging of the i-th token (take i-th token
in tagger output and return its tag); $ref(i)$ denotes the tag assigned to the i-th
token in the reference corpus.

Following the suggestion of Radziszewski and Acedański (2012), we also provide the values of an additional statistic called *accuracy upper bound*, reflecting
a hypothetical upper limit of tagging accuracy. The upper bound is the virtual
tagging accuracy under the (false) assumption that each token from the reference corpus that was not explicitly present in tagger output due to unexpected

tokenisation would be correctly tagged. This statistic may be formalised as (5). The underlying idea is that some of the actual changes in tokenisation are practically more important, while others are less; the lower and upper bounds define a range where the actual tagging accuracy lies, regardless of which changes in tokenisation we decide to penalise. Note that the accuracy lower bound is the measure recommended for tagger comparison.

$$Acc_{upper} = \frac{|\{i : tag(i) = ref(match(i)), i \in match\}| + |\{i : 0 < i \leq N \wedge i \notin match\}|}{N}$$

(5)

Note that both statistics operate on the level of tokens, which include words, but also punctuation, sequences of digits etc. The accuracy figures are counted including all the tokens without any distinction between correctly tagging a word and a punctuation mark or another string.

Our evaluation has been performed using the data from NCP 1.0 (1 215 513 tokens) and 10-fold cross-validation. Strictly speaking, we used exactly the same data (including the same division into training and test folds, the same version of the morphological analyser) as used for the experiments reported by Radziszewski and Acedański (2012), hence the results are fully comparable.

We report values of accuracy lower and upper bound (Acc_{lower}, Acc_{upper}), but also, accuracy lower bound measured separately for words known and unknown to the morphological analyser (Acc_{lower}^K and Acc_{lower}^U, respectively).

Both versions of WMBT, as well as the algorithm proposed here, assume that the input has been tokenised and morphologically analysed. To obtain this structure from plain text we use the MACA software (Radziszewski and Śniatowski 2011) and the configuration named `morfeusz-nkjp-official-guesser`, as suggested in Radziszewski and Śniatowski (2011).

Table 1 shows the values of *accuracy lower bound* and other measures for the following taggers:

PANTERA with default parameter values (threshold value of 6). The tagger was run against plain text to take advantage of its heuristics to solve segmentation ambiguities.

WMBT — the basic algorithm as proposed in Radziszewski and Śniatowski (2011).

WMBT+u — the modified algorithm with handling unknown words. Also, the training data were subjected to morphological reanalysis beforehand.

WCRFT — the algorithm proposed here. Reanalysis of training data was employed as well.

The improvement over PANTERA and WMBT+u is statistically significant (as measured in accuracy lower bound, using paired *t*-test with 95% confidence). The improvement in accuracy lower bound in comparison with both WMBT variants is mirrored in accuracy upper bound increase, which was expected as the same toolchain was employed for tokenisation and morphological analysis. It may be noticed that the proposed CRF-based algorithm exhibits slightly higher error rate for unknown words than the WMBT+u.

Table 1. Accuracy measures obtained during evaluation on NCP 1.0

Tagger	Acc_{lower}	Acc_{upper}	Acc_{lower}^{K}	Acc_{lower}^{U}
WMBT	87.50%	87.82%	89.78%	13.57%
PANTERA	88.79%	89.09%	91.08%	14.70%
WMBT+u	89.71%	90.04%	91.20%	41.45%
WCRFT	90.34%	90.67%	91.89%	40.13%

The performance of the tagger is about 1.5 times faster that that of WMBT+u. We recorded 269 tokens per second for WMBT+u and 408 tokens per second for WCRFT on an Intel Core i7 machine. These figures does not include using MACA, although its overhead is minor (the employed MACA configuration results in over 18k tokens output per second).

The training of WCRFT is unfortunately much slower: training ten models (one for each training fold) using three concurrent processes on a server with Intel Xeon 2.67GHz CPU took 7 days.

7   Conclusion

We have shown that application of Conditional Random Fields together with tiered tagging and a rich feature set allows to achieve improvement on state-of-the-art results in tagging Polish. The practical value of the work presented here lies in the implemented tagger, WCRFT. The tagger will be used internally in the SyNaT project, but is also publicly available under GNU LGPL. What is more, a dedicated web-service is being developed at the moment, which will facilitate using the tagger.

A practical disadvantage of the tagger is its relatively low tagging speed. Most likely it could be made considerably faster by merging together tiers corresponding to some attribites. Initial experiments show that the training data for a couple of less important attributes are almost 'empty', that is only a small percentage of tokens assign any value to the attributes. This is due to the characteristics of the tagset: the set of attributes whose value must be given depends on the grammatical class. What is more, many of the ambiguities are already resolved after dealing with grammatical class, number, gender and case (this was the reason why disambiguation in TaKIPI stopped here, remaining infrequent ambiguities at output). This intuition could be exploited to boost the tagger performance by reducing the number of tiers.

It would also be interesting to test the tagger on data from another Slavic languages, e.g. Croatian or Slovene. We consider adding support for MULTEXT-East tagsets (Erjavec 2012) to Corpus2 library for this purpose.

Acknowledgement. This work was financed by the National Centre for Research and Development (NCBiR) project SP/I/1/77065/10 ("SyNaT").

References

Acedański, S.: A Morphosyntactic Brill Tagger for Inflectional Languages. In: Loftsson, H., Rögnvaldsson, E., Helgadóttir, S. (eds.) IceTAL 2010. LNCS, vol. 6233, pp. 3–14. Springer, Heidelberg (2010)

Brill, E.: A simple rule-based part of speech tagger. In: Proceedings of the Third Conference on Applied Natural Language Processing, pp. 152–155. Association for Computational Linguistics, Morristown (1992)

Cohn, T.: Scaling conditional random fields for natural language processing. PhD thesis, Department of Computer Science and Software Engineering, University of Melbourne, Australia (2007)

Erjavec, T.: MULTEXT-East: morphosyntactic resources for Central and Eastern European languages. Language Resources and Evaluation 46(1), 131–142 (2012)

Hajič, J., Krbec, P., Květoň, P., Oliva, K., Petkevič, V.: Serial combination of rules and statistics: A case study in Czech tagging. In: Proceedings of the 39th Annual Meeting on Association for Computational Linguistics, pp. 268–275. Association for Computational Linguistics (2001)

Kudo, T.: CRF++: Yet another CRF toolkit (2005), User's manual and implementation available at http://crfpp.googlecode.com/svn/trunk/doc/index.html

Kuta, M.: Tagging and Corpus based Methods for improving Natural Language Processing of Polish. PhD thesis, Wydział Elektrotechniki, Automatyki, Informatyki i Elektroniki, Akademia Górniczo-Hutnicza, Kraków (2010)

Lafferty, J., McCallum, A., Pereira, F.: Conditional random fields: Probabilistic models for segmenting and labeling sequence data. In: Proceedings of the Eighteenth International Conference on Machine Learning, ICML 2001 (2001)

Lehnen, P., Hahn, S., Ney, H., Mykowiecka, A.: Large-scale Polish SLU. In: Interspeech, Brighton, UK, pp. 2723–2726 (2009)

Marcińczuk, M., Janicki, M.: Optimizing CRF-Based Model for Proper Name Recognition in Polish Texts. In: Gelbukh, A. (ed.) CICLing 2012, Part I. LNCS, vol. 7181, pp. 258–269. Springer, Heidelberg (2012)

Piasecki, M., Godlewski, G.: Effective Architecture of the Polish Tagger. In: Sojka, P., Kopeček, I., Pala, K. (eds.) TSD 2006. LNCS (LNAI), vol. 4188, pp. 213–220. Springer, Heidelberg (2006)

Przepiórkowski, A.: The IPI PAN Corpus: Preliminary version. Institute of Computer Science, Polish Academy of Sciences, Warsaw (2004)

Przepiórkowski, A.: The IPI PAN Corpus in numbers. In: Vetulani, Z. (ed.) Proceedings of the 2nd Language & Technology Conference, Poznań, Poland (2005)

Przepiórkowski, A.: A comparison of two morphosyntactic tagsets of Polish. In: Koseska-Toszewa, V., Dimitrova, L., Roszko, R. (eds.) Representing Semantics in Digital Lexicography: Proceedings of MONDILEX Fourth Open Workshop, Warsaw, pp. 138–144 (2009)

Przepiórkowski, A., Woliński, M.: A flexemic tagset for Polish. In: Proceedings of Morphological Processing of Slavic Languages, EACL 2003 (2003)

Przepiórkowski, A., Górski, R.L., łaziński, M., Pęzik, P.: Recent developments in the National Corpus of Polish. In: Proceedings of the Seventh International Conference on Language Resources and Evaluation, LREC 2010, Valletta, Malta. ELRA (2010)

Przepiórkowski, A., Murzynowski, G.: Manual annotation of the National Corpus of Polish with Anotatornia. In: Goźdź Roszkowski, S. (ed.) The Proceedings of Practical Applications in Language and Computers, PALC 2009, Frankfurt, Germany. Peter Lang (2009)

Przepiórkowski, A., Woliński, M.: The unbearable lightness of tagging: A case study in morphosyntactic tagging of Polish. In: Proceedings of the 4th International Workshop on Linguistically Interpreted Corpora (LINC 2003), EACL 2003 (2003)

Radziszewski, A.: Treatment of unknown words in WMBT. Wrocław University of Technology (2012),
http://nlp.pwr.wroc.pl/redmine/projects/wmbt/wiki/Guessing

Radziszewski, A., Acedański, S.: Taggers Gonna Tag: An Argument against Evaluating Disambiguation Capacities of Morphosyntactic Taggers. In: Sojka, P., Horák, A., Kopeček, I., Pala, K. (eds.) TSD 2012. LNCS, vol. 7499, pp. 81–87. Springer, Heidelberg (2012)

Radziszewski, A., Pawlaczek, A.: Large-Scale Experiments with NP Chunking of Polish. In: Sojka, P., Horák, A., Kopeček, I., Pala, K. (eds.) TSD 2012. LNCS, vol. 7499, pp. 143–149. Springer, Heidelberg (2012)

Radziszewski, A., Śniatowski, T.: Maca — a configurable tool to integrate Polish morphological data. In: Proceedings of the Second International Workshop on Free/Open-Source Rule-Based Machine Translation (2011)

Radziszewski, A., Śniatowski, T.: A memory-based tagger for Polish. In: Proceedings of the 5th Language & Technology Conference, Poznań (2011)

Radziszewski, A., Wardyński, A., Śniatowski, T.: WCCL: A morpho-syntactic feature toolkit. In: Proceedings of the Balto-Slavonic Natural Language Processing Workshop. Springer (2011)

Sutton, C., McCallum, A.: An introduction to conditional random fields. In: Foundations and Trends in Machine Learning (2011)

Tufiş, D.: Tiered Tagging and Combined Language Models Classifiers. In: Matoušek, V., Mautner, P., Ocelíková, J., Sojka, P. (eds.) TSD 1999. LNCS (LNAI), vol. 1692, pp. 28–33. Springer, Heidelberg (1999)

Vidová-Hladká, B.: Czech Language Tagging. PhD thesis, Uniwersytet Karola, Wydział Matematyki i Fizyki, Praga (2000)

Wallach, H.M.: Conditional random fields: An introduction. Technical Report MS-CIS-04-21, Department of Computer and Information Science, University of Pennsylvania, USA (2004)

Liner2 — A Customizable Framework for Proper Names Recognition for Polish

Michał Marcińczuk, Jan Kocoń, and Maciej Janicki

Wrocław University of Technology, Wrocław, Poland
michal.marcinczuk@pwr.wroc.pl, janekkocon@gmail.com, macjan@o2.pl

Abstract. In the paper we present a customizable and open-source framework for proper names recognition called Liner2. The framework consists of several universal methods for sequence chunking which include: dictionary look-up, pattern matching and statistical processing. The statistical processing is performed using Conditional Random Fields and a rich set of features including morphological, lexical and semantic information. We present an application of the framework to the task of recognition proper names in Polish texts (5 common categories of proper names, i.e. *first names*, *surnames*, *city names*, *road names* and *country names*). The Liner2 framework was also used to train an extended model to recognize 56 categories of proper names which was used to bootstrap the manual annotation of KPWr corpus. We also present the CRF-based model integrated with a heterogeneous named entity similarity function. We show that the similarity function added to the best configuration improved the final result for cross-domain evaluation. The last section presents NER-WS — a web service for proper names recognition in Polish texts utilizing the Liner2 framework and the model for 56 categories of proper names. The web service can be tested using a web-based demo available at `http://nlp.pwr.wroc.pl/inforex/`.

Keywords: Proper Names Recognition, Information Extraction, Liner2, Proper Name Similarity Function, NER Web Service.

1 Introduction

Named entity recognition (NER) is one of the information extraction tasks. The goal of NER is to identify and classify multiword expressions in running text which refer to some objects from the world. Proper names are one of the classes of named entities. Proper names always denote unique (to some extent, because proper name can be also ambiguous) objects from the word.

Robust recognition of proper names is very important in many tasks from the natural language processing (NLP) domain, i.e. information extraction from medical documentation [1], text anonymization [2], machine translation [3] or forensic linguistics [4].

Recognition of named entities is a well studied task in the case of English [5]. However, only a few researches were conducted so far for Polish. Works for other

R. Bembenik et al. (Eds.): *Intell. Tools for Building a Scientific Information*, SCI 467, pp. 231–253.
DOI: 10.1007/978-3-642-35647-6_17   © Springer-Verlag Berlin Heidelberg 2013

Slavic languages are not numerous either; Czech [6], Bulgarian [7] and Ukrainian [8]. Majority of approaches to NER for Polish are based on manual construction of rules, cf., applications in [1, 9–13], machine anonymization [2] and machine translation [3]. Only a few preliminary works were presented on the application of Machine Learning methods to NER, e.g. Memory Based Learning in [14], Decision Trees C4.5 and Naïve Bayes in [15], Conditional Random Fields in [16].

NER is not a trivial task and cannot be effectively performed using naive methods like simple dictionary look-up or pattern matching (see [17]). However, usage of external language resources can be very important to make the NER robust. In the paper we present a NER method which combines statistical approach with various language resources like wordnet, lexicons (proper names and keywords), rules and tools like similarity function for proper names. As the result of the research we constructed a customizable framework for NER for Polish called Liner2, which is also presented in the paper.

The paper is organised as following. Section 2 contains the description of Liner2 framework including its architecture, available modes, supported feature categories and chunkers. Section 3 presents a lexicon of proper names called NELexicon. In section 4 we present two applications of Liner2 framework and NELexicon. In section 5 we present and discuss an approach to improve the lexicon recall by applying a similarity function for proper names. Section 6 presents a web service for proper name recognition build on top of the Liner2 framework.

2   Liner2 Framework

This section presents the technical and usability aspects of Liner2 framework. The technical aspect refers to the framework architecture and division into components (Section 2.1). The usability aspect refers to available functions and configuration options (Sections from 2.2 to 2.5).

2.1   Architecture

Liner2 is divided into 5 components (see Figure 1). They are:

- **main controller** — it controls the data flow between components. The controller behavior depends on Liner2 running mode (see Section 2.2),
- **data reader** — reads the data from file or standard input and transforms them into internal structure (see Section 2.3),
- **feature space generator** — extends the feature space by generating new features according to given configuration (see Section 2.4),
- **chunker controller** — creates a set of chunkers arranged into a pipeline (see Section 2.5),
- **result writer** — writes processed data to a file or computes and prints evaluation of the chunking result (see Section 2.3).

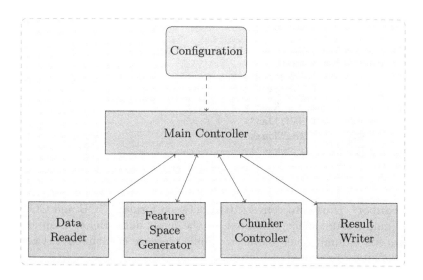

Fig. 1. Liner2 framework architecture

2.2 Modes

Liner2 can be run in several modes suitable for different applications. Single document can be easily processed in `pipe` mode. Multiple documents and interactive processing can be done in `batch` mode. Feature generation can be done in `convert` mode. Quick evaluation can be performed using `eval` or `evalcv` mode. A brief summary of all modes is presented in Table 1. The following subsections describe them in details.

Table 1. List of available Liner2 running modes

Mode	Brief description
batch	console interactive mode
convert	convert text from one format to another
eval	perform evaluation on given data
evalcv	perform n-fold cross validation on given data
pipe	processe single document in pipe-style
train	construct and serialize chunkers

Mode *batch* is used to process multiple documents, test Liner2 in interactive manner and run Liner2 as subprocess. After loading data user is prompted to enter text to process. The text can be entered in CCL, IOB and plain format (see Section 2.3). For plain format Liner2 requires *maca* for morphological analysis and optionally *wmbt* for tagging. Below is a sample interaction with Liner2 in the `batch` mode:

```
> $liner2 batch -ini model.ini -i plain -o tuples -maca -
# Loading, please wait...
# Enter a sentence and press Enter.
# To finish, enter 'EOF'.
> Jan Nowak mieszka w Krakowie.
(0,2,PERSON_FIRST_NAM,"Jan")
(3,7,PERSON_LAST_NAM,"Nowak")
(16,23,CITY_NAM,"Krakowie")
```

Mode *convert* is used to convert data between formats (IOB and CCL) and also to dump feature space generated for the input data. Below is an example of feature space dump generated for a sample sentence (`features.txt` contains feature definition presented in Figure 2).

Input (content of *input.xml* file):

```xml
<?xml version="1.0" encoding="UTF-8"?>
<!DOCTYPE cesAna SYSTEM "xcesAnaIPI.dtd">
<chunkList xmlns:xlink="http://www.w3.org/1999/xlink">
<chunk id="ch1" type="p">
  <sentence>
  <tok>
   <orth>Pan</orth>
   <lex disamb="1"><base>Pan</base><ctag>subst:sg:nom:m1</ctag></lex>
   <lex disamb="1"><base>pan</base><ctag>subst:sg:nom:m1</ctag></lex>
  </tok>
  <tok>
   <orth>Marek</orth>
   <lex disamb="1"><base>Marek</base><ctag>subst:sg:nom:m1</ctag></lex>
   <lex disamb="1"><base>marek</base><ctag>subst:sg:nom:m1</ctag></lex>
  </tok>
  <tok>
   <orth>Groszek</orth>
   <lex disamb="1"><base>groszek</base><ctag>subst:sg:nom:m3</ctag></lex>
  </tok>
  <tok>
   <orth>jest</orth>
   <lex disamb="1"><base>być</base><ctag>fin:sg:ter:imperf</ctag></lex>
  </tok>
  <tok>
   <orth>prezesem</orth>
   <lex disamb="1"><base>prezes</base><ctag>subst:sg:inst:m1</ctag></lex>
  </tok>
  <tok>
   <orth>Company</orth>
   <lex disamb="1"><base>company</base><ctag>subst:sg:nom:m1</ctag></lex>
  </tok>
  <tok>
   <orth>SA</orth>
   <lex disamb="1"><base>sa</base><ctag>subst:sg:nom:f</ctag></lex>
  </tok>
  </sentence>
</chunk>
</chunkList>
```

Command:

```
liner2 convert -i ccl -f input.xml -o iob -ini features.txt
```

Output:

```
-DOCSTART CONFIG FEATURES orth base ctag syn hyp1 class case number gender pattern
-DOCSTART FILE ch1
Pan      Pan     subst:sg:nom:m1    Pan      Pan     subst nom  sg m1   UPPER_INIT 0
Marek    Marek   subst:sg:nom:m1    Marek    Marek   subst nom  sg m1   UPPER_INIT 0
Groszek  groszek subst:sg:nom:m3    groszek  groszek subst nom  sg m3   UPPER_INIT 0
jest     być     fin:sg:ter:imperf  być      być     fin   null sg null ALL_LOWER  0
prezesem prezes  subst:sg:inst:m1   prezes   głowa   subst inst sg m1   ALL_LOWER  0
Company  company subst:sg:nom:m1    company  company subst nom  sg m1   UPPER_INIT 0
SA       sa      subst:sg:nom:f     sa       sa      subst nom  sg f    ALL_UPPER  0
```

Mode *eval* is used to evaluate the performance of a model on a given corpus.

Command:

```
liner2 eval -i ccl -f input.xml -ini features.txt \
    -chunker c1:CHUNKER -use c1
```

Output (fragments):

```
(...skipped...)

Sentence #4924 from /nlp/corpora/pwr/wikinews4//annotated/0099843.txt

Text  : Budynek powstaje przy ulicy Władysława Reymonta w pobliżu skrzyżowania
        z  ulicą Józefa Chłopickiego ,  naprzeciwko cmentarza żydowskiego .
Tokens: 1_____ 2_____ 3___ 4____ 5_____ 6_____ 7 8_____ 9_____
        10 11___ 12____ 13_____ 14 15_____ 16_____ 17_____ 18

Chunks:
  TruePositive ROAD_NAM [5,6] = Władysława Reymonta
  TruePositive ROAD_NAM [12,13] = Józefa Chłopickiego
  FalsePositive PERSON_FIRST_NAM [5,5] = Władysława
  FalsePositive PERSON_LAST_NAM [6,6] = Reymonta
  FalsePositive PERSON_FIRST_NAM [12,12] = Józefa
  FalsePositive PERSON_LAST_NAM [13,13] = Chłopickiego

(...skipped...)

Annotation       &  TP &  FP &  FN & Precision & Recall  & F$_1$  \\
\hline
ROAD_NAM         &  11 &  10 &  45 &  52,38% &  19,64% &  28,57% \\
PERSON_LAST_NAM  & 824 &  30 & 813 &  96,49% &  50,34% &  66,16% \\
COUNTRY_NAM      & 1504 & 92 & 251 &  94,24% &  85,70% &  89,76% \\
PERSON_FIRST_NAM & 825 &  28 & 287 &  96,72% &  74,19% &  83,97% \\
CITY_NAM         & 492 & 166 & 173 &  74,77% &  73,98% &  74,38% \\
\hline
*TOTAL*          & 3656 & 326 & 1569 &  91,81% &  69,97% &  79,42%
```

Mode *evalcv* is used to evaluate the performance of a model using n-fold cross validation on given set of documents. The documents must be a priori divided into n folds. The output contains evaluation for every fold and summary for all folds. The results are presented in the same way as in mode `eval`.

Mode *pipe* is used to process a single document. The document can be read form a file (CCL, IOB or plain format) or from standard input and saved to a file (CCL, IOB or tuple format) or printed to standard output.

Mode *train* is used to train and serialize chunkers defined with the `-chunker` parameter.

2.3 Input/Output Formats

Liner2 handles following data formats:

Plain text can be processed by Liner2 after providing the -maca parameter. The parameter value can be: - if *maca* is installed in the system or PATH to a `maca-analyze` program.

CCL is an XML-based format used by the *maca* tool [18]. Document can be read and write in CCL format.

```xml
<?xml version="1.0" encoding="UTF-8"?>
<!DOCTYPE cesAna SYSTEM "xcesAnaIPI.dtd">
<chunkList xmlns:xlink="http://www.w3.org/1999/xlink">
  <chunk id="ch1" type="p">
    <sentence>
     <tok>
        <orth>Ala</orth>
        <lex disamb="1"><base>Ala</base><ctag>subst:sg:nom:f</ctag></lex>
     </tok>
     <tok>
        <orth>mieszka</orth>
        <lex disamb="1"><base>mieszkać</base><ctag>fin:sg:ter:imperf</ctag></lex>
     </tok>
     <tok>
        <orth>w</orth>
        <lex disamb="1"><base>w</base><ctag>prep:loc:nwok</ctag></lex>
     </tok>
     <tok>
        <orth>Krakowie</orth>
        <lex disamb="1"><base>Kraków</base><ctag>subst:sg:loc:m3</ctag></lex>
     </tok>
     <ns/>
     <tok>
        <orth>.</orth>
        <lex disamb="1"><base>.</base><ctag>interp</ctag></lex>
     </tok>
    </sentence>
  </chunk>
</chunkList>
```

IOB is a format used in CoNLL-2003 shared task named entity recognition[1]. In this format every token is represented as a set of space-seperated values in a single row. Sentences are separated with empty lines. Every chunk is encoded using two symbols: `B-NAME` and `I-NAME`, where `NAME` is a chunk category, `B-NAME` (*begins*) is assigned to the first token and `I-NAME` (*inside*) to the other tokens of the chunk. Tokens which do not belong to any chunk are marked with `O` symbol (*outside*).

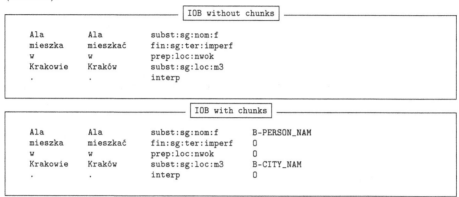

```
┌──────────────────────────  IOB without chunks  ──────────────────────────┐
│                                                                          │
│    Ala         Ala         subst:sg:nom:f                                 │
│    mieszka     mieszkać     fin:sg:ter:imperf                             │
│    w           w           prep:loc:nwok                                  │
│    Krakowie    Kraków      subst:sg:loc:m3                                │
│    .           .           interp                                         │
│                                                                          │
└──────────────────────────────────────────────────────────────────────────┘
```

```
┌──────────────────────────  IOB with chunks  ──────────────────────────┐
│                                                                        │
│    Ala         Ala         subst:sg:nom:f        B-PERSON_NAM           │
│    mieszka     mieszkać     fin:sg:ter:imperf     O                     │
│    w           w           prep:loc:nwok         O                      │
│    Krakowie    Kraków      subst:sg:loc:m3       B-CITY_NAM             │
│    .           .           interp                O                       │
│                                                                        │
└────────────────────────────────────────────────────────────────────────┘
```

Tuple format is used to display result of chunking in a compact way.

```
┌──────────────────────────  Tuple format  ──────────────────────────┐
│                                                                    │
│    (1,3,PERSON_NAM,"Ala")                                          │
│    (11,18,CITY_NAM,"Krakowie")\end{verbatim}                       │
│                                                                    │
└────────────────────────────────────────────────────────────────────┘
```

2.4   Feature Generator

Liner2 contains a stand-alone feature generator which can generate 21 types of features. The features can be grouped into four categories, i.e. orthographic, morphological, lexicon-based and wordnet-based features.

1. **Orthographic features**
   - **orth** — a word itself, in the form in which it is used in the text,
   - *n*-**prefix** — *n* first characters of the encountered word form, where $n \in \{1, 2, 3, 4\}$. If the word is shorter than *n*, the missing characters are replaced with '_'.
   - *n*-**suffix** — *n* last characters of the encountered word, where $n \in \{1, 2, 3, 4\}$. If the word is shorter than *n*, the missing characters are replaced with '_'. We use prefixes to fill the gap of missing inflected forms of proper names in the gazetteers.
   - **pattern** — encode pattern of characters in the word:
     - ALL_UPPER — all characters are upper case letters, e.g. "NASA",

---

[1] Web page: `http://www.cnts.ua.ac.be/conll2003/ner/`

- ALL_LOWER — all characters are lower case letters, e.g. "rabbit"
- DIGITS — all character are digits, e.g. "102",
- SYMBOLS — all characters are non alphanumeric, e.g. "-_-"',
- UPPER_INIT — the first character is upper case letter, the other are lower case letters, e.g. "Andrzej",
- UPPER_CAMEL_CASE — the first character is upper case letter, word contains letters only and has at least one more upper case letter, e.g. "CamelCase",
- LOWER_CAMEL_CASE — the first character is lower case letter, word contains letters only and has at least one upper case letter, e.g. "pascalCase",
- MIXED — a sequence of letters, digits and/or symbols, e.g. "H1M1".

  - **binary orthographic features**, the feature is 1 if the condition is met, 0 otherwise. The conditions are:
    - *(word) starts with an upper case letter,*
    - *starts with a lower case letter,*
    - *starts with a symbol,*
    - *starts with a digit,*
    - *contains upper case letter,*
    - *contains a lower case letter,*
    - *contains a symbol*
    - *contains digit.*

    The features are based on filtering rules described in [17], e.g., first names and surnames start from upper case and do not contain symbols. To some extent these features duplicate the *pattern* feature. However, the *binary features* encode information on the level of single characters, while the aim of the *pattern* feature is to encode a repeatable sequence of characters.

2. **Morphological features** — are motivated by the NER grammars which utilise morphological information [9]. The features are:
   - **base** — a morphological base form of a word,
   - **ctag** — morphological tag generated by tagger,
   - **part of speech**, **case**, **gender**, **number** — enumeration types according to tagset described in [19].

3. **Lexicon-based features** — one feature for every lexicon. If a sequence of words is found in a lexicon the first word in the sequence is set as B and the other as I. If word is not a part of any dictionary entry it is set to O.

4. **Wordnet-base features** — are used to generalise the text description and reduce the observation diversity. The are two types of these features:
   - **synonym** — word's synonym, first in the alphabetical order from all word synonyms in Polish Wordnet. The sense of the word is not disambiguated,
   - **hypernym** n — a hypernym of the word in the distance of n, where $n \in \{1, 2, 3\}$

The feature space is defined using a set of `-feature` parameters. The generated features can be used by any chunker, like CRF or rule-based (see Section 2.5 for chunkers description). Figure 2 presents a sample definition of feature space.

```
-feature orth
-feature base
-feature ctag
-feature syn
-feature hyp1
-feature class
-feature case
-feature number
-feature gender
```

Fig. 2. Sample set of features definition

```
-template t1:orth:-2:-1:0:1:2
-template t1:base:-2:-1:0:1:2
-template t1:syn:-2:-1:0:1:2
-template t1:hyp1:-2:-1:0:1:2
-template t1:hyp2:-2:-1:0:1:2
-template t1:hyp3:-2:-1:0:1:2
-template t1:class:-1:0:1
-template t1:case:0
-template t1:gender:-2:-1:0:1:2
```

Fig. 3. Sample template definition

2.5   Sequence Chunking Methods

Liner2 offers several methods for sequence chunking, which can be divided into five categories:

1. **Supervised statistical models** — perform chunking on the basis of statistial model created on a training corpus.
2. **Rule-based chunkers** — perform chunking on the basis of rules.
3. **Gazetteer-based chunkers** — perform chunking on the basis of lexicons.
4. **Iterative** — the chunking process is performed in several iterations. The result from one iteration is used in the next iteration.
5. **Ensamble** — perform majority voting or union of other chunkers.

Most of the chunkers can be defined from command line or in configuration file and do not require any modifications in the source code. The only exception is the *heuristic* chunker, which requires to encode the rules as Java functions. Table 2 presents a summary of all available chunkers and the following subsections describe them in details.

The chunkers are defined using the -`chunker` parameter as following:

```
-chunker chunker_name:chunker_description
```

where `chunker_name` is an unique chunker name among all defined chunkers. The `chunker_description` defines the chunker. Format descriptions for all chunkers are presented in the following subsections.

Table 2. List of available sequence chunking methods

Type	Symbol	Brief description
statistical	`crfpp-train`	train and serialize a CRF model
statistical	`crfpp-load`	deserialize an existing CRF model
rule-based	`heuristic`	set of rules defined as Java functions
rule-based	`wccl`	rules written in WCCL
lexicon-based	`dict-compile`	compile and serialize an unambigous dictionary look-up
lexicon-based	`dict-load`	deserialize an unambigous dictionary look-up
lexicon-based	`dict-full-compile`	compile and serialize a full dictionary look-up
lexicon-based	`dict-full-load`	deserialize a full dictionary look-up
iterative	`adu`	run given chunker twice, after first iteration update lexicon-base features
ensamble	`ensamble`	combine chunker results by applying a majority voting or merging

Supervised Statistical Chunker

```
crfpp-train:p=P:template=T:(iob|ccl|data)=DATA:model=FILE
crfpp-load:PATH
```

Liner2 offers a trainable CRF-based chunker. The chunker utilises an existing open source implementation of the Conditional Random Fields called CRF++[2]. The chunker can be defined in two ways: `crfpp-train` — trains the model from scratch or `crfpp-load` — loads existing model.

Training model from scratch

```
crfpp-train:p=P:template=T:(iob|ccl|data)=DATA:model=FILE
```

where:

- **P** is a number of threads used to train the model (this parameters is passed to the CRF++),

---

[2] Web page: `http://crfpp.googlecode.com/svn/trunk/doc/index.html`

- **T** is a name of features template used to train the model. The feature template is defined in the configuration file and is transformed to a format accepted by CRF++ (see Figure 3),
- **DATA** is a path to a corpus used to train. The corpus can be defined directly by providing path to a file (`iob=PATH` or `ccl=PATH`) or by a name defined with `-corpus` parameter (`data=NAME`).
- **FILE** is a file name where the trained model will be saved. The saved model can be loaded using the `crfpp-load` chunker.

Loading existing model

```
crfpp-load:PATH
```

where **PATH** is a path to a serialized CRF model trained and saved with `crfpp-train` chunker.

Rule-based

```
heuristic:comma-separated-list-of-rules
```

This module chunks the text using Java-coded rules. Every rule is implemented as a Java function. The module requires changes in the Liner2 source code. The rules have access to complete feature space defined with the `-feature` parameters (see Section 2.4).

Sample rules implemented in the Liner2:

- **city** — set of rules which recognize city-related patterns, e.g. "*postal_code upper_case*",
- **general-ign-dict** — if given token's part of speech is *ign* (unknown) and any of the lexicon-based features has value *B* then mark the token with a symbol refering to the lexicon name,
- **general-camel-base** — if given token starts from an upper case letter and is marked by one lexicon-based feature as *B* then mark the token with a symbol refering to the lexicon name,
- **person** — set of rules which recognize patterns like "Mr. *first_name surname*", "J. K. *surname*", etc.,
- **road** — if given token starts from upper case letter, previous token is present in lexicon of road name prefixes and following token is number, then mark the token as road name.
- **road-prefix** — if given token is present in lexicon of road names and previous token is present in lexicon of road name prefixes then mark the token as road name,

```
wccl:WCCL_FILE
```

This chunker requires an external tool called *wccl*[3]. The chunker creates annotations on the basis of rules written in WCCL format. A sample rule, that mark person name is presented below.

```
apply(
  match(
    optional(
      is('person_position_full'),
      optional( text(',') ),
    ),
    optional( is('person_title') ),
    is('person_first_nam_gaz'),
    optional( is('person_first_nam_gaz') ),
    is('person_last_nam_gaz'),
    optional( is('person_suffix') )
  ),
  actions(
    mark(:3, 'PERSON_FIRST_NAM'),
    mark(:4, 'PERSON_FIRST_NAM'),
    mark(:5, 'PERSON_LAST_NAM'),
    mark(:3, :5, 'PERSON_NAM')
  )
)
```

Lexicon-based — includes two types of chunkers: dict and dict-full. dict marks all unambiguous proper names that are not common words. dict-full marks every proper name that is present in the lexicon.

```
dict-compile:dict=DICT:common=COMMON:model=BIN
dict-load:BIN
```

where:

- **DICT** is a path to a lexicon of proper names,
- **COMMON** is a path to a list of common words,
- **BIN** is a path to a file where the compiled dictionary will be saved.

dict-compile is used to compile a lexicon from scratch and save it in a file. dict-load is used to load compiled lexicon from a file.

```
dict-full-compile:dict=DICT:model=BIN
dict-full-compile:dict=DICT
dict-full-load:BIN
```

---

[3] Web page: http://nlp.pwr.wroc.pl/redmine/projects/joskipi/wiki.

3 Lexicon of Proper Names

For the need of lexicon-based chunkers we have prepared a large lexicon of proper names called NELexicon[4]. We have processed several resources from the Internet containing lists of proper names of different categories. The NELexicon contains 1.4 millions unique entries "name:category" and 1.37 millions of unique names among all categories. Detailed statistics of NELexicon are presented in Table 3.

Table 3. Statistics of proper names in NELexicon

Count	Annotation	Count	Annotation	Count	Annotation
866	bay	8519	inst_org	371378	person_last
10	bridge	1930	island	61	political_
163	canal	2052	lake		party
749	cape	1358	landscape	221	powiat
259	cave	3	mausoleum	385	province
107285	city	660	mountain	2725	river
14	coast	460	mountain	40862	road
415970	company	516	nation	131	sea
48	continent	65	oasis	1469	square
108	desert	11	ocean	272	strait
2	district	419039	organization	1137	subdivision
115	fortification	322	park	72	swamp
46	glacier	154	peninsula	47	temple
8	graveyard	66	periodic	22	title
327	institution	22434	person_first	71	waterfall

4 Liner2 Applications

4.1 Recognition of Basic Proper Names

Liner2 has already been used to construct and evaluate a model for recognition 5 basic categories of proper names in Polish texts, i.e. first names, surnames, names of cities, countries and roads. The model was evaluated in two ways — single domain evaluation following 10-fold cross validation on a Corpus of Stock Exchange Reports (CSER) and cross-domain evaluation on a Corpus of Economic News from Wikinews (CEN). The model is a cascade of following chunkers (Figure 4 presents the chunkers configuration):

- CRF-based tagger (the complete list of features is presented in [20] and [21]),
- unambiguous dictionary look-up for country and city names,
- set of Java-coded rules.

---

[4] Web page: http://nlp.pwr.wroc.pl/en/tools-and-resources/nelexicon

```
-ini {INI_PATH}/features.ini
-chunker c1:crfpp-load:{INI_PATH}/crf_cicling.bin
-chunker c2:dict-load:{INI_PATH}/dict.bin:types=COUNTRY_NAM,CITY_NAM
-chunker c3:heuristic:person,road,road-prefix,city
-chunker en:ensamble:c1+c2+c3
-chunker f:adu:en
-use en
```

Fig. 4. Liner2 configuration for proper name recognition model

In [20] we presented a basic model and in [21] we proposed and evaluated several ways of improving the performance of the model. The final configuration including all optimizations obtained in total 95.08% of precision and 96.07% of recall what is very high result (see Table 4 for detailed results). In the cross-domain evaluation the model obtained lower results on the level of 91.44% of precision and 70.53% of recall (see Table 5 for detailed results), but still acceptable performance.

Table 4. 10-fold cross validation on CSER

Annotation	TP	FP	FN	Precision	Recall	F-measure
first names	654	13	29	98.05%	95.75%	96.89%
surnames	636	17	52	97.40%	92.44%	94.85%
countries	441	36	32	92.45%	93.23%	92.84%
cities	1947	128	34	93.83%	98.28%	96.01%
roads	377	16	19	95.93%	95.20%	95.56%
All	4055	210	166	95.08%	96.07%	95.57%

4.2 Proper Names Bootstrapping

Liner2 was also used to support the process of manual annotation of KPWr corpus [22] with 56 categories of proper names. The model was trained on 400 documents manually annotated by linguists using the configuration used for the basic model. The extended model was then run on another 400 unannotated documents. The results of automatic proper names annotation were manually verified by linguists. The verification was performed using a dedicated perspective in the Inforex system [23] (see Figure 5). For every recognized annotation linguist could choose to accept the annotation, change the proper name category (if the annotation boundary was recognized properly but the category was incorrect), discard the annotation (if both the annotation boundary and category were incorrect) or leave the evaluation for later (for example, if not sure at the moment).

Table 5. Cross domain evaluation on CEN

Annotation	TP	FP	FN	Precision	Recall	F-measure
first names	827	27	285	96.84%	74.37%	84.13%
surnames	831	31	806	96.40%	50.76%	66.51%
countries	1524	92	231	94.31%	86.84%	90.42%
cities	492	186	173	72.57%	73.98%	73.27%
roads	11	9	45	55.00%	19.64%	28.95%
All	3685	345	1540	91.44%	70.53%	79.63%

The results of verification were stored in the database and used to measure the performance of bootstrapping. By applying this procedure we added more than 1700 new annotations. Evaluating the correctness of annotation boundaries and proper name categories, the model obtained 71% of precision and 42% of recall. When considering only the correctness of annotation boundaries, the model obtained 93% of precision and 54% of recall. This means, than more than half of annotations were automatically added to the documents and 71% of them had correct category. Only the other 30% required correction of the category and 46% of annotations had to be added manually from scratch.

5   Improving the Lexicon Recall

To improve the recall of lexicon-based features were applied a similarity function for named entities. The following subsections present the concept of heterogeneous named entity similarity function, the ways how the function could be combined with existing models for proper names recognition and evaluation of the presented approaches.

5.1   Introduction

The basic application of the heterogeneous named entity similarity function (henceforth *NamEnSim*) is to find for an input word w_I its lemma or an morphological word form in *NELexicon* if it exists. In practice, similar word set P returned by similarity function for w_I should contain its proper lemma w_L and the decision function value for all other pairs: $\langle w_I, w_O \rangle$ where $w_O \in P$ and $w_O \neq w_L$ should be lower than for $< w_I, w_L >$ [24]. This task is different than morphological guessing, e.g. [25] which is aimed at generation of lemmas for unknown words on the basis of an *a tergo* index.

Named entities similarity function is based on several string metrics, combined into a complex one using Logistic Regression. Similar approach as aggregation of different string metrics in named-entity similarity task (using Supported Vector Machine) was presented in [26, 27]. A complex similarity function trained on the reduced set of single metrics (*NamEnSim5*) described in [24] was used as the additional feature in CRF-based model for proper name recognition in Polish texts.

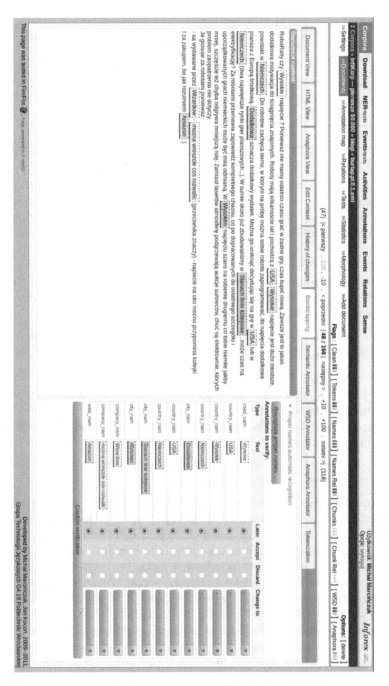

Fig. 5. Perspective for bootstrapping verification in the Inforex system

5.2   Evaluation of CRF-Based Model

In evaluation process we used two corpora: CEN and CSER. Cross-domain evaluation on CEN trained on CSER was performed in 3 variants of different feature configuration and three versions of dictionaries representing search space for dictionary/similarity features:

- Variants:
  - IC – *Initial Configuration*, top features configuration for CRF-based model, described in [21].
  - SC – *Similarity Configuration*, dictionary features from IC are replaced by similarity features.
  - BC – *Best Configuration*, similarity features are added to IC.
- Versions:
  - NB – *No Bases*, search space for dictionary/similarity features contains only proper names from NELexicon.
  - B – *Bases*, base forms (lemmas) of NER (IC+NB configuration) results (only False Negative – unrecognized proper names) are added to NB.
  - BF – *Base and Forms*, unrecognized proper names (NER IC+NB False Negative results) added to B.

Table 6 contains the results of cross-domain evaluation on CEN trained on CSER for different test variants and versions.

Table 6. Results of evaluation for different test variants and versions

Variant	Version	TP	FP	FN	Precision [%]	Recall [%]	F-measure [%]
NB	IC	3681	346	1544	91.41	70.45	79.57
NB	SC	3509	338	1716	91.21	67.16	77.36
NB	BC	3729	329	1496	**91.89**	**71.37**	**80.34**
B	IC	3998	347	1227	92.01	76.52	83.55
B	SC	3698	340	1527	91.58	70.78	79.84
B	BC	4031	328	1194	**92.48**	**77.15**	**84.12**
BF	IC	4107	348	1118	92.19	78.60	84.86
BF	SC	3733	345	1492	91.54	71.44	80.25
BF	BC	4141	333	1084	**92.56**	**79.25**	**85.39**

The analysis of the results showed that the best evaluation results can be obtained by adding similarity features to the top features configuration (baseline), described in [21]. Replacing the dictionary features from the baseline with similarity features does not improve the results in the cross-domain evaluation. The best improvement (comparing to the baseline IC results in each test variant) can be observed with search space, extended by adding base forms of proper names, which were not recognized using baseline configuration (B variant). For each variant the difference between best configuration and initial configuration is: *NB – 0.77%, B – 0.91%, BF – 0.64%*.

6   NER-WS Web Service

The NER-WS web service makes the Liner2 available through Web. It utilises basic web services technologies [28]: SOAP protocol for data exchange and WSDL description language.

The service's implementation is based on a three-layer architecture, that was earlier successfully applied for other Natural Language Processing web services, e.g. the TaKIPI tagger [29]. This architecture was proved to be robust and well-scalable.

6.1   Usage

The web service is accessed through the WSDL description file, which contains URL addresses of endpoints, defines available functionalities and structure of request and answers, that are used for communication between the client and the server.

NER-WS has an *asynchronous* model of communication, i.e. it doesn't keep the connection with the client up within the time of computation. Instead, it allows two types of client requests, which are immediately answered and the connection is terminated:

- **Annotate** — client sends text to annotate, the WebService returns a unique *request identifier*,
- **GetResult** — client sends the request identifier, the WebService returns status of the request (one of: QUEUED, PROCESSING, READY, ERROR, FINISHED) and the result or error code if available).

The probing of WebService with GetResult requests (e.g. once per second) should be implemented on the client side.

6.2   Architecture

NER-WS has a three layer architecture (see Figure 6): the *SOAP server layer* communicates with clients, accepting requests and sending back responses. The texts to process, as well as processing results, are stored by the *database layer*. Text annotating is done by the *daemon layer*, which consists of many liner2 daemons, running parallelly.

This architecture was chosen because of the following advantages:

- **asynchronous communication** — the connection with client need not to be kept up during the computation,
- **distributed processing** — the daemons may run on many different machines and communicate with other layers through network,
- **parallel processing** — many requests may be handled simultaneously,,
- **scalability** — the number of daemons running may be adjusted to the needs and available resources; an increase of the number of daemons increases the processing speed and capacity of the WebService.

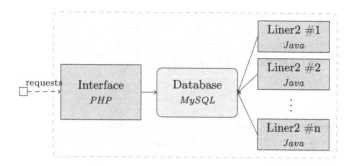

Fig. 6. NER-WS architecture

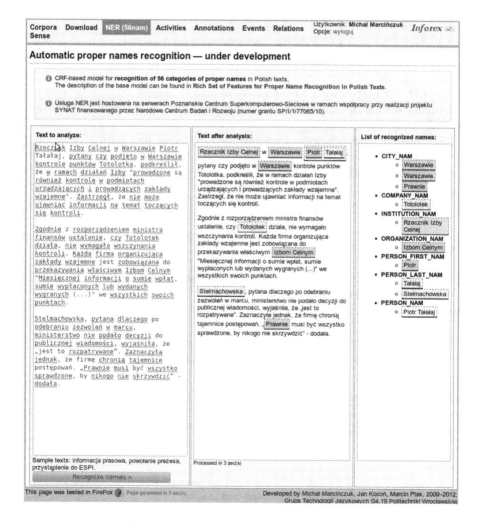

Fig. 7. Graphical interface using NER-WS

An additional mechanism was implemented to improve the server-side robustness: the daemons, that are inactive for a long period of time (e.g. because they were abnormally terminated or have lost connection to the server) are automatically detected and disconnected from the system.

6.3 Demo

A mockup instance of NER-WS is hosted at the Poznań Supercomputing and Networking Center (PSCN)[5]. The web service can be tested using a web-based application[6] that is a part of Inforex system (see Figure 7).

7 Summary

In the paper we presented a customizable framework for proper names recognition called Liner2. The framework was used to construct a model for recognition 5 basic categories of proper names and was also applied to bootstrap the annotation of KPWr corpus with 56 categories of proper names. We also presented a preliminary results of combining the Liner2 with similarity function for proper names.

The Liner2 framework was developed mainly for proper names recognition task for Polish. However, many of the existing components can be used for other sequence labelling tasks, like syntactic chunks recognition. Most of the component are also language independent. The other, with some effort, could be also adopted to other languages than Polish.

License and Access

Liner2 — is available on a free license (GPL) and can be downloaded from the following page: http://nlp.pwr.wroc.pl/liner2,

Liner2 models — can be also downloaded from http://nlp.pwr.wroc.pl/liner2,

NER-WS demo — a web-based application utilizing NER-WS is available at http://nlp.pwr.wroc.pl/inforex/index.php?page=ner,

Acknowledgements. Work financed by NCBiR NrU.: SP/I/1/77065/10 (project SyNaT[7]) and Innovative Economy Programme project POIG.01.01.02-14-013/09 (project NEKST[8]).

---

[5] Home page: http://www.man.poznan.pl/online/en/
[6] Web page: http://nlp.pwr.wroc.pl/inforex/index.php?page=ner
[7] Web page: http://www.synat.pl
[8] Web page: http://www.ipipan.waw.pl/nekst/

References

1. Mykowiecka, A., Kupść, A., Marciniak, M., Piskorski, J.: Resources for Information Extraction from Polish texts. In: Proceedings of the 3rd Language & Technology Conference: Human Language Technologies as a Challenge for Computer Science and Linguistics (LTC 2007), Poznań, Poland, October 5-7 (2007)
2. Graliński, F., Jassem, K., Marcińczuk, M.: An Environment for Named Entity Recognition and Translation. In: Màrquez, L., Somers, H. (eds.) Proceedings of the 13th Annual Conference of the European Association for Machine Translation, Barcelona, Spain, pp. 88–95 (2009)
3. Graliński, F., Jassem, K., Marcińczuk, M., Wawrzyniak, P.: Named Entity Recognition in Machine Anonymization. In: Kłopotek, M.A., Przepiórkowski, A., Wierzchoń, A.T., Trojanowski, K. (eds.) Recent Advances in Intelligent Information Systems, pp. 247–260. Academic Publishing House Exit (2009)
4. Marcińczuk, M., Zaśko-Zielińska, M., Piasecki, M.: Structure Annotation in the Polish Corpus of Suicide Notes. In: Habernal, I., Matoušek, V. (eds.) TSD 2011. LNCS, vol. 6836, pp. 419–426. Springer, Heidelberg (2011)
5. Marrero, M., Sánchez-Cuadrado, S., Lara, J.M., Andreadakis, G.: Evaluation of Named Entity Extraction Systems. Research in Computing Science 41, 47–58 (2009)
6. Kravalová, J., žabokrtský, Z.: Czech named entity corpus and SVM-based recognizer. In: Proceedings of the 2009 Named Entities Workshop: Shared Task on Transliteration, pp. 194–201. ACL, Suntec (2009)
7. Osenova, P., Kolkovska, S.: Combining the Named-entity Recognition Task and NP Chunking Strategy for Robust Pre-processing. In: Proc. of the 1st Workshop on Treebanks and Linguistic Theories, Sozopol, Bulgaria, pp. 167–182 (2002)
8. Katrenko, S., Adriaans, P.: Named Entity Recognition for Ukrainian: A Resource-Light Approach. In: Proceedings of the Workshop on Balto-Slavonic Natural Language Processing: Information Extraction and Enabling Technologies, pp. 88–93. ACL, Prague (2007)
9. Piskorski, J.: Extraction of Polish named entities. In: Proceedings of the Fourth International Conference on Language Resources and Evaluation, LREC 2004, pp. 313–316. Association for Computational Linguistics, Prague (2004)
10. Piskorski, J.: Named-Entity Recognition for Polish with SProUT. In: Bolc, L., Michalewicz, Z., Nishida, T. (eds.) IMTCI 2004. LNCS (LNAI), vol. 3490, pp. 122–133. Springer, Heidelberg (2005)
11. Urbańska, D., Mykowiecka, A.: Multi-words Named Entity Recognition in Polish texts. In: SLOVKO 2005 – Third International Seminar Computer Treatment of Slavic and East European Languages, Bratislava, Slovakia, pp. 208–215 (2005)
12. Abramowicz, W., Filipowska, A., Piskorski, J., Węcel, K., and Wieloch, K.: Linguistic Suite for Polish Cadastral System. In: Proceedings of the LREC 2006, Genoa, Italy, pp. 53–58 (2006) ISBN 2-9517408-2-4
13. Savary, A., Piskorski, J.: Lexicons and Grammars for Named Entity Annotation in the National Corpus of Polish. In: Kłopotek, M.A., Marciniak, M., Mykowiecka, A., Penczek, W., Wierzchnoń, S.T. (eds.) Intelligent Information Systems, Siedlce, pp. 141–154 (2010) ISBN 978-83-7051-580-5

14. Marcińczuk, M., Piasecki, M.: Pattern Extraction for Event Recognition in the Reports of Polish Stockholders. In: Proceedings of the International Multiconference on Computer Science and Information Technology, Wisła, Poland, October 15-17, vol. 2, pp. 275–284 (2007) ISSN 1896-7094

15. Marcińczuk, M.: Pattern Acquisition Methods for Information Extraction Systems. Master's thesis, Blekinge Tekniska Högskola, Sweden (2007)

16. Savary, A., Waszczuk, J.: Narzedzia do anotacji jednostek nazewniczych. In: Przepiórkowski, A., Bańko, M., Górski, R.L., Lewandowska-Tomaszczyk, B. (eds.) Narodowy Korpus Języka Polskiego [Eng.: National Corpus of Polish], pp. 225–252. Wydawnictwo Naukowe PWN, Warsaw (2012)

17. Marcińczuk, M., Piasecki, M.: Statistical Proper Name Recognition in Polish Economic Texts. Control and Cybernetics (2011)

18. Radziszewski, A., Śniatowski, T.: Maca: a configurable tool to integrate Polish morphological data. In: Proceedings of FreeRBMT 2011, Barcelona, Spain (2011)

19. Przepiórkowski, A.: The IPI PAN Corpus: Preliminary version. Institute of Computer Science. Polish Academy of Sciences, Warsaw (2004)

20. Marcińczuk, M., Stanek, M., Piasecki, M., Musiał, A.: Rich Set of Features for Proper Name Recognition in Polish Texts. In: Bouvry, P., Kłopotek, M.A., Leprévost, F., Marciniak, M., Mykowiecka, A., Rybiński, H. (eds.) SIIS 2011. LNCS, vol. 7053, pp. 332–344. Springer, Heidelberg (2012)

21. Marcińczuk, M., Janicki, M.: Optimizing CRF-Based Model for Proper Name Recognition in Polish Texts. In: Gelbukh, A. (ed.) CICLing 2012, Part I. LNCS, vol. 7181, pp. 258–269. Springer, Heidelberg (2012)

22. Broda, B., Marcińczuk, M., Maziarz, M., Radziszewski, A., Wardyński, A.: KPWr: Towards a Free Corpus of Polish. In: Calzolari, N., Choukri, K., Declerck, T., Doğan, M.U., Maegaard, B., Mariani, J., Odijk, J., Piperidis, S. (eds.) Proceedings of the Eighth International Conference on Language Resources and Evaluation (LREC 2012), Istanbul, Turkey. European Language Resources Association, ELRA (2012) ISBN 978-2-9517408-7-7

23. Marcińczuk, M., Kocoń, J., Broda, B.: Inforex — a web-based tool for text corpus management and semantic annotation. In: Calzolari, N., Choukri, K., Declerck, T., Doğan, M.U., Maegaard, B., Mariani, J., Odijk, J., Piperidis, S. (eds.) Proceedings of the Eighth International Conference on Language Resources and Evaluation (LREC 2012), Istanbul, Turkey. European Language Resources Association, ELRA (2012) ISBN 978-2-9517408-7-7

24. Kocoń, J., Piasecki, M.: Heterogeneous Named Entity Similarity Function. In: Sojka, P., Horák, A., Kopeček, I., Pala, K. (eds.) TSD 2012. LNCS(LNAI), vol. 7499, pp. 223–231. Springer, Heidelberg (2012)

25. Piasecki, M., Radziszewski, A.: Polish Morphological Guesser Based on a Statistical A Tergo Index. In: Proceedings of the International Multiconference on Computer Science and Information Technology — 2nd International Symposium Advances in Artificial Intelligence and Applications (AAIA 2007), pp. 247–256 (2007)

26. Cohen, W.W., Ravikumar, P., Fienberg, S.E.: A Comparison of String Distance Metrics for Name-Matching Tasks. In: Proceedings of IJCAI 2003 Workshop on Information Integration, pp. 73–78 (2003)

27. Cohen, W.W., Ravikumar, P., Fienberg, S.E.: A Comparison of String Metrics for Matching Names and Records. In: Proceedings of the KDD 2003 Workshop on Data, Washington, DC, pp. 13–18 (2003) ISBN 3-540-29754-5

28. Newcomer, E.: Understanding Web Services: XML, WSDL, SOAP and UDDI. Pearson (2002)
29. Broda, B., Marcińczuk, M., Piasecki, M.: Building a Node of the Accessible Language Technology Infrastructure. In: Calzolari, N., Choukri, K., Maegaard, B., Mariani, J., Odjik, J., Piperidis, S., Rosner, M., Tapias, D. (eds.) Proceedings of the Seventh Conference on International Language Resources and Evaluation (LREC 2010), Valletta, Malta, May 19-21 (2010) ISBN 2-9517408-6-7

Chapter IV

Ontology-Based Systems

Theoretical and Architectural Framework for Contextual Modular Knowledge Bases[*]

Krzysztof Goczyła, Aleksander Waloszek, Wojciech Waloszek, and Teresa Zawadzka

Gdańsk University of Technology, Poland
{kris,alwal,wowal,tegra}@eti.pg.gda.pl

Abstract. The paper presents the approach aimed at building modularized knowledge bases in a systematic, context-aware way. The paper focuses on logical modeling of such knowledge bases, including an underlying SIM metamodel. The architecture of a comprehensive set of tools for knowledge-base systems engineering is presented. The tools enable an engineer to design, create and edit a knowledge base schema according to a novel context approach presented elsewhere by the authors. It is explained how a knowledge base built according to SIM (Structured-Interpretation Model) paradigm is processed by a prototypical reasoner Conglo-S, which is a custom version of widely known Pellet reasoner extended with support for modules of ontologies called tarsets (also introduced elsewhere under the name of conglomerates). The user interface of the system is a plug-in to Protégé ontology editor that is a standard tool for development of Semantic Web ontologies. Possible applications of the presented framework to development of knowledge bases for culture heritage and scientific information dissemination are also discussed.

Keywords: knowledge base, modularization, OWL API, ontology editor, reasoned.

1   Introduction

Knowledge bases grow bigger and bigger. The vast amount of information that has to be stored and managed in an intelligent way motivated intensive research in the field of management of large knowledge bases. One of the hot topics in this field is modularization of knowledge bases. In this chapter we present a method of modularization that we invented and intend to use in the SyNaT project (see the footnote of this page). The method is based on the notion of tarset (previously called "conglomerate") that enables a system to manage a knowledge bases in pieces that themselves are (much smaller) knowledge bases. Tarsets can be seen as analogs to single relations or groups of relations in a relational database, that can be handled

[*] This work has been partially supported by the Polish National Centre for Research and Development (NCBiR) under Grant No. SP/I/1/77065/10 by the strategic scientific research and experimental development program: „Interdisciplinary System for Interactive Scientific and Scientific-Technical Information - SyNaT".

R. Bembenik et al. (Eds.): *Intell. Tools for Building a Scientific Information*, SCI 467, pp. 257–280.
DOI: 10.1007/978-3-642-35647-6_18   © Springer-Verlag Berlin Heidelberg 2013

separately (e.g. as views), or as a whole. Upon the structure of tarsets different logical structures of a knowledge base can be built. In this chapter we show how the contextual approach to design and management of a knowledge base can be realized using the tarset-based approach in order to obtain a semantically modularized knowledge base.

This chapter is organized as follows. In the Section 2, that is the main contribution of this paper, we present theoretical background of our approach, with recall of basic notions that lie behind contexts and tarsets. We also introduce the concept of couplers that enable a system to interpret different logical dependencies between tarsets (e.g. contextual dependencies). This sections contains also an ontological metamodel for a tarset-aware knowledge base. In Section 3 we present implementation aspects of our framework: an ontology editor especially suited to create and edit contextual ontologies, an important component of our framework that is the Conglo-S module which is a tarset-aware Description Logic reasoner that is able to interpret tarset algebra operators, and then the architecture of the framework that we build to implement the ideas presented in the previous sections. Both: Conglo-S and the editor use widely known open-source tools, extending them with facilities needed to support our approach. In Section 4 we discuss our attitude to application of the framework in the SyNaT project. Section 5 gives a comparison with other approaches to modularization of knowledge bases, which justifies our approach, and presents directions for future work.

2   Basics: Underlying SIM Metamodel

In this section we present our vision of what a modular knowledge base really is and how it should be utilized. Theoretical foundation for this is the theory of tarsets, presented in details in [1], [2], [3]. In the first subsection we recall some basic elements of this theory. In the next subsection we introduce the notion of a modular knowledge base schema. Then we present the SIM (Structured-Interpretation Model) and discuss some problems connected with designing SIM knowledge bases.

We assume familiarity of the reader with basic assumptions of Description Logics. From among less widely used terms we exploit the notion of *vocabulary* (aka. signature), which is a triple (C, R, I) consisting of sets of names of concepts (C), roles (R), and individuals (I) respectively. We assume that every Tarski-style interpretation of a vocabulary assigns each concept name a subset of a domain, each role a set of pairs of domain elements, and each individual name an element of a domain.

2.1   Theory of Tarsets

The theory of tarsets was already presented [4]. Here we recall only some of its properties important for the rest of the paper. A knowledge base is defined as a freely chosen subset of the set of all possible tarsets $\mathbf{K}$. An element of this set, a tarset M, is defined as a pair (S, W) of a vocabulary S and a set W containing all intended models of S. Using $S(M)$ and $W(M)$ we describe the two parts of a tarset M. "Intended models" mean the models of S which are allowed by the creator of a given tarset.

Tarsets are semantic modules and are defined in a way which disregards the exact form of a language (like DL). They focuses only on (Tarski-style) interpretations. However, the only known way to describe them is a language. So we say that a tarset satisfies a particular sentence α, denoted $M \vDash \alpha$, iff $\forall \mathcal{I} \in W(M)$: $\mathcal{I} \vDash \alpha$. Now we can regard a tarset M as a representation of an ontology, defined as a set of sentences, whose all of the sentences are satisfied by every interpretation contained in $W(M)$.

The tarsets contained by the set $\mathbf{K}$ are related to each other. There exists a way to describe these relationships. The theory defines an algebra of tarsets allowing to cover some of them. We can perceive the operations of the algebra as predicates describing relations between tarsets or as functions producing new tarsets from old ones. The list of the operations is as follows:

Intersection:	$M_1 \cap M_2 = (S(M_1) \cup S(M_2), W(M_1) \cap W(M_2))$
Union:	$M_1 \cup M_2 = (S(M_1) \cup S(M_2), W(M_1) \cup W(M_2))$
Difference:	$M_1 - M_2 = (S(M_1) \cup S(M_2), W(M_1) - W(M_2))$
Renaming:	$\rho_\gamma(M) = (\gamma(S(M)), \gamma(W(M)))$
Projection:	$\pi_S(M) = (S, \{\mathcal{I}|S: \mathcal{I} \in W(M)\})$
Selection:	$\sigma_\alpha(M) = (S(M), \{\mathcal{I} \in W(M): \mathcal{I} \vDash \alpha\})$

Intersection, union and difference are obvious. Renaming ρ_γ involves a function γ giving a new name a' to an element a of $S(M)$ and assigning all interpretations of a to a' in $W(M)$. Projection π_S changes the vocabulary $S(M)$ into S and cuts the interpretation in $W(M)$ in the way that all relationships between original concepts, roles, and individuals whose names remain in the vocabulary are preserved. Selection σ_α leaves in $W(M)$ only the interpretation satisfying the sentence α.

Using the algebra we can manipulate our knowledge base by producing new modules (or choosing them from $\mathbf{K}$) and in this way changing the state of the base.

2.2   Knowledge Base Schema

From the engineer's point of view such a definition of a knowledge base, as presented above, is insufficient. A properly defined engineer's methodology should allow to constrain users in their activities during all stages of a system's life. There are some essential questions a designer should answer during the design stage:

- Which modules are removable, and which are not?
- Which modules are unchangeable?
- What relationships exist between modules?
- Which relationships are removable, and which are not?
- How to create new modules and new relationships?

All above questions are very important but they take into account only the technical aspect of the knowledge base dynamics. Equally important is the semantic aspect: how those activities correspond to the semantics of the modeled problem:

- What does a module to be added represent?
- What is the meaning of a relationship between two modules?
- What do our actions mean (from semantic point of view) when we add or remove a module/relationship?
- Why some of the modules/relationships are removable and some are not?
- Why some of the modules should be unchangeable?

To address these problems we decided to define a knowledge base schema and a knowledge base instance. Generally speaking, a knowledge base instance is its state. Every time a user changes something in a knowledge base he/she produces its new instance. A knowledge base schema is a set of constraints imposing restrictions on users' activities. From another point of view a schema is a description which categorizes modules and relationships between modules. By creating this description a designer explains semantics of modules and relationships.

2.2.1 Expressions, Equalities and Inequalities

In order to make clear what a schema is, it is necessary to introduce some new notions. The first is a tarset algebra expression. The definition bases on the fact that the operations of the algebra produce new tarsets from old ones.

Definition 1 (*tarset algebra expression*)
An *expression of the tarset algebra* for the domain D is a formula of the form: $F, G := F \cup G \mid F \cap G \mid F - G \mid \pi_S(F) \mid \rho_\gamma(F) \mid \sigma_\alpha(F) \mid d$, where S, γ, α and d respectively mean a vocabulary, a renaming function, a sentence and an element of D. ∎

The simplest choice of D is a subset of $\mathbf{K}$. It is the only situation where we can perceive an expression as a complex operation producing a tarset as a result.

The other domains are also very useful, e.g. a set of tarset variables. A tarset variable can be a name chosen from a set of allowed variable names (the set is defined by a specific implementation), which has to be assigned a value from $\mathbf{K}$ during the instantiation process. The symbol $|v|$ means a value assigned to the variable v.

In order to define equality and inequality of tarsets, we assume that every tarset is reduced. A *reduced tarset* is a tarset whose vocabulary and set of models are also reduced. A *reduced vocabulary* $[S(M)]$ is a vocabulary $S(M)$ modified by removing all words with unrestricted meaning accordingly to all interpretations from $W(M)$ (i.e. such words x for which $M = \pi_{S(M)}(\pi_{S(M) - \{x\}}(M))$). Then a *reduced set of models* $[W(M)]$ is produced from $W(M)$ by projection to the reduced vocabulary.

Such an assumption facilitates comparing tarsets because it allows for neglecting insignificant differences between vocabularies. A greater tarset is simply a tarset with greater set of models (i.e. $M_1 > M_2 \Leftrightarrow W(M_1) \supseteq W(M_2)$). Moreover, with this assumption, it is easy to define an *empty tarset*, which is the only one in the space of reduced tarsets and has the form $M_0 = \{\varnothing, \varnothing\}$.

Now we can define *predicative expressions*: tarset equality and inequality:

Definition 2 (*tarset equality and inequality*)
An *algebraic equality* of tarsets for a domain D (a tarset equality) is a formula of the form: $F = G$, and an *algebraic inequality* of tarsets is a formula of the form $F > G$, $F < G$, $F \geq G$ or $F \leq G$, where F, G are expressions of the tarset algebra for the domain D. ∎

In general a choice of a domain D is arbitrary as long as its elements refer to tarsets (the domain may, apart from trasets, embrace e.g. tarset functions, or tarset variables). In the most straightforward situation, when $D \subseteq \mathbf{K}$, the satisfaction of a predicative expression depends only on the relationship between tarsets being the results of F and G. The way of satisfying expressions for other domains should be independently defined.

2.2.2   Tarset Knowledge Base, Schema and Instance

It is clear that the old definition of knowledge base is not sufficient to define rules allowing to control the contents of an instance. In order to achieve this goal the definition of a knowledge base should be enriched with new elements. In this subsection we redefine the notion of a knowledge base, taking into consideration its static (a schema) and dynamic (an instance) aspects. The form of the definition is strongly determined by the kinds of possible changes we can distinguish:

1. Changes of the contents of a tarset:
    (a) vocabulary changes, i.e. adding new words (general or individual names); in order to keep monotonicity we do not expect removing words,
    (b) semantic changes, i.e. adding new axioms describing relations between general names from the vocabulary,
    (c) factual changes, i.e. accepting new facts about individual names from the vocabulary.
2. Adding and/or removing tarsets.
3. Adding or removing relationships between tarsets.

These forms of changes will be discussed in next subsections.

Changes of the contents of a tarset
This kind of change is equivalent to addition of a new axiom to a tarset. It is not an addition itself, as a tarset is not a syntactic entity and does not contain sentences. Thus when we speak about "adding a sentence" α, we mean something like executing the operation of selection σ_α. As a result we get a new set of interpretations, that is a new tarset from the set $\mathbf{K}$. According to the old definition the result of such a change is a new knowledge base: a new set of tarsets. The conclusion is that it is much better to perceive a knowledge base as a set of tarset variables than a set of tarsets. The described change would not alter this set but only an assignment of (at least one of) the variables.

The decision that a knowledge base is a set of tarset variables allows us to define a new kind of elements controlling the contents of tarsets.

Definition 3 (*coupler*)
Let V be a set of tarset variables. A *coupler over* V is a predicative expression (tarset equality or inequality) for the domain V. ∎

A coupler s is satisfied iff the predicative expression is satisfied after mapping all tarset variables into elements of V. We assume that in a knowledge base the only accepted changes are the changes satisfying all its couplers. The couplers represent the aforementioned relationships between tarsets.

Adding or removing tarsets or relationships between tarsets
This kind of changes is much more difficult to control but is the most essential from the modular knowledge bases point of view. Adding or removing a tarset is connected with the change of the set of tarset variables. This change, on the other hand, imposes a change of the set of couplers. But the last change has also be under control. Thus there is a need to introduce new elements: types of tarsets and types of couplers. The purpose of these elements is to categorize groups of tarsets and couplers w.r.t their properties and to take control over this kind of changes.

Types of tarsets make a structure ordered by the inheritance relation. This structure is called a *hierarchy of tarset types*:

Definition 4 (*hierarchy of tarset types*)
A *hierarchy of tarset types* is a strict partial order $(T_K, \lhd)$, where T_K is called a *set of tarset variable types*, and $\lhd$ is called the *inheritance relation*. We say that t_1 *extends* (*inherits from*) t_2 iff $t_1 \lhd t_2$ ($t_1, t_2 \in T_K$). ∎

It is possible to put a hierarchy of tarset types on a given set of tarset variables.

Definition 5 (*establishing types*)
Let V_K be a finite set of tarset variables. *Establishing types by putting a hierarchy* $(T_K, \lhd)$ *on the set* V_K consists in setting a function $f_{in}: T_K \longrightarrow \mathcal{P}(V_K)$, where $\mathcal{P}(V_K)$ is the powerset of V_K. The function f_{in} takes into account transitivity of the relation $\lhd$ in the way that $\forall t_1, t_2 \in T_K \ \forall v \in V_K \ ((t_1 \lhd t_2 \wedge v \in f_{in}(t_2)) \to v \in f_{in}(t_1)))$. Every variable v assigned to $t \in T_K$, i.e. such that $v \in f_{in}(t)$, is called a *variable of the type t*. f_{in} is called a *result of establishing types*. ∎

It is worth to note that tarset types are types of tarset variables. Two variables of different types may be assigned the same value.

The other kind of types, coupler types, are used to categorize couplers.

Definition 6 (*coupler types*)
Let V_K be a finite set of tarset variables with types established by a structure $(T_K, \lhd, f_{in})$. A *coupler type over* V_K is a predicative expression (tarset equality or inequality) for a domain of slots L. A *slot* $l \in L$ is a pair (v, t), where $t \in T_K$ and $v \in V_K$ or v is a variable, whose range of assignment is $f_{in}(t)$. ∎

A given coupler s is of the type t_s iff it is a formula resulting of the substitution of all slots (v, t) in t_s by v, where $v \in V_K$, or by |v| otherwise.

Having coupler types defined, we can introduce the notion of tarset types. A tarset type consists of constraints describing the requirements against tarsets conforming to the type. We can formulate the definition of tarset type constraints as follows:

Definition 7 (*tarset type constraints*)
Let T_S be a set of coupler types. *Tarset type constraints* over T_S are constraints of the form:

cardinality constraint for a given tarset type t is a triple (t_s, l, N), where $t_s \in T_S$, l is a chosen slot, and N is a subset of the set of the natural numbers;

editing constraint for a given tarset type t is a value from the set $\{noTerminology, noFacts, noInstances\}$. ∎

Now we are ready to define a schema and an instance of a knowledge base. A schema is its fixed part; we assume that by changing a schema we define a new knowledge base.

Definition 8 (*tarset knowledge base schema*)
A *tarset knowledge base schema* is a structure $(K_S, V_{KS}, (T_K, \lhd), f_{inS}, T_S, C, f_o, w_{KS})$, where:

- K_S — subset of **K**,
- V_{KS} — a set of tarset variables,
- $(T_K, \lhd)$ — a hierarchy of tarset types,
- f_{inS} — the result of establishing $(T_K, \lhd)$ on V_{KS},
- T_S — a set of coupler types,
- C — a set of tarset type constraints over T_S,
- f_C — a function mapping a subset of C to every type from T_K; f_c takes into account transitivity of the relation $\lhd$, mapping to every type all constraints assigned to the inherited types in such a way that $t_1 \lhd t_2 \Rightarrow f_C(t_1) \subseteq f_o(t_2)$,
- w_{KS} — a partial function $V_{KS} \longrightarrow K_S$.

All the sets mentioned above are finite. ∎

The elements K_S and V_{KS} allow to predefine a part of an instance during the designing stage. In other words, every schema can contain predefined tarset variables and tarsets (constants). As members of a schema they cannot be removed.

Unlike a schema, which is a fixed part of a base, an instance is its temporary state. Every change generates a new state and a new instance comes into existence (or is chosen from the universe of all possible instances).

Definition 9 (*tarset knowledge base instance*)
An *instance of a tarset knowledge base* $\Sigma = (K_S, V_{KS}, (T_K, \lhd), f_{inS}, T_S, C, f_C, w_{KS})$ is a structure $(\Sigma, (K, V_K, S, w_K), f_{in}, f_{sin})$, where:

- K — subset of **K**,
- V_K — a set of tarset variables,
- S — a set of couplers,

- w_K — a function $V_K \longrightarrow K$,
- f_{in} — the result of establishing $(T_K, \lhd)$ on V_K,
- f_{sin} — a function $S \longrightarrow T_S$.

All the elements of the structure have to satisfy the following conditions, called the *schema adjustment conditions*:

- *the condition of constants preservation*: $K \supseteq K_S$,
- *the condition of variables preservation*: $V_K \supseteq V_{KS}$,
- *the condition of constant values preservation*: $w_K \supseteq w_{KS}$,
- *the condition of proper tarsets typing*: $f_{in} \supseteq f_{inS}$,
- *the condition of couplers satisfaction*: all the couplers from the set S have to be satisfied,
- *the condition of proper couplers typing*: for every $s \in S$: s has to be of the type $f_{sin}(s)$,
- the condition of constraints preservation: for every tarset type t and every tarset variable v from $f_{in}(t)$, all the constraints from the set $f_C(t)$ have to be satisfied, i.e.:
  - every cardinality constraint (t_s, l, N) is satisfied, when the number of couplers from S, created by substitution of every l by v and such that $f_{sin}(s) = t_s$, belongs to N,
  - every editing constraint E is satisfied.   ∎

Strict mathematical description of the meaning of an editing constraint E would take too much space. Simplifying, we can say that if $E = noInstances$ then it is forbidden to create new tarsets of a given type (i.e. $f_{in}(t) = f_{inS}(t)$); if $E = noTerminology$ then adding any terminological axioms to a tarset of a given type is forbidden; if $E = noFacts$ then adding new facts is forbidden.

2.3   Ontological Description of the Metamodel

The mathematical metamodel of a tarset knowledge base presented above has been described in the ontological way The ontology is expressed in the OWL DL language and was edited using Protégé (http://protege.stanford.edu/). It consists of three modules: *Algebraization*, *SchemaBasicTBox* and *InstanceBasicTBox*. The modules are standard, flat ontologies connected with commonly used option called *import*.

The module *Algebraization* contains definitions of the elements representing the algebraic and predicative expressions. There are concepts allowing to describe main entities, like *SKBPredicativeExpression* or *SKBExpression*, the operators (e.g. *SKBIntersection*, *SKBUnion*, *SKBProjection*, etc.), operands (e.g. *SKBExpression Variable*) and some auxiliary entities (among others: *AxiomSet*, *NameSet* and operations on them). The relations between concepts are described by 12 roles, e.g.: *hasSKBOperand* (domain: *SKBOperation*, range: *SKBExpression*), *isRestrictedBy* (domain: *SKBSelection*, range: *AxiomSet*), etc. This module is very important as it is the essential part of coupler type definitions.

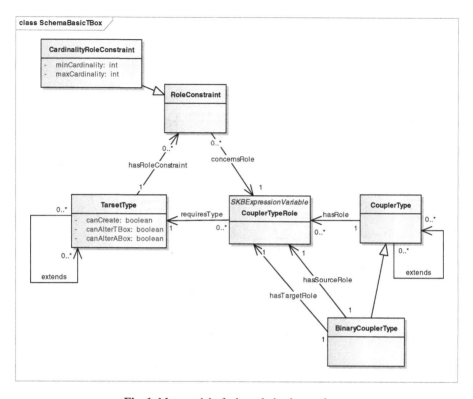

Fig. 1. Metamodel of a knowledge base schema

The module *SchemaBasicTBox* defines the metamodel of a tarset knowledge base schema. It is shown in Fig. 1. Classes represent concepts and associations represent roles. The main concepts are *TarsetType* and *CouplerType*. The former is extended by *BinaryCouplerType*, because such a type is expected to be used most frequently. As we see, the concept *CouplerTypeRole* inherits from the concept *SKBExpressionVariable* and, in this way, implements also the notion of a slot. There is one more role not depicted in this diagram: it is called *isRealizedBy* and connects the concept *CouplerType* with *SKBPredicativeExpression*. This connection is essential for the definition of a coupler type.

The module *InstanceBasicModule* is focused on the notions related to knowledge base instances. It imports *SchemaBasicTBox* as every element of an instance has to be connected with the corresponding element of a schema. The concepts *Tarset* and *Coupler* are connected with *ConceptType* and *CouplerType* respectively with the role *hasType*. The role *assigns* connects *RoleAssignment* with *CouplerTypeRole*. Figure 2 presents the metamodel of a knowledge base instance. From this figure it is easy to see what is the reason for defining *BinaryCouplerType*. The concept *BinaryCoupler* is directly connected with the concept *Tarset* without mediation of the concept *RoleAssignment*. Thanks to this, the schema of an instance is much simpler.

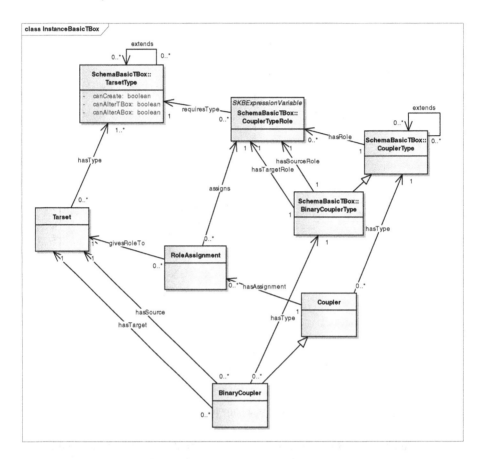

Fig. 2. Metamodel of a knowledge base instance

The main idea behind using the ontological form of a description is to deliver a basis for practical solutions for future implementations. We assume that every implementation of tarset knowledge base will also be described ontologically as an extension of the metamodel. Figure 3 shows a possible architecture.

The block named "Basic terminology" is the metaontology itself. The ontologies inside form the common part of all knowledge bases and are available as an internet resource. Every knowledge base (the blocks "Knowledge Base 1", "Knowledge Base 2", etc.) has its own schema (the fixed part) and instance (the dynamic part), both described ontologically. The schema imports the basic terminology. It may be extended with new general terms, but by doing this one cannot change the semantics of the metaontology. Such an extension is called "conservative extension" (see [5]). Any extension of the terminology here has only a local meaning because it is assumed that the tarset reasoning process is aware only of the language defined by the basic terminology.

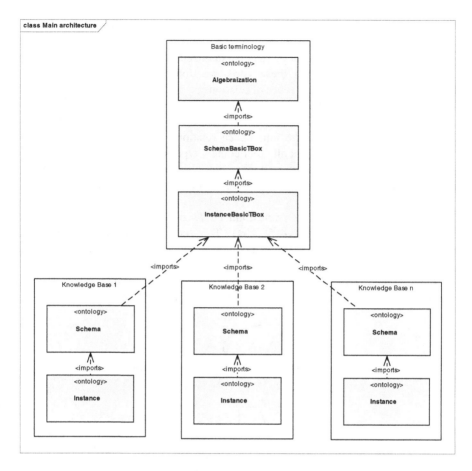

Fig. 3. Architecture of ontological descriptions of tarset knowledge bases

The instance ontology consists of facts (is an ABox), constantly changes (yet preserving monotonicity), and imports the schema ontology. Every tarset and every coupler is represented as an individual, but their contents are kept in separate ontologies, pointed by special annotations. It is convenient to describe the contents of a tarset in the form of sentences using the notion of $\mathcal{L}$-(2)-**D**-representation (see the next part of the paper). This kind of representation assumes that every module is represented by a set of sets of sentences. It is convenient to keep every set of sentences in a separate ontology.

2.4   SIM Contextualized Knowledge Bases

In [4] we described shortly the SIM methodology so now we discuss only the most important of its properties needed to understand the rest of the paper. More information may be found in [6][7][8].

SIM (Structured-Interpretation Model) is an extension of Description Logic formalism allowing to define modular ontologies of a specific form. As a normal DL ontology it consists of a terminology (a contextualized TBox) and a world description (a contextualized ABox). Both these parts are divided into modules which are connected to each other by three types of relations.

A *contextualized TBox* is a poset ($\{T_i\}_{i \in I}$, $\trianglelefteq$) with a single least element T_m. The indexed set $\{T_i\}$ contains terminological modules (TBoxes), called context types. Context types are ordered by the inheritance relation $\trianglelefteq$. The T_m is called the *root context type* and is the ancestor of all other context types in the structure.

The semantics of such a structure impose that every successor inherits all axioms defined in its ancestors. It is equivalent to importing the contents of the "less" module to the "greater" module.

A *contextualized ABox of the contextualized TBox* ($\{T_i\}_{i \in I}$, $\trianglelefteq$) is a structure that contains three elements: ($\{A_j\}_{j \in J}$, f_{inst}, $\lessdot$). The set $\{A_j\}$ contains modules with descriptions of facts (ABoxes). These modules are called context instances. f_{inst} is a function $\{A_j\} \longrightarrow \{T_i\}$, called *instantiation function*, assigning every context instance to a single context type. This connection also resembles an OWL import, since a context instance imports all the contents from its context type and its ancestors.

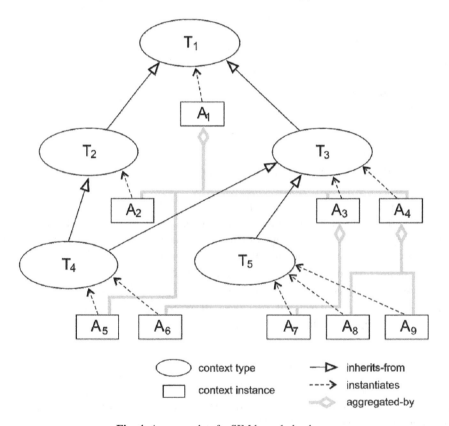

Fig. 4. An example of a SIM knowledge base

The last element of the structure is the aggregation relation $\ll$. This relation is also a partial order with exactly one least element.

The both relations ($\unlhd$ and $\ll$) and the function f_{inst} determine the paths of flow of conclusions during reasoning. In the case of aggregation relation the flow has the direction from aggregated to aggregating instances. All sentences must be taken into account, but due to the fact that the attached TBox is more general some information must be reinterpreted in more general terms. Figure 4 depicts three aggregating context instances: A_1, A_3 and A_4.

It is worth stressing that if levels of generality (introduced by context types) are properly chosen, we can aggregate information from context instances holding contradictory assertions without making the knowledge base inconsistent.

The main principle of SIM structures is that the higher level of hierarchy, the bigger set of individuals is described in more general way.

Properties of the SIM method allow us to achieve one more aim. It is easy to establish the boundary between a schema (a fixed part of a knowledge base) and an instance (a changeable part). In [8] this problem is discussed more precisely. Here, not going into details, we can say that a *SIM knowledge base schema* is a contextual TBox with so called *admissible places of aggregation* pointing out places where instances of a given context type should be aggregated. On the other hand a *SIM knowledge base instance* is a contextual ABox built regarding to the requirements imposed by the schema.

2.5 SIM Modules as Tarsets

To explain the correlation between the two described above models we have to refer to the analogy with relational databases. While designing a new database, we first prepare an E-R diagram, in which we introduce relationships between entity sets. Each kind of relationships may be expressed in relational algebra, but they are only a fraction of all kind of relationships expressible in this way. In other words, during the high level design we discard variety of possibilities offered by the algebra and focus on selected, most important, kinds of relationships.

We perceive the role of a SIM schema similarly as this of E-R. The model captures the most important factors connected with contextual modeling (like e.g. change of perspective) and, as such, is an ideal candidate for providing a core set of relationships between semantic modules.

It is very important that all the constraints imposed by SIM models are expressible in tarset algebra. This allowed us to create a uniform, general system of defining knowledge base schemas, in which we treat SIM elements as a vital part of every tarset knowledge base (depicted in Fig. 3 as Knowledge Base 1..n).

In practice the task has been carried out by development of an additional OWL ontology named *SIMMetaschema*. This ontology has been integrated within the described framework. For each kind of SIM relation (inheritance, instantiation, aggregation) it introduces an appropriate binary coupler type. The tarsets themselves are divided into two types: context types, and context instances.

Figure 5 presents an excerpt from the ontology, where the individual *inheritance* is defined as a type of coupler. Indeed, inheritance is a binary coupler requiring tarsets of the type "context type" for its both roles (ancestor and descendant). The coupler is described by an inequality with two slots (for the two roles) and the inequality itself is a formula $D \leq D \cap A$, which simply means that all the axioms from an ancestor are "imported" by a descendant. All the other required inequalities are defined in a similar fashion.

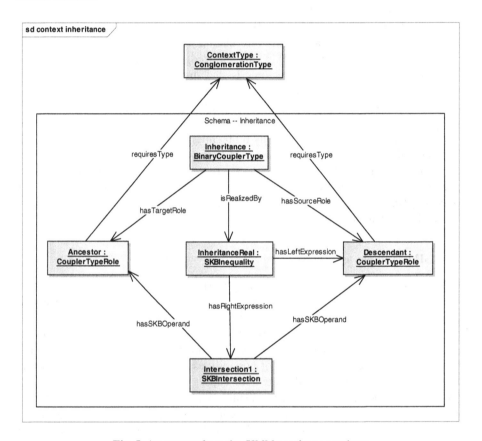

Fig. 5. An excerpt from the *SIMMetaschema* ontology

The achieved result is notable, as on the one hand it indicates that the tarset algebra is powerful enough to fully represent the SIM model, and on the other hand SIM introduces important relations that are very useful for organizing the design of knowledge bases of various kinds in a systematic way.

The resulting framework integrates both methods and allows for exploiting their strengths. We assume that a typical way of use of the framework is to design a "backbone" of a knowledge base only with use of SIM relations, and then augmenting it in the next steps with more specialized couplers. Such course of actions would be similar to creating an E-R backbone design of a database, and then preparing specialized triggers in order to fulfill more sophisticated requirements.

This framework constitutes a base for a set of tools addressed to designing modular knowledge bases described in the next section of this chapter.

3   The Tools

For the theoretical framework presented above to be realized in software practitioners' environments, a set of supporting tools is needed. This section is devoted to presentation of such a set of tools. We stress that some crucial components of this set are required to fulfill needs of knowledge-based systems engineers; these are: a context-aware modular ontology editor (Section 3.1), an appropriate reasoner (Section 3.2), and a general architecture of the framework with use-case analysis (Section 3.3).

3.1   Protégé Plugin

During the SyNaT project we strive to achieve the aim of developing a suite of tools allowing for designing and using SIM knowledge bases. For the first task we developed an extension (in the form of a set of plugins) in for a well-known and widely used ontology editor – Protégé.

While designing an editor interface of contextual plug-ins, we aimed at obtaining an interface as similar to standard Protégé editor interface as possible. Thus, while presenting this interface, names of consequent elements of interface are taken from standard OWL editor. The terms "window" and "tab" are defined analogically as in operating system. Additionally, we use the term "view" as one of the elements of the specified tab.

The standard set of tabs, normally used in the OWL editor, is supplemented by a new tab (Structure Tab) containing elements allowing to navigate and edit the main structures of a SIM knowledge base. The functionality of the standard set is unchanged: it allows to view and edit OWL ontologies. As SIM modules are, in general, ontologies, the set is also useful for editing their contents. In some cases, though, some possibilities are restricted (e.g. inserting assertions into TBox modules has been prohibited).

The Structure tab help the user with creating the ontological description of the knowledge base (cf. Sec 2.3) by hiding its details and presenting the structure of a SIM knowledge base. The view named Context Type Hierarchy shows the hierarchy of context types in the similar way to the Class Hierarchy view showing the hierarchy of concepts. Selecting any context type causes the hierarchy of context instances being instances of selected context type to be displayed in the view named Context Instances. Additional view allows for designing several knowledge bases within one instance of Protégé editor, and choosing one as an active base.

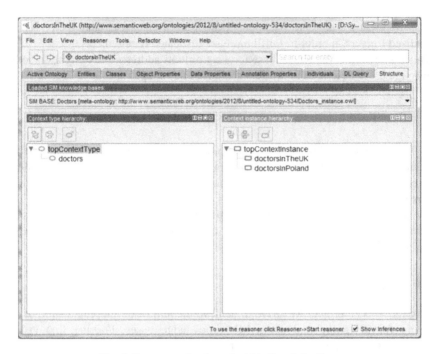

Fig. 6. Structure tab selected within Protégé editor

3.2   Reasoning with Conglo-S

Conglo-S inference engine (its name being a shortcut for Conglomerated Sentencer) is a reasoner integrated with Protégé system. It is designed for processing tarset knowledge bases with use of one standard OWL reasoning services (currently, by default, it uses Pellet [12] as its internal reasoner, though it is possible to configure it—with use of Preferences Tab—to use any of reasoners integrated with Protégé). Conglo-S reasoner is an extension of a previously developed tool, and the extension is towards handling schemas: i.e. module variables and couplers.

3.2.1 Sentential Representation of Tarsets

In order to meet its basic requirements, Conglo-S needs to convert the semantic modules to a form which will be readable by standard reasoning systems. For this purpose we exploit the theory of *sentential representation* of tarsets.

It is obvious that there are situations when a single module may be represented with a set of sentences expressed in some language $\mathcal{L}$. For such representation one must be able to create exactly the set of satisfying interpretations as for the original module. The notion of $\mathcal{L}$-representability formalizes the idea (by $Sig(O)$, where O is a set of sentences, we understand a vocabulary of all terms used in the sentences):

Definition 10 ($\mathcal{L}$-representability and $\mathcal{L}$-representation)

A set of sentences $O \subseteq \mathcal{L}$ is an $\mathcal{L}$-representation of a tarset M iff $Sig(O) \subseteq \mathbf{S}(M)$ and $\mathbf{W}(M) = \{\mathcal{I}: \mathcal{I} \vDash O\}$.

Tarset M is $\mathcal{L}$-representable iff there exists a set of sentences $O \subseteq \mathcal{L}$ being its $\mathcal{L}$-representation.   ∎

While some operations (like ∩) preserve the $\mathcal{L}$-representability (i.e. while performed on $\mathcal{L}$-representable modules, they produce a module which is also $\mathcal{L}$-representable), it is not always the case, as it can be seen in the following example.

Consider a module $M = M(\{A \sqsubseteq B\}) \cup M(\{B \sqsubseteq A\})$. While it seems that it may be represented with an empty set of sentences (because from M we cannot draw any sensible conclusion which can be expressed with a single sentence), in fact it cannot, because M carries some subtle information, which can be seen by performing $M \cap M(\{A \sqcap \neg B(a), \neg A \sqcap B(b)\})$: in the result, we obtain the empty module.

The second source of trouble with representing tarsets in sentential form is connected with the operation of projecting. It may turn out that the reduced signature of the module is "not enough" to represent all the interrelationships between remaining terms. A simple example of this is the module $M = \pi_{\{A, B\}}(M(\{A \sqsubseteq B, A(a)\}))$. The original module contained an assertion $A(a)$ whose effect on the set of models was that it excluded all the interpretations with empty A (and, consequently, B). There is no way of expressing such a constraint while using only the names A and B.

The above discussion indicates that the notion of $\mathcal{L}$-representability has to be extended at least towards handling alternative sets of models (like in the case of the union) and towards handling projection. The first problem can be solved by using multiple sets of sentences instead of a single one. Each set can now determine a separate set of models, and can be treated as a one of several "alternatives". The second problem is tackled with by introducing a set of "auxiliary" names from a special set $\mathbf{D}$ which can be used regardless of a signature of a module. The simultaneous use of both proposed solutions leads to the new notion of $\mathcal{L}$-(2)-$\mathbf{D}$-representability.

Definition 11 ($\mathcal{L}$-(2)-D-representabiliy)

Tarset M is $\mathcal{L}$-(2)-$\mathbf{D}$-representable iff there exists a (finite) set of sets of sentences $S = \{O_i\}_{i \in [1..k]}$, $k \in N$, each $O_i \subseteq \mathcal{L}$, being its $\mathcal{L}$-(2)-$\mathbf{D}$-representation (denoted $M \sim_{\mathcal{L}\text{-}(2)\text{-}\mathbf{D}} S$), which means that $\bigcup_{i \in [1..k]} Sig(O_i) \subseteq \mathbf{S}(M) \cup \mathbf{D}$ and $\mathbf{W}(M) = (\bigcup_{i \in [1..k]} \{\mathcal{I}: \mathcal{I} \vDash O_i\})|\mathbf{S}(M)$.   ∎

It can be shown that with such a definition of $\mathcal{L}$-(2)-$\mathbf{D}$-representability the core of algebraic operations preserves this feature. An outline of the proof is presented below:

1. If $M_1 \sim_{\mathcal{L}\text{-}(2)\text{-}\mathbf{D}} \mathcal{S}_1$, $\mathcal{S}_1 = \{O_{1:i}\}_{i \in [1..k]}$ and $M_2 \sim_{\mathcal{L}\text{-}(2)\text{-}\mathbf{D}} \mathcal{S}_2$, $\mathcal{S}_2 = \{O_{2:j}\}_{j \in [1..m]}$, then $M_1 \cup M_2 \sim_{\mathcal{L}\text{-}(2)\text{-}\mathbf{D}} \mathcal{S}_1 \cup \mathcal{S}_2$.

2. If $M_1 \sim_{\mathcal{L}\text{-}(2)\text{-}\mathbf{D}} \mathcal{S}_1$, $\mathcal{S}_1 = \{O_{1:i}\}_{i \in [1..k]}$ and $M_2 \sim_{\mathcal{L}\text{-}(2)\text{-}\mathbf{D}} \mathcal{S}_2$, $\mathcal{S}_2 = \{O_{2:j}\}_{j \in [1..m]}$, then $M_1 \cap M_2 \sim_{\mathcal{L}\text{-}(2)\text{-}\mathbf{D}} \{O_{1:i} \cup O_{2:j} : i \in [1..k], j \in [1..m]\}$.

3. If $M \sim_{\mathcal{L}\text{-}(2)\text{-}\mathbf{D}} \mathcal{S}$, $\mathcal{S} = \{O_i\}_{i \in [1..k]}$, then $\pi_\mathbf{S}(M) \sim_{\mathcal{L}\text{-}(2)\text{-}\mathbf{D}} \{\gamma_{\mathbf{S}(M) - \mathbf{S} \to \mathbf{D}}(O_i) : i \in [1..k]\}$; where by $\gamma_{\mathbf{S}(M) - \mathbf{S} \to \mathbf{D}}$ we understand a function renaming terms not included in $\mathbf{S}$ to terms from $\mathbf{D}$.

4. If $M \sim_{\mathcal{L}\text{-}(2)\text{-}\mathbf{D}} \mathcal{S}$, $\mathcal{S} = \{O_i\}_{i \in [1..k]}$, then $\rho_\gamma(M) \sim_{\mathcal{L}\text{-}(2)\text{-}\mathbf{D}} \{\gamma(O_i) : i \in [1..k]\}$.

Various reasoning tasks can be performed with use of $\mathcal{L}$-(2)-**D**-representation. For instance checking the consistency of the tarset can be performed by checking the consistency of each of the sets in the representation. Consequently, to check if the module entails a given axiom α, one has to check if every set of the representation entails α.

3.2.2 Module Variables and Couplers in Conglo-S

Conglo-S is extended by functions to manage a knowledge base schema. Apart from administering "contents" of each module, which is done in a way sketched in the previous section, Conglo-S is able to handle schemas: module variables with contents changing over time, and couplers, which bind together specific variables with a relationship described by an algebraic inequality.

A user of the system is presented with an interface which allows for creating, updating, and dropping modules. Originally each module is perceived by the user as an ontology (or a set of sentences). Update operations comprise of adding new sentences and retracting the previously added ones. The sentences provided by a user forms the "core" part of each module variable in the knowledge base.

The schema of the knowledge base may be altered by addition (or by dropping) a coupler. Conglo-S currently allows only for using couplers in the form of $M < E$, where M is a single module variable, and E is an algebraic expression which may contain all the defined variables.

Conglo-S interprets couplers in a fix-point fashion. It means that, after each update the contents of every module variable is recalculated by "firing" every coupler (by performing the assignment $M := E$), and the operation is repeated until the fix-point is reached. While the "core" module contents is preserved, for the sake of enabling updates, the contents visible to the user is in fact the result of performing algebraic operations contained in the couplers.

3.3   Architectural Considerations

The tools presented in the previous subsections were developed in a way which allows for using them in different software configurations. It is possible to use them also outside of Protégé. The two factors that support this feature are: use of standard OWL API and component design with use of OSGi framework.

In fact different configurations, i.e. collaboration between standard (non context-aware) and SIM tools, were analyzed during very early stages of development. The comparison of various configurations is presented in Table 1 (see next page). In the *Ontology* column it is specified if user works with standard ontology or SIM knowledge base. In the *Editor* column it is specified if user works with Protégé without extensions (Standard) or works with Protégé with SIM plug-in (Context-aware). The next column *Reasoner* defines if user works with standard reasoner (Standard; e.g. Pellet [9]) or Conglo-S reasoner (Context-aware). The last column *Comment* contains remarks on a specific configuration.

The set of SIM tools works correctly in all the enlisted configurations. It was achieved by integration with OWL API, especially by the exploit of flow of events provided by *OWLOntologyManager*.

Table 1. Comparison of various configurations of collaboration between standard and context-aware tools

No.	Ontology	Protégé	Reasoner	Comment
1	Standard	Standard	Standard	-
2	Standard	Standard	Context-aware	Context-aware reasoner can cooperate with standard ontologies. It works analogically to the first configuration.
3	Standard	Context-aware	Standard	Standard ontology can be edited by context-aware plug-in, analogically as for non-extended Protégé.
4	Standard	Context-aware	Context-aware	Works analogically to the second and third configurations.
5	Context-aware	Standard	Standard	Context-aware ontology can be edited in Protégé only by expert users, familiar with ontological description of the metamodel (see Sec. 2.3 and 2.5). Reasoner provides sound (but not complete) conclusions.
6	Context-aware	Standard	Context-aware	Context-aware ontology can be edited in Protégé only by expert users familiar with ontological description of the metamodel (see Sec. 2.3 and 2.5). Reasoner provides sound and complete conclusions.
7	Context-aware	Context-aware	Standard	Context-aware ontology is edited with use of designated views. Reasoner provides sound conclusions but not complete ones.
8	Context-aware	Context-aware	Context-aware	Complete set of context-aware tools

OWLOntologyManager is a standard OWL API interface so, as it was previously mentioned before, we can also integrate SIM tools into systems that are devoid of Protégé components. An example of such a system is a Conglo-S console, which function as a demonstrator of tarset algebra for users, and allows for formulating commands in dedicated query/command language developed in ANTLR [13] (cf. Fig. 7).

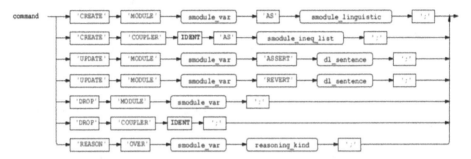

Fig. 7. Grammar for console command in Conglo-S (from ANTLR-works tool [14])

Further possibilities of interoperability stem from using OSGi component-based framework. This decision allows us to easily embed the tools into any system exploiting this standard (including Eclipse RCP) and is important for the further plans of deployment of the tools into SYNAT project YADDA platform [16].

4   Possible Applications for Cultural Heritage

This section contains a discussion about possible applications of the introduced ideas within the SyNaT system designed for facilitation of management and use of digital objects connected with cultural heritage.

4.1   Improving the Efficiency of Reasoning

Let us consider use of SIM model for improving the efficiency of inference from a knowledge base organized in accordance with CIDOC CRM. The motivation behind this work is strictly connected with work (see [15]) of PCSS (Poznan Supercomputing and Networking Centre; a partner in SyNaT). PCSS prepared a rich source of linked data on the basis of the contents of D-Libra system: a system designated for indexing library resources in digital form.

PCSS puts a great effort into adjusting the data in compliance with the requirements of CIDOC. The resulting base is very large and can be processed only by reasoners performing variants of *pD** inference, like OWLIM. Such inference is attractive because of computational properties, but it is not as exhaustive as standard DL reasoning.

Loading the base in its original form to a standard reasoner is an infeasible task. Tests performed by us indicate that for a terminology of such complexity an efficiency barrier occurs at about 3000 individuals. Because the base is several orders of magnitude larger, modularization seems to be a viable solution.

We propose to introduce two-dimensional contextual breakdown of the structure of the base. A "vertical" breakdown would consists in separating the terminology into conceptual units embracing: time-space information, events, agents, and objects. A "horizontal" breakdown would concern the description of the world (context instances) and would consist in separating information objects coming from different libraries.

We plan to perform experiments in order to answer the question whether such decomposition would be enough to perform full OWL reasoning (with Open World Assumption), and what would be the cost of combining knowledge from different contexts in order to answer more comprehensive queries.

4.2   Introduction of Contextual Reasoning to Library Catalogs

Another very interesting stream of our work on contextualization of the SyNaT knowledge base is connected with library catalogs. Books (and other digital objects) are currently organized in accordance with a special catalog of terms (hierarchical thesaurus) called KABA. Use of KABA gives great advantages in unifying library information records, but it also has some disadvantages.

KABA in the present form cannot be treated as a viable source of terminological information for inference engines. One of the reasons is the fact that it carries only loose relationships between terms. KABA distinguishes between broader and narrower words and expression, but this hierarchy should not be confused with concept taxonomy. For example, *a lighthouse* has a broader term *rescue missions*, while naturally there is no subsumption between the terms. However, to some extent, we may say that a book about lighthouses carries some knowledge potentially useful for carrying out rescue missions.

Our aim here is to introduce contextual method of knowledge base organization to distinguish different fragments of KABA hierarchy, and to determine which parts of it and under what assumptions may be treated as proper taxonomies. This is an ambitious task because most probably it would require us to exploit some kind of meta-description (in order to differentiate between various parts of KABA) and to deploy a method of converting parts of PCSS knowledge base ABox containing KABA terms to context types.

While the difficulty of the task seems to be vast, the potential advantages are very appealing. KABA constitutes a large portion of PCSS knowledge base (close to the number of 300,000 individuals), and using it for improving semantic search seems very natural. Meta-knowledge about how to relate different categories of queries to different portions of the hierarchy would greatly facilitate and improve the process of answering. Moreover, the specific fragments of KABA may be used as starting-point ontologies for describing various fields of scientific knowledge. Furthermore, out of the necessity, KABA has to be maintained on the current basis (as it is used for indexing new papers, journals, and books) and should keep pace with the development of science and technology, so that such ontologies may be automatically or semi-automatically kept up-to-date.

5   Related and Further Work

The approach presented in the paper is not the only one concerning knowledge base modularity. In fact there are several notable methods in the field, with probably most prominent ε-Connections [17] and Distributed Description Logics [18]. In the both approaches authors take as a starting point a pair of ontologies and relate their terms with use of special syntactic constructions: axiom with linking relations in the case of ε-Connections, and bridge rules in the case of DDL.

Another very significant stream of works connected with modularity is the research concerning conservative extensions ([5], [19]). While potential applications of the theory seem to be very broad, the most notable studies in the area concern ontology evolution (preparing new versions and assessing their influence on possible prior extensions), and improvement of reasoning performance (with use of automated division of an ontology into modules).

Also within the research area there were formulated proposals of algebras for ontologies. One of the examples is described in [20]; without going into details, the authors postulate treatment of ontologies as graphs and try to use an algebra of graphs including mainly set-theoretic operations (like intersection, union etc.).

Summarizing the above short analysis we should clearly state that probably every element of the proposed here framework can be related to ongoing research. While we are aware of importance of such comparison (and, in fact, performed it, e.g. in [21]), we stress the fact that we primarily perceive the strength of the framework in its wholeness and uniformity. We introduce relatively simple mathematical apparatus and consequently build over it additional elements, driven by the idea that modularity should be considered as one of the fundamental factors in ontology engineering. Through such course of work we managed to integrate several dissipated areas of interest (like ontology evolution, integration, querying, and design) into one consequential stream of works, which would hopefully lead to creation of a model and a suite of tools for knowledge bases comparable in maturity and flexibility to relational databases.

The holistic framework presented in this chapter, although mature from the theoretical point of view, is currently under development. Our work concentrates on preparing a basic set of tools, which would allow us go through a process of designing a tarset knowledge base, as a proof concept for our approach and as an example of practical application of presented theories.

To sum up: Our main goal is to create a set of conceptual and development tools that will make development of knowledge-based systems as natural for a software engineer as it is now in the case of data-based systems. This follows the idea that we call "New Knowledge Engineering Initiative" (NKEI) that we promote among researchers and practitioners. To this end, we, among others, plan to implement a subset of KQL [22] as a suitable and semantically adequate language for querying modularized knowledge bases. In parallel, further theoretical work will be conducted. Among others, we plan to investigate other types of dependencies between tarsets than those that support context-aware knowledge-bases. This would enable us to construct modularized knowledge bases with definable modalities, according to the needs of complicated reality that we live in and that knowledge engineering must face.

References

1. Goczyla, K., Waloszek, A., Waloszek, W.: S-modules - Approach to Capture Semantics of Modularized DL Knowledge Bases. In: Proc. of KEOD 2009, pp. 117–122 (2009)
2. Goczyla, K., Waloszek, A., Waloszek, W.: A Semantic Algebra for Modularized Description Logics Knowledge Bases. In: Proc. of DL 2009. CEUR-WS, vol. 477 (2009)
3. Goczyła, K., Waloszek, A., Waloszek, W.: Algebra of Ontology Modules for Semantic Agents. In: Nguyen, N.T., Kowalczyk, R., Chen, S.-M. (eds.) ICCCI 2009. LNCS(LNAI), vol. 5796, pp. 492–503. Springer, Heidelberg (2009)
4. Goczyła, K., Waloszek, A., Waloszek, W., Zawadzka, T.: Modularized Knowledge Bases Using Contexts, Conglomerates and a Query Language. In: Bembenik, R., Skonieczny, L., Rybiński, H., Niezgodka, M. (eds.) Intelligent Tools for Building a Scient. Info. Plat. SCI, vol. 390, pp. 179–201. Springer, Heidelberg (2012)
5. Lutz, C., Walther, D., Wolter, F.: Conservative extensions in expressive description logics. In: Proc. of IJCAI 2007, pp. 453–459 (2007)
6. Goczyla, K., Waloszek, A., Waloszek, W.: Contextualization of a DL knowledge base. In: Proc. of DL 2007, pp. 291–299 (2007)
7. Goczyla, K., Waloszek, A., Waloszek, W.: Hierarchical partitioning of ontology space into contexts (in Polish). In: Kozielski, S., Małysiak, B., Kasprowski, P., Mrozek, D. (eds.) Bazy Danych. Nowe Technologie - Architektura, Metody Formalne i Zaawansowana Analiza Danych, pp. 247–260. Wydawnictwa Komunikacji i Łączności, Warszawa (2007)
8. Waloszek, A.: Hierarchical contextualization of Knowledge Bases (in Polish). Doctoral dissertation, Gdańsk University of Technology (2010)
9. Pellet: OWL 2 Reasoner for Java, http://clarkparsia.com/pellet
10. Goczyla, K., Waloszek, A., Waloszek, W.: A Semantic Algebra for Modularized Description Logics Knowledge Bases. In: Proc. of DL 2009, pp. 1–12 (2009)
11. The OWL API, http://owlapi.sourceforge.net
12. Sirin, E., Parsia, B., Grau, B.C., Kalyanpur, A., Katz, Y.: Pellet: A Practical OWL-DL Reasoner. Journal of Web Semantics 5(2), 51–53 (2007)
13. Parr, T.: The Definitive ANTLR Reference: Building Domain-Specific Languages. Pragmatic Bookshelf (2007)
14. Bovet, J.: ANTLR Works: The ANTLR GUI Development Environment, http://www.antrl.org/works (accessed June 11, 2012)
15. Mazurek, C., Sielski, K., Stroiński, M., Walkowska, J., Werla, M., Węglarz, J.: Transforming a Flat Metadata Schema to a Semantic Web Ontology: The Polish Digital Libraries Federation and CIDOC CRM Case Study. In: Bembenik, R., Skonieczny, L., Rybiński, H., Niezgodka, M. (eds.) Intelligent Tools for Building a Scient. Info. Plat. SCI, vol. 390, pp. 153–177. Springer, Heidelberg (2012)
16. About YADDA, http://yaddainfo.icm.edu.pl (accessed June 11, 2012)

17. Kutz, O., Lutz, C., Wolter, F., Zakharyaschev, M.: E-connections of abstract description systems. Artificial Intelligence 156(1), 1–73 (2004)
18. Borgida, A., Serafini, L.: Distributed Description Logics: Assimilating Information from Peer Sources. In: Spaccapietra, S., March, S., Aberer, K. (eds.) Journal on Data Semantics I. LNCS, vol. 2800, pp. 153–184. Springer, Heidelberg (2003)
19. Grau, B.C., Horrocks, I., Kazakov, Y., Sattler, U.: Modular Reuse of Ontologies: Theory and Practice. J. of Artificial Intelligence Research (JAIR) 31, 273–318 (2008)
20. Mitra, P., Wiederhold, G.: An Ontology-Composition Algebra. In: Handbook on Ontologies, pp. 171–216. Springer (2004)
21. Goczyla, K., Waloszek, A., Waloszek, W., Zawadzka, T.: Analysis of Mapping within S-module Framework. In: International Conference on Knowledge Engineering and Ontology Development, KEOD 2011 (2011)
22. Goczyła, K., Piotrowski, P., Waloszek, A., Waloszek, W., Zawadzka, T.: Terminological and Assertional Queries in KQL Knowledge Access Language. In: Pan, J.-S., Chen, S.-M., Nguyen, N.T. (eds.) ICCCI 2010, Part III. LNCS(LNAI), vol. 6423, pp. 102–111. Springer, Heidelberg (2010)

Hermeneutic Cognitive Process and Its Computerized Support

Janusz Granat, Edward Klimasara, Anna Mościcka, Sylwia Paczuska,
Mariusz Pajer, and Andrzej P. Wierzbicki

National Institute of Telecommunication, Szachowa 1, 04-894 Warsaw, Poland
{J.Granat,E.Klimasara,A.Moscicka,S.Paczuska,M.Pajer,
A.Wierzbicki}@itl.waw.pl

Abstract. The paper discusses the role of tacit knowledge (including emotive and intuitive knowledge) in cognitive processes, recalls an evolutionary theory of intuition, academic and organizational knowledge creation processes, describes diverse cognitive processes in a research group, but especially concentrates on individual hermeneutic processes and processes of knowledge exchange while distinguishing their hermeneutic and semantic role. The functionality and main features of the PrOnto system are shortly recalled, and the relation of changes in individualized ontological profile of the user to her/his hermeneutical perspective is postulated. An interpretation of such changes for both hermeneutic and semantic aspects of knowledge exchange is presented. Conclusions and future research problems are outlined.

Keywords: hermeneutic circle, hermeneutic spiral, hermeneutic perspective, supporting hermeneutic processes, hermeneutic versus semantic understanding.

1 Introduction

The process of human interpretation of a text, called hermeneutic process, is one of basic cognitive processes and has been analyzed for many centuries; however, most of ontological engineering does not take into account the characteristic features of this process. Thus, the paper concentrates first on the properties of the hermeneutic process, independent whether it is classically understood as a *hermeneutic circle* or in a more contemporary way as a *hermeneutic spiral*. The difference between these two interpretations is whether the phenomenon of intuition is understood as supernatural (in the classical interpretations) or natural, evolutionary motivated (in the contemporary interpretation). Independent from this distinction, it is agreed that the hermeneutic process is a-logical, is based on using a special intuitive mode of human perception called *hermeneutic horizon* or *hermeneutic perspective*. This hermeneutic perspective defines important aspects of human personality and is responsible for the diversity of human tastes and ideas.

Moreover, while the classical interpretation of hermeneutic processes limits them to humanistic research or even postulates that hermeneutics defines the specificity of

R. Bembenik et al. (Eds.): *Intell. Tools for Building a Scientific Information*, SCI 467, pp. 281–304.
DOI: 10.1007/978-3-642-35647-6_19   © Springer-Verlag Berlin Heidelberg 2013

humanistic science [6], more recently it was observed [24, 25] that any science – including exact sciences and technological sciences - relies on an interpretation of scientific texts and thus on hermeneutic processes. Obviously, hermeneutic perspectives of mathematicians are different than those of technologists and those of humanists, but even inside of each such group there is a great diversity of hermeneutic perspectives.

Therefore, any attempt to support cognitive processes of a group of researchers involving interpretation of texts should take into account this diversity of hermeneutic perspectives. The authors of this paper studied this problem using a system PrOnto that applies some concepts from ontological engineering to support highly individualized selection of texts based actually on an hermeneutic perspective. It is shown that recording some aspects of behavior of a user of the PrOnto system it is possible to note changes in her/his hermeneutic perspective. These changes might be interesting and useful to her/his co-workers in a research group. Exchange of information about highly individualized ontological profiles of researchers might, therefore, serve two purposes. One, typically assumed as dominant in ontological engineering, is to support *semantic understanding and rapport,* that is, a deepened common (group) understanding of concepts used in communication. This is a very important aspect of analyzing texts, however, not the only one. Another, necessary in hermeneutic processes, is to support *hermeneutic understanding and intuition,* that is, a deepened individual interpretation of a text or a research idea. We concentrate on the second purpose while being aware that typical ontological engineering concentrates on the first purpose. Even if there is a large body of literature concerning user-oriented ontology formation, see, e.g., [8, 9], it assumes an automatic formation of an ontological model of a user, which can help at most in a semantic rapport but cannot help in hermeneutic understanding because of its a-logical character.

However, before going into detail of this study, we recall first some basic features of cognitive processes, because they contain important a-logical elements, usually undervalued by ontological engineering and semantic studies.

2   The Role of Tacit Knowledge in Cognitive Processes

In a contemporary understanding of cognitive processes, a fundamental role plays *tacit knowledge* [17, 19], which is preverbal, difficult to express in speech, emotional and intuitive, a-logical. This concerns in particular *hermeneutic processes,* see [6, 24, 25], that is, processes of interpreting knowledge in the form of a text. It is tacit knowledge, intuitive and emotional, that allows for an individual interpretation of a text and – immediately or after a shorter or longer period of *gestation* – for a creative generation of ideas stimulated by the text.

The importance of gestation period results from a fundamental difference between knowledge expressed by words, logical and rational, and tacit knowledge, preverbal and unconscious. We need time for an *immersion* of a new rational knowledge into our unconscious or subconscious, emotional and intuitive tacit knowledge, in order to create a new idea (this creative effect has many names: *illumination, enlightenment, aha, eureka, abduction*). This was stressed by a fundamental book of brothers Dreyfuses [5], but also by an evolutionary theory of intuition [23, 27], rational in the sense of Karl Popper [20].

According to this theory intuition consists of preverbal, holistic, unconscious (or subconscious, or quasi-conscious) processing of signals from environment and own memory, motivated by life-long experience and imagination, historically remaining from preverbal stage of human evolution. Quasi-conscious action utilizes *tacit knowing* [19] such as walking, riding bicycle or driving cars. Signals from environment encompass all their forms – *immanent perception,* by all senses, while the sense of vision was probably dominant for humans (before the evolutionary development of speech) and for primate apes. A part of this evolutionary theory of intuition consists of a techno-informational substantiation of the power of intuition through a comparison of informational and computational complexity of speech and vision, with the conclusion that *a picture is worth at least ten thousand words*. This explains *the power of intuitive reasoning* – and also the fact that historically, from Plato, Descartes, Locke, Kant, many philosophers referred to visual aspects of intuitive perception of truth. On the other hand, such evolutionary theory does not say that intuition leads infallibly to truth; our judgments are *mesocosmic biased* [22], and for all examples of Kantian *a priori synthetic judgments* (true, according to Kant, because of the power of our intuition) we can find conditions, in which such judgments cease to be true.

From the evolutionary theory of intuition we can derive a model of a creative intuitive decision process, different than the classical analytical model of a decision process. Herbert Simon [21] defined essential phases of a decision process as *intelligence, design and choice;* later [14] another important phase, *implementation* was added. However, an intuitive decision process has different phases [23]:

1) Recognition, which often starts with a subconscious feeling of uneasiness. This feeling is sometimes followed by a conscious identification of the type of the problem.

2) Deliberation or analysis; for experts, a deep thought deliberation suffices, as suggested by the Dreyfuses [5]. Otherwise any tools of analysis or an analytical decision process is useful - with intelligence and design but suspending the final elements of choice.

3) Gestation; this is an extremely important phase - we must have time to forget the problem in order to let our subconscious work on it.

4) Enlightenment (illumination, aha, eureka, abduction); the expected eureka effect might come but not be consciously noticed; for example, after a night's sleep it is simply easier to generate new ideas (which is one reason why group decision and brainstorming sessions are more effective if they last at least two days).

5) Rationalization; in order to communicate our decision to others we must formulate our reasons verbally, logically, and rationally. This phase can be sometimes omitted if we implement the decision ourselves.[1]

---

[1] The word *rationalization* is used here in a neutral sense, without necessarily implying self-justification or advertisement, though they are often actually included. Note the similarity of this phase to the classical phase of *choice*.

6) Implementation, which might be conscious, after rationalization, or immediate and even subconscious.

Especially important are the a-rational phases of gestation and enlightenment. They rely on utilizing the enormous potential of our mind on the level of preverbal processing: if not bothered by conscious thought, the mind might turn to a task previously specified as most important but forgotten by the conscious ego. Especially in the Far East, there exist many cultural institutions supporting gestation and enlightenment. The advice of *emptying your mind, concentration on emptiness or beauty, forgetting the prejudices of an expert* stemming from Zen meditation or Japanese tee ceremony can be interpreted as useful methods of supporting the functioning of our unconscious part of mind.

The shortly signalized here role of tacit knowledge in cognitive processes implies that these processes have especially individual character: tacit knowledge is a determinant of personality, thus any attempt to make uniform models of it, or to corporately privatize this knowledge (to subordinate it to the goals of a market corporation) have aspects of new slavery. If we start to make automatic model of a user, we should actually first ask her/his permission. Therefore, the theme of this paper should be approached with caution: we can distinguish some common or group features in cognitive processes, we should not, however, go too far in attempts to make uniform models of them.

2.1   Organizational Processes of Knowledge Creation

The understanding how to create knowledge for the needs of an organization, especially a market corporation, was increasing gradually during the second half of XX Century. Historically, the first of such methods appeared rather early (Osborn 1957, [18]); it was the method of *brainstorming,* much later fully formalized and described as a *DCCV spiral of brainstorming* [12, 13].

Brainstorming has many definitions, its name suggests an intensive inspiration and group generation of new ideas, a kind of group enlightenment (illumination, aha, eureka, abduction). However, the term *brainstorming* obtained a specific meaning after the book *Applied Imagination* [18]; *brainstorming is a group process of creating new ideas with postponing an evaluation of their value.* Later it was observed that the method of brainstorming can also be applied individually, since its essence relies on generation and recording new ideas with postponing their evaluation, although obviously we can generate more ideas in a group and competition provides for a specific positive feedback in the group process.

This phase of brainstorming was called *divergent,* or *divergent production.* The rules of this divergent phase are following:

1) The goal of brainstorming in the divergent phase is the creation of a large number of ideas, not necessarily the best ideas;
2) The evaluation of ideas (whether they are good or bad, utopian or not, etc.) should be postponed;

3) Unusual ideas are welcomed;
4) Further development or modifications of already proposed ideas are also welcomed.

Brainstorming has many advantages and disadvantages, see [13]. The basic disadvantage is its inconsistence: after the divergent phase it is necessary to switch to a divergent phase, to selection and choice of ideas, and this collides psychologically with the attitude of "let all flowers bloom" of the first phase. In other words: who should be responsible for the selection of ideas – the entire group or the organizer of brainstorming? Who owns ideas generated during brainstorming? Despite these drawbacks (which can be overcome by defining several variants of the process), brainstorming became one of most often used methods of solving problems and generating ideas in industrial and other organizations, in *organizational knowledge creation*, even if it has much less applications in *academic knowledge creation,* see below. Thus, brainstorming is the oldest and broadest applied organizational cognitive process, earlier than other such processes described below. The first applications of brainstorming occurred in NASA in relation to planning cosmic research.

There were many attempts to define a general model of brainstorming, see [12, 13], but the essential phases of such process are as follows, represented as transitions in the model of *DCCV brainstorming spiral* in Fig. 1:

A) *Divergent thinking (Divergence),* as in the divergent phase described above;
B) *Convergent thinking (Convergence),* selection and choice of ideas;
C) *Crystallization of ideas (Crystallization),* a more detail development of ideas (in particular, analytical, since the initial phases have essentially an intuitive character);
D) *Verification of ideas (Verification),* that can apply diverse methods (learning by doing, a *debate* discussed below, etc.).

An interpretation of the model from Fig. 1 as a *spiral* results from the fact that a repetition of the process of brainstorming can only increase the number and improve the quality of generated ideas – knowledge is not lost by its intensive use. Significant in this model are not only *transitions*, but also their interpretation as routes connecting the *nodes* of this network-like model. The nodes are *individual intuition* (the source of divergent transition), *group intuition* (individual ideas are presented to the group), *group rationality* (*Convergence* results in group rationalization of individual ideas), *individual rationality* (*Verification* can be interpreted as improving individual intuition by an attempt to verify an idea).

Another method of knowledge creation for the needs of today and tomorrow, younger than brainstorming but the oldest between diverse methods that emerged in the last decade of XX Century, is *Shinayakana Systems Approach* [16], developed by Nakamori and Sawaragi and related to interactive decision support. *Shinayakana* is a Japanese term expressing a combination of the hardness of a sword and the elasticity of a willow withe; it stresses a synthesis of so called *soft* and *hard* systems analysis.

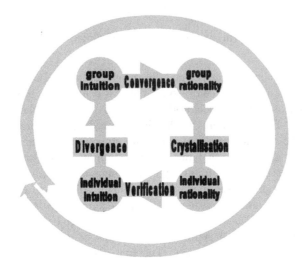

Fig. 1. DCCV *brainstorming spiral* (Kunifuji et al. [12], [13])

Parallel, in management science, another theory was developed also by Japanese authors – Nonaka and Takeuchi [17]. Without the limitations resulting from the discourse between soft and hard systemic approaches[2], these authors were first to propose and algorithmic process to create knowledge. This theory has a revolutionary importance, even if it concerns only the creation of small increases of utilitarian knowledge, ready to be used in organizations functioning on economic markets. It was first to stress clearly the role of a group in processes of knowledge creation

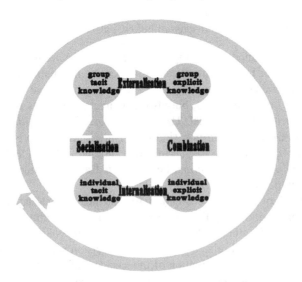

Fig. 2. SECI *spiral* of organizational knowledge creation (Nonaka i Takeuchi [17])

---

[2] Soft systemic approaches deny the applicability of algorithmic methods to social problems.

(brainstorming uses a group, but originally did not stress its role), it also assumes clearly the use of a-rational *tacit knowledge* (Nonaka and Takeuchi generalized thus the concept of *tacit knowing* of Polanyi [19]), necessary for knowledge creation.

This theory, very popular today in management science, is usually expressed in the form of the *SECI spiral* illustrated by Fig. 2. This spiral contains four transitions between four nodes on two axes; the axes are called *epistemological dimension* composed of *explicit knowledge* and *tacit knowledge,* and the *social dimension[3]* that contains an *individual* and a *group.* Consecutive transitions are:

1) *Socialization,* in which tacit knowledge of individuals is transformed into tacit knowledge of the group (this is characteristic for the Far East character of this method: Japanese stay after work to drink together and informally discuss problems related to work);

2) *Externalization,* in which group tacit knowledge is rationalized and transformed into group explicit, rational knowledge;

3) *Combination,* in which group explicit knowledge is transformed into individual explicit knowledge;

4) *Internalization,* in which explicit individual knowledge is transformed into tacit knowledge (e.g., by practical implementation of explicit knowledge, increasing the intuition of an individual).

Each application of knowledge – as illustrated by many examples described by Nonaka and Takeuchi - can only increase it, hence *SECI spiral* leads to organizational knowledge creation.

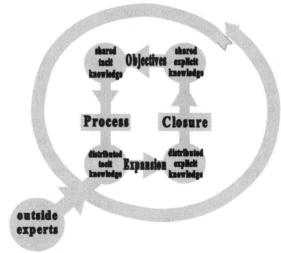

Fig. 3. OPEC *spiral* of knowledge creation in an Anglo-Saxon type of organization (Gasson [7])

---

[3] Originally called *ontological dimension* by Nonaka and Takeuchi, but this creates misunderstandings related to diverse interpretations of the word *ontology.* Similarly, Nonaka and Takeuchi used the term *knowledge conversion* instead of a more general term *transition* (because knowledge is not lost when used, it is not converted, only transformed).

Because of great international impact of the theory of Nonaka and Takeuchi, many competing theories were created, particularly in the USA. One of them is the theory of Gasson [7], called *OPEC spiral* and illustrated on Fig. 3.

The nodes on the network considered by Gasson, even if differently named, are practically identical with the nodes from Fig. 2. The transitions between them have different direction and character: these are *Objectives, Process, Expansion,* and *Closure.* For a more detailed analysis see [24, 25]; these transitions have a typical Anglo-Saxon character (starting with a joint discussions of goals, *objectives,* common for the Western culture, while *socialization* of Nonaka and Takeuchi has Japanese character).

Despite some essential differences between three creative processes illustrated by Figures 1, 2, 3, they all have an *organizational character:* they are motivated by group interest and the results of these processes are the property of the group, of an organization.

2.2   Academic Processes of Knowledge Creation

Quite different character have typical processes of *academic* knowledge creation. They are motivated by an individual interest of the researcher, even if they could be also supported by a group, and the results of these processes are usually individual property – except in cases when they are published to become a part of intellectual heritage of humanity. We present here shortly models [24, 25] of two such academic processes of creating knowledge – *debate and experiments,* and in more detail a model of the third one – *hermeneutics,* understood here broader than in its classical, humanistic understanding, as an art of interpretation of any texts or other elements of the intellectual heritage of humanity, no matter whether they concern theology, human sciences, exact sciences or even technology.

Debate is a cognitive process that uses – approximately – the same nodes as *SECI spiral* or *OPEC spiral,* but with different transitions and different interpretation. This process of normal creating knowledge at universities is well known and easy to recognize in the *EDIS spiral* in Fig. 4.

An individual researcher, due to his intuition, has an idea – of a smaller or larger significance – and the illumination alone is not sufficient, it is necessary to convert this idea into words or equations; let us call this transition *Enlightenment.* A group supports the researcher primarily through securing for him a forum for discussing her/his ideas – the more discerning discussion, the better the support of the group; this transition is called *Debate.* These two stages are well known, but now comes a new element, resulting from the evolutionary theory of intuition. This theory indicates that a deeper, more penetrating debate will be achieved if we give the group time for reflection, for gestation of comments, an *Immersion* of group rationality in group intuition. This implies a practical conclusion, *the principle of double debate:* discussion of new ideas should be repeated e.g. after a few days or after a week, if we want to support the researcher not only with full explicit knowledge, but also with full tacit knowledge and intuition of the debating group. After obtaining the comments of the group, the individual researcher makes a choice – *Selection* – between these comments, and we know well that such choice is not rational, occurs on the intuitive level.

Fig. 4. EDIS *spiral of debate* (Wierzbicki and Nakamori [24])

We could ask why to describe in the form of abstractive, heuristically approximate models such cognitive processes that are well known to us? The answer is simple: even in such well known process as *Debate* it was possible to draw additional important conclusions, such as *the principle of double debate.* We can also ask what was simplified or neglected in such a model? For example, normal academic processes of creating knowledge reach also into rational, emotional and intuitive intellectual heritage of humanity [24, 25] and end often with a publication; thus, the *EDIS spiral of debate* describes only a small part of such processes.

In experimental science, beside discussion we check our ideas through experiments. The spiral describing experimental knowledge creation is a modification of the *EDIS spiral* with debate substituted by an experiment. Such modified *EEIS spiral of experiments* is shown in Fig. 5.

The transition *Experiment* denotes here simply experimental verification. However, each experimental researcher knows well that raw experimental data are not worth much, it is necessary to interpret them – in the transition *Interpretation,* which has a character of an immersion of the raw data into the experimental intuition of the researcher, based on her/his experience. A *Selection* occurs here as well, this time between these aspects of experimental data that have most important impacts on the development of researcher's ideas. Experiments have often individual character, although larger experiments are obviously organized and realized by a group.

Hermeneutics – initially understood as an art of interpreting the Bible, later of interpreting any texts – was for a long time understood as a typically humanistic activity, even as a distinguishing mark of human sciences [6]. However, some great philosophers of XX Century treated hermeneutics broader, as applicable to all sciences.

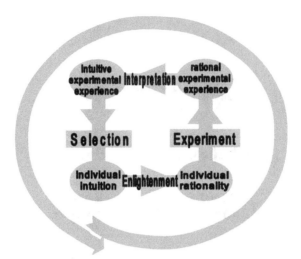

Fig. 5. EEIS *spiral of experiments* (Wierzbicki and Nakamori [24])

After their example, we want to treat here hermeneutics as the art of interpretation of any selected elements of the intellectual heritage of humanity, thus a part of any cognitive process, in particular an academic one, including a collection of texts (or films, etc.) important for a research theme, their interpretation and reflection on them. Hermeneutic creation of knowledge is described by *EAIR hermeneutic spiral,* see Fig. 6. This spiral is equivalent to the *hermeneutic circle,* only in a naturalistic interpretation: closed not through transcendence, as postulated by Gadamer [6], but through the power of natural human intuition.

If we have a new idea due to a personal *abduction* or *illumination (Enlightenment),* we search for related materials in libraries or in Internet and subject the materials to a rational *Analysis.* But this does not suffice for their hermeneutic capture; we must submit them to an *Hermeneutic Immersion* into our unconscious mind and intuition, transmit them into our intuitive perception. This immersion is possible, as suggested by the evolutionary theory of intuition, through letting our unconscious mind time for *Reflection* that might result in new ideas. We stress that this is not only a description of the work of a humanist in a *hermeneutic circle:* a technological researcher, working on new technical constructions, performs an inspection of research materials in the same way.

A thorough understanding of the hermeneutic spiral is a necessary element for a good support in text analysis, unfortunately omitted usually in logical or semantic analysis, ontological engineering or artificial intelligence. This is also an element of fundamental importance for such projects as SYNAT. Therefore, we add here some comments to the subsequent transitions in Fig. 6. An individual idea resulting from *Enlightenment* can guide us in searching for texts or other materials in a digital library. The *Analysis* of this idea can be obviously supported by the tools of semantic analysis or ontological engineering. We must remember, however, that such analysis is only a part of the actual cognitive process: the result of such analysis, a penetrating

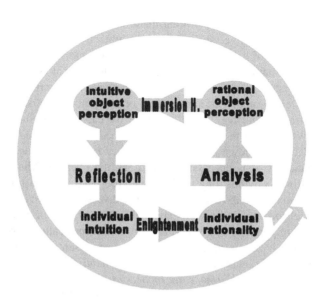

Fig. 6. EAIR *hermeneutic spiral* (Wierzbicki and Nakamori [24])

rational perception of the research object, must be *Immersed* into our *hermeneutic perspective,* we must let interpret this result by our unconscious mind. Such *intuitive perception of the research object* is the foundation of *Reflection* as a method of enriching our intuition that might result in new ideas.

The hermeneutic spiral is the most individual process of knowledge creation. We should stress again that almost all academic micro-processes of knowledge creation are motivated individually, even if they assume a participation of a group, as in debate: the group only supports the improvement of individual ideas, the goal is individual scientific degree or publication. On the other hand, all organizational micro-processes of knowledge creation – brainstorming, *SECI spiral, OPEC spiral* etc. – are motivated by the interests of the group: it is assumed that the knowledge thus created will be the property of the organizers of the process or the organization (even if this assumption is often unstated and the members of the group might use the intellectual heritage of humanity and some elements of the hermeneutic process). This might be the main reason why the well known process of brainstorming did not find a broader applications in academic knowledge creation. This means, however, that academic and organizational processes of knowledge creation are essentially different, which might be one of the main reasons of the difficulties and delays in transferring knowledge from academia to industry.

Nevertheless, hermeneutic processes have also important group aspects; it concerns, in particular, the concept of *hermeneutic perspective* or *hermeneutic horizon.* As stressed by Zbigniew Król (see e.g. [11]), hermeneutic horizon - interpreted in mathematics as the collection and interpretation of axioms that are assumed intuitively to be true – is not absolutely permanent, but is relatively permanent and common for the society of mathematicians in long historical periods. Therefore, hermeneutic horizon is not fully subjective, nor is it an intersubjective result of a discourse, but rather it is an element of a *long duration historical structure* as defined by Fernand Braudel [1], e.g., in mathematics it results from the doctrine of educating mathematicians in a given historical period. This means, however, that professional groups – such as lawyers, telecommunication engineers or computer scientists – have specific *group hermeneutic perspectives,* usually distinct for each group. Such group hermeneutic perspectives can be to some extent researched, *reconstructed,* in such a way that Zbigniew Król has shown the *reconstruction of hermeneutic horizon* of ancient Greek mathematicians: it is necessary then to ask the question what truth is considered to be intuitively obvious for a professional group in a given historical period.

However, in a given group there is also a large diversity of *individual hermeneutic perspectives.* Such internal convictions about the truth of certain assumptions about the world and the importance of given values are sometimes called *horizontal convictions;* they are decisive for a personality and have an essential impact on cognitive processes (see, e.g., [15]). This means, on the other hand, that such internal stages of the hermeneutic process, as *Hermeneutic Immersion* or *Reflection* are deeply individual and the chances of their computerized support are scarce. Nevertheless, any information even indirectly characterizing such processes might be useful for the work of a research group.

3   Cognitive Processes in a Research Group

Here we limit the range of analysis to a relatively small research group, using (not necessarily only academic) micro-processes of creating knowledge, with a particular attention paid – because of general interest of digital libraries in the project SYNAT – to individual or group studies of literature of a research object, in particular to hermeneutic processes. These means either individual hermeneutic processes, as described above generally when discussing the *EAIR spiral,* or group hermeneutic processes, consisting of exchange of information and knowledge obtained in individual hermeneutic processes. Before we describe the possibilities of their computerized support, it is necessary to recall fundamental functions of the PrOnto system.

3.1   System PrOnto as a Tool for Supporting Research Group

The research of the stage A15 of the SYNAT project in the National Institute of Telecommunications concerned the construction of an individualized interface of an user of a digital library for finding interesting texts in large text repositories. Similar interfaces can be used for network search, this leads, however, to additional problems. Many organizations have already large text repositories, a vast number of texts is accessible in Internet, but the problem of finding interesting texts in a given collection of them still does not have a satisfactory solution. This is because typical search engines in Internet are commercially oriented and classify texts according to their own, internal criteria, while a user would like to be sovereign in her/his choice, thus obtain a ranking list classified according to own intuitive criteria. Moreover, it is sometimes not necessary to search in entire network, a given digital library might be sufficient, including a text repository and possibly also other materials (e.g., multimedia, provided they are described by tags and comments forming their verbal metadata).

This work was based on a prototype of the system PrOnto, developed earlier in a PBZ project *Teleinformatic Services and Networks of Next Generation: Technical, Application and Market Aspects,* and improved for the needs of SYNAT project. System PrOnto supports the work of a research group (Virtual Research Community, VRC) and contains a *radically personalized user interface.* The radical personalization of the interface results from the assumption that research preferences of the user cannot be formalized logically or probabilistically (according to the evolutionary theory of intuition described earlier and the related estimate that at most 0.01% of neurons in our brain are used in a logical, analytical way, see also [23, 24, 25]). Therefore, the interface should preserve and support the intuitive character of user choices, in fact − support rather hermeneutic understanding than semantic rapport, even if the analytical part of the system exploits the tools of ontological engineering. The latter choice results from the conviction that the organization of knowledge in an ontological structure is much more effective in representing knowledge for a VCR than relying only on keywords and key phrases. Ontological concepts and relations between them support intuitive knowledge classification in a way that can help in individual or group research. Individual profiles of interests, obtained this way, can be also used for supporting knowledge sharing in VCR.

Such radical personalization, however, contradicts the paradigmatic approach to ontological engineering, that treats ontology always as an expression of shared, common knowledge, a (very useful) tool of supporting semantic rapport, e.g., as the summary of all knowledge accumulated in the Web, see e.g. [4]. Even in user-centric approaches to ontology construction, see, e.g., [8, 9], the goal is to construct a model of a user that has a semantic value and does not serve to support hermeneutic understanding. We should stress again, therefore, that even if system PrOnto uses the concepts and tools of ontological engineering, it uses them with a different goal than their traditional treatment; *the goal is supporting the work and document search for an individual researcher or a small research group, while concentrating on hermeneutic aspects of research process.* In this sense we speak about a radical personalization of the interface, even if the tools of ontological engineering are used in this interface.

PrOnto model supports a group of researchers (VRC) through functionalities serving an individual researcher or group cooperation. This model includes:

1) A radically personalized ontological model, called personal ontological profile of the user, composed of three layers:
a) The layer of intuitive, a-logical *concepts* **C***;* the *radical personalization* consist in treating those concepts as intuitive, hermeneutic, personal creations of the user and in avoiding too far reaching logical interpretations of these concepts (e.g., the concept *Markov chains* can actually mean *these aspects that are now interesting for me in the theory of Markov chains*). We also refrain from too far reaching automation of transforming such concepts, although we allow an intuitive definition of relations between these concepts.

b) The layer of *keyword phrases* **K** (that are subject to semantic and logical analysis with the use of all tools of ontological engineering);

c) The layer of *relations between concepts and keyword phrases* **f**: **CxK→R** (in the initial version of PrOnto we used only simple weighting coefficients subjectively defined by the user, but precisely this layer can be diversely interpreted and extended).

The basic version of PrOnto system assumed only one type of an index of semantic importance $h(d,c)$ of a document $d \in D$ (in a given set of documents, e.g. a repository) for a concept $c \in C$ defined by the user:

$$h(d, c) = \sum_{k \in K} f(c,k) \cdot g(d,k) \qquad (1)$$

where $k \in K$ denotes a keyword or key phrase, $g(d,k)$ is a result of indexing the semantic importance of a document d with respect to the key phrase k, using the tools of ontological engineering (e.g., the classical index *TF-IDF*, *Term Frequency – Inverse Document Frequency*), while $f(c,k) \in F$ denotes an importance coefficient defined subjectively by the user in her/his personalized ontological profile of interests. However, in further work we included diverse interpretations of these relations and thus diverse indexes of resulting semantic importance $h(d,c)$.

2) *Document repository* **D**, containing documents interesting for the user or entire group of them (VRC) in the form of full text or a network link to such text;

3) *A method of search and ranking of documents in the repository* for an individual user based on her/his radically personalized ontological profile (many methods are possible and the model of a user does not uniquely define such a method);

4) *An agent of network search* (so called *hermeneutic agent*) that performs search in all accessible network – usually with help of accessible search engines – for new documents in order to enrich the repository, including a ranking method and(or) a decision rule;

5) *Functionalities supporting an effective exchange of knowledge between users* that can enrich PrOnto system either for an individual or for group user. Such functionalities include:

a) cataloguing documents for a group of users (VRC);

b) supporting research collaboration in the group (information about new documents judged as interesting by some users, etc.);

c) search for similarities in user interests, etc.

Since *concepts* are treated in PrOnto system as highly personalized, hermeneutic and intuitive, we should not give them far reaching logical interpretations. For example, we cannot demand that they constitute a consistent logical classification of key phrases, since a user might relate several concepts to a given key phrase. On the other hand, we assume that the user might relate her/his concepts by an acyclic graph structure. The concepts in PrOnto are similar, in a sense, to *tags*, annotations to documents defined intuitively by an user, with the distinction that concepts in PrOnto describe not a document, but the preferences and profile of the user.

Because of such personalized and highly intuitive character of concepts in PrOnto, we might use *fuzzy logic* (that was devised precisely for representing fuzzy intuitive relations) to describe and interpret relations between concepts or between concepts and key phrases. For example, suppose that key phrases are interpreted as fuzzy logical product *'and'* relations between keywords in these phrases; but we might also group key phrases in groups K_l, $l = 1,... L$, inside which fuzzy logical *'or'* relation is assumed (in particular, we might assume that such relation holds between all key phrases, in which case $l = 1$, $K_l = K$). However, suppose between groups K_l again fuzzy logical product *'and'* relation is assumed. Suppose we interpret importance coefficients $f(c,k)$ as fuzzy membership values. Then Eq. (1) that defines a measure of importance of a document $d \in D$ with respect to a given concept $c \in C$ as a function $h(d, c)$ can be modified as follows[4]:

$$\mathbf{h}(d, c) = \min_{l = 1,... L} \max_{k \in K_l} f(c,k) \; g(d,k) \tag{2}$$

Since the subdivision of the set of key phrases into groups K_l might be too onerous for the user, we might assume the simplest case $l = 1$, $K_l = K$, in which case Eq. (2) modifies to:

$$\mathbf{h}(d, c) = \max_{k \in K} f(c,k) \; g(d,k) \tag{3}$$

This way of defining the importance of a document for a concept we shall assume to be basic for fuzzy interpretations of concepts, because the opposite interpretation (using fuzzy logical *'and'* relation between all key phrases) would be equivalent to interpreting the set of key phrases as one long key phrase, with the corresponding equation:

$$\mathbf{h}(d, c) = \min_{k \in K} f(c,k) \; g(d,k) \tag{4}$$

which would most probably lead to extreme or erroneous interpretations (e.g., if some $f(c,k)$ are correctly defined to be zero, when a key phrase is meaningless for a concept).

The above equations define the importance of a document for a concept; but how to define the importance of a document for entire ontological profile? If we just assume fuzzy logical *'or'* relation between all (equally important)[5] concepts, then the corresponding measure of importance $\mathbf{h}_C(d)$ can be expressed as

$$\mathbf{h}_{Cor}(d) = \max_{c \in C} \max_{k \in K} f(c,k) \; g(d,k) \tag{5}$$

We can as well assume fuzzy relation *'and'* between the concepts, but this assumption has a smaller applicability because of possible extreme interpretations as described above:

$$\mathbf{h}_{Cand}(d) = \min_{c \in C} \max_{k \in K} f(c,k) \; g(d,k) \tag{6}$$

---

[4] We use here the simplest fuzzy operators for *and* and *or*. For other forms of such operators see, e.g., [8].

[5] We might assume initially equal importance of all concepts, since differences in their importance can be expressed by the user by assigning higher or lower levels of importance coefficients $f(c, k)$ uniformly for a concept deemed to be more or less important. See, eg., [26].

The measure of importance $\mathbf{h}_C(d)$ in either version can be used then in ranking documents for the user. However, the issue of ranking can be approached not only from the perspective of fuzzy logic; equally, or even more adequate might be approaches resulting from the theory of multiple criteria decision making.

In such approaches, we interpret concepts $c \in \mathbf{C}$ as *criteria of choice*. We shall distinguish then at least two cases: the criteria can be *compensable* (i.e. a large value of one criterion can compensate a small value of other criteria), or the criteria can be *essential* (i.e. all criteria should have reasonably large values). In the latter, but also in the former case, it is reasonable not to compare absolute values of all criteria, but their values compared to statistical means, as in the concept of *objective ranking* [24]. This is because it might happen that a set of key phrases related to a given concept is occurring rarely in all documents, thus all documents will have small importance measures for the concept; however, it is important to find documents that have importance larger than the statistical mean. For this reason, we compute also statistical means (average importance of all documents for a given concept):

$$\mathbf{h}_{av}(c) = \Sigma_{d \in \mathbf{D}} \, \mathbf{h}(d, c) \, / \, |\mathbf{D}| \tag{7}$$

where $\mathbf{h}(d, c)$ is computed as in (3) and $|\mathbf{D}|$ is the number of documents in the repository. If we use such statistical means both for the case of interpreting concepts as *compensable criteria* (of equal importance), and for the case of interpreting them as *essential criteria*, we obtain the following measures of importance of a document $d \in \mathbf{D}$ for entire set of concepts $\mathbf{C}$:

$$\mathbf{h}_{Ccom}(d) = \sum_{c \in C} (\mathbf{h}(d,c) - \mathbf{h}_{av}(c)) \cdot \tag{8}$$

$$\mathbf{h}_{Cess}(d) = \min_{c \in \mathbf{C}} (\mathbf{h}(d, c) - \mathbf{h}_{av}(c)) + \varepsilon \Sigma_{c \in \mathbf{C}} (\mathbf{h}(d, c) - \mathbf{h}_{av}(c)) \tag{9}$$

where $\mathbf{h}(d, c)$ is also computed as in (3) and the coefficient $\varepsilon > 0$ in (9) indicates a compromise between interpreting the relations between concepts as a fuzzy logical *'and'* operation and interpreting them as compensable criteria (for $\varepsilon = 0$, Eq. (9) would be almost equivalent to (6)).

We decided to test experimentally these four cases, while interpreting the concepts:

    a) As many compensable criteria of choice, Eq. (8) – method COMP;
    b) As many essential criteria of choice, Eq. (9) – method ESS;
    c) As fuzzy logical concepts subject to a logical sum *'or'* operation, Eq. (5) – method OR;
    d) As fuzzy logical concepts subject to a logical product *'and'* operation, Eq (6) – method AND.

We have thus at least four different variants of semantic importance indexes of a document with respect to a given ontological profile of a user. As reported in [2, 3], we performed many empirical tests of such indexes and related methods of ranking documents.

The tests were performed with a group of five users that have downloaded a repository of over four hundred of documents. Table 1 presents the results of ranking of first

20 of these documents for the ontological profile of user 1, while applying four methods (for the method ESS interpreting concepts as essential criteria, $\varepsilon = 0.5$ was used). Rank 20+ means that according to the given method, this document would be placed outside of 20 first documents. Rank [0.0] means that according to the given method, the measure of importance of this document would be zero (and thus the document would be not only 20+, but also not ranked at all). Note that this occurs for most of 20 first documents ranked according to method COMP, if method AND is used. This confirms our theoretical expectations described earlier: method AND results in extreme interpretations. This does not mean that the method is totally useless: note that two documents in Table 1 were ranked by method AND, one (on ranking place 16 according to method COMP) ranked by AND on place 1.

Table 1. Ranking of documents by diverse methods for user 1

Document title \ Ranking method	COMP	ESS	OR	AND
OntologyMatching.org	1	1	1	[0.0]
Web Ontology Language…	2	2	2	[0.0]
From SHIQ and RDF to OWL: The Making...	3	3	3	[0.0]
OWL DL vs. OWL Flight: Conceptual...	4	4	4	[0.0]
Three Theses of Representation in...	5	5	5	[0.0]
Semantic Mediawiki: A User-Oriented...	6	6	6	[0.0]
The 3Cs of Knonledge Sharing	7	9	17	[0.0]
Taking QuickPlace to the next level...	8	10	11	[0.0]
Knowledge Management - Empolis	9	11	7	[0.0]
QuizRDF: Search Technology for Semantic Web	10	8	14	13
Knowledge Management in a Research Organization...	11	12	8	[0.0]
Scalable Semantic Web Data Management...	12	13	9	[0.0]
Genea: Schema-Aware Mapping of Ontologies...	13	14	10	[0.0]
decoi2009rewerse.pdf	14	15	12	[0.0]
Knowledge management…	15	16	13	[0.0]
Towards Peer-to-Peer Semantic Web...	16	7	20+	1
Knowledge management system...	17	17	15	[0.0]
Collaborative Knowledge Sharing	18	18	20+	[0.0]
Semantic Alignment of Business Processes	19	19	20+	[0.0]
Techniki informacyjne dla wnioskowania...	20	20	16	[0.0]

Similar conclusions can be drawn from Table 2 that presents the ranking of first 20 documents, but according to the ontological profile of user 2.

Table 2. Ranking of documents by diverse methods for user 2

Document title \ Ranking method	COMP	ESS	OR	AND
Distributional hypothesis ...	1	1	1	[0.0]
ACL Anthology	2	2	2	[0.0]
Construction Grammar website	3	3	3	[0.0]
OntologyMatchnig.org	4	4	6	[0.0]
The Emile Program	5	5	4	[0.0]
el.org - FCG publications	6	6	5	[0.0]
pinto.pdf	7	8	7	[0.0]
Roberto Navigli - Publications	8	7	20	[0.0]
A Framework for Understanding and...	9	9	8	[0.0]
Ontology Research and Development Part 2...	10	10	15	[0.0]
metamodel.com - What are the differences...	11	12	9	[0.0]
Fluid construction grammar...	12	14	10	[0.0]
Three Theses of Representation...	13	13	11	[0.0]
Construction Grammar For Kids	14	18	12	[0.0]
A Bottom-Up Strategy for Enterprise...	15	15	13	[0.0]
Heterogeneous Ontology Structures for...	16	16	14	[0.0]
Ontology (information science)...	17	17	17	[0.0]
Ontology Research and Development Part 1...	18	11	20+	1
On Accepting Heterogeneous Ontologies...	19	19	16	[0.0]
Collaborative Ontology Construction...	20	20	18	[0.0]

However, *the most important conclusion* results from a comparison of these two tables: *even if the interests of these two users are similar* (both are co-authors of the PrOnto system), *the results of ranking are quite different* – only one document (OntologyMatching.org) is repeated in both Tables. Thus, the ontological profiles specified by them represent rather the differences of their personalities, their hermeneutic perspectives, than the semantic coherence of their interests. *This empirical fact confirms our theoretical conviction that the ontological profiles of users should be radically personalized when expressing their hermeneutic interests.* Thus, PrOnto system with its highly personalized user profile might give results that would be difficult to interpret semantically, but are useful for support of hermeneutic processes.

Both Tables suggest that the three first methods of ranking (COMP, ESS, OR) give similar results on the first places of ranking – even if these results are different for different users. For example, user 1 obtained identical ranking by these three methods for his first 6 documents, user 2 – identical ranking by these three methods for his first 3 documents and rather similar ranking for next 4 documents. This is not a drawback, but suggests rather good robustness of ranking; lower ranking places are obviously different for different methods.

The last ranking method, AND, prefers documents that are related (through key phrases) to all concepts in user's ontological profile; this property can be interpreted both as an advantage and an disadvantage, depending on the needs of the user. Thus, we should give the user the possibility of defining her/his own compromise between a multiplicatory method (AND) and an additive method (all three methods COMP, ESS, OR are partly additive). Such a compromise is offered by the ESS method through the possibility of choosing the parameter ε.

In Table 3, we present the results of ranking obtained by the COMP method, AND method – as two methods from the opposite ends of the range – and the in-between method ESS with the parameter ε changing from the value 1.0 (close to COMP) to the value 0.001 (close to but slightly different from AND).

Table 3. Ranking of documents by diverse methods for user 2, change of the coefficient ε for ESS (upper part COMP presents 10 documents ranked as first by COMP method, lower part AND presents 8 documents ranked as first by AND method, numbers denote places in ranking, sign - denotes a very far place in ranking)

Document title \ Ranking method ($\varepsilon = ...$)	COMP	ESS 1.0	ESS 0.5	ESS 0.1	ESS 0.05	ESS 0.04	ESS 0.01	ESS 0.001	AND
COMP:									
Distributional hypothesis...	1	1	1	1	1	2	-	-	-
ACL Anthology	2	2	2	2	4	5	-	-	-
Construction Grammar …	3	3	3	3	5	6	-	-	-
OntologyMatchnig.org	4	4	4	4	3	3	27	-	-
The Emile Program	5	5	5	7	8	22	-	-	-
el.org - FCG publications	6	6	6	8	21	-	-	-	-
pinto.pdf	7	7	8	9	10	16	-	-	-
Roberto Navigli - Publications	8	8	7	6	6	4	3	4	-
A Framework for …	9	9	9	10	14	23	-	-	-
Ontology Research … Part 2	10	10	10	11	19	28	-	-	-
AND:									
Ontology Research … Part 1	18	12	11	5	2	1	1	1	1
42.pdf	-	-	-	-	-	-	24	24	2
Text Onto Miner...	-	-	25	12	7	7	2	2	3
Ontology Creation Process...	-	-	-	-	27	29	25	25	4
Complexity Analysis...	-	-	-	-	-	-	-	-	5
D3.3-Business...	-	-	-	-	-	-	-	-	6
Mining meaning...	-	-	-	-	24	20	4	3	7
State od The Art...	-	-	-	-	-	-	-	-	8

The results in Table 3 suggest that method ESS with ε > 0.1 gives results practically identical to method COMP. Method ESS with ε < 0.04 gives results similar to, nevertheless different than method AND. The reason for this distinction is that method ESS analyzes differences from average values, while method AND analyzes absolute values. Therefore, if the average value $\mathbf{h}_{av}(c)$ is small for a given concept c, documents d with sufficiently high values of $\mathbf{h}(d, c)$ - $\mathbf{h}_{av}(c)$ are ranked high by method ESS, while they might be ranked low by method AND because of low absolute value of $\mathbf{h}(d, c)$; in such situation; AND method would rank high only documents with exceptionally high values of $\mathbf{h}(d, c)$. It appears, therefore, that method AND has an essential drawback of neglecting the averages $\mathbf{h}_{av}(c)$ and method ESS with small values of parameter ε might be preferable. We conclude that a sufficiently universal method might be ESS with parameter ε interactively set by the user; this conclusion will be tested in future with diverse users.

On the background of these descriptions and results of empirical studies of PrOnto system we can return to the question: how to support individual and group hermeneutic processes?

3.2 Individual Hermeneutic Processes

What is characteristic for a hermeneutic process of a researcher using PrOnto system? Generally, this process is similar to that described by the *EAIR spiral*. The user has an idea or a research theme and uses the system PrOnto for finding texts that might be useful in a further development of this idea or theme. Her/his ontological profile was constructed precisely for that – it does not express her/his full research interests, only that what is interesting for her/him now, in relation to the idea or theme that is actual for her/him. Therefore, the changes of personalized ontological profile might be frequent.

These changes might be related to a new interpretation of the idea or theme, resulting from a full cycle of *EAIR spiral*. We can try to register such changes and summarize several subsequent individual profiles into an integrated ontological profile; there exist for that appropriate tools, also developed for the PrOnto system. Such integrated ontological profile can in such a case represent a holistic, multi aspect interpretation of the idea or theme currently developed.

However, the changes of an individualized ontological profile might also result from a substantive change of the research problem or theme, or an emergence of a new, different idea. The question is how to interpret an integrated, holistic ontological profile in such a case. If somebody would claim that in such a way – perhaps after a long observation and many cycles of *EAIR spiral* – it is possible to characterize entire hermeneutic perspective of the user, such a claim would be erroneous. Firstly, an individual hermeneutic perspective of a researcher is always related to the hermeneutic perspective of the entire professional group that she/he belongs to, and the latter is a long duration historical structure, not necessarily accessible by keywords or key phrases. Secondly, each individualized ontological profile is related to at most two stages of the hermeneutic spiral – the currently researched idea and its rational analysis – and does not have any direct relation to hermeneutic immersion and reflection.

Therefore, the registration of changes in an individualized ontological profile, even for the purpose of supporting a single user, has to be enhanced by additional information that the user might (though not obligatorily) specify for a better interpretation of these changes. We could ask her/him (for a choice and response) two essential reasons for a change of this profile:

1) A change of interpretation of the old research theme;
2) A change of the old research theme to a new one (we could also ask for a name of the new research theme and how far the new theme differs from the old one).

We could also ask, whether:

a) To register (save) her/his old individualized ontological profile (it is a kind of intellectual property of the user and its further application requires her/his approval);
b) To integrate her/his old individualized ontological profile with other profiles registered earlier.

Saving the answers to such questions helps to document at least some aspects of the cognitive process of the user. We can further enrich the documentation, e.g., by saving not only the old individualized ontological profile, but also the results of resulting ranking of research materials, etc.

3.3   Group Hermeneutic Processes

As described earlier, group cognitive processes are not only hermeneutic processes, also processes of debate, experiments, brainstorming, etc. Group hermeneutic processes, concerned with an interpretation of research materials (whether in text or multimedia form), might be related to the processes of debate and *EDIS spiral*. However, PrOnto system was not constructed to support the processes of debate; it was only assumed that the exchange of information about individual ontological profiles together with resulting rankings of documents in a given repository or digital library might be interesting for other users working together on a research theme. This function, together with a possibility of integration of ontological profiles for the entire research group, is the basis of the group use of the system PrOnto.

An information exchange about individual ontological profiles together with resulting rankings of interesting documents in a given repository might be:

1) Spontaneous, i.e. initiated by an individual user who wants to share with designated colleagues the results of her/his analyses;
2) Automatic, initiate by the system after each use, addressed to all registered members of given research group.

Spontaneous exchange should be always admissible, but the application of automatic exchange should be discussed and decided by the entire group, together with a definition of its limits (unlimited automation is always dangerous). Note that, according to our distinction of supporting semantic rapport and of hermeneutic understanding, an exchange of individual ontological profiles can have double effects: obviously, it can be helpful in semantic rapport, but most interesting is how far it might be helpful in deepening hermeneutic interpretation and understanding. Further development of PrOnto system should not only include tools for organization of a research group empowered to use the system, automation of information exchange in this group and options of limitations of this automation, but also test such tools with the distinction of their semantic and hermeneutic aspects..

Another question is a group integration of individualized ontological profiles – either after their single use, or integrated after several modifications by the same user. Here also important are the problems whether such integration has more semantic or more hermeneutic impacts, whether it should be spontaneous or automatic, but it is more difficult to define the limitations of automatic integration. Spontaneous integration might be initiated by two or more users, asking other members of the research group to join them. Any automatic integration, even concerning a separate research theme, must be discussed and voted by the entire group, with the right of any user to keep her/his ontological profiles private.

Future development of PrOnto system might include tools for a spontaneous integration of individual ontological profiles as well as tools supporting a debate and voting inside a research group.

4   Conclusions

System PrOnto, although it is oriented mainly to support an individual hermeneutic process – searching for research sources (texts and multimedia elements) in a given repository or electronic library, then analyzing and interpreting them – can be also used for supporting group cognitive processes of knowledge exchange. For this purpose, it must be accordingly enhanced not only by tools supporting a group integration of individual ontological profiles, but also distinguishing semantic and hermeneutic aspects of such integration and supporting limitations of exploiting such enhancements decided jointly by the group. Such limitations result from the fact that cognitive processes are based to a large extent on the use of tacit knowledge, preverbal and a-logical, hence the tools based on a logical text analysis can only partially support cognitive processes.

This negative observation is understood here in a positive sense, to stimulate thinking how far the tools of ontological engineering – and other tools of computer science – can be used to support deep cognitive (especially hermeneutic) processes, whether and how it is possible to use computers and networks to support the psychology of the depth? Even if currently a negative answer seems to prevail, it is not determined that such an answer must be absolutely true.

References

[1] Braudel, F.: Civilisation matérielle, économie et capitalisme, XV-XVIII siècle. Armand Colin, Paris (1979)

[2] Chudzian, C., Granat, J., Klimasara, E., Sobieszek, J., Wierzbicki, A.P.: Wykrywanie wiedzy w dużych zbiorach danych: przykład personalizacji inżynierii ontologiczne (Telekomunikacja i Techniki Informacyjne), vol. (1-2), pp.s.3–s.28. Instytut Łączności - Państwowy Instytut Badawczy, Warszawa (2011)

[3]   Chudzian, C., Klimasara, E., Paterek, A., Sobieszek, J., Wierzbicki, A.P.: Personalized Search Using Knowledge Collected and Made Accessible: A Model of Ontological Profile of the User and Group in Pronto System. Konferencja Projektu SYNAT, Warszawa, lipiec (2011a)

[4]   Curtis, J., Baxter, D., Cabral, J.: On the Application of the Cyc Ontology to Word Sense Disambiguation. In: Proceedings of the Nineteenth International FLAIRS Conference, Melbourne Beach, FL, pp. 652–657 (May 2006)

[5]   Dreyfus, H., Dreyfus, S.: Mind over Machine: The Role of Human Intuition and Expertise in the Era of Computers. Free Press (1986)

[6]   Gadamer, H.-G.: Warheit und Methode. Grundzüge einer philosophishen Hermeneutik. J.B.C. Mohr (Siebeck), Tübingen (1960)

[7]   Gasson, S.: The management of distributed organizational knowledge. In: Sprague, R.J. (ed.) Proceedings of the 37th Hawaii International Conference on Systems Sciences. IEEE C.S. Press (2004)

[8]   Gauch, S., Chaffee, J., Pretschner, A.: Ontology-Based User Profiles for Search and Browsing. User Modeling and User-Adapted Interaction: The Journal of Personalization Research, Special Issue on User Modeling for Web and Hypermedia Information Retrieval (2003)

[9]   Golemati, M., Katifori, A., Vassilakis, C., Lepouras, G., Halatsis, C.: Creating an Ontology for the User Profile: Method and Applications. In: First IEEE International Conference on Research Challenges in Information Science (RCIS), Morocco (2007)

[10]   Kacprzyk, J.: Linguistic Summaries of Data Using Fuzzy Logic. International Journal of General Systems 30(2), 133–154 (2001)

[11]   Król, Z.: Matematyczny platonizm i hermeneutyka (Mathematical Platonism and hermeneutics). Wydawnictwo IFiS PAN, Warszawa (2006)

[12]   Kunifuji, S., Kawaji, T., Onabuta, T., Hirata, T., Sakamoto, R., Kato, N.: Creativity support systems in JAIST. In: Proceedings of JAIST Forum 2004: Technology Creation Based on Knowledge Science, pp. 56–58 (2004)

[13]   Kunifuji, S., Kato, N., Wierzbicki, A.P.: Creativity Support in Brainstorming. W Wierzbicki, A.P., Nakamori, Y. (red) Creative Environments, op.cit., str. 93-126 (2007)

[14]   Lewandowski, A., Wierzbicki, A.P. (eds.): Aspiration Based Decision Support Systems. Springer, Heidelberg (1989)

[15]   Michalska, A.: Niezbywalność założeń metafizycznych w nauce. Motycka, W.A. (ed.) Nauka a Metafizyka, Wyd. IFIS PAN, Warszawa, str. 207-227 (2009)

[16]   Nakamori, Y., Sawaragi, Y.: Shinayakana systems approach in environmental management. In: Proceedings of 11th World Congress of IFAC, vol. 5, pp. 511–516. Pergamon Press, Tallin (1990)

[17]   Nonaka, I., Takeuchi, H.: The Knowledge-Creating Company. How Japanese Companies Create the Dynamics of Innovation. Oxford University Press, New York (Polish translation: Kreowanie wiedzy w organizacji, Poltext 2000) (1995)

[18]   Osborn, A.F.: Applied imagination. Scribner, New York (1957)

[19]   Polanyi, M.: The tacit dimension. Routledge and Kegan, London (1966)

[20]   Popper, K.R.: Objective Knowledge. Oxford University Press, Oxford (1972)

[21]   Simon, H.A.: Models of Man. Macmillan, New York (1957)

[22]   Vollmer, G.: Mesocosm and Objective Knowledge: On Problems Solved by Evolutionary Epistemology. In: Wuketits, F.M. (ed.) Concepts and Approaches in Evolutionary Epistemology. D. Reidel Publishing Co., Dordrecht (1984)

[23]   Wierzbicki, A.P.: On the role of intuition in decision making and some ways of multicriteria aid of intuition. Multiple Criteria Decision Making 6, 65–78 (1997)

[24] Wierzbicki, A.P., Nakamori, Y.: Creative Space: Models of Creative Processes for the Knowledge Civilization Age. Springer, Heidelberg (2006)

[25] Wierzbicki, A.P., Nakamori, Y. (eds.): Creative Environments: Issues of Creativity Support for the Knowledge Civilization Age. Springer, Heidelberg (2007)

[26] Wierzbicki, A.P.: The problem of objective ranking: foundations, approaches and applications. Journal of Telecommunications and Information Technology 3, 15–23 (2008)

[27] Wierzbicki, A.P.: Techne$_n$: Elementy niedawnej historii technik informacyjnych i wnioski naukoznawcze (Techne$_n$: Elements of recent history of information technologies and epistemological conclusions). Komitet Prognoz "Polska 2000 Plus" przy Prezydium PAN oraz Instytut Łączności (PIB), Warszawa (2011)

Associations between Texts and Ontology[*]

Anna Wróblewska[1], Grzegorz Protaziuk[1], Robert Bembenik[1],
and Teresa Podsiadły-Marczykowska[2]

[1] Institute of Computer Science, Warsaw University of Technology,
Nowowiejska 15/19, 00-665 Warszawa, Poland
{A.Wroblewska,G.Protaziuk,R.Bembenik}@ii.pw.edu.pl
[2] Institute of Biocybernetics and Bioengineering,
Trojdena 4, 02-109 Warszawa, Poland
tpodsiadly@ibib.waw.pl

Abstract. Intelligent automatic text processing methods require linking between texts (written in natural languages) and concepts representing semantics of particular words and phrases. Concepts are supposed to be independent from languages and can be expressed as ontologies. The mapping of concepts to a specific language may be expressed as a linguistic layer of a given ontology. The paper presents a model of lexical layer that establishes relations between terms (words or phrases) and entities of a given ontology. Moreover, the layer contains representations not only of terms but also their contexts of usage. It also provides associations with commonly used lexical knowledge resources, such as WordNet, Wikipedia and DBPedia.

Keywords: Ontology lexical layer, ontology localization, text mining, discriminators.

1 Introduction

Ontologies hold knowledge about particular domains of interest. They are used for sharing common understanding about the domains what makes communication between various beings (e.g. computer programs, agents) possible. Ontologies have richer internal structure than taxonomies and are not only a kind of representation of knowledge. Ontologies refer to a description of a given part of the world by using a specific defined vocabulary and a set of explicit assumptions concerning intended meaning of words from the vocabulary. A set of assumptions is usually expressed in a first-order logical theory, where vocabulary words are names of concepts (classes) and relations. Such understating of ontology leads to distinguishing two layers of it, namely: a conceptual (semantic, core) layer (CL) (often represented by a semantic

---

[*] This work is supported by the National Centre for Research and Development (NCBiR) under Grant No. SP/I/1/77065/10 by the Strategic scientific research and experimental development program: "Interdisciplinary System for Interactive Scientific and Scientific-Technical Information".

R. Bembenik et al. (Eds.): *Intell. Tools for Building a Scientific Information*, SCI 467, pp. 305–321.
DOI: 10.1007/978-3-642-35647-6_20   © Springer-Verlag Berlin Heidelberg 2013

network), defining semantics and a linguistic[1] layer (LL) defining vocabulary used for expressing concepts in a given language included in an ontology.

To a large extent notions seem to be language-independent, so we can assume that a semantic layer is independent from any natural language. On the other hand, in many cases knowledge is expressed and described in a natural language. Thus there is a natural need to link texts with semantics.

A linguistic layer provides such associations between unstructured texts and semantics (defined with ontologies) (Fig. 1). This layer may be also perceived as a mapping of entities of conceptual layer (concepts, relations, axioms) to terms (words or phrases). The mappings can be considered as vertical relationships between a core layer of ontology and domain texts. Many LL may be defined for a given conceptual layer so texts expressed in many different languages can be mapped to one semantic layer.

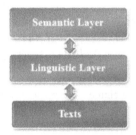

Fig. 1. Layers of knowledge representation

A linguistic layer may be realized by different dictionaries or other specialized frameworks designed particularly for this purpose and providing required linguistic representation. Some frameworks emphasize the attachment of linguistic information to an ontology. The linguistic information includes syntactic behavior of terms (sub-categorization frames), morphology, dependency information – the full or partial parse tree. The frameworks are: LEMON, LMF, LexInfo etc. Almost all the designed frameworks include the attachment of lexical semantics (defined to some extent), e.g. WordNet, SKOS, LIR, LEMON

In this paper we present the new version of LEXO – a special structure for a lexical layer. Our approach takes into account first and foremost lexical semantics and mutual co-occurrence of terms in texts. The first version of our lexical layer (LEXO, LEXical layer of Ontology) was presented in [1]. In this paper we propose the more mature second version of LEXO.

In the following sections the scope and principal contributions of this document are: an analysis of issues in associating texts and ontologies and an overview of the state-of-the-art for linguistic frameworks that we have taken as starting point in our work (section 2), a short listing of the requirements for our model of lexical layer,

---

[1] Or lexical layer in approaches with a simplified representation of linguistic aspects.

main assumptions and definition of key notions (section 3), a description of the structure of LEXO v.2. and outlining improvements regarding the previous version of LEXO (section 4), a use cases (examples) for our lexical layer (section 5). At the end we give conclusions in section 6.

2   Associating Texts with Ontologies

2.1   Polysemes and Synonyms

In natural languages there are a lot of terms that have several different meanings but spelling the same (polysemes and homographs). In WordNet average polysemy including monosemous words is 1.52 (for English WordNet) [2] and 1.56 (for Polish WordNet) [3]. The average polysemy excluding monosemous words is 2.89 and 2.78, for English and Polish WordNet respectively. In English Wikipedia there are above 82.5 thousand disambiguation pages. Each of them lists different meanings for one term used as a name of article (a Wikipedia concept) [4].

Various meaning of a polysemous term may be expressed and differentiated in a lexical layer. We can define a context of occurrence of the term in a given meaning (*TCtx*) in texts and linked it to different concepts represented in an ontology (Fig. 2).

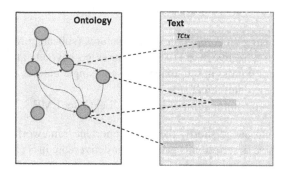

Fig. 2. Associations between an ontology and a text: circles inside semantic layer – the ontology concepts, terms (utterance of meaning) – rectangles in the text, term neighborhood (context, *TCtx*) - the larger rectangle. A term may refer to many concepts (polysemy). It is also possible that many terms refer to the same concept (synonymy).

Terms that have different spellings but the same meaning (synonyms) can be linked to each other by using appropriate relations. It is worth to mention that most words are synonyms only in particular contexts. They can be defined as partial synonyms [5]. The phenomenon in [6] was described as: "multidimensionality occurs when a concept can be seen from more than one perspective and can therefore be classified and designated in more than one way based on the different characteristics that it possesses". For example, one meaning of "skoczek" (a jumper, a knight in Polish) is synonymous with one meaning of a word "figura" (figure) only when it means a knight - a figure in chess. However, the most popular meaning of "skoczek" is a jumper or an athlete, and "figura" means a statue or a very important person.

When *TCtx* is defined one can express such subtle variations. In WordNet there is a similar idea of synsets as a sets of words defining one meaning. There are often examples and descriptional definitions of the synsets. However, there are no information about frequency of co-occurrence of a given term with words from sysnsets.

2.2   Frameworks

Several suitable linguistic frameworks that might be applied for building a linguistic layer were described in [7]. They are as follows:

- LMF (Linguistic Markup Framework) [8] emphasizing morphosyntactic information and semantic interpretations of words (or more complex lexical constructions),
- SKOS (Simple Knowledge Organization Systems) - a W3C Recommendation for the representation of concept schemes,
- LIR (Linguistic Information Repository) [9] – a subset of lexical and terminological description elements that account for the linguistic realization of a domain ontology in different natural language,
- LingInfo – another representation of linguistic information with ontology elements; focus on multilingual terms, lexical information, morpho-syntactic decomposition,
- LexOnto focusing on complex lexical structures (subcategorization frames, lexical entries for adjectives etc.). The structures go beyond "atomic" entities (e.g. single labels or words) at the lexical side and single entities (classes, properties, instances) at the ontological side.
- LexInfo [10] – a combination of LingInfo, LexOnto, LMF.

In the sequel we introduced shortly the most important frameworks from our point of view. More extensive description is given in our deliverable in SYNAT project [11].

The latest development in the domain of lexical models is LEMON (Lexicon Model for Ontologies) [12][13][14]. LEMON was designed to represent lexical information about words and terms relative to ontology, and enabling exchange of data on the Semantic Web. LEMON is based on previous frameworks such as LMF, LIR and LexInfo, but it has less complex structure.

LMF is an ISO standard (namely ISO 12620 Data Categories) for representing lexical resources in XML/UML. Its main purpose is to provide a common framework for modeling and representing lexical information, including morphological, syntactic, and semantic aspects in order to permit data exchange and interoperability. The description of an LMF entry [15] is very extended and detailed , but the model does not attempt to establish any connection with domain ontologies.

LIR is a model inspired by LMF. It enables association of lexical information with OWL ontologies. The main goal of LIR is to provide a model allowing enriching a conceptual part of an ontology with a lexico-cultural layer. It is focused on multilingual aspects and on capturing specific variants of terms such as abbreviations, short forms, acronyms, transliterations, etc. LIR put no emphasis on part-of-speech, morphological information, syntactic decomposition of complex terms, and contextual information.

LexInfo is another model for linguistic grounding of ontologies, allowing to associate linguistic information with respect to any level of linguistic description and expressivity to elements in ontology. LexInfo has been implemented in OWL. The LexInfo project identified and realized five key requirements of a lexicon-ontology model: (i) separation of the lexicon and ontology (the ontological layer and lexica must be interchangeable); (ii) representation of structured linguistic information (it should be possible to represent linguistic descriptions, e.g., part of speech); (iii) representation of the syntactic behavior of its entries; (iv) representation of morphological decomposition; (v) the lexicon should be reusable for different ontologies.

Despite clearly stated goals, the incorporation of LIR and LMF models in LexInfo was the source of some problems. The incorporation of LIR in LexInfo resulted in a large number of categories for parts of speech, which may lead to difficulties in adapting them to non-European languages. LexInfo is also very verbose, due to the way it adapts LMF.

In 2012 a LEMON model was proposed as the next standard for exchanging lexicon resources on the Web. LEMON inherits a lot from the predecessors. However, it concentrates on the following goals: conciseness, descriptiveness, modularity, RDF-based implementation. Conciseness means using as few classes and definitions as needed. Descriptiveness means that LEMON can be extended in different ways to handle different purposes, i.e., terminological variation, morpho-syntactic description, translation memory exchange. Moreover, LEMON uses external sources for the majority of its definitions. Modularity means that LEMON divides into a number of modules and it is not necessary to implement the entire model to create a functional lexicon. RDF-based implementation enables sharing on the semantic web.

The core of LEMON consists of the following six main entities: (i) *Lexicon*: the object representing the lexicon as a whole, lexicon objects in *lemon* are assumed to be monolingual; (ii) *Lexical Entry*: an entry in a lexicon is a container for one or several forms and one or several meanings of a lexeme; all forms of an entry must be realized with the same part of speech, homonyms are treated as separate lexical entries; (iii) *Lexical Form*: an inflectional form of an entry; the entry must have one canonical form and may have any number of other forms; (iv) *Lexical Sense*: a sense object links the lexical entry to the *reference* used to describe its meaning; (v) *Component*: a lexical entry may be broken up into a number of components; (vi) *Representation*: a given lexical form may have several orthographic and phonetic representations.

To sum up, LEMON maintains a high degree of compatibility with LMF, LIR and LexInfo. It focuses on compactness and high expressivity what allows representing large amount of linguistic information. Nevertheless, it does not emphasize variety of direct links with a core layer of ontology. We can only associate instances of *Sense* class with concepts of a domain ontology.

3   Motivation and Assumptions

Up to date, in designed lexical layers there are little emphasis on representation and usage of contexts of terms (*i.e.* their neighborhood in a written language) associated with particular lexico-semantic meanings of terms. In some frameworks, it is taking into consideration only as a possibility to represent parse trees of particular sentences when a given term occurs (as it is in LEMON).

We define the context with regard to text data mining approach and lay less emphasis on linguistic information. We reuse the well-known idea "a word is characterized by the company it keeps" [16]. The context *TCtx* of a chosen term *T* in a given meaning that is being defined (i.e. the core term of *TCtx*) contains terms being in any lexico-semantic relation to *T* as well as terms occurring frequently in a neighborhood of *T*. A neighborhood can be defined as a sentence or a paragraph or any words laying in texts in particular chosen distance from the described term *T*.

Another important drawback of exiting solutions is laying little emphasis on defining a way of linking elements from a lexical layer with elements from a conceptual layer and contexts of such connections (OCtx) . The most extended solution in this aspect is LIR. However, we add more flexible and detailed definition of the interface.

3.1   LEXO – Purpose and Requirements

Generally, in our approach we have looked at ontologies from the perspective of semantic processing (automatic or semi-automatic) of texts written in a natural language. In contrast to the aforementioned approaches to realizing a lexical layer (described in section 2), LEXO focuses less on a morphological specification of a given language. It rather concentrates on relationships between words or phrases which may be extracted from texts, and which are useful for distinguishing between different meaning of a given term in a given texts. It is dedicated first and foremost to applications associated with semantic analysis of texts.

The goal of the design of LEXO was to create a structure satisfying the following main requirements:

1. storing information about single words as well as multiword expressions;
2. providing representation[2] of different and/or various meanings of terms (words or phrases);
3. describing a context of a term in a given meaning (*TCtx*) in order to make it possible to determine the given meaning based on its context taken from text in which that word or phrase occurs;
4. representing general linguistic relations between terms and relations valid only in a given context;

---

[2] In LEXO a meaning of a word is only indicated and associated with co-occurring terms, whereas its definition should be provided in the conceptual layer which LEXO refers to.

5. representing context (*OCtx*) of associations of objects from LEXO with elements from a core layer of a given ontology;
6. being usable for knowledge engineers and linguists as well as in automatic approaches.

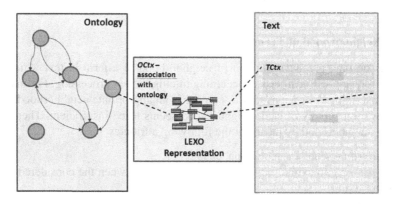

Fig. 3. Linking a context of a term (*TCtx*) with an ontology using LEXO representation

Summing up, our idea of LEXO (shown in Fig. 3) emphasizes two notions: (i) a context of a meaning of a term (*TCtx*) providing information needed for determining a lexicalized semantic meaning in which a given term occurs in a text; (ii) a context of connections with a conceptual layer of a given ontology (*OCtx*) providing additional information about texts for which associations between elements from a lexical layer and elements from a semantic layer are valid.

3.2   Key Notions

Before we present the design of LEXO in detail we introduce the key concepts under-lying LEXO, namely discriminant terms, a neighborhood of term, and neighbor terms.

3.3   Discriminant Terms

A meaning of a word in some text may be determined by analyzing a context of its occurrence. Such context may be naturally created from words and phrases surround-ing an analyzed word in some unit of a text (e.g. in sentence or paragraph). Thus, intuitively a discriminant for a given word is a term which unequivocally determines some meaning of that word. If the analyzed term and the word occur together in a unit of a text, a meaning of the word can be determined based only on that term (in most cases). Such approach was successfully applied in [17] for finding various meanings of a word.

In our approach we allow composed discriminants e.g. discriminants composed of two or more terms. The formal definition of discriminants is provided below.

Definition 1. A given not-empty set of terms $\{t_1, t_2,.. ,t_n\}$ is called a *discriminant* of a term t if it occurs together with term t in a unit of text then it unequivocally determines some meaning i of term t.

Definition 2. A given discriminant $d = \{dt_1, dt_2,.., dt_n\}$ is a *proper discriminant* if for any $dt_i \in d$ the set $d' = d - \{t\}$ is not a discriminant.

3.4   Neighborhood of Term

There is no one, precise definition of neighborhood of a term in text which can be successfully applied in all tasks concerning intelligent text processing. For instance, in the discovering proper nouns task a sentence is an appropriate neighborhood, whereas in discovering synonyms such neighborhood seems to be insufficient. Therefore, we define a neighborhood by means of the following attributes:

- *granularity* – sentence, paragraph, chapter, etc.;
- *distance* – e.g. maximal number of words in text between the considered term and another term;
- *side* - left or right.

3.5   Neighbor Terms

In LEXO we define a neighbor terms with respect to a particular meaning of a term, not with regard to a term.

Definition 3. A word or phrase is a neighbor term for a given term T in a given meaning if it occurs frequently enough in a neighborhood of that term used in that meaning in texts from a domain of interest.

Of course, for calculating frequency only one chosen type of neighborhood is used. A needed frequency of occurrence depends on several aspects, such as: frequency of occurrence of the term, cumulative size of analyzed text, a size of neighborhood, etc. Note that such definition allows that a determinant of a term may not be its neighbor term.

4   The Structure of LEXO

The general conceptual schema of LEXO in UML is presented in Fig. 4. The structure consists of six main classes: *Term, LexicalEntry, ContextEntry, LexicalMeaning, UsageExample, LexicalKnowledgeSource,* and *DomainTerminology*. Three association classes represent attributes of relations between main classes. They are: *OntologyRelationContext, MeaningOccurence, TermContextRelation. ConceputalLayer* class represents all elements (e.g. concepts, instances, properties) of a conceptual layer with which elements from LL may be associated. The most important classes and relations are described below.

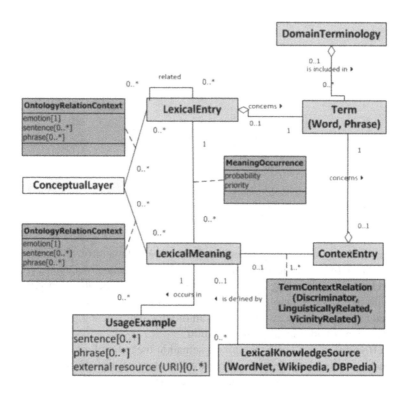

Fig. 4. The conceptual schema of LEXO

Term represents a single word or a phrase from a natural language (Fig. 5). A single word is written in the *lexicalForm* attribute of *Word* class (a subclass of *Term*). *POS* (the part of speech) is another attribute of *Word* class. Additionally, we can indicate a canonical form of a given word (represented also as an instance of *Word* class) includes through *has_canonical_form* relation. *Phrase* class (another subclass of *Term*) represents a phrase from a natural language composed of two or more words. Thus an instance of *Phrase* consists of a sequence of *Word* instances. The place of a word in the sequence are given by an attribute *position* of *PhrasePart* aggregation relation between *Word* and *Phrase* classes.

LexicalEntry class depicts a term which meaning is determined in *LexicalMeaning* class (Fig. 4). Representing a term by an instance of *LexicalEntry* is realized by *concerns* binary relation between *LexicalEntry* and *Term* classes.The different and/or various meanings of a given term are expressed by relations with different *LexicalMeaning* objects. The *MeaningOccurrence* relation is used for assigning to a *LexicalEntry* object one or more its lexicalized meanings. The relation has two attributes: *probability* indicating probability of occurrence in texts of a term represented by a *LexicalEntry* object in a given meaning and *priority* showing the position of a given meaning in the sequence of lexicalized meanings ordered by probability of occurrence of the term in a given meaning.

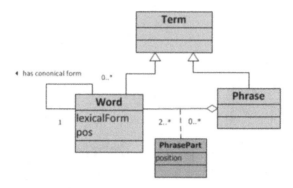

Fig. 5. The representation of terms

Any *LexicalEntry* object may be connected with an element (or elements) from a conceptual layer (e.g. notion, relation): 1) directly - by means of *OntologyRelation-Context* relation (representing *OCtx*) or 2) through an associated *LexicalMeaning* object with the same relation. Possible direct linguistic relations between terms represented as *LexicalEntry* instances (e.g. antonymy, equivalency) may be expressed directly by means of additional r*elated* relation.

LexicalMeaning represents a lexicalized semantic meaning of a given *LexicalEntry* object. Such meaning is defined foremost by a context of a given word/phrase. The context (*TCtx*) is described by related *ContexEntry* objects (defined in the sequel) associated with the *LexicalMeaning* instance by using *TermContextRelation* relation. Subclasses of *TermContextRelation* represent different kinds of relations between *LexicalMeaning* and *ContextEntry* instances (Fig. 6). Despite the subclasses are represented differently in UML schema (an association class, a class, an N-ary link), we implemented them in OWL2 and all of them can be modeled as a class. The subclasses of *TermContextRelation* are as following:

- *LingusticallyRelated* – modeling linguistic relations between a word in a given lexicalized meaning with some other words. The relation has the two following attributes: *linguistic_relation* (*synonymy*, *hyponym*, and *hypernym*) and *parameter* –a generic parameter for all linguistic relations. It can be used e.g. for storing information about degree of similarity in the synonymy relation.
- *Discriminator* – indicates a phrase/word (or set of them) which if occurs together with a given word/phrase in texts unequivocally determines its meaning. The *discriminator* relation is modeled by means of auxiliary class *Discriminator*. The *type* attribute has assigned one of the two following values: *AND* which indicates that all *ContextEntry* objects included in a *Discriminator* object constitute one discriminator, and *OR* which indicates that each *ContextEntry* object included in a *Discriminator* object constitutes one discriminator of the lexicalized meaning.
- *VicinityRelated* modeling relations between a term with a given lexicalized meaning and other words/phrases which occur frequently in a certain neighborhood of that term. The neighborhood is a parameter of the relation and is defined by *Neighborhood* class.

Neighborhood is an auxiliary class (Fig. 6) used for defining a neighborhood of a given term/phrase. A neighborhood is defined by means of three attributes: *granularity* – (sentence, paragraph, chapter), *distance* (e.g. maximal number of words between two phrases), and *side* (left or right). *ContextEntry* represents a word/phrase included in a context of a given term, which indicates a certain lexicalized semantic meaning of a given word/phrase. The way of representation is identical as in *LexicalEntry* class.

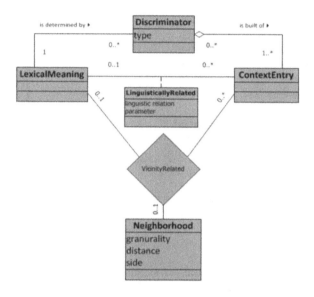

Fig. 6. Relations between *LexicalMeaning* and *ContextEntry* classes. *Discriminator, LinguisticallyRelated, VicinityRelated* are subclasses of *TermContextRelation*.

Furthermore, a lexical meaning may be defined by using external resources such as WordNet, Wikipedia or DBPedia (Fig. 4). Such resources are represented in the schema by *LexicalKnowledgeSource* class, which is a superclass for classes dedicated to a particular resource (*WordNet, DBpedia,* and *Wikipedia*). The classes gather links to definitions existing in the dedicated resources (Fig. 7). One instance of *LexicalKnowledgeSource* subclass can include many links because in the existing resources there may be no one definition for a given lexicalized meaning defined in LEXO.

Additionally, examples of usage of a given word/phrase in a given meaning may be associated with a *LexicalMeaning* object with the use of *UsageExample* class and *occurs_in* relation (Fig. 4).

LexicalMeaning is connected with an element (or elements) from a conceptual layer by means of *OntologyRelationContext* relation.

Additionally, we define *DomainTerminology* class to associated any terms loosely coupled with the conceptual layer. They are not associate directly with any notion of the CL but generally with the whole CL).

The more precise description and implementation of LEXO is presented in our deliverable [18] and the model of LEXO implemented in OWL2 is available in [19].

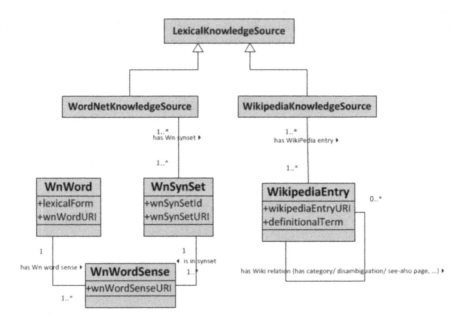

Fig. 7. Additional definition of lexicalized meaning in other resources, e.g. WordNet and Wikipedia

4.1   Improvements Regarding the Previous Version of LEXO

The first version of LEXO was introduced in [1]. First of all in LEXO v.1. we have four different definitions of one particular meaning of a term, which may be associated with that term. The main definition is created by means of a set of terms associated with the described term in the lexicalized meaning. The lexicalized meaning can be associated with other terms in context by means of lexical relations (e.g. antonymy, synonymy, equivalence, broader or narrower terms) or frequent proximity. Other meaning are links to exact definition in commonly used lexical knowledge sources, e.g. Wikipedia or WordNet.

In LEXO v.2. we define one lexical meaning (an instance of *LexicalMeaning* class) for one meaning of a given term. A lexical meaning may be associated with various definitions from lexical resources. The definition may be more extended: it may be a set of definitions from one lexical resource, e.g. a set of Wikipedia pages, a set of WordNet synsets.

The second of the important differences between versions of LEXO is introducing a notion of *OCtx* context that allow describing the association between elements of our lexical layer and entities from a given conceptual layer of an ontology.

The next change in LEXO is the application of the notion of discriminators as it was defined before in this paper.

Moreover, there are a lot of changes (in comparison to the previous version) less important and more implementation oriented, e.g. a definition of a phrase as a series of words with an index of occurrence.

5   LEXO – Usage Example

One word may have many meanings. For example "process" may relate to a computer program, a scientific process or a random/stochastic process. To check the possibility to describe lexicalized senses we decided to incorporate words and phrases dedicated to these three meaning into LEXO.

5.1   The Exemplary Method of LEXO Populating

We have selected definitions of these three meanings of "process" from Wikipedia disambiguation page (http://en.wikipedia.org/wiki/Process) and defined the lexical model LEXO based on the first paragraph from pages representing the chosen concepts (if the paragraph comprises more than 3 sentences, we take into consideration only the first three sentences). We collected only words being nouns (NN), verbs (VB) if they have only one sense in WordNet, adjectives (JJ) in their base forms. Adjectives are added to the context only when they precede the analyzed word (in our case the word is "process"). A pattern for incorporating adjectives to LEXO is /(JJ)+NN/. We also added phrases that are marked in the analyzed article as hyperlinks to other Wikipedia articles. We add only the whole phrases, not separate words constituting the phrases. We incorporate all the terms (words and phrases) into LEXO model as terms associated with a vicinity property to the particular described lexical meaning (Wikipedia concept). Additionally we determine discriminators, checking if any candidate term does not occur in other Wikipedia articles (related to different meanings of the word "process"). The discussed words, related terms and relations with the meaning have been collected in Table 1.

5.2   A Fragment of Populated LEXO

Given a domain ontology of information technology and services we can define a lexical layer for the ontology containing different expressions associated with a computer process, a demon etc. In our example we have a word "process" in LL and a concept "demon" in CL. We determine a lexicalized meaning for the word "process" and terms related to it in LL. Then we associate the meaning with the concept "demon" by means of OntologyRelationContext relation.

Table 1. Neighbor terms in different definitions of the word "process" in Wikipedia

Different meanings of word *process*	Related terms (abbreviations in this column stand for: PH – phrase, NN - noun, JJ – adjective, RB – adverb)	Relation with the meaning (abbreviations in this column stand for: D - discriminator, V - vicinity relation)
process (computing)	Program, NN,	D
	Instance, NN,	D
	Computer program, PH	D
	Computing, NN	D
	Code, NN	D
	Activity, NN	D
	Operating system, PH	D
	Threads of execution (thread), PH	V
	System, NN	V
	Execution, NN	D
	Instruction, NN	D
	Concurrently, RB	D
process (science)	Sequence, NN	V
	Change, NN	V
	Method, NN	V
	Science, NN	V
	Event, NN	V
	Transformation, NN	D
	Object, NN	D
	Substance, NN	D
	Organism, NN	D
	body, NN	D
	Scientific method, PH	D
stochastic process	Stochastic, JJ	D
	Probability theory, PH	D
	Random, JJ	D
	Deterministic, JJ	D
	Collection, NN	V
	Variable, NN	D
	Random variable, PH	D
	Evolution, NN	V
	Value, NN	V
	System, NN	V
	Time, NN	D
	Counterpart, NN	D
	Deterministic system, PH	D

Thus, we create an instance of *LexicalMeaning* (namely *LM.process_computing*) for the particular meaning of the word *"process"* in computing. The instance *LM.process_computing* is associated with some related words and phrases, e.g. a word *"system"* and a phrase *"operating system"*, which are instances of *Word* class and *Phrase* class respectively. Consequently, we have three instances of *Word* class, i.e. *W.process, W.system, W.operating* and one instance of *Phrase* class, i.e. *PH.operating_system* shown in Fig. 8. Because a phrase is a sequel of words we need two instances of *PhrasePart* class to define *"operating system"* phrase, i.e. *PHP.os1* and *PHP.os2*.

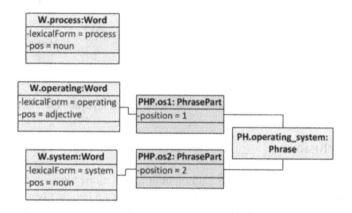

Fig. 8. Instances for definitions of terms: "process", "system", "operating system"

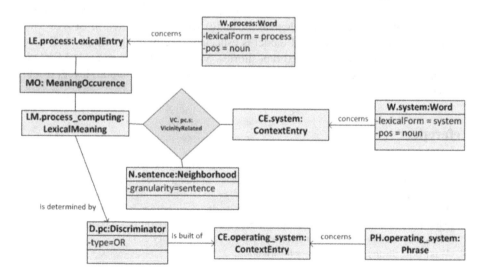

Fig. 9. A lexicalized meaning of "process" word and associated terms

To define lexicalized meaning of process word (in the domain of computing) we should create an instance of *LexicalEntry, MeaningOccurence* and *LexicalMeaning* (as it was mentioned before). To associate terms with the meaning we should make instances of relations: *Discriminator* and *VicinityRelated* and appropriate *ContextEntry* instances for the related terms. All the mentioned before and needed instances are shown in Fig. 9.

Having the lexicalized meaning of *process* word we can associate it with the conceptual layer and the proper notion (*demon* class in CL) giving the context with means of *OntologyRelationContext* instance (Fig. 10).

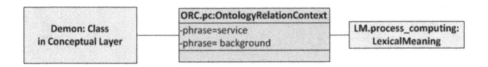

Fig. 10. An example of an association between the lexical layer and the conceptual layer

6   Conclusions

In this paper we have investigated a problem of associations between texts and ontologies. We have outlined LEXO - the special structure of a lexical layer for ontologies, which is seen as a bridge between terms from a natural language and notions defined in an ontology. In this model we emphasize linguistic meanings of words or phrases and the linguistic relations between them rather than other linguistic property of words such as morphology or syntax. We have defined formally the key constructs (such as discriminant, context of lexical meaning) which are applied in LEXO. This approach causes that the proposed model is especially useful in algorithms dedicated to discovering knowledge from text repositories. Moreover, the structure of LEXO is independent of a conceptual layer and it may be used for different core layers of ontologies.

References

1. Wróblewska, A., Protaziuk, G., Bembenik, R., Podsiadły-Marczykowska, T.: LEXO: a Lexical Layer for Ontologies - Design and Building Scenarios. Studia Informatica 33(2B(106)), 173–186 (2012)
2. English WordNet statistics, http://plwordnet.pwr.wroc.pl/wordnet/stats
3. Maziarz, M., Piasecki, M., Szpakowicz, S.: Approaching plWordNet 2.0. In: Proceedings of the 6th Global Wordnet Conference, Matsue, Japan (January 2012)
4. Wikipedia disambiguation pages information,
   http://en.wikipedia.org/wiki/Wikipedia:Disambiguation_pages_
   with_links/The_Daily_Disambig

5. Montiel-Ponsoda, E., Aguado-de-Cea, G., McCrae, J.: Representing Term Variation in lemon. In: 9th International Conference on Terminology and Artificial Intelligence, pp. 47–50 (2011)
6. Fernández-Silva, S., Freixa, J., Cabré, M.T.: A proposed method for analysing the dynamics of cognition through term variation. Terminology 17(1), 49–73 (2001)
7. Buitelaar, P., Cimiano, P., Haase, P., Sintek, M.: Towards Linguistically Grounded Ontologies. In: Aroyo, L., Traverso, P., Ciravegna, F., Cimiano, P., Heath, T., Hyvönen, E., Mizoguchi, R., Oren, E., Sabou, M., Simperl, E. (eds.) ESWC 2009. LNCS, vol. 5554, pp. 111–125. Springer, Heidelberg (2009)
8. Francopoulo, G., George, M., Calzolari, N., Monachini, M., et al.: Lexical markup framework (LMF). In: Proceedings of the Fifth International Conference on Language Resource and Evaluation, LREC 2006 (2006)
9. Montiel-Ponsoda, E., Aguado de Cea, G., Gomez-Perez, A., et al.: Modelling multilinguality in ontologies. In: Proceedings of the 21st International Conference on Computational Linguistics (COLING). Linguistic Information Repository (2008),
   `http://mayor2.dia.fi.upm.es/oeg-upm/index.php/en/downloads/63-lir`
10. Cimiano, P., Buitelaar, P., McCrae, J., Sintek, M.: LexInfo: A Declarative Model for the Lexicon-Ontology Interface. Web Semantics: Science, Services and Agents on the World Wide Web 9(1), 29–51 (2011)
11. State of the Art on Ontology and Vocabulary Building & Maintenance Research And Applications Solutions for System and Domain Ontologies, Technical Report B12 in SYNAT project, Institute of Computer Science, WUT (March 2011)
12. LEMON ontology, `http://www.monnet-project.eu`
13. McCrae, J., Aguado-de-Cea, G., Buitelaar, P., Cimiano, P., Declerck, T., Gómez Pérez, A., Gracia, J., Hollink, L., Montiel-Ponsoda, E., Spohr, D., Wunner, T.: The Lemon Cookbook, `http://www.monnet-project.eu/Monnet/Monnet/English/Navigation/LemonCookbook`
14. McCrae, J., Spohr, D., Cimiano, P.: Linking Lexical Resources and Ontologies on the Semantic Web with Lemon. In: Antoniou, G., Grobelnik, M., Simperl, E., Parsia, B., Plexousakis, D., De Leenheer, P., Pan, J. (eds.) ESWC 2011, Part I. LNCS, vol. 6643, pp. 245–259. Springer, Heidelberg (2011)
15. LMF specification, `http://www.tagmatica.fr/lmf/iso_tc37_sc4_n453_rev16_FDIS_24613_LMF.pdf`
16. Firth, J.R.: A synopsis of linguistic theory 1930-1955. In: Studies in Linguistic Analysis, pp. 1–32. Philological Society, Oxford (1957)
17. Rybiński, H., Kryszkiewicz, M., Protaziuk, G., Kontkiewicz, A., Marcinkowska, K., Delteil, A.: Discovering Word Meanings Based on Frequent Termsets. In: Raś, Z.W., Tsumoto, S., Zighed, D.A. (eds.) MCD 2007. LNCS (LNAI), vol. 4944, pp. 82–92. Springer, Heidelberg (2008)
18. Lexical layer for the system ontology. Technical Report B12 in SYNAT project, Institute of Computer Science, WUT (March 2012)
19. LEXO model in OWL2, `http://wizzar.ii.pw.edu.pl/passim-ontology/lexo.owl`

SYNAT System Ontology: Design Patterns Applied to Modeling of Scientific Community, Preliminary Model Evaluation*

Anna Wróblewska[1], Teresa Podsiadły-Marczykowska[2], Robert Bembenik[1],
Henryk Rybiński[1], and Grzegorz Protaziuk[1]

[1] Institute of Computer Science, Warsaw University of Technology,
Nowowiejska 15/19, 00-665 Warszawa, Poland
{A.Wroblewska,R.Bembenik,H.Rybinski}@ii.pw.edu.pl
[2] Institute of Biocybernetics and Bioengineering
Trojdena 4, 02-109 Warszawa, Poland
tpodsiadly@ibib.waw.pl

Abstract. The paper presents the extended version of the SYNAT system ontology, used design patterns, modeling choices and preliminary evaluation of the model. SYNAT system ontology was designed to define semantic scope of the SYNAT platform. It covers concepts related to scientific community and its activities i.e.: people in science and their activities, scientific and science-related documents, academic and non-academic organizations, scientific events and data resources, geographic notions necessary to characterize facts about science as well as classification of scientific topics. In its current version SYNAT system ontology counts 472 classes and 296 properties, its consistency was verified using Pellet and HermiT reasoners.

Keywords: Ontology building, ontology design patterns, semantic modeling, scientific community.

1 Introduction

Ontology is usually perceived in computer science as a method of representing static knowledge about given part of the world. The aim of an ontology design is, first of all, functional. Typically, ontologies are used for sharing knowledge and common understanding about particular domain of interest which makes communication between various beings possible (e.g. human users, computer programs (agents)) and unambiguous. Ontology refers to a description of a given part of the world using a specific vocabulary and a set of explicit assumptions concerning intended meaning of

---

* This work is supported by the National Centre for Research and Development (NCBiR) under Grant No. SP/I/1/77065/10 by the strategic scientific research and experimental development program: "Interdisciplinary System for Interactive Scientific and Scientific-Technical Information".

R. Bembenik et al. (Eds.): *Intell. Tools for Building a Scientific Information*, SCI 467, pp. 323–340.
DOI: 10.1007/978-3-642-35647-6_21   © Springer-Verlag Berlin Heidelberg 2013

words from the vocabulary. A set of assumptions is usually expressed in a first-order logical theory, where vocabulary words are names of concepts (classes) and relations. Ontology is not only a kind of representation nor taxonomy. Ontologies have richer internal structure than taxonomies, and incorporate common vision (to some people, agents, programs and so on) about knowledge that is (to be) represented.

The SYNAT system ontology (also referred as the SYNAT ontology or system ontology) aims to support the SYNAT project content storage and sharing platform for academia, education and open knowledge society. The main goals of ontology application in the platform are assistance in validation and automatic acquisition of knowledge about scientific community from heterogeneous resources. The resources include, for example, metadata repositories of scientific publications (i.e. DBLP, Ci teSeer, and different university resources), researches' homepages and conference pages. The structuring of the information in the system will be based on the system ontology. The functionality of modules supporting knowledge acquisition requires that a search module and all sources of information share the common understanding of concepts used in queries. It can be achieved by means of ontology.

In the following sections of this paper the scope and principal contributions of this document are: a general view of the SYNAT system ontology (section 2), an extensive presentation of attributes and characteristics of the main conceptualized entities (section 3), an analyses of ontology design patterns and their application to the SYNAT system ontology (section 4), listing of criteria and aspects of ontology quality and preliminary mainly descriptive assessment of the SYNAT system ontology (section 5). In the last sections we give an overview of future works (section 6) and then we summarize our work (section 7).

2   Assumptions of the SYNAT System Ontology and Knowledge Reuse

The SYNAT system ontology describes facts about academic and science community and provides knowledge about: people in science, scientific documents, academic and non-academic organizations, scientific events and information resources.

The preliminary version of the SYNAT system ontology and architecture of search module of ontology knowledge base was introduced in [1]. The model of the current version of the SYNAT ontology in OWL2 language is available at [2][1].

The preliminary version of the system ontology has been assessed against the informational needs of other participants of SYNAT project: employees of the Main Library of Warsaw Technical University, the National Library of Poland, the Institute of Information and Library Science of Jagiellonian University in Cracow, the Faculty of Mathematics, Informatics and Mechanics of University of Warsaw. The result of assessment revealed that there is a need for more detailed and enlarged descriptions of notions in nearly all the modules of the ontology. A detailed description of the required changes in the SYNAT system ontology is available in our deliverable [3].

---

[1] The first two authors of paper are main contributors of the SYNAT system ontology, and their contribution is similar.

The current version of the ontology[2] (Fig. 1) is composed of five main modules (sub-taxonomies, at the highest level of class hierarchy), characterized below:

- *InformationResource* - definitions of different kinds of data resources being the result of scientific activity or used in science community divided into three following classes: various documents, multimedia and collections of other resources;
- *Agent* - definitions of agents (causative bodies) - different types of organizations, persons in science and groups (of people or organizations) that act in science community;
- *Project* – describing scientific projects;
- *Event* – representing events associated with scientific activities (Fig. 2), e.g. conferences, seminars, workshops, lectures, competitions etc.;
- *Characteristic* – specification of concepts used to describe class attributes in the remaining (previously mentioned) modules. This module is the most complex and important and is presented in detail in section 3, it contains notions necessary to detailed description of the recognized area of scientific community.

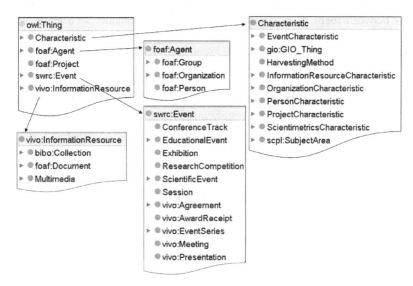

Fig. 1. Five main modules of the SYNAT system ontology. Arrows in the figure point to the panels presenting structure and content of five main ontology submodules and submodules of the Event module. In case of reusing other ontologies, we show the source namespaces of concepts.

There are other ontologies that can be used to model scientific community and its activities, i.e. SWRC [4], FOAF [5], VIVO [6], ESWC2006 Conference ontology [7], Science Ontology [8], SIOC [9], Dublin Core [10], BIBO[3] [11]. Ontologies that have the closest scope to our SYNAT system ontology are: FOAF, SWRC and VIVO.

---

[2] In the following we present ontologies implemented in OWL-DL language based on SHOIN[(D)] description logics.

[3] Extended descriptions of the reused ontologies are provided in our first deliverable [3].

The *Friend-of-a-Friend ontology* (FOAF) provides simple characteristic of people and social (scientific) groups. It does not describe dependence on time and technology, but it concentrates on inter-dependences between its classes. Basic notions of FOAF, such as Agent, Group, Person and Project are reused in the SYNAT system ontology.

The *Semantic Web Research Community Ontology* (SWRC, published in 2004) is one of the first attempts to provide formal definitions of key notions describing a typical research community and relations within the community. The limited scope of included domain notions makes it impossible to model for example scientific or science-based resources, find a link between such scientific event as a conference and related proceedings or represent broad range of characteristics of organizations involved in scientific life. Moreover, SWRC does not use modeling patterns for roles played by a person in scientific community. We reuse some SWRC notions in definition of Event and Document subtypes.

VIVO ontology has been built as part of VIVO project finalized in 2011 [6]. VIVO ontology describes people, organizations, and activities involved in scientific research. Wide range of VIVO notions allows for to some extent fine-grained explication of concepts characterizing scientific community and also for characterization of relations between them. It is possible, among others, to model web resources and relations between any kind of documents and their topics within VIVO. Despite those advantages, the way of constructing VIVO ontology taxonomy structure can represent a barrier for effective knowledge reuse. The VIVO authors decided to represent closely related notions as distinct non-related classes: for example roles of persons in scientific life are modeled as subclasses of a class *vivo:role*, as subclasses of a class

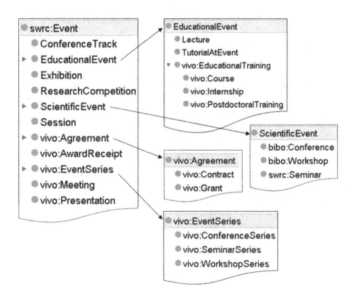

Fig. 2. The Event module modeling notions associated with scientific events. Arrows in the figure point to the panels presenting structure and content of submodules of the Event module.

vivo:position and subclasses of a class *Person*. Generally, in VIVO model there are some cases of semantically tied notions that are scattered throughout the hierarchy without a link to a more general class (e.g. geographic notions: *geo:area, vivo:Address, vivo:Location*), such way of modeling makes knowledge reuse problematic. Nevertheless, we reuse selected VIVO notions to model some personal role types, document types and event subtypes.

Summing up, the scopes of aforementioned ontologies only partially overlap our needs. Therefore we did not import whole ontologies into the SYNAT ontology, although, we reused the selected concepts defined in the above mentioned ontologies[4] clearly using their source namespaces in the SYNAT system ontology (see Fig. 1 and Fig. 2).

3   Ontology Characteristic Module – The Result of Conceptualization

Module *Characteristic* is composed of nine submodules: *EventCharacteristic, HarvestingMethod, InformationResourceCharacteristic, OrganizationCharacteristic, PersonCharacteristic, ProjectCharacteristic, ScientimetricsCharacteristic*, and two submodules containing imported ontologies. The two incorporated ontologies are: geographical and scientific subject classifications, which have been created separately from the system ontology. The function of all the above listed submodules is to gather and structure notions used to characterize and define classes included in the rest of the main modules. Particularly, the module *Characteristic* contains:

- modules dedicated to characterize the other parts of the ontology;
- imported geographical ontology (represented by *gio:GIO_Thing* class);
- scientific subject classification ontology *SCPL* (represented by *scpl:SubjectArea* class);
- module *ScientimetricsCharacteristic* for scientimetric measures for individual scientists and for scientific organizations;
- catalogue of methods used for data harvesting (represented by *HarvestingMethod* class).

The selected, most interesting submodules are presented in the subsequent sections.

3.1   Information Resource and Its Characteristics

The most difficult notions to characterize are those associated with various information resources. The basic types of the resources are modeled as subclasses of *InformationResource* class (Fig. 3). However, the distinction between the form and the content of information resources is unclear and sometimes impossible to make. Therefore, the other types of resources and their content types are modeled as characteristics in *ContentFormType* (Fig. 4).

---

[4] Some of them were extended.

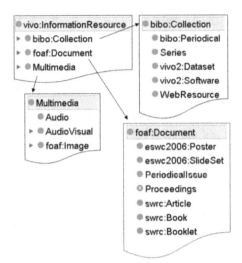

Fig. 3. The *InformationResource* module modeling notions associated with the basic types of information resources. Arrows in the figure point to the panels presenting structure and content of submodules of the *InformationResource* module.

Fig. 4. A fragment of *InformationResourceCharacteristic* submodule. The central panel presents general structure of the submodule and its subclasses are indicated with arrows.

InformationResourceCharacteristic submodule (Fig. 4) contains notions that allow a user to describe in detail such facts about data resources as: kinds of published and unpublished documents, their general content, assessment of the usability in science, different kinds of articles (form the perspective of their contents), relations between documents (i.e. citing and referencing), thematic scope of documents using keywords and related subject area.

The module also allows describing web resources (see Fig. 5) and information about collections of documents such as journals, forums, book series, edited books (with editors and multiple authors).

3.2   Organization Characteristics

The science related organizations (universities, laboratories) play an important role in scientific community. They can be characterized by means of basic data such as: name, acronym, address, location, organization home page, etc. More interesting organization description contains information about areas of conducted research or educational activity, their composition (departments, laboratories), organization activity profile and associations between organizations.

It is almost impossible to build a clear and accurate hierarchy of different kinds of organizations. Therefore, we decided to collect the terminology used to describe organization profile, activity and other necessary attributes.

The submodule *OrganizationCharacteristic* collects terminology which allow a user to characterize academic and non-academic organizations module (see Fig. 6). The system ontology allows to create instances of institutions supporting science such as libraries, archives, museums, foundations, associations, etc.

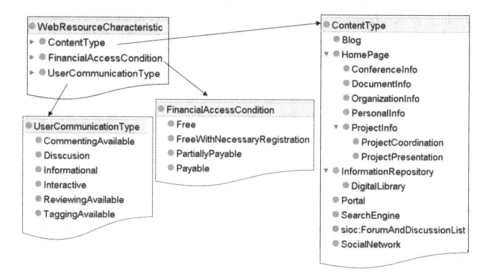

Fig. 5. A fragment of *WebResourceCharacteristic* submodule. The top left panel presents general structure of the submodule and its subclasses indicated with arrows.

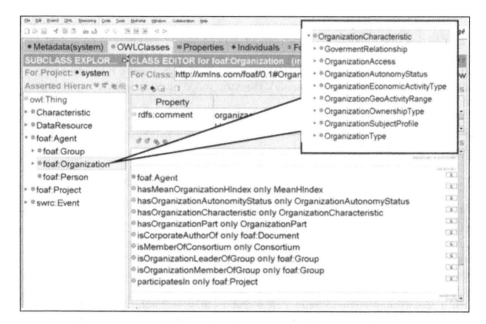

Fig. 6. Necessary conditions for *foaf:Organization* class and its characteristics

3.3   Person Characteristics

The *PersonCharacteristic* submodule (see panel A in Fig. 8) allows for describing a scientist using various aspects: basic information (email, homepage, phone, etc.), education aspects (academic degrees, subject domains, titles of theses, educational organization, when the degree was obtained, supervisors), research interests (authored documents, publications, blogs, patents), participation in projects (a role in a project), teaching (topics of lectures for students and PhD students, conducted seminars, presentations, invited lectures, tutorials and workshops), function and roles at work, affiliations (current and past), reviewing in journals, theses supervised (types of theses, e.g. Master, PhD and titles of theses), conference profile (member of committees, reviewing topics, chair roles), professional connections with other scientists (co-authors, collaborators in projects or grants, supervision, leader, etc.).

3.4   Project Characteristics

ProjectCharacteristic module groups notions for describing such aspects of scientific projects as: organization of financing and sponsoring, place of work, participants, project type, scientific events associated with academic community (e.g. seminars, conferences etc.), geographical information, documents artifacts (papers, slide sets, tutorials, etc.), organizational data (sessions, chairs, committees), organizations that finance and sponsor a project.

3.5   Incorporating Geographical Information and Science Classification into SYNAT System Ontology

Geographic Information Objects ontology, designed for BOEMIE project [12], was adapted for the SYNAT system ontology needs and incorporated into the *Characteristic* module. The main goal of BOEMIE project was to use sophisticated algorithms to extract semantics from multimedia content (audio, video, images and text) in the athletics domain. The GIO ontology describes geopolitical areas, points of interest (e.g. sports, tourism, transport, eating, healthcare objects), athletic routs, a georeference object (location with its data properties: Cartesian coordinates, geographic coordination system parameters, postal address) etc. We extended the GIO ontology with concepts regarding spoken and written languages, geographical scopes and their relation to geopolitical areas (Fig. 7).

Science Classification i.e. (*SCPL*) ontology was created as the autonomic model, it is based on a new 3-level classification of scientific topics (areas, domains and disciplines) elaborated by the Polish Ministry of Science and Education. For the use in the SYNAT project resulting SCPL ontology was imported in SYNAT ontology and represented in *Characteristic* module as *scpl:SubjectArea* class.

The classification of scientific topics is difficult and ambiguous task, it can be done using different criteria. We envisage a possibility to use another science classification,

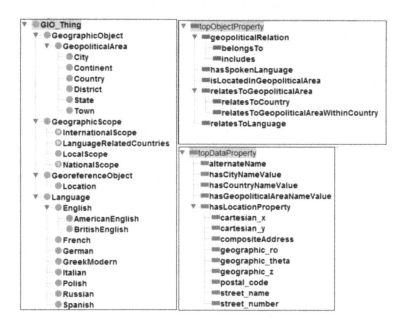

Fig. 7. The GIO SYNAT ontology is composed of: 4 general modules - *GeographicObject, GeographicScope, GeoreferenceObject, Language* (left panel), object properties representing relations between objects and data properties representing names of geographic objects and properties of a location object

but they should have links to the introduced SCPL ontology. Possible linkage are: equivalency; being sub-concepts (narrower and broader concepts); different kinds of relations: similar research methods, similar data for processing, similar types of results, documents, similar reasons for performing research, etc.; linkage with ontology annotations, ε-connections language, conglomerates and their couplers (as it was mentioned before).

The two above described ontologies, GIO SYNAT ontology and SCPL ontology were incorporated to SYNAT ontology using *owl:import* constructor. Using this constructor does not automatically assure the consistency of the resulting ontology, though our ontology including imports is consistent (please refer to Section 5.3). To modularize future versions of the SYNAT system ontology properly and systematically mechanisms that allow storing and reasoning having distinct ontology modules are needed. Those mechanisms may be based on coupling between distinct ontologies: ε-connections [13] (provided only in Swoop, the editor for educational use) and more advanced representation dedicated to modularized ontologies: conglomerates and their couplers [14][15].

4   Ontology Design Patterns Applied to SYNAT Ontology

The notion of Design Patterns (DPs) comes from object-oriented programming [16], where they are defined as abstract solutions to common modeling problems. The same concept can be applied to ontology engineering: an Ontology Design Pattern (ODPs) solves a given general modeling problem and can be applied every time that the problem appears in different ontologies [17].

ODPs are defined as reusable modeling solutions for a recurrent ontology design problems. They make ontology construction process faster and more reliable, the use of ODPs helps to create reusable ontologies and to enable reasoning [17]. By using ODPs a structure of ontology and structures of modeled entities can be made explicit, the resulting modular models are more useful for reuse. The ODPs are actually an emerging method of ontology construction, but there is a lack of well documented and sufficiently detailed examples how to apply them to real and complete modeling use cases. In [17] several types of ODPs are presented including:

- *Logical ODPs* that provide solutions to solve problems with lacking OWL expressivity e.g., expressing n-ary relations in OWL;
- *Architectural ODPs* that describe the overall internal and external structure of an ontology that is convenient with respect to a specific ontology-based task or application, e.g. a certain DL family;
- *Content ODPs* that are small ontologies addressing a specific modeling issues; they can be reused by importing them into an ontology under development;
- *Presentation ODPs* that provide good practices for e.g. naming conventions;
- *Reasoning ODPs* that are applications of logical patterns oriented towards obtaining certain reasoning results; they are based on the behavior implemented in a reasoning engine;

- *Reengineering ODPs* provide designers with solutions to the problem of transforming a conceptual model, which can be a non-ontological resource, into a new ontology.

4.1   Logical ODPs in the Construction of SYNAT Ontology

OWL gives only logical language constructs, but does not provide any guidelines concerning their use to solve actual modeling tasks and decisions[5] during ontology development. Moreover in semantic web languages OWL and RDF binary relations link two individuals, or an individual and a value, so it is impossible to directly model relations of higher arity.

Modeling real world entities in their context nearly always requires the association of additional info with a binary relation, or to model a network of relations between several individuals[6]. In both above mentioned cases, the use of logical ODP called n-ary relations (described in [18]) is the solution recommended by W3C[7] for ontology modeling problems.

Logical ODPs fulfilled our needs for realistic and effective modeling of composed entities in SYNAT system ontology, they were applied during SYNAT ontology construction to define important notions such as *Person*, *WebResource*, *Project* or *Organization*.

The class *Person* is the most illustrative example of the concept modeling using logical ODPs, an n-ary relation, in SYNAT system ontology. Subsequent paper subsection presents in more details the modeling of the class *Person*, other ontology classes represented as n-ary relations are also mentioned.

4.2   Modeling Classes in SYNAT Ontology as N-ary Relations

A person in science, a researcher, can play several roles in life of scientific community. He or she: attends scientific events or is involved in projects, has not only personal data but can has several work positions at scientific or publishing organizations and can be active in many research fields, is author of papers and books, is also a reviewer of papers and books, is theses supervisor, has achieved scientific degrees and scored work results. The majority of concepts necessary to describe the *Person* concept are defined in the *PersonCharacteristic* module of the ontology. The realistic modeling of the class *Person* requires the creation of the network linking ontology elements representing above listed notions, i.e. the creation of so-called n-ary relation with no

---

[5] Modeling something as a class or an object property is an important decision during ontology development. The modeling options depend on the assumed ontology applications.

[6] N-ary relation modeling the network of relations between several individuals is often called n-ary relation with no distinguished participant. Its use does not require the creation of an additional class, as it is the case when modeling the relation with additional attributes [18].

[7] By the World Wide Web Consortium, not only the main international standards organization, but also organization engaged in education.

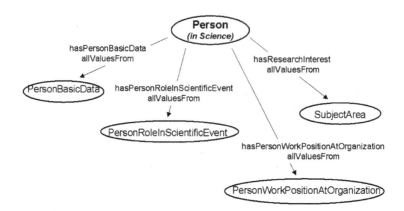

Fig. 8. The schema of the n-ary relation modeling class *Person*. The n-ary relation is linking individuals playing different roles in class *Person* definition. For the sake of the schema clarity only some selected relations and fillers are presented.

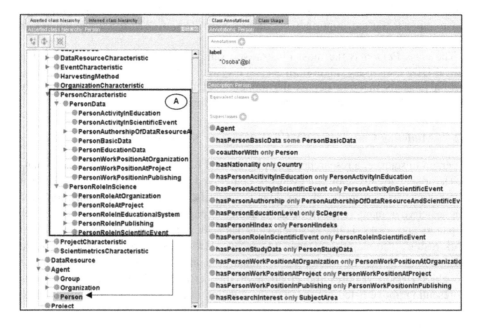

Fig. 9. The editing of the class *Person* in Protégé [19] ontology editor. In the left panel – a part of SYNAT ontology hierarchy, submodule *PersonCharacteristic* is marked with letter A. In the right panel properties of the class *Person* are modeled as relations of the n-ary relation. Most of the relations fillers are subclasses of the class *PersonCharacteristic*.

distinguished participants [18][8]. The simplified[9] schema of the network representing class *Person* is presented in Fig. 8. The editing of the definition of the composed class *Person* in Protégé ontology editor is presented in Fig. 9.

5   Evaluation of SYNAT Ontology

Even though the ontology community noticed the need for ontology evaluation no comprehensive and global approach to this problem has been proposed so far [20]. Authors publishing in the field agree that the evaluation should be systematic and should accompany the ontology not just at the design time but throughout its lifespan. In this section we discuss ontology evaluation methods and criteria and present how the SYNAT ontology adheres to them.

5.1   Automatic Ontology Evaluation

In the literature one can find references to software tools supporting ontology evaluation. Most of the tools are either dated or their development is discontinued. [21] and [22] mention two tools that actually deal with ontology evaluation: OntoAnalyser and OntoClean. OntoAnalyser is a tool that uses rules and focuses on evaluation of ontology properties, in particular language conformity and consistency. OntoAnalyser was realized as a plug-in to OntoEdit (which is now part of a commercial ontology tool). OntoClean is an approach towards the formal evaluation of ontologies, as it analyses the intensional content of concepts [23]. OntoClean enables the formal analysis of concepts and taxonomic relationships based on the the philosophical notions *rigidity*, *unity*, *dependency* and *identity* (known as OntoClean meta-properties).

The application of OntoClean consists of two main steps [24]. First, all concepts are tagged with regards to the OntoClean meta-properties. Second, the tagged concepts are checked against predefined constraints, with constraint violations indicating potential misconceptualisations in the subsumption hierarchy. Although OntoClean is well documented in numerous publications, and its importance is widely acknowledged, it is still used rather infrequently due to the high costs for application. Several tools supporting the OntoClean methodology have been developed and integrated into ontology editors such as ODEClean for WebODE, OntoEdit or Protégé. The main obstacle for applying OntoClean is the manual tagging of concepts with the correct meta-properties which requires substantial efforts of highly experienced ontology engineers. Manual tagging of an ontology consisting of 266 concepts can take an expert from a week up to a month [24]. Such an evaluation is thus a long and expensive process. [23] and [24] proposed AEON - a tool which automatically tags concepts with appropriate OntoClean meta-properties. The tool however is not distributed as a plugin for a particular ontology editor but is rather a prototype available in the form of a source code, which makes using it a challenging task.

---

[8] Logical patterns described in defining N-ary relations on the Semantic Web 2006, Use Case 3: N-ary relation with no distinguished participant.

[9] I.e. not showing, for the sake of the clarity, all of the relation participants.

5.2   Ontology Evaluation Criteria

In [21] the authors propose to evaluate ontologies according to the following criteria: consistency, completeness, conciseness, expandability and sensitiveness. The authors of [25] mention some of the criteria proposed in [21] and complement them with some additional ones, like accuracy, computational efficiency and organizational fitness/commercial accessibility.

The criteria listed above have the following meaning [21,25]:

- Consistency – refers to whether it is possible to obtain contradictory conclusions from valid input definitions. A given definition is consistent if and only if the individual definition is consistent and no contradictory sentences can be inferred from other definitions and axioms. A given definition can be individually consistent or inferentially consistent.

- Completeness. Incompleteness is a fundamental problem in ontologies, even more when ontologies are available in an open environment such as the Semantic Web. We cannot prove either the completeness of ontology or the completeness of its definitions, but we can prove both the incompleteness of an individual definition and thus deduce the incompleteness of an ontology, and the incompleteness of an ontology if at least one definition is missing in the established reference framework. Ontology is thus complete when (i) all that is supposed to be in the ontology is explicitly stated in it, or can be inferred, and (ii) each definition is complete.

- Conciseness. An ontology is concise: (i) if it does not store any unnecessary or useless definitions, (ii) if explicit redundancies between definitions of terms do not exist, and (iii) if redundancies cannot be inferred from other definitions and axioms.

- Expandability refers to the effort required to add new definitions to an ontology and more knowledge to its definitions without altering the set of well-defined properties already guaranteed.

- Sensitiveness relates to how small changes in a definition alter the set of well-defined properties already guaranteed.

- Accuracy refers to whether the axioms comply to the expertise of one or more users and whether the ontology captures and represents correctly aspects of the real world.

- Computational efficiency relates to how easily and successfully reasoners can process the ontology and how fast the usual reasoning services (satisfiability, instance classification, querying, etc.) can be applied to the ontology.

- Organizational fitness/commercial accessibility refers to the ease of deployment of the ontology within the organization, whether ontology-based tools within the organization put constraints upon the ontology, whether the process for creating the ontology used was proper, if the ontology was certified (if that is required), whether it meets legal requirements, is easy to access and aligns to other ontologies already in use.

5.3   Evaluation of the SYNAT Ontology

Applications utilizing SYNAT ontology are currently under development. It is thus not yet possible to fully evaluate the ontology. Below we present technical evaluation according to the criteria listed in Section 5.2.

- Consistency. The syntactical consistency of the SYNAT ontology has been verified using two reasoners, Pellet and HermiT. Based on the verification definitions and restrictions of all concepts are consistent and can have instances, thus the whole SYNAT ontology is consistent.
- Completeness. We asked employees of two libraries: Main Library of Warsaw University of Technology and Polish National Library, and also employees of the Institute of Information and Library Science of Jagiellonian University in Cracow to verify the content of ontology definitions, especially those related to descriptions of web information resources. In their expert opinion the ontology in its first version had some deficiencies, which we corrected and repaired in the current version of the ontology.
- Conciseness. SYNAT ontology was designed to avoid redundancies (reasoner does not detect equivalent classes based on notions' definitions) and was verified by the above-mentioned experts.
- Expandability. In the *Characteristic* module one can add various attributes (any number of hierarchical attributes) of the main classes (*Agent, InformationResource, Event, Project*) defined in the ontology.
- Sensitiveness. This evaluation of the ontology according to this criterion is difficult, as this criterion is ambiguous, as it does not define exactly what a small change is. If a small change is understood by adding an attribute to a class – the ontology is not sensitive to such changes. Thus the SYNAT ontology ensures that small changes in a definition does not alter the set of well-defined properties already guaranteed.
- Accuracy. Throughout the process of creation of the ontology all the decisions concerning the aspect of extending the ontology with particular classes or class properties was consulted and discussed with the potential users. The ontology reflects their everyday needs and is thus accurate with regard to the definition of this criterion.
- Computational efficiency. This criterion has been verified using reasoners (Pellet and HermiT). Classification times for those reasoners are short and comparable (0.343 sec and 0.287 sec respectively) and thus computational efficiency of the SYNAT ontology can be judged satisfying.
- Organizational fitness/commercial accessibility. At this moment this criterion is hard to verify. It will become possible once the ontology becomes part of a working system.

6   Future Work

In the future we plan to extend the SYNAT system ontology in cooperation with the group from the National Library of Poland working on SYNAT metadata. The other ontologies that can be reused in our project are Semantic Publishing and Referencing (SPAR) ontologies [26,27]. The SPAR ontologies, FaBiO, CiTO, BiRO and C4O can be useful for extensive and detailed modeling of bibliographic objects, bibliographic records and references, citations, citation counts, citation contexts and their relationships to relevant sections of cited papers, and the organization of bibliographic records and references into bibliographies, ordered reference lists and library catalogues.

Other models that can be reused and extended in the SYNAT domain are: geographical ontologies (as it was mentioned before), other science classifications [3]. As it was mentioned before, it is possible to use various science classifications, however they should have links to the introduced SCPL ontology. Moreover, the SYNAT system ontology should be flexibly associated with other domain ontologies encapsulating concepts from particular scientific domains or disciplines.

Important extension of the SYNAT scope is to make references to CIDOC Conceptual Reference Model (CRM) used in SYNAT by Poznan Supercomputing and Networking Center [28]. The CIDOC CRM provides definitions and a formal structure for describing the implicit and explicit concepts and relationships used in cultural heritage documentation [29]. The model can provide the "semantic glue" needed to mediate between different sources of cultural heritage information, such as that published by museums, libraries and archives. The two models - ours and CIDOC CRM – are mutually complement. Ours is dedicated to semantically define social network of scientific community and their products. The other (CIDOC) concentrates on proper documentation of cultural heritage.

The above described ontology extensions and associations require to solve the most important problem in the current SYNAT system ontology i.e. the mechanism of effective linking with other ontologies. Currently used, standard *owl:import* constructor does not allow to properly modularize various ontologies and to store them separately with properly defined relations between ontology modules.

We plan to incorporate and adjust the SYNAT system ontology using the most advanced solution to store modularized knowledge bases i.e. conglomerates and their couplers and a tool S-Pellet to reason with the use of conglomerates [14,15].

Other important tasks in the projects are associated with the use of SYNAT system ontology as metadata in acquisition tasks and knowledge base used for this purpose.

7   Summary

In the paper we presented SYNAT system ontology which allows for precise and wide modeling of the academic/scientific community. It formalizes concepts related to the scientific community and its activities i.e.: people in science and their activities, scientific and science-related documents, academic and non-academic organizations, scientific events and data resources, geographic notions necessary to characterize facts about science and classification of scientific topics. In its current version SYNAT system ontology counts 472 classes and 296 properties, its consistency was verified using HermiT reasoner.

The ontology was at first designed to model a scope of the system for scientific community data search and acquisition. This approach enables to model all aspects and roles of a person in science. Relations between persons and publications, persons and projects make it possible to analyze cooperation between people. It is also possible to search experts in given domains and analyze the academic social network. Inclusion of the geographical ontology in the ontology enables localization of ontology instances, which is an entry point for conducting various spatial analyses of the scientific community (using e.g. spatial data mining methods). Moreover, application of design patterns to the SYNAT system ontology makes its structure more clear and prepared to reuse.

References

1. Wróblewska, A., Podsiadły-Marczykowska, T., Bembenik, R., Protaziuk, G., Rybiński, H.: Methods and Tools for Ontology Building, Learning and Integration – Application in the SYNAT Project. In: Bembenik, R., Skonieczny, L., Rybiński, H., Niezgodka, M. (eds.) Intelligent Tools for Building a Scient. Info. Plat. SCI, vol. 390, pp. 121–151. Springer, Heidelberg (2012)
2. SYNAT system ontology, http://wizzar.ii.pw.edu.pl/passim-ontology/
3. System and Domain Ontologies in PASSIM project. Design of the system for building and development ontologies. Technical Report B12 in SYNAT project, Institute of Computer Science, WUT (September 2011)
4. Sure, Y., Bloehdorn, S., Haase, P., Hartmann, J., Oberle, D.: The SWRC Ontology - Semantic Web for Research Communities. In: Bento, C., Cardoso, A., Dias, G. (eds.) EPIA 2005. LNCS (LNAI), vol. 3808, pp. 218–231. Springer, Heidelberg (2005); SWRC model available at, http://ontoware.org/swrc/
5. FOAF Vocabulary Specification 0.98. Namespace Document - Marco Polo Edition (August 9, 2010), http://xmlns.com/foaf/spec/
6. Krafft, D.B., Cappadona, N.A., Caruso, B., Corson-Rikert, J., Devare, M., Lowe, B.J.: & VIVO Collaboration: VIVO: Enabling National Networking of Scientists. In: WebSci10: Extending the Frontiers of Society On-Line, Raleigh, NC, April 26-27 (2010), VIVO model http://vivoweb.org/download
7. ESWC Conference ontology, http://www.eswc2006.org/rdf/
8. Science Ontology, http://protege.stanford.edu/ontologies/ontologyOfScience/ontology_of_science.htm
9. SIOC ontology, http://rdfs.org/sioc/spec/
10. Dublin Core, http://www.cs.umd.edu/projects/plus/SHOE/onts/dublin.html
11. BIBO Ontology, http://bibotools.googlecode.com/svn/bibo-ontology/trunk/doc/index.html
12. Paliouras, G., Spyropoulos, C.D., Tsatsaronis, G. (eds.): Knowledge-Driven Multimedia Information Extraction and Ontology Evolution. LNCS, vol. 6050. Springer, Heidelberg (2011)
13. Grau, B.C., Parsia, B., Sirin, E.: Working with Multiple Ontologies on the Semantic Web. In: McIlraith, S.A., Plexousakis, D., van Harmelen, F. (eds.) ISWC 2004. LNCS, vol. 3298, pp. 620–634. Springer, Heidelberg (2004)

14. W. Waloszek: Implementing Sentential Representation of S-modules in S-Pellet. In: BDAS 2012, vol. 33(2A (105)). Silesian University of Technology Press, Gliwice (2012)

15. Goczyła, K., Waloszek, A., Waloszek, W.: Algebra of Ontology Modules for Semantic Agents. In: Nguyen, N.T., Kowalczyk, R., Chen, S.-M. (eds.) ICCCI 2009. LNCS, vol. 5796, pp. 492–503. Springer, Heidelberg (2009)

16. Gamma, E., Helm, R., Johnson, R., Vlissides, J.: Design Patterns: Elements of Reusable Object-Oriented Software. Professional Computing Series. Addison-Wesley (1995)

17. Gangemi, A.: Ontology design patterns. In: Staab, R. (ed.) Handbook of Ontologies, International Handbooks on Information Systems, 2nd edn. Springer (2009)

18. N-ary relations ODP, http://www.w3.org/TR/swbp-n-aryRelations/

19. Protégé editor, http://protege.stanford.edu

20. Gangemi, A., Catenacci, C., Ciaramita, M., Lehmann, J.: Modelling Ontology Evaluation and Validation. In: Sure, Y., Domingue, J. (eds.) ESWC 2006. LNCS, vol. 4011, pp. 140–154. Springer, Heidelberg (2006)

21. Gomez-Perez, A.: Ontology Evaluation. In: Staab, S., Studer, R. (eds.) Handbook on Ontologies, pp. 251–274. Springer (2003)

22. Aruna, T., Saranya, K., Bhandari, C.: A Survey on Ontology Evaluation Tools. In: Process Automation, Control and Computing, PACC (2011)

23. Völker, J., Vrandečić, D., Sure, Y.: Automatic Evaluation of Ontologies (AEON). In: Gil, Y., Motta, E., Benjamins, V.R., Musen, M.A. (eds.) ISWC 2005. LNCS, vol. 3729, pp. 716–731. Springer, Heidelberg (2005)

24. Völker, J., Vrandecic, D., Sure, Y., Hotho, A.: AEON – An approach to the automatic evaluation of ontologies. Journal of Applied Ontology 3(1-2), 41–62 (2008)

25. Vrandecic, D.: Ontology Evaluation. In: Staab, S., Studer, R. (eds.) Handbook on Ontologies, pp. 293–314. Springer (2009)

26. SPAR ontologies, http://opencitations.wordpress.com/2010/10/14/introducing-the-semantic-publishing-and-referencing-spar-ontologies/

27. Shotton, D.: CiTO, the Citation Typing Ontology. Journal of Biomedical Semantics 1(Suppl. 1), S6 (2010), http://dx.doi.org/10.1186/2041-1480-1-S1-S6

28. Mazurek, C., Sielski, K., Stroiński, M., Walkowska, J., Werla, M., Węglarz, J.: Transforming a Flat Metadata Schema to a Semantic Web Ontology: The Polish Digital Libraries Federation and CIDOC CRM Case Study. In: Bembenik, R., Skonieczny, L., Rybiński, H., Niezgodka, M. (eds.) Intelligent Tools for Building a Scient. Info. Plat. SCI, vol. 390, pp. 153–177. Springer, Heidelberg (2012)

29. Crofts, N., Doerr, M., Gill, T., Stead, S., Stiff, M.: Definition of the CIDOC Conceptual Reference Model, 5.0.2 edn. (2010), http://www.cidoc-crm.org/docs/cidoc_crm_version_5.0.2.pdf

Chapter V
Text Mining

Hierarchical, Multi-label Classification of Scholarly Publications: Modifications of ML-KNN Algorithm

Michał Łukasik[1], Tomasz Kuśmierczyk[1],
Łukasz Bolikowski[1], and Hung Son Nguyen[2]

[1] Interdisciplinary Centre for Mathematical and Computational Modelling,
University of Warsaw, Warsaw, Poland
{m.lukasik,t.kusmierczyk,l.bolikowski}@icm.edu.pl
[2] Faculty of Mathematics, Informatics and Mechanics,
University of Warsaw, Warsaw, Poland
son@mimuw.edu.pl

Abstract. One of the common problems when dealing with digital libraries is lack of classification codes in some of the documents. In the following publication we deal with this problem in a multi-label, hierarchical case of Mathematics Subject Classification System. We develop modifications of ML-KNN algorithm and show how they improve results given by the algorithm on example of Springer textual data.

Keywords: document classification, multi-label classification, hierarchical classification, ML-KNN, YADDA2 software platform.

1 Introduction

Document classification is an old problem and does not require a computer to be solved. A good example is a library, in which categories are assigned to books. A problem that occurs with manual approach is scalability. Automatic text classification is considered since 1960s [1].

Nowadays, document classification is a common problem. Sebastiani [1] brings up such examples as: document indexing, document filtering, meta-data extraction, word sense disambiguation, creating hierarchical catalogue of Internet websites. This list can be extended with analysis of emotions expressed by a text's author [2]. An important problem which appears when dealing with text corpora is assigning classification codes to documents, based on previously classified documents.

In this paper we inspect an established multi-label classification algorithm: ML-KNN. We show a problem that might occur when dealing with noisy data and develop a new KNN-based algorithm that is more resistant to noise. We also use an established method for dealing with hierarchical classification problem and join the method with ML-KNN modifications. We show on real data how new algorithms perform better than ML-KNN.

R. Bembenik et al. (Eds.): *Intell. Tools for Building a Scientific Information*, SCI 467, pp. 343–363.
DOI: 10.1007/978-3-642-35647-6_22   © Springer-Verlag Berlin Heidelberg 2013

The rest of this work is organised as follows: in section 2 we formally define what a classification problem is. We show different classification measures for multi-label classification and specify a measure for hierarchical classification. In section 3 we review literature on multi-label and hierarchical classification. Section 4 contains a detailed description of ML-KNN algorithm. We show a problem, which might occur when working with noisy data using this algorithm. We propose and describe novel modifications of ML-KNN which are not prone to this specific problem. Section 5 describes data, on which we tested our algorithms. Section 6 shows what experimental settings have been taken, whereas section 7 contains the results for data. We finish our work with summary and propositions of future work which might improve the algorithms.

2 Problem Statement

In this section we formalize a classification problem of documents. We consider an example of such problem in section 5.

2.1 Classification Problem

In the classification problem of scientific documents a set of documents D is considered. Furthermore, k attribute functions are defined, each mapping a document into a value from some domain:

$$\forall_{i \in 1, \cdots, k} : a_i : D \rightarrow D_{a_i} \tag{1}$$

Let $Q = \{q_1, \cdots, q_n\}$ be the set of n labels describing documents from set D. We can then specify a function $K : D \times Q \rightarrow \{0, 1\}$, such that $K(d, q) = 1 \Leftrightarrow$ q describes document d. Let $K(d)$ be the set of labels describing d.

The solution to the classification problem is creating a function $K' : D \times Q \rightarrow \{0, 1\}$, which is as similar in a given sense to K as possible. It is done based on some finite set of training documents $D_{train} \subset D$.

2.2 Evaluation

Testing is evaluating, how similar function K' is to K. It is based on comparing values $K'(d)$ and $K(d)$ returned for documents d from some set of documents $D_{test} \subset D$.

Different approaches to evaluation exist in the literature. It is worth noting, that many labels may be assigned to a single document, which brings even more complexity to the problem. We have shown some of the existing evaluation methods below [3].

Accuracy (defined in equation (2)) measures classification quality, not distinguishing errors resulting from choosing too many labels from errors resulting from not choosing the label that should be chosen.

$$Accuracy(K, K', D_{test}) = \frac{1}{|D_{test}|} \sum_{d_{test} \in D_{test}} \frac{|K(d_{test}) \cap K'(d_{test})|}{|K(d_{test}) \cup K'(d_{test})|} \tag{2}$$

Let P and R (defined by equations: (3) and (4)) be: Precision and Recall evaluated for a document. Precision is the ratio of correct decisions made by a classifier to all the labels that have been chosen. Recall is the ratio of correct decisions made by a classifier to all the labels that describe a document.

$$P(K, K', d_{test}) = \frac{|K(d_{test}) \cap K'(d_{test})|}{|K'(d_{test})|} \tag{3}$$

$$R(K, K', d_{test}) = \frac{|K(d_{test}) \cap K'(d_{test})|}{|K(d_{test})|} \tag{4}$$

Based on P and R, analogous variables can be specified for the whole data set: these are the arithmetic means of measures calculated for single documents (equations: (5) and (6)).

$$Precision(K, K', D_{test}) = \frac{1}{|D_{test}|} \sum_{d_{test} \in D_{test}} P(K, K', d_{test}) \tag{5}$$

$$Recall(K, K', D_{test}) = \frac{1}{|D_{test}|} \sum_{d_{test} \in D_{test}} R(K, K', d_{test}) \tag{6}$$

F-measure is a popular classification measure, which deals with a problem of imbalanced label representation. F-measure for a single document is defined as a harmonic mean of Precision and Recall. In equation (7) F-measure is defined as an arithmetic mean of such variables calculated for each of the documents.

$$F\text{-}measure(K, K', D_{test}) = \frac{1}{|D_{test}|} \sum_{d_{test} \in D_{test}} 2\frac{P(K, K', d_{test})R(K, K', d_{test})}{P(K, K', d_{test}) + R(K, K', d_{test})} \tag{7}$$

The last 2 measures we list show how wrong the classifier was in the evaluation process. Hamming Loss (8) returns number of labels, for which incorrect answer has been given, averaged over all documents.

$$Hamming\text{-}loss(K, K', D_{test}) = \tag{8}$$
$$\frac{1}{|D_{test}|} \sum_{d_{test} \in D_{test}} \frac{|(K(d_{test}) - K'(d_{test})) \cup (K'(d_{test}) - K(d_{test}))|}{|Q|}$$

Subset Zero-One loss (9) returns amount of documents for which at least one error has been made in the classification process.

$$Zero\text{-}One\text{-}loss(K, K', D_{test}) = \frac{1}{|D_{test}|} \sum_{d_{test} \in D_{test}} K(d_{test}) \neq K(d_{test}) \tag{9}$$

2.3   Label Dependencies

In the classification problem dependencies among the labels may be specified. They can be given in a form of a relation: $R \subset Q \times Q$. An example of such relation is as follows: $q_1 R q_2 \Leftrightarrow q_1$ is a sub-category of q_2. In such a case, labels with their dependencies form a graph. In general, each label may have a few parental labels. When adding a constraint that each label can have only one parental node, categories form a tree. Only the case of a balanced tree will be considered in this paper.

In a problem stated in such a way one of 2 possibilities can occur. The first possibility is that each of the labels assigned has to be a leaf node in a category tree. The second possibility is the opposite. In this paper we will consider the first option.

In the literature, many approaches to evaluating hierarchical classifiers exist. The common idea behind most of the evaluation methods is that the closer the labels in the category graph are, the less punishment in the evaluation process they should bring when the mistake is made for the other label. There is no established method for hierarchical evaluation yet [4].

In this paper, we evaluate the classifier by comparing functions K and K' in the following way: for each height h of nodes in the tree of labels (where maximum height is taken by the leaves, and minimum value of 0 by the root) project all the leaf nodes to their ancestor nodes of height h. Such projected labels can be compared in the traditional way, for example using F-measure. We will receive as many results as there are levels in the tree. Because we consider only the case of balanced tree, where only leaf nodes are assigned, we will not encounter a problem that an assigned label does not have an ancestor node of a given height.

3 Previous Work

In this section we review methods for solving multi-label and hierarchical classification problems.

3.1 Multi-label Classification

Multi-label classification methods can be divided into 2 groups: problem transformation methods and algorithm adaptation methods. Problem transformation methods are algorithms that decompose a multi-label problem into one or more single-label problems. Algorithm adaptation methods are about extending an established single-label algorithm to deal with multiple labels.

Problem Transformation Methods. One of the simple approaches to the multi-label classification is creating a single-label classifier for each label, which discriminates between a label and all other labels. There are $|Q|$ classifiers, each of which is trained on whole data set. At classification time each classifier answers to the question whether a given label should be assigned to an object or not.

Another approach is about training $|Q|^2 - |Q|$ classifiers, which discriminate between all pairs of labels: q_1 and q_2. The set of positive samples consists of objects to which label q_1 has been assigned, whereas the set of negative samples contains only objects with label q_2 assigned. Comparing to previous approach, there are more classifiers to be trained and less training examples for each of them.

It is worth noting that each of the methods mentioned above assumes independence between the labels. An approach to multi-label classification which does not make such an assumption is about creating a new set of labels Q', which contains the power set of the set Q. The problems with this approach are: small number of training samples for each of the new categories and exponential number of new labels.

Read et al describe a method called classifier chains in [5], which assigns labels one after another, at each step using information about labels assigned this far. A problem how to determine the order of labels is solved by randomly choosing several options.

Zhang proposes in [6] an algorithm based on creating a Bayesian network, which models dependencies between the labels. Information about the dependencies is fetched from correlations between errors given by single-label classifiers.

Algorithm Adaptation Methods. There are various algorithm adaptation methods for multi-label classification. Clare and King described in [7] a modification of C4.5 algorithm, with appropriately modified formula for entropy calculation.

There exist modifications of Ada-Boost approach which allow multi-label classification, that are not transformation based [8]. The idea is very similar to the idea behind basic Ada-Boost approach.

A popular approach in Multi-label classification based on algorithm adaptation is Multi-label KNN, introduced in [9]. It is a Bayesian classifier based on distance features, calculated on neighbours from the training set. The classifier is described in details in next section.

3.2 Hierarchical Classification

In case, when labels form a tree, using the information about the dependencies might increase the efficiency of a classifier. There exist various approaches to how to use such information [10] [4].

First approach is called the flat method, which is about ignoring the hierarchical dependencies and just using some standard classification algorithm.

Another approach is about dividing nodes by their distance from the root. Each set of nodes is then treated as a separate flat classification problem.

The most popular approach is about creating one classifier per node of a label tree [10]. Each of the classifiers is trained on appropriately narrowed data set. There are different subclasses of this approach. There might be a single-label classifier in each node, returning the truth value whenever a class describes a given object. There might also be multi-label classifiers in parental nodes. In such approach a classifier can choose multiple labels from node's children.

The last category of hierarchical classification listed in [10] is the Big-Bang approach. It is about training a single classifier for the whole hierarchy, which is somehow built in the algorithm. The arguments for using this approach are savings in time and space complexity. There has not been made much research about this kind of classifiers [10]. An example of the Big-Bang approach is: casting the hierarchical classification into a multi-label problem, saving the information about the hierarchy by adding the labels that are parents of those describing objects [11]. The post-processing step enforces consistency with the hierarchy.

4   ML-KNN

ML-KNN (Multi-Label KNN)[9] is a popular multi-label classification algorithm [3]. It uses 2 popular approaches to classification [12]: Naive Bayes and KNN.

Naive Bayes is an algorithm, which is popular because of its efficiency: in case of simple features it uses only linear time for training.

KNN is an algorithm, which achieves efficiency close to the best classifiers. In case of problems, for which Bayes Error Rate equals 0, 1NN algorithm converges to the optimal classifier as the training data becomes larger [12].

It is worth noting, that already in 1998 Joachims noticed some serious arguments for using SVM for text classification [13]. However, SVM depends heavily on solving a quadratic programming problem, which makes it a computationally demanding task. At the same time, KNN-based algorithms can be efficiently implemented, for example using k-d trees. When working with big text corpora it is therefore worth considering more efficient methods than SVM, such as ML-KNN.

In this section we describe in detail ML-KNN algorithm and inspect its nature. We point at a possible problem when working with ML-KNN on real data. We deal with this problem, developing novel modifications of ML-KNN that do not increase asymptotic time complexity.

4.1   Basic Algorithm

Let us use the notation defined in section 2 and moreover let us define:

- S_x - neighbourhood of object x, e.g. its k nearest neighbours (where k is earlier defined)
- $S_x(q)$ - number of occurrences of label $q \in Q$ among the objects from S_x

Let H_q be an event, that a given object belongs to class q, and let $\neg H_q$ be the opposite event. Let $E_{S_x(q)}$ be an event, that an object has $S_x(q)$ neighbours belonging to class q.

Category q is being assigned to a given object, if $P(H_q|E_{S_x(q)}) > P(\neg H_q|E_{S_x(q)})$. Bayes theorem states, that this inequality is equivalent to the following:

$$P(E_{S_x(q)}|H_q)P(H_q) > P(E_{S_x(q)}|\neg H_q)P(\neg H_q) \tag{10}$$

It is possible to estimate variables from the inequality (10) using the training set.

The training algorithm for ML-KNN is shown in listing: Algorithm 1 (D is the training set, Q is the label set, m is its size, K is the known classification of the training objects, k is the neighbourhood size, s is the smoothing parameter). It uses 2 arrays: c and c', both of size $|Q| \times (k+1)$. Their purpose is explained below.

Algorithm 1. ML-KNN(D, Q, m, K, k, s)

1: Initialize 2-dimensional arrays c and c', both of size $|Q| \times (k+1)$
2: **for** $q \in Q$ **do**
3:     $P(H_q) = \frac{s + \sum_{x \in D} K(x,q)}{2s + m}$
4:     $P(\neg H_q) = 1 - P(H_q)$
5: **end for**
6: **for** $x \in D$ **do**
7:     S_x = find k nearest objects to x in D
8:     **for** $q \in Q$ **do**
9:         $i =$ how many times class q occurs among objects in S_x
10:         **if** $K(x, q)$ **then**
11:             increment $c[q][i]$
12:         **else**
13:             increment $c'[q][i]$
14:         **end if**
15:     **end for**
16: **end for**
17: **for** $q \in Q$ **do**
18:     **for** $i \in 0..k$ **do**
19:         $P(E_i|H_q) = \frac{s + c[q][i]}{s(k+1) + \sum_{p \in \{0..k\}} c[q][p]}$
20:         $P(E_i|\neg H_q) = \frac{s + c'[q][i]}{s(k+1) + \sum_{p \in \{0..k\}} c'[q][p]}$
21:     **end for**
22: **end for**
23: **return** $\quad \forall_{q \in Q} P(H_q), \quad \forall_{q \in Q} P(\neg H_q), \quad \forall_{q \in Q} \forall_{i \in \{0, \cdots, k\}} P(E_i|H_q),$
    $\forall_{q \in Q} \forall_{i \in \{0, \cdots, k\}} P(E_i|\neg H_q)$

The algorithm works as follows. First, a-priori probabilities for each label occurrence are calculated. This is performed by calculating occurrences of categories in the training set. Next, in the double-nested loop, values $c[q][i]$ are calculated. They denote, how many times the following situation occurs: object belonging to class q has exactly i neighbours, which belong to class q. Similarly, $c'[q][i]$ can be evaluated. They correspond to situations, when object not belonging to class q has exactly i neighbours belonging to class q. In the end, the posterior probabilities are calculated using values from arrays c and c'.

Classification of an object is implemented as the inequality (10) defines.

Each category is considered separately. Therefore, label independence is assumed.

The ML-KNN algorithm depends on values calculated in the arrays c and c'. It can be noticed, that when smoothing parameter s equals 0, the algorithm is equivalent to comparing counts $c[q][i]$ and $c'[q][i]$ (for given i) and choosing class

q iff $c[q][i] > c'[q][i]$ (in [9] it was not stated). This can be shown by the following series of equivalent inequalities:

$$P(E_{S_x(q)}|H_q)P(H_q) > P(E_{S_x(q)}|\neg H_q)P(\neg H_q) \tag{11}$$

$$\frac{c[q][S_x(q)]}{\sum_{p\in\{0..K\}} c[q][p]} \frac{\sum_{p\in\{0..K\}} c[q][p]}{m} > \frac{c'[q][S_x(q)]}{\sum_{p\in\{0..K\}} c'[q][p]} \frac{\sum_{p\in\{0..K\}} c'[q][p]}{m} \tag{12}$$

$$c[q][S_\star(q)] > c'[q][S_x(q)] \tag{13}$$

4.2   Threshold ML-KNN

Let us consider the following situation: 2 objects x_1 and x_2 are given for classification and the following inequality holds for them: $S_{x_1}(q) < S_{x_2}(q)$ for some label q. In such case, ML-KNN algorithm allows the following to happen: $P(H_q|E_{S_{x_1}(q)}) > P(\neg H_q|E_{S_{x_1}(q)})$ and at the same time $P(H_q|E_{S_{x_2}(q)}) < P(\neg H_q|E_{S_{x_2}(q)})$. It means, that classifier can learn to assign category q to object x_1 with small number of neighbouring objects described by category q and at the same time not to assign category q to object with big number of neighbours described by class q. Such a situation has been observed when analyzing data described in chapter 5. At the same time, the classifiers efficiency was low.

Example of data, where such situation should be allowed is shown in figure 1. Nevertheless, intuitively this phenomenon corresponds to noise. Therefore, it seems reasonable to consider a modification of ML-KNN, which does not allow such situations to happen.

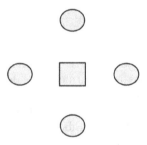

Fig. 1. An example, where situation described should be allowable. When considering neighbourhood of size 4, each of the circles has only 3 neighbouring circles, whereas the square has 4 neighbouring circles.

In order to achieve this, we propose the following modification. Instead of estimating the probabilities shown in inequality (10), threshold number of neighbours p can be chosen for each category, such that $K(x,q) = 1 \Leftrightarrow S_x(q) > p$. In the training data it is possible that such a value does not exist. Common situation is such as shown in table 1. It shows, that in almost all cells non-zero values exist. Nevertheless, intuitively the threshold in the case of data shown in table 1 should be chosen for p=1, because $\forall_{t>1} : c[l] > c'[l]$ and $\forall_{t\leq1} : c[l] \leq c'[l]$. It is less obvious, what the threshold should be like in case of data shown in table 2, because: $c[1] < c'[1]$, $c[2] > c'[2]$, $c[3] < c'[3]$ and $c[4] > c'[4]$.

Table 1. Example table with counts for category q

neighbours count	c	c'
0	0	100
1	20	40
2	30	24
3	10	8
4	8	1

Table 2. Example table with counts for category q: complicated situation

neighbours count	c	c'
0	0	100
1	20	40
2	30	21
3	10	14
4	8	1

We propose to choose threshold by maximizing the F-measure. Let us use the following notation:

- FN (false negatives), number of objects incorrectly classified as not belonging to class q. These are the samples incrementing the count $c[q][i]$ for $i \leq p$.
- TP - (true positives), number of objects correctly classified as belonging to class q. These are the samples incrementing the count $c[q][i]$ for $i > p$.
- TN - (true negatives), number of objects correctly classified as not belonging to class q. These are the samples incrementing the count $c'[q][i]$ for $i \leq p$.
- FP - (false positives), number of objects incorrectly classified as belonging to class q. These are the samples incrementing the count $c'[q][i]$ for $i > p$.

Now, the threshold p can be chosen in such a way, that F-measure is maximized. F-measure is calculated using the following formula:

$$F1 = \frac{2PR}{P + R} \tag{14}$$

In formula (14), P means precision and R means recall. They are calculated in the following way: $P = \frac{TP}{TP+FP}$, $R = \frac{TP}{TP+FN}$.

Complete training algorithm for Threshold ML-KNN is shown in listing: Algorithm 2. It uses 2 arrays: c and c', both of size $|Q| \times (k+1)$. Furthermore, an array p of size $|Q|$ and an auxiliary variable $best_{f1}$ are used. Their meanings are explained below.

The algorithm works as follows. In lines 3 - 13 counts c and c' are calculated in a similar way as in case of ML-KNN. Next, for each category q a threshold value $p[q]$ maximizing F-measure is being chosen. F-measure is calculated using values from arrays c and c', as explained above. An array p is returned.

Algorithm 2. THRESHOLD-ML-KNN(D, Q, K, k, s)

1: Initialize 2-dimensional arrays c and c', both of size $|Q| \times (k+1)$
2: Initialize 1-dimensional array p of size $|Q|$ and floating point variable $best_{f1}$
3: **for** $x \in D$ **do**
4:     $S_x =$ find k nearest objects to x in D
5:     **for** $q \in Q$ **do**
6:         $i =$ how many times class q occurs among objects in S_x
7:         **if** $K(x, q)$ **then**
8:             increment $c[q][i]$
9:         **else**
10:             increment $c'[q][i]$
11:         **end if**
12:     **end for**
13: **end for**
14: **for** $q \in Q$ **do**
15:     $p[q] = -1$
16:     $best_{f1} = -1$
17:     **for** $i \in 0..k$ **do**
18:         $FN = \sum_{l \in \{0, \cdots, i-1\}} c[q][l]$
19:         $TP = \sum_{l \in \{i, \cdots k\}} c[q][l]$
20:         $TN = \sum_{l \in \{0, \cdots, i-1\}} c'[q][l]$
21:         $FP = \sum_{l \in \{i, \cdots k\}} c'[q][l]$
22:         $F1 =$ Calculate F-measure based on FN, TP, TN and FP
23:         **if** $F1 > best_{f1}$ **then**
24:             $p[q] = i$
25:             $best_{f1} = F1$
26:         **end if**
27:     **end for**
28: **end for**
29: **return** p

4.3   Ensemble Threshold ML-KNN

Choice of value for parameter k is a problem that appears each time a KNN based algorithm is used. The way how Threshold ML-KNN has been defined allows to cope with the problem by using different values at the same time.

Algorithm 3. ENSEMBLE-THRESHOLD-ML-KNN-TRAIN
$(D, Q, K, k\text{-}list, s)$

1: Initialize 3-dimensional arrays c and c', both of size $k\text{-}list_{len} \times |Q| \times (\max(k\text{-}list) + 1)$
2: Initialize 2-dimensional arrays p i $best_{f1}$, both of size $k\text{-}list_{len} \times |Q|$
3: **for** $x \in D$ **do**
4:     $S_x = $ find $\max(k\text{-}list)$ nearest objects to x in D
5:     **for** $q \in Q$ **do**
6:         **for** $k_j \in k\text{-}list$ **do**
7:             $i = $ how many times class q occurs among first k_j objects in S_x
8:             **if** $K(x, q)$ **then**
9:                 increment $c[j][q][i]$
10:            **else**
11:                increment $c'[j][q][i]$
12:            **end if**
13:        **end for**
14:    **end for**
15: **end for**
16: **for** $k_j \in k\text{-}list$ **do**
17:    **for** $q \in Q$ **do**
18:        $p[j][q] = -1$
19:        $best_{f1}[j][q] = -1$
20:        **for** $i \in 0..k_j$ **do**
21:            $FN = \sum_{l \in \{0, \cdots, i-1\}} c[j][q][l]$
22:            $TP = \sum_{l \in \{i, \cdots k_j\}} c[j][q][l]$
23:            $TN = \sum_{l \in \{0, \cdots, i-1\}} c'[j][q][l]$
24:            $FP = \sum_{l \in \{i, \cdots k_j\}} c'[j][q][l]$
25:            $F1 = $ Calculate F-measure based on FN, TP, TN and FP
26:            **if** $F1 > best_{f1}[j][q]$ **then**
27:                $p[j][q] = i$
28:                $best_{f1}[j][q] = F1$
29:            **end if**
30:        **end for**
31:    **end for**
32: **end for**
33: **return** $p, best_{f1}$

A few Threshold ML-KNN classifiers may be constructed simultaneously, not making the asymptotic time complexity larger.

After finding k nearest neighbours of an object in a sorted order, it is possible to find j nearest neighbours out of them efficiently, for $j \leq k$. Therefore, for a list of neighbour sizes $k\text{-}list = [k_1, k_2, \cdots, k_n]$ it is enough to find a number of $\max(k\text{-}list)$ nearest neighbours once in order to calculate the parameters needed to train a number of Threshold ML-KNN classifiers (each of them corresponds to a value $k_j \in k\text{-}list$). When training Threshold ML-KNN classifiers, we can make use of already fetched nearest neighbours. Such a solution allows for savings in time complexity, because the most time consuming part of training is finding nearest neighbours.

After the training process, the estimated F-measure values can be used to choose the best Threshold ML-KNN for each class.

In the listing: Algorithm 3 the training algorithm for Ensemble Threshold ML-KNN is shown. It makes use of the following data structures:

- 3-dimensional arrays: c and c', both of size: k-$list_{len} \times |Q| \times (\max(k\text{-}list)+1)$ (k-$list_{len}$ denotes length of the list),
- 2-dimensional arrays: p and $best_{f1}$, both of size: k-$list_{len} \times |Q|$,

First, counts c and c' are calculated. After calculations, $c[j][q][i]$ shows how many times the following situation occurs: object belonging to class q has exactly i neighbours out of the closest k_j nearest neighbours (where k_j belongs to list k-$list$), which belong to class q. On the other hand, $c'[j][q][i]$ shows how many times object not belonging to class q has exactly i neighbours out of the closest k_j neighbours, which belong to class q.

Next, arrays p and $best_{f1}$ are calculated. After calculations, $p[j][q]$ shows, what minimum number of neighbours out of the closest k_j neighbours should be described by class q in order to assign class q to an object. $best_{f1}[j][q]$ shows what F-measure value has been achieved for such threshold. Based on these arrays the best value $k_q \in k$-$list$ in terms of achieved F-measure can be chosen for each class q.

5   Data Description

In this section we describe data on which we tested algorithms described in previous sections. The classifiers have been evaluated on Springer data[1], which has been made available to the ICM[2].

5.1   General Description

Data made available consists of 1342882 records, describing consecutive scientific papers. Each record is described by meta-data such as: list of authors, title, abstract, keywords. Full list of fields has been shown in table 3.

Each record is described by a list of labels. There are different categorization systems in data, such as: MSC[3], PACS[4] (Physics and Astronomy Classification Scheme) etc. In table 4 we show basic statistics describing categorization systems.

The categories are very rare in the corpus. Therefore, in the evaluation process we consider MSC only.

---

[1] http://www.springer.com/
[2] http://www.icm.edu.pl/
[3] http://www.ams.org/mathscinet/msc/msc2010.html
[4] http://publish.aps.org/PACS

Table 3. Fields describing the records

Field	Description
an	Unique identifier.
py	Publication year.
ti	Title.
ut	Keywords.
ab	Abstract. Consisting of 1-4 short sentences describing the paper.
au	List of authors.
jy	Year of journal publication.
mc	MSC classification tags.
jp	Journal publisher.
ps	Page numbers in the journal.
jt	Journal title.
uv	Affiliations of authors.
vl	Volume of a journal.
jc	ISSN number of a journal.

Table 4. Statistics about various categorization systems in the corpus

Categorization	No. of documents	No. of occurrences	No. of distinct codes
ZDM	297	783	125
PACS	13715	38639	3970
CLC	8536	8536	2495
QICS	37	66	40
MSC	20275	54410	5130
JEL	7927	22349	860

5.2 MSC Codes

MSC (Mathematics Subject Classification) is the classification system for documents on mathematics. It consists of more than 5000 categories, each represented by a string of 2, 3 or 5 characters. Each document can be described using more than one category (in such case, first category is considered as most important). Categories form a hierarchy, in which each category is represented by a string of 2 or 3 characters and is divided into subcategories. In picture 2 we show part of the MSC tree. Each node corresponds to some field of mathematics. In the example shown, 60 corresponds to probability theory and stochastic processes, 60E to distribution theory, and 60E15 to inequalities and stochastic orderings.

5.3 Analysis of Record Content

There are many fields containing information concerning the publications in the meta-data. It seems that some introduce relevant information (such as keywords and title) and some do not (such as page numbers).

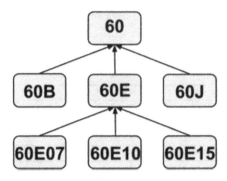

Fig. 2. Part of an MSC category tree

It can be stated, that the most important fields are textual ones: title, abstract and keywords. List of authors can also contain relevant information, since if a person published in some field, then it is very possible that he or she is still going to work on it.

In table 5 we showed how many records contain all fields from given sets. We excluded fields, which do not seem to introduce important information for choosing the topic of a paper.

After counting fields occurrences it can be noticed, that number of records containing all the textual information (title, abstract and keywords) is not much smaller then number of records containing MSC codes.

5.4   Data Filtering

In picture 3 we show a histogram showing number of codes with consecutive occurrence numbers. It can be noticed, that there is a big number of codes with very few occurrences. Training a classifier on a small number of training samples is a hard task.

Because of low number of occurrences of some codes, we performed data filtering. We left only these codes, which appeared at least 30 times and at the same time have not been the only subcategory of their parental category. The criterion has been applied recursively to codes of 2nd and 3d level. As a result, some codes have been left without any labels, therefore they have been removed from the corpus. Finally, 240 different codes for the bottom level remained. We got 9180 records.

5.5   Data Characteristics

Tsoumakas in [3] describes measures which describe complexity of the data set: *label cardinality* and *label density*.

Label cardinality shows, how many labels on average are being assigned to a record. In our case the value is 1.56.

Table 5. Number of records containing subsets of fields

ab	1105609
au	1289023
jt	14959
mc	20275
py	1342065
ti	1281409
ut	851240
ab au mc py ti	20155
ab au mc py ut	17977
ab au mc ti ut	17959
au mc py ti ut	18017
au jt mc py ti ut	470
ab au jt mc py ti	502
ab au jt mc py ut	468
ab au jt mc ti ut	468
ab au mc py ti ut	17959
ab jt mc py ti ut	468
ab au jt mc py ti ut	468
ab au jt mc py ti ut	468

Label density describes, what part of all the labels is on average assigned to a document. The value for the data is 0.65%.

The corpus can also be described by measures describing, how homogeneous labels assigned to a document are on average. In table 6 we showed in how many documents there are at least 2 similar codes (similar in the hierarchical sense) and at least 2 different codes.

6   Experimental Settings

We decided to use only textual fields in the experiments. We joined keywords, title and abstract of each record to form a single text describing each document. We removed stop words and projected words to their base forms using Porter stemmer. We then performed a popular text analysis technique called TF-IDF [12]. This way, we got vectors of numbers describing each document.

In each algorithm, we used cosine distance, which is a good measure to distinguish between high dimensional objects, such as texts described by TF-IDF vectors [12].

As for ML-KNN algorithms, we used the following settings for the parameters:

- smoothing parameter (s): 1
- k parameter: 5
- *k-list* parameter: [3, 5, 8]

As was the case in [9], we also noticed, that value for parameter k does not influence the relative performance of the algorithms.

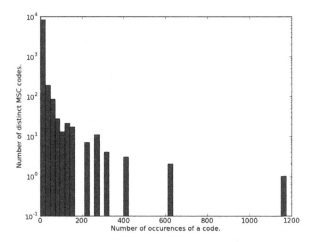

Fig. 3. Number of MSC codes of a given number of occurrences in the corpus before filtering

Table 6. The percentage amount of records with at least 2 similar (different) codes. The criterion for similarity is being subcategory of the same category.

Property	Percentage
Contains at least 2 sub-codes of the same highest level category	35.39%
Contains at least 2 sub-codes of different highest level categories	13.66%
Contains at least 2 sub-codes of the same 2nd level category	29.59%
Contains at least 2 sub-codes of different 2nd level categories	24.78%

7   Experimental Results

In this section we show results for classifiers described earlier.

We have evaluated the following classifiers: ML-KNN, Threshold ML-KNN and Ensemble Threshold ML-KNN, where the 2 latter algorithms are our modifications of ML-KNN algorithm (they have been described in section 4).

Furthermore, we have evaluated modifications, where we put each of the 3 listed classifiers into parental nodes of the hierarchical structure of labels. This is one of the most popular approaches to hierarchical classification, as stated in section 3. We list tested algorithms in table 7.

The evaluation has been performed on data described in section 5, prepared as stated in section 6. 5-fold cross validation has been performed.

We used measures described in section 2.2. Furthermore, we used a method for evaluating hierarchical classification problems, described in section 2.3. Results for each consecutive level of the label tree are shown in tables: 8, 9, 10.

Table 7. Evaluated classifiers. We separately test flat approaches and hierarchical approaches. Novel methods that have been proposed in this paper have been highlighted.

Approach	Base Algorithm	Modification	Ensemble Modification
Flat	ML-KNN	Threshold ML-KNN	Ensemble Threshold ML-KNN
Hierarchical	Hierarchical ML-KNN	Hierarchical Threshold ML-KNN	Hierarchical Ensemble Threshold ML-KNN

Table 8. Evaluation measures for results projected to the 1st level of labels

Classifier	Accuracy	Precision	Recall	F-measure	Hamming Loss	Subset 0/1 Loss
ML-KNN	21.30%	23.08%	21.35%	21.90%	0.38%	80.49%
Threshold ML-KNN	44.36%	47.48%	45.23%	45.68%	0.30%	59.55%
Ensemble Threshold ML-KNN	63.33%	66.59%	72.38%	67.43%	0.29%	48.59%
Hierarchical ML-KNN	44.62%	48.09%	45.91%	46.20%	0.32%	60.03%
Hierarchical Thresh. ML-KNN	50.95%	54.73%	51.75%	52.47%	0.28%	53.55%
Hierarchical Ens. Thresh. ML-KNN	66.59%	70.66%	71.33%	69.54%	0.25%	42.06%

It can be noticed, that ML-KNN algorithm yields low efficiency. This is a result of problems introduced in chapter 4. Much better results given by proposed modifications support this hypothesis.

Ensemble Threshold ML-KNN is the only classifier which does not use hierarchical information and that exceeds 50% in terms of accuracy in table 8. Comparing to ML-KNN, each of the modifications (Threshold ML-KNN and Ensemble Threshold ML-KNN) gives much better results in terms of all used measures.

As for the results given by the hierarchical approaches, it can also be noticed, that each of the modifications exceeds the simple Hierarchical ML-KNN algorithm in terms of all used measures. The best results are given by the most advanced classifier: Hierarchical Ensemble Threshold ML-KNN. What is interesting is that the hierarchical approach introduces low improvement (a few percent in table 8 and even less in table 10) comparing to the flat approach. In some works similar observation has been made: use of hierarchy brings small improvement in terms of classification correctness [12].

8   Future Work

There are many directions, in which our work can be improved. We listed some of the possibilities below.

Table 9. Evaluation measures for results projected to the 2nd level of labels

Classifier	Accuracy	Precision	Recall	F-measure	Hamming Loss	Subset 0/1 Loss
ML-KNN	18.40%	21.52%	18.51%	19.44%	0.45%	84.59%
Threshold ML-KNN	36.49%	41.76%	37.85%	38.67%	0.40%	69.72%
Ensemble Threshold ML-KNN	49.42%	54.19%	61.72%	55.14%	0.49%	66.44%
Hierarchical ML-KNN	35.27%	40.69%	37.38%	37.76%	0.44%	71.78%
Hierarchical Thresh. ML-KNN	40.91%	46.95%	42.78%	43.53%	0.41%	66.59%
Hierarchical Ens. Thresh. ML-KNN	51.83%	57.57%	61.75%	57.00%	0.46%	62.41%

Table 10. Evaluation measures for results directly on leaf nodes

Classifier	Accuracy	Precision	Recall	F-measure	Hamming Loss	Subset 0/1 Loss
ML-KNN	13.83%	18.87%	14.03%	15.45%	0.59%	90.61%
Threshold ML-KNN	25.12%	32.60%	27.71%	28.43%	0.60%	83.92%
Ensemble Threshold ML-KNN	31.74%	37.14%	46.55%	38.56%	0.84%	85.82%
Hierarchical ML-KNN	24.78%	32.63%	27.11%	28.04%	0.63%	84.03%
Hierarchical Thresh. ML-KNN	26.60%	34.32%	30.39%	30.45%	0.66%	83.82%
Hierarchical Ens. Thresh. ML-KNN	31.94%	37.37%	48.62%	38.92%	0.96%	85.34%

- In our work, we assumed a bag of words model, which in turn assumes independence between the words. What can be done to improve this is to try extracting semantic information from the text.
- We excluded non-textual attributes from data. It seems that information about authors can be very useful to categorize documents.
- We used TF-IDF algorithm to generate weights. There exist other algorithms (such as LSA or LDA), which may improve the results.
- The most popular approach to make use of hierarchical structure between the labels has been chosen. One can try other solutions.
- We narrowed our work to a single algorithm (ML-KNN). We could try use other approaches and join them with our algorithms. For example we could form a hierarchy of different classifiers, where our algorithm deals with the first classification step, and more difficult situations to decide are delegated to more effective classifiers (such as SVM).
- Evaluating hierarchical classification is not a trivial problem and it can be further explored.

9   Summary

In this paper we dealt with a problem of multi-label hierarchical classification of documents. We chose to work on ML-KNN and supported our decision with efficiency of this algorithm. Analysis of ML-KNN algorithm has been performed. It turns out, that when there is noise in data, the algorithm fits to it very much. We proposed modifications of ML-KNN, which help to deal with this issue.

The problem has been noticed when working on data, on which algorithms have been evaluated. It turns out, that a situation, where document's neighbourhood is not very stable is not rare. The algorithm's ability to learn about such instabilities causes degradation in classification.

What is worth mentioning is that our modifications, apart from giving better results, sustain low computational cost of the algorithm. This is important, because as we pointed out earlier, it is one of the biggest strengths of this approach.

Acknowledgements. This work is supported by the National Centre for Research and Development (NCBiR) under Grant No. SP/I/1/77065/10 by the Strategic scientific research and experimental development program: "Interdisciplinary System for Interactive Scientific and Scientic-Technical Information."

References

1. Sebastiani, F.: Machine learning in automated text categorization. ACM Computing Surveys 34(1), 1–47 (2002)
2. Melville, P., Gryc, W., Lawrence, R.D.: Sentiment analysis of blogs by combining lexical knowledge with text classification. In: Proceedings of the 15th ACM SIGKDD International Conference on Knowledge Discovery and Data Mining, KDD 2009, pp. 1275–1284. ACM, New York (2009)
3. Tsoumakas, G., Katakis, I.: Multi-label classification: An overview. IJDWM 3(3), 1–13 (2007)
4. Costa, E., Lorena, A., Carvalho, A., Freitas, A.: A review of performance evaluation measures for hierarchical classifiers. In: Drummond, C., Elazmeh, W., Japkowicz, N., Macskassy, S. (eds.): Evaluation Methods for Machine Learning II: Papers from the AAAI-2007 Workshop, AAAI Technical Report WS-07-05, pp. 1–6. AAAI Press (July 2007)
5. Read, J., Pfahringer, B., Holmes, G., Frank, E.: Classifier Chains for Multi-label Classification. In: Buntine, W., Grobelnik, M., Mladenić, D., Shawe-Taylor, J. (eds.) ECML PKDD 2009, Part II. LNCS, vol. 5782, pp. 254–269. Springer, Heidelberg (2009)
6. Zhang, M.L., Zhang, K.: Multi-label learning by exploiting label dependency. In: Proceedings of the 16th ACM SIGKDD International Conference on Knowledge Discovery and Data Mining, KDD 2010, pp. 999–1008. ACM, New York (2010)
7. Clare, A.J., King, R.D.: Knowledge Discovery in Multi-label Phenotype Data. In: Siebes, A., De Raedt, L. (eds.) PKDD 2001. LNCS (LNAI), vol. 2168, pp. 42–53. Springer, Heidelberg (2001)
8. Zhu, J., Rosset, S., Zou, H., Hastie, T.: Multi-class adaboost. Technical report (2005)

9. Zhang, M.L., Zhou, Z.H.: Ml-knn: A lazy learning approach to multi-label learning. Pattern Recognition 40(7), 2038–2048 (2007)
10. Silla, C., Freitas, A.: A survey of hierarchical classification across different application domains. Data Mining and Knowledge Discovery 22, 31–72 (2011), 10.1007/s10618-010-0175-9
11. Kiritchenko, S., Matwin, S., Nock, R., Famili, A.F.: Learning and Evaluation in the Presence of Class Hierarchies: Application to Text Categorization. In: Lamontagne, L., Marchand, M. (eds.) Canadian AI 2006. LNCS (LNAI), vol. 4013, pp. 395–406. Springer, Heidelberg (2006)
12. Manning, C.D., Raghavan, P., Schtze, H.: Introduction to Information Retrieval. Cambridge University Press, New York (2009)
13. Joachims, T.: Text categorization with support vector machines: learning with many relevant features. In: Nédellec, C., Rouveirol, C. (eds.) ECML 1998. LNCS, vol. 1398, pp. 137–142. Springer, Heidelberg (1998)
14. Sylwestrzak, W., Rosiek, T., Bolikowski, L.: YADDA2 – Assemble Your Own Digital Library Application from Lego Bricks. In: Proceedings of the 2012 ACM/IEEE Joint Conference on Digital Libraries (2012)

A   Implementation in the YADDA2 Architecture

Results of research presented in this paper are currently implemented as a module in the SYNAT system. SYNAT is a strategic project commissioned by the Polish National Centre for Research and Development, with the goal of building "Interdisciplinary System for Interactive Scientific and Scientific Technical Information." YADDA2 framework, developed at ICM UW, is a core part of that system.

YADDA2 [14] has a two-tier architecture, with base services tier providing generic fuctionalities independent of the type of content being processed, and application tier where business logic and user interfaces are located. YADDA2 facilitates creation of several types of products:

- stand-alone repositories with a web front-end and a publication application in the back-end;
- repository federations containing multiple autonomous collections, accessed through a central front-end;
- publication data warehouses aggregating content from multiple repositories in order to provide long-term preservation of data and access for researchers and analysts.

Several configurable components are already implemented and are ready to be used, for example: meta-data and content storage, full-text indexing, batch processing engine, relational index, user annotation service. Results of this research are being implemented as yet another reusable module, providing hierarchical, multi-label classification tailored for scholarly publications.

The classification module is intended to be part of back-end work-flows for improving meta-data quality and enriching it with inferred information. In a typical setting, one back-end process fetches from a storage all the documents that are already classified using codes from a given classification scheme, passes them to the module in question in order to train it and places results of the training in a storage. Another back-end process fetches all documents from a given domain lacking codes from a given classification scheme, pipes them to the module for classification (configured to use the results of an earlier training) and updates document meta-data using output from the classifier.

Comparing Hierarchical Mathematical Document Clustering against the Mathematics Subject Classification Tree

Tomasz Kuśmierczyk[1], Michał Łukasik[1],
Łukasz Bolikowski[1], and Hung Son Nguyen[2]

[1] Interdisciplinary Centre for Mathematical and Computational Modelling,
University of Warsaw, Warsaw, Poland
{t.kusmierczyk,m.lukasik,l.bolikowski}@icm.edu.pl
[2] Faculty of Mathematics, Informatics and Mechanics,
University of Warsaw, Warsaw, Poland
son@mimuw.edu.pl

Abstract. Mathematical publications are often labelled with Mathematical Subject Classification codes. These codes are grouped in a tree-like hierarchy created by experts. In this paper we posit that this hierarchy is highly correlated with content of publications. Following this assumption we try to reconstruct the MSC tree basing on our publications corpora. Results are compared to the original hierarchy and conclusions are drawn.

Keywords: documents clustering, documents representation, clustering similarity, measures of agreement, Mathematics Subject Classification.

1 Introduction

1.1 Research Problem

There are several established classification schemes for scholarly literature, for example: Mathematics Subject Classification (MSC), Physics and Astronomy Classification Scheme (PACS), Journal of Economic Literature (JEL) Classification System, ACM Computing Classification System, or a much broader Dewey Decimal Classification. All these systems are created by human experts (rather than generated by an algorithm), all are hierarchical, and many undergo periodical updates which result in minor-to-moderate differences between revisions.

In this research, we are primarily interested in building algorithms that would, as far as it is possible, recreate a classification system for a given domain. In a wider sense, we are interested in studying the process that governs the development of such classification systems, in particular, understanding which features of the classified documents have the largest impact on the final hierarchy. While most of our theoretical work is applicable to any hierarchical classification scheme, our experiments are conducted on the 2000 revision of Mathematics Subject Classification.

R. Bembenik et al. (Eds.): *Intell. Tools for Building a Scientific Information*, SCI 467, pp. 365–392.
DOI: 10.1007/978-3-642-35647-6_23 © Springer-Verlag Berlin Heidelberg 2013

This paper is structured as follows. In the remainder of this section we briefly summarize the MSC 2000 system, similarity as it is understood in computer science, and the data set used in our experiments. In Section 2 we investigate approaches to measuring structural similarity of objects, we outline state-of-the-art, present our original ideas and analyze the results of our experiments. In Section 3 we focus on similarity of documents. Section 4 describes methodology of evaluation of similarity matrices, and Section 5 presents various experiments related to reconstructing MSC 2000 hierarchy. The last sections contain summary and conclusions.

1.2   MSC Codes

MSC codes[1] are a hierarchical system for multi-tagging of mathematical documents. It was created by experts from Mathematical Reviews and Zentralblatt MATH[2]. There are two slightly different version of codes: MSC2000 and MSC2010.

In MSC hierarchy there are three levels of codes:

- leaves (denoted as L) - named with 5 characters (2 digits + letter or special character '-' + 2 digits) - for example: 05C05 means 'Trees'
- middle level (denoted as M) - named with 3 characters - for example: 05C means 'Graph theory'
- higher level (denoted as H) - named with 2 characters - for example: 05 means 'Combinatorics'

Special character '-' is used for special purpose documents (instructional exposition, proceedings etc.).

In MSC every single document can have one or more codes assigned. First code is the most important and is called 'primary'. Subsequent are called 'secondary'. What is more, not only leaf-codes can be assigned but also codes from upper levels.

1.3   Similarity

Similarity is an intuitive and subjective concept. Many different approaches to this idea exist in psychology.

One of the earliest and the one that has most in common with computer science is mental distance approach [18]. In this approach objects are represented as points within the space and similarity is represented by some distance function.

The second most influential approach is featural approach [19] (in formalism closely related to Jaccard Coefficient). In this method objects are represented as sets of features. Similarity is then measured by comparing two sets of features. It increases with the number of common features and decreases with the number of differences. This approach deals with several psychological aspects of

---

[1] http://www.ams.org/mathscinet/msc/msc2010.html
[2] http://www.zentralblatt-math.org/zbmath/

similarity but has several disadvantages e.g. an assumption that commonalities and differences are independent.

In computer science, properties of similarity measures are formulated closely to the featural approach [11]:

F1) similarity is a value in [0, 1]
F2) similarity reaches its maximum when two objects are identical
F3) the more differences two objects have, the less similar they are
F4) the more commonalities two objects share, the more similar they are
F5) $1.0 - similarity$ has metric properties apart from triangle inequality

During our work we dealt with similarity of objects of different kinds e.g. documents and elements localized in different structures. Details are described in further sections.

1.4   Data Description

We used 13,609 documents tagged with MSC2000 codes. Documents originated from following digital libraries:

- ZentralBlatt-MATH[3]
- CEDRAM[4]
- NUMDAM[5]

Every document was represented by an abstract, keywords and a title that were merged into single list of words (apart from bigram calculation where bigrams are calculated before the merge). Length statistics (in words; after filtering - see section 5.1) of these fields are shown in table 1. Some symptoms of preprocessing problems can be found. Especially it is rather uncommon to have abstract containing over 35 thousands of words. The most probable explanation is that during extraction process some parts of document were glued to the abstract. Similar situations can happen in real, fully automatic systems. Due to this fact we decided to leave data after preprocessing without further modifications.

Table 1. Length statistics of documents

field	min	max	avg	std
abstract	11	35522	509.23	493.66
keywords	7	374	77.17	39.13
title	10	318	64.59	34.74
merged	47	35846	651.00	505.39

---

[3] http://www.zentralblatt-math.org/zbmath/
[4] http://www.cedram.org/
[5] http://www.numdam.org/?lang=en

In our experiments we decided to consider only typical leaf codes composed of 5 letters and leave out special codes (with special character '-' instead of one letter in the name of a code). It should not influence our results much, as codes of different types account for less than 5% of all codes.

We also filtered out the codes that occurred less often than 10 times as a primary code. In the end, we had 345 non-special leaf-codes (level L). At level M of MSC tree we had 144 codes. Each of these codes had, on average 2.40, children (std=2.33). 76 codes had only single child and the maximum number of children was 13. On the H-level of MSC tree we had 37 codes out of which 15 had only a single child. Average number of children at this level was 3.89 (std=3.38) and the maximum number of children was 12. Statistics presented above show how diverse MSC subtree is and how complicated is the problem of its reconstruction.

The filtering left only $10,201$ documents in corpus: $7,575$ with primary code assigned and $7,032$ with at least one secondary code. Every leaf-code occurred as a primary code on average 21.96 times (with standard deviation 19.48; max=185) and as a secondary code 29.54 times (with standard deviation 24.01; max=204). What is worth noting, every code appeared at least once as a secondary code.

Before filtering, single document had averagely 2.98 codes assigned. After filtration, an average of only 1.74 codes was left (standard deviation = 0.88). The maximum noticed number of codes per document was 10. Every document had averagely 0.74 primary codes (std=0.44; before filtering=1.00) and 1.00 secondary codes (std=0.89; before filtering=1.98).

2   Similarity of Structures

To compare two structures (e.g. original MSC hierarchy and the reconstructed tree) we decided to adapt pairwise comparisons (in psychology called paired comparisons). In classical clusterings comparisons this group of methods is described as counting of pairs of elements (other two are: information-theoretical mutual information and summation of set overlaps). It was extended for purpose of comparing fuzzy clusterings. Later, we adapted it for hierarchical structures and introduced simple formalism.

Having two elements (e.g. leaves of MSC tree or publications) l_i and l_j, the bonding (introduced in [3]) between them is described by the function:

$$B_{ij} = b_T(l_i, l_j) \in [0, 1] \qquad (1)$$

Index T denotes the structure in which leaves are positioned. It means that bonding in different structures T and T' can be different. Indexes i and j always denote the row and the column in a matrix.

Bonding measures how close two elements are to each other. Complementary value:

$$C_{ij} = 1 - B_{ij} \in [0, 1] \qquad (2)$$

measures how much two elements (indexed by i and j) are separated according to structure T.

Having bondings (and complementary values) of two elements l_i and l_j in structures T and T' we need to decide how much these values 'agree'. It is performed by another function τ:

$$\Theta(\beta, \beta')_{ij} = \tau(\beta_{ij}, \beta'_{ij}) \in [0,1] \tag{3}$$

where: β, β' can be either B, B' or C, C' matrices for T and T'. For example $\Theta(B, B')_{ij}$ measures how much two structures 'agree' on how much elements l_i and l_j should be bonded.

Using four matrices:

$$\Theta(B, B'), \Theta(B, C'), \Theta(C, B'), \Theta(C, C')$$

we calculate four coefficients:

$$a = h(\ \Theta(B, B')\ )$$
$$b = h(\ \Theta(B, C')\ )$$
$$c = h(\ \Theta(C, B')\ )$$
$$d = h(\ \Theta(C, C')\ )$$

where h is a function that aggregates values from matrices. It can be interpreted as summarizing over all pairs and therefore can be implemented as:

$$h(X) = \sum_{i,j>i} X_{ij}$$

Derived coefficients measure how much two structures 'agree' (a, d) and 'disagree' (b, c). Using them we can adapt similarity indexes designed for typical clusterings comparisons. The most common is the *Rand index* $\in [0,1]$ [15]:

$$RI = \frac{a+d}{a+b+c+d}$$

$$RI = \frac{a+d}{|LL|} \tag{4}$$

$$RI = \frac{|LL|-(b+c)}{|LL|}$$

where number of pairs of elements ($n = |L|$ stands for number of elements):

$$|LL| \equiv \frac{n(n-1)}{2}$$

The measure strongly depends on the number of clusters in the clustering structure [12] (we showed that this property is preserved for hierarchical structures). It was shown [5] that for some kinds of structures its value increases up to 1.0 with number of clusters. What is more, it was shown that RI for two random clusterings is not a constant. For these reasons *Rand Index* was modified to *Adjusted Rand Index* [3]:

$$ARI' = \frac{2(ad-bc)}{c^2+b^2+2ad+(a+d)(c+b)} \in [-1,1]$$

$$ARI = \frac{ARI'+1}{2} \in [0,1] \tag{5}$$

Strong critic [4] is also given to RI for equal treatment of a and d. In some situations d dominates over a what can be a serious problem [8]. To overcome this problem *Jaccard coefficient* was introduced:

$$JI = \frac{a}{a+b+c} \in [0,1] \qquad (6)$$

Presented above methods of similarity measurement depend on two functions: $b_T(l_i, l_j)$ and $\tau(\beta_{i,j}, \beta'_{i,j})$. Their selection changes the properties of measures.

In section 1.3 we assumed that $1.0 - similarity$ should have metric's properties (apart from triangle inequality). For above indexes (apart from *Rand Index*), it is not known what $b_T(l_i, l_j)$, $\tau(\beta_{i,j}, \beta'_{i,j})$ to use and how to modify them to fulfill this condition. Especially, when comparing T to itself we can obtain similarity lower than 1.0. Examples of such behaviour can be found in [7].

In [7] authors modified *Rand Index* in a way that $1.0 - RI$ is a metric (apart from the condition: $d(T, T') = 0 \implies T = T'$ which may not be fulfilled). During our research we showed that, assuming $\tau(a, b) \equiv min(a, b)$ and taking second formula for RI in equation 4, their result can be shown in our formalism. Assumption for $1 - RI$ to be (almost) metric is that $1 - b_T(l_i, l_j)$ must be a metric.

In the context of MSC-like trees our proposition for $b_T(l_i, l_j)$ is to use the common fraction ($\in [0,1]$) of two paths from root to leaves. In this case $b_T(l_i, l_j) = 0$ when two leaves have only root in common and $b_T(l_i, l_j) = 1$ iff $l_i = l_j$. Other metric properties were also proven.

An example of such measure is shown in the figure 1. For leaves l_1 and l_3 common fraction of paths has length q. For l_1 and l_2 it has the length $p + q$.

This measure is well-defined for trees where all leaves have the same depth. If we want to generalise to all trees, for two leaves: l_i and l_j, we have two fractions of paths f_i and f_j. Having f_i and f_j we can use $b_T(l_i, l_j) = F(f_i, f_j)$ where F can be *average*, *min*, *max* etc.

An alternative to 'path fraction' approach can be cosine-like measure:

$$b_T(l_i, l_j) \equiv 1 - \frac{2 \cos^{-1}\left(\frac{v_i}{|v_i|} \cdot \frac{v_j}{|v_j|}\right)}{\pi} \in [0,1] \qquad (7)$$

where v_i, v_j are membership vectors assigned to l_i and l_j. Their length is equal to the number of nodes at the medium (M) level in MSC tree. The value of an m-th component in this vector describes the affiliation of the m-th element to a (M-level) node:

- 2.0 ⇔ element and node have common prefix of length 3 (for example '03A')
- 1.0 ⇔ element and node have common prefix of length 2 (for example '03')
- 0.0 ⇔ element and node have no common prefix

2.1   Experiment 1

To show behaviour of our indexes in different situations we tested them with randomly generated trees. We took $n = |L| \sim 350$ MSC leaves. These leaves describe part of MSC tree. Out of them we built 100 trees.

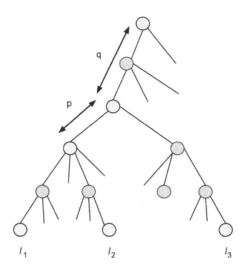

Fig. 1. Common fraction of paths from root to leaves

In the first step, out of range $[n^{0.25}, n^{0.75}]$ we randomly selected number (m) of nodes at M level. Then, every leaf was assigned to one of the m nodes. In the next step, we randomly selected h out of range $[m^{0.25}, m^{0.75}]$ and assigned middle-level nodes to high-level nodes. In such procedure we generated single random tree. This procedure was repeated 100 times. In the end, we obtained 100 random trees.

In the picture 2 there are values of different indexes. We compared random trees and part of the original MSC tree. Prefixes of indexes stands for different configurations of b_T and τ:

- Hf - b_T - path fraction, $\tau(a, b) = min(a, b)$
- Hm - b_T - formula 7, $\tau(a, b) = min(a, b)$
- Bf - b_T - path fraction, $\tau(a, b) = ab$
- Bm - b_T - formula 7, $\tau(a, b) = ab$

Conclusions:

- ARI does not depend on number of nodes and is the most stable
- JI is also very stable but slightly decreases with number of nodes
- RI is very unstable and strongly increases with number of clusters
- τ does not influence much results (Hm-RI and Bm-RI give almost equal results; the same for Hf-RI and Bf-RI)
- b_T does not change order (plots for Hm-RI and Hf-RI are just translated and scaled; the same for Bm-RI and Bf-RI)

In the end, Bf-ARI and Hf-ARI seem to be the best measures. They take constant value (~ 0.5) for random tree no matter how much tree has nodes. The calculation of a common path is also more intuitive and faster than the calculation

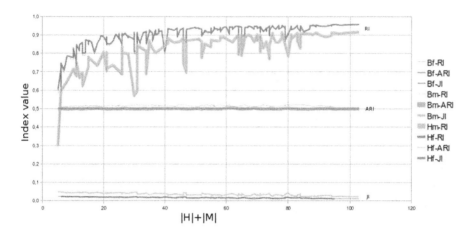

Fig. 2. Indexes' values for random trees

of a membership vectors. Having $n = |L|$ leaves overall complexity is $O(D \times n^2)$ where D stands for height of trees. To select between these two measures we designed further experiments.

2.2 Experiment 2

In the figure 3 behaviour of two indexes: Bf-ARI, Hf-ARI is shown. For each index 10 runs of an experiment is shown. There is also black, thick line that represents an average over all runs. In the single run we randomly selected nodes from H level and split them according to its child nodes. Every node from H level was replaced with c_x nodes, where c_x is a number of children of the node $x \in H$. Each new node has just single child (one of the previous children of x).

An analysis of figure 3 reveals that for comparing MSC subtree (described in the section 1.4) to itself (0 modifications) values of the indexes are different and smaller than 1.0 (0.71 for Bf-ARI and about 0.76 for Hf-ARI). This behaviour is consistent with description from the beginning of the section 2. Another conclusion is that both indexes decrease monotonically (in every single run) with number of such modifications. It is consistent with the intuition.

2.3 Experiment 3

In the figure 4 another experiment's results are shown. For each index 10 runs of an experiment is shown. In the single run we randomly selected nodes from H level and merged their child nodes. For every node from H-level we took all its children and merged them into single child node. In the end, every node from H-level had only single child node.

An analysis of figure 4 reveals that Bf-ARI behaves counter-intuitively for such modifications what makes it useless for further use.

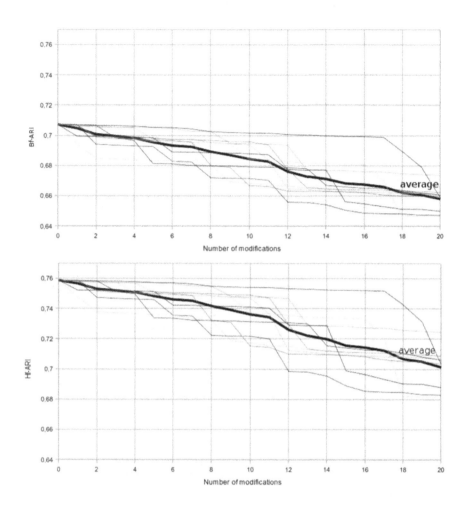

Fig. 3. Indexes' behaviour for splitting nodes at H level

2.4   Measure Selection

An analysis of the behaviour of different indexes showed that the best, out of described indexes, is Hf-ARI. This index is resistant to change of number of nodes (section 2.1), to 'pushing' nodes down (section 2.2) and up (section 2.3). The index have also relatively low computational complexity ($O(D \times n^2)$).

For the tree described in the section 1.4 the lower bound of Hf-ARI is equal to 0.5 (value for random trees) and the upper bound is equal to 0.759 (value for a comparison of the tree to itself). This two values state the range in which we expect to operate.

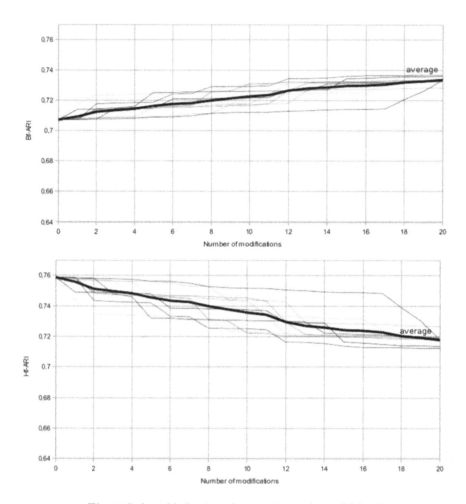

Fig. 4. Indexes' behaviour for merging nodes at M level

3   Similarity of Documents

Modern techniques of similarity determining split into two groups [2] [11]:

- content-based
- link-based

We focused on the first group.

For the reason that dealing with structure in documents is very demanding task, *bag-of-words model* is applied. In this model single document is represented as a set of pairs: (term, count). To compare such representations several measures was developed. One of the most successful [10] is *Ratio Model* [19] that is related to the first model described in section 1.3:

$$sim_{tv}(d_i, d_j) = \frac{|d_i \cap d_j|}{|d_i \cup d_j|} \tag{8}$$

where d_i - bag of terms (words, bigrams etc.) representing i-th document.

More advanced approaches apply *vector space model* [17] (related to the second model from section 1.3) in which documents are represented as vectors of numbers. Document d_i is thus represented by a vector $\boldsymbol{w_i}$. While vectors are often sparse, cosine-like measurements are used [12]:

$$sim_{cos}(d_i, d_j) = 1 - \frac{2\cos^{-1}(\frac{\boldsymbol{w_i}}{|\boldsymbol{w_i}|} \cdot \frac{\boldsymbol{w_j}}{|\boldsymbol{w_j}|})}{\pi} \tag{9}$$

The most popular method of constructing vector representation of documents is $TF \times IDF$ weighting scheme. $TF \times IDF$ belongs to wider group of methods called $LW \times GW$ where LW stands for local weight and GW for global weight. For every term t in document d single weight is generated:

$$\boldsymbol{w}^t = LW_{t,d} \times GW_t \tag{10}$$

Intuitively, weight should be higher if term is more important for document (e.g. occurs many times) but lower if is not very characteristic (e.g. occurs in many documents) (for further description see [12]). In $TF \times IDF$:

$$TF_{t,d} = \frac{d^t}{|d|} \tag{11}$$

$$IDF_t = log(\frac{N}{N^t}) \tag{12}$$

where:

- d^t - number of occurrences of t in document d
- $|d|$ - number of all terms in document d
- N - number of documents in corpus
- N^t - number of documents having term t

There are many doubts about $TF \times IDF$ scheme. Particularly, TF grows linearly with the number of term occurrences but psychologically the occurrence is more important than count. For example, it does not make big difference whether term t occurred 4 or 6 times. To overcome this problem another local weight can be introduced:

$$WF_{t,d} = \begin{cases} 1 + log(1 + TF_{t,d}) & TF_{t,d} > 0 \\ 0 & otherwise \end{cases} \tag{13}$$

Also for IDF there exist many replacements e.g. ENT [13] was reported as particularly efficient [10]:

$$ENT_t = 1 + \frac{\sum_d p_{t,d} \log(p_{t,d})}{log(N)} \tag{14}$$

$$p_{t,d} = \frac{TF_{t,d}}{GF_t} \qquad (15)$$

where GF_t stands for t frequency in whole corpus.

The most advanced techniques (ESA, LSA, LDA etc.) that calculate vector representations try to discover some semantic behind documents. In this group the dominating approach is LSA (for details see [12]) in which original dimensions are linearly combined into new ones. Then, only the most 'informative' dimensions are kept. New dimensions are believed to be latent 'topics' of documents. The method is close to dimensionality reduction techniques.

LSA is calculated using Singular Value Decomposition. In our experiments we used *Gensim* implementation [16] because of its scalability and stream processing mode.

3.1   MSC Leaves Similarity

In our experiments we performed clustering (reconstruction of a MSC hierarchy) of MSC leaves. We assumed that leaves' similarity can be computed basing on the similarity of tagged documents. We considered several strategies of aggregation documents' similarity into MSC leaves similarity.

The first strategy is to consider only primary tags: MSC leaf is represented as a set of documents tagged with it at first place. Sets are disjoint. To estimate similarity between two leaves l_i and l_j aggregating function is calculated:

$$sim_{pr}(l_i, l_j) = A(\{sim(d_k, d_l) : primary(d_k) = l_i, primary(d_l) = l_j\}) \qquad (16)$$

We considered $A \equiv average$ (denoted avg) and $A \equiv max$ (single linkage, denoted $single$). Complete linkage in this situation is pointless as in two sets of documents there are always two with similarity equal to 0.

The second strategy is to consider both: primary and secondary tags. It means that sets of documents overlap. In such case we assigned to every tag l_i of a document d_k weight: $e_{k,i}$. Now, similarity between l_i and l_j is calculated as a weighted average:

$$sim_{sec}(l_i, l_j) = \frac{1}{Z} \sum_{l_i \in tags(d_k), l_j \in tags(d_l)} (e_{k,i} + e_{l,j}) \cdot sim(d_k, d_l) \qquad (17)$$

$$Z = \sum_{l_i \in tags(d_k), l_j \in tags(d_l)} e_{k,i} + e_{l,j} \qquad (18)$$

Primary codes have always weight 1.0. Secondary codes can have constant weight $e_{k,i} = 0.5$ (strategy denoted as $avg - e0.5$) or $e_{k,i} = 0.75$ ($avg - e0.75$) or weight dependent on number of secondary codes assigned to particular document:

$$e_{k,i} = \frac{C}{|secondary(d_k)|} \qquad (19)$$

where $C = 0.75$ (denoted as $avg - s0.75$) or $C = 0.5$ (denoted as $avg - s0.5$).

4   Similarity Matrices Evaluation

Typical approach in clustering [9] is to perform evaluation in the end of the process - after clustering. For this strategy, in our case, having similarity matrices we should cluster using one of the clustering algorithms and then compare results.

Although, in practical applications, where the goal is to tune clustering process as much as it is possible, it is reasonable, this strategy does not give knowledge about similarity matrices themselves. To deal with this problem we decided to evaluate different features and similarity matrices just before clustering.

There are several intuitive assumptions about clustering algorithms. The most common and the most intuitive is that closer object should be joined more likely than distant. According to this rule, formal conditions can be set up [1]. Having three elements: l_i, l_j, l_k the requirement that l_i, l_j should be merged more likely than l_i, l_k and l_j, l_k can be written as inequalities:

$$sim(l_i, l_j) > sim(l_i, l_k) \tag{20}$$

$$sim(l_i, l_j) > sim(l_j, l_k) \tag{21}$$

In worst case number of conditions is bounded by $O(n^3)$ where $n = |L|$ - number of elements. That makes this method useful only if n is small (no bigger than ~ 1000).

Though its shortcomings, the approach has big advantage: can be interpreted as a simple thought experiment. Having n elements, we take three of them and show to an expert. He selects two the most similar or says that he does not know. If he knows, we check in our similarity matrix if the similarity conditions are held. The procedure is repeated for each tuple of size three. In the end, we have fraction of situations when we guessed correctly. Situations when the expert did not know do not count. For random similarity matrix result should be about 50%. For ideal data about 100%.

In our case, expert answers are read directly from MSC tree. The longer common prefix two elements have, the more similar should be. For example: $30A01$ and $30A02$ are the more similar than $30A01$ and $30B01$. Situations when elements have common prefix of equal length is treated as experts' answer: 'do not know'.

5   Experiments

To perform various kinds of experiments efficiently we designed modular framework. Sample configuration of the framework is shown in the figure 5. The data in most of components is processed in a stream. According to different input and output data type we have components of following types:

- document in/document out: documents' filtering, n-grams construction
- document in/vector out: term counts per document generation
- vector in/data model out: $LW \times GW$/LSA models construction

- vector and model in/vector out: $LW \times GW$/LSA vectors calculation
- vector in/similarity matrix out: Tversky/Cosine-like similarity calculation
- similarity matrix in/structure out: hierarchical/k-medoids clustering
- two structures in/similarity index value out: similarity indexes calculation

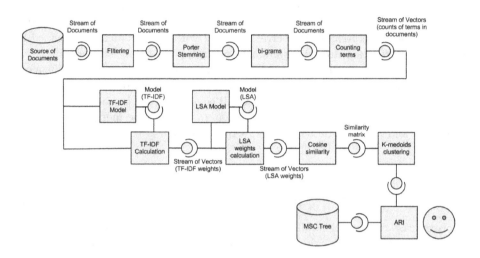

Fig. 5. Sample configuration of the experiment

In different experiments some of the components were removed or replaced with another. For example $TF \times IDF$ can be replaced with $WF \times ENT$ or the component that calculates bigrams can be removed. The table 2 presents considered configurations of the framework.

Each column in the table represents single step in reconstruction process. First column describes possible representation in which documents could be prepared. Second - possible similarity measures. For bigrams and words we used Tversky measure (equation 8). In other cases Cosine-like measure was applied (equation 9). The third column shows possible similarity aggregation methods (for details see section 3.1). The last two columns are related to process of clustering of MSC leaves (section 5.6).

As it can be easily seen, most of the processing steps work on documents. Only in the last part (clustering/reconstruction of the tree, similarity index calculation) MSC-related data was considered.

5.1   Preprocessing

Every document in the corpus was preprocessed. Whole interpunction, brackets, numbers etc. were replaced with spaces. Then letters were changed to lower case. Also single letters and common words (stopwords from the list[6]) were removed.

---

[6] http://www.textfixer.com/resources/common-english-words.txt

Table 2. Possible configurations of the experiment

Representation (section 3)	Similarity (section 3)	Aggregation (section 3.1)	Clustering (section 5.6)	Linkage (section 5.6)
Words Words-TFIDF Words-WFENT Words-TFIDF-LSA Words-WFENT-LSA Bigrams Bigrams-TFIDF Bigrams-WFENT Bigrams-TFIDF-LSA Bigrams-WFENT-LSA	Tversky (sim_{tv}) Cosine-like (sim_{cos})	average (avg) single (max) weighted average (avg-xYY)	hierarchical 3-level hierarchical 3-level k-medoids	average-linkage (avg) single-linkage (single) complete-linkage (complete)

In brackets numbers of connected sections are given.
In further text 'Words' is default and omitted.

In the next step, words were stemmed using Porter algorithm [14] (believed to be the best stemming algorithm [6]).

5.2   Similarity Matrices Evaluation

Using the method described in section 4 we evaluated several variants of similarity matrices for MSC leaves.

We considered documents representations listed in the table 2. To calculate similarity between two documents we used techniques from section 3 (either Tversky for Words/Bigrams or Cosine-like in other cases). In the end, we used one of the strategies from section 3.1 to obtain MSC-leaves' similarity matrix.

An evaluation of different similarity matrices is shown in figures 6 and 7. The results of different variants of weighting strategies in aggregation (section 3.1) were consistent so we decided to show an average value of them (denoted as Avg-xYY)

An analysis of the figures 6 and 7 showed that the best results (above 94% fulfilled conditions) were obtained for documents represented as bigrams with WF-ENT weighting scheme and LSA applied. Overall conclusions for the above figures are following:

- bigrams overcome representation as a bag of single words
- TF-IDF performs worse than WF-ENT
- single-linkage strategy used for aggregation of similarity is the weakest one: averaging gives better results

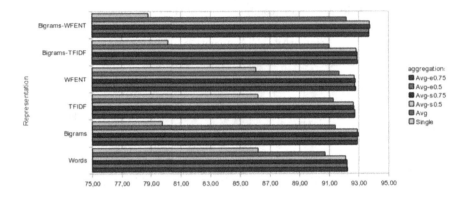

Fig. 6. Evaluation results for different MSC similarity matrices

- considering both primary and secondary codes improves results
- optimal number of dimensions in LSA depends on weighting scheme and it
  is hard to find any rule for its selection

5.3 Extracted Topics

In tables 3, 4, 5, 6 LSA topics for different configurations (weighting schemes)
are shown. Top ten topics is listed for every configuration. For every topic terms
with highest weights are shown. Terms are sorted with weights so the leftmost
are the most important.

The results show several problems in our data. First of them are TEX-tags (e.g.
sb, sp, ąl) that occurred in some documents. Another is that, some documents
contain parts in other than English languages (e.g. French, German).

It can be easily seen that $WF \times ENT$ scheme managed much better than
$TF \times IDF$. In the table 3 we see that TEX-tags were filtered out from the most
important terms in the most influential topics. Also non-english words occur only
in 4-th topic. Even better situation is shown in the table 4. Both: non-english
words and TEX-tags were almost totally filtered out from top ten topics. It is
important to remember that these 'bad terms' were not completely removed.
They were just pushed down to less important topics or given lower weights.

5.4 Linkage and Similarity Aggregation

Figure 8 shows comparison of different linkage (single/complete/average) and
aggregation similarity (see section 3.1) strategies. The results only for 3-level
hierarchical clustering (section 5.6) are shown but for other configurations are
similar. Because results for different variants of weighted average aggregation
method (section 3.1) were almost the same we decided to show an average of
them (denoted as $avg - xYY$).

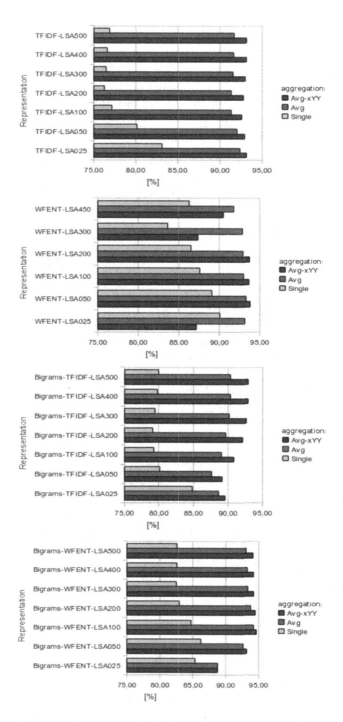

Fig. 7. Evaluation results for different MSC similarity matrices (LSA representations)

Table 3. Top ten LSA topics for $WF \times ENT$ weighting scheme

1	-0.105*problem + -0.103*equat + -0.094*function + -0.093*gener + ...
2	0.987*obituari + 0.080*public + 0.076*jan + 0.064*reiterman + ...
3	0.197*equat + 0.196*solut + 0.151*problem + 0.141*nonlinear + ...
4	0.185*une + 0.183*de + 0.181*un + 0.173*la + 0.168*sur + ...
5	0.190*algorithm + 0.157*numer + 0.145*method + 0.144*comput + ...
6	0.293*process + 0.226*random + 0.192*brownian + 0.167*stochast + ...
7	-0.200*manifold + 0.137*number + 0.131*integ + 0.124*bound + ...
8	0.162*word + -0.156*field + 0.155*algorithm + 0.154*languag + ...
9	-0.200*schrödinger + -0.190*word + -0.170*languag + 0.139*error + ...
10	0.164*solut + -0.153*asymptot + 0.151*equat + -0.146*schrödinger + ...

Table 4. Top ten LSA topics for $WF \times ENT$ weighting scheme on bigrams

1	0.670*list-public + 0.179*public-item + 0.013*comput-scientist + ...
2	0.759*public-item + -0.517*obituari + 0.349*list-public + ...
3	-0.984*order-determin + -0.048*physic-requir + -0.028*expect-uniqu + ...
4	0.482*differenti-ideal + 0.461*modul-field + 0.137*ring-modul + ...
5	-0.123*differenti-ideal + -0.112*modul-field + 0.110*math-zbl + ...
6	0.966*show-impli + 0.072*defin-notion + 0.038*theorem-survey + ...
7	0.653*preview-zbl + 0.653*see-preview + 0.255*extens-doubl + ...
8	0.531*applic-constitut + 0.456*inequ-base + 0.445*base-tetrahedra + ...
9	0.323*applic-constitut + 0.279*inequ-base + 0.259*base-tetrahedra + ...
10	-0.829*ring-proof + -0.471*present-formal + -0.065*eingeführt-und + ...

Table 5. Top ten LSA topics for $TF \times IDF$ weighting scheme

1	-0.471*sb + -0.264*sp + -0.148*omega + -0.123*ąl + -0.114*group + ...
2	0.705*sb + 0.281*sp + -0.125*equat + -0.105*solut + ...
3	0.508*de + 0.216*la + 0.199*le + 0.182*group + 0.179*est + ...
4	0.328*group + -0.290*de + -0.177*solut + -0.174*omega + ...
5	0.595*omega + -0.271*sb + 0.160*ąl + 0.150*text + 0.146*partial + ...
6	0.355*process + 0.257*brownian + 0.234*random + 0.187*measur + ...
7	-0.650*ąl + 0.377*group + -0.206*categori + 0.133*subgroup + ...
8	0.487*ąl + -0.273*manifold + -0.230*surfac + 0.175*categori + ...
9	-0.387*group + 0.242*curv + 0.168*number + 0.159*alpha + -0.153*lie + ...
10	-0.406*omega + 0.235*alpha + 0.199*lambda + 0.193*equat + ...

An analysis of the figure 8 leads to the following conclusions:

- the weakest aggregation methods is *single*
- an information about secondary codes, in most cases, improves results
- complete-linkage is the worst strategy for clustering
- for single-linkage results are similar for different representations (apart from the best one: Bigrams-WFENT-LSA)

Table 6. Top ten LSA topics for $TF \times IDF$ weighting scheme on bigrams

1	-0.581*sb-sb + -0.369*sp-sb + -0.281*sb-sp + -0.143*bbfr-sp + ...
2	0.421*finit-element + 0.256*error-estim + -0.245*sb-sb + ...
3	0.387*finit-element + 0.327*sb-sb + 0.226*element-method + ...
4	-0.496*navier-stoke + -0.424*stoke-equat + 0.262*finit-element + ...
5	-0.516*brownian-motion + -0.269*local-time + -0.234*random-walk + ...
6	0.370*navier-stoke + 0.322*stoke-equat + -0.257*bbfr-sp + 0.210*sb-sb + ...
7	0.263*de-la + 0.140*dan-le + 0.128*bbfr-sp + 0.117*sur-le + ...
8	-0.494*sb-sb + 0.277*sp-sb + 0.219*sb-sp + 0.218*al-sb + ...
9	0.535*random-walk + -0.288*differenti-equat + -0.254*brownian-motion + ...
10	-0.595*al-sb + 0.278*lie-algebra + 0.242*lie-group + ...

5.5 Representations and Similarity

Results in the figure 8 can be interpreted as a comparison of different representations against aggregation and linkage strategies. Such analysis of figure 8 leads to following conclusions:

- Bigrams-WFENT-LSA is the best representation among considered
- the use of LSA improves results (it is especially apparent for complete-linkage)

For single-linkage it is hard to derive any consistent rules. Different representations lead to different results. For average-linkage we can assume that bigrams and more advanced representations generally give better results but there are some exceptions.

5.6 Clustering Method

In our experiments we considered two typical clustering approaches. First was hierarchical clustering. Second was k-medoids.

In hierarchical clustering we tested two strategies: either we compared MSC hierarchy to the binary tree (no modification in clustering results) or we compared MSC hierarchy to the tree reduced to just three levels (3-level hierarchical). Reduction was obtained by testing possible splits (number of nodes at M and H level; we tested values differing by 10) and selecting one with the highest Hf-ARI. Obtained value can be interpreted as an approximation of the upper bound for the clustering method.

For k-medoids clustering we used similar strategy. We tested possible values of k on two levels: M and H. Results of clustering from M level were used to compute input similarity matrix for clustering on H level. To aggregate similarity we tested three linkage method: single/complete/average. Because k-medoids method is non-deterministic the Hf-ARI value was averaged for 5 runs of clustering. It is worth mentioning here that differences between results in each run were very small and could not influence our conclusions.

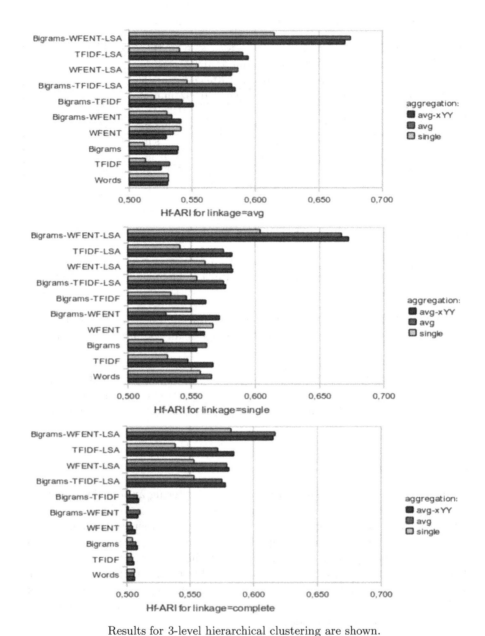

Results for 3-level hierarchical clustering are shown.
$Avg - xYY$ stands for an average of $avg - e0.5$, $avg - s0.5$, $avg - e0.75$, $avg - s0.75$.
For LSA different number of topics were considered and the best (the one with the
highest Hf-ARI value) was chosen for every configuration.

Fig. 8. Comparison of different linkage and similarity aggregation strategies against
documents' representations

Figure 9 presents Hf-ARI values for different clustering methods in different configurations. Three, described previously, clustering methods were considered. It is clear that the best result are obtained for hierarchical clustering with reducing to three levels (3-level hierarchical). For k-medoids and pure hierarchical clustering (binary tree) results are much worse. In some situations k-medoids obtains higher Hf-ARI value what shows that this measure strongly prefers hierarchies with the same numbers of levels.

5.7   Number of LSA Topics

In our experiments we tested LSA representation of documents. For LSA we tested different weighting schemes and different numbers of topics (25, 50, 100, 150, 200, 250, 300, 350, 400, 450, 500). Table 7 presents what were the best values for different configurations. The column number four presents what weighting scheme was used and whether bigrams were used or not. In fifth column the best found Hf-ARI value is shown. The last column presents number of topics assigned to the highest Hf-ARI value.

Table 7. The best number of topics for LSA representation in different configurations

Aggregation	Linkage	Clustering	Representation	Hf-ARI	Num topics
avg-e0.5	avg	3-level hierarchical	Bigrams-TFIDF-LSA	0.671	500
avg-e0.5	avg	3-level hierarchical	Bigrams-WFENT-LSA	0.680	300
avg-e0.5	avg	3-level hierarchical	TFIDF-LSA	0.664	350
avg-e0.5	avg	3-level hierarchical	WFENT-LSA	0.669	150
avg-e0.5	avg	3-level kmedoids	Bigrams-TFIDF-LSA	0.576	150
avg-e0.5	avg	3-level kmedoids	Bigrams-WFENT-LSA	0.575	25
avg-e0.5	avg	3-level kmedoids	TFIDF-LSA	0.586	25
avg-e0.5	avg	3-level kmedoids	WFENT-LSA	0.587	200
single	single	3-level hierarchical	Bigrams-TFIDF-LSA	0.614	300
single	single	3-level hierarchical	Bigrams-WFENT-LSA	0.613	50
single	single	3-level hierarchical	TFIDF-LSA	0.600	25
single	single	3-level hierarchical	WFENT-LSA	0.649	25
single	single	3-level kmedoids	Bigrams-TFIDF-LSA	0.551	25
single	single	3-level kmedoids	TFIDF-LSA	0.537	25
single	single	3-level kmedoids	WFENT-LSA	0.550	25
single	single	hierarchical	Bigrams-WFENT-LSA	0.548	100

Generally, we can not give any simple advice for choice of number of topics. The best value strongly depends on representation and clustering algorithm what is consistent with observation from section 5.2. Anyway, some dependencies can be observed.

For 3-level kmedoids clustering low number of topics is preferred (especially for aggregation=single and linkage=single). For most configurations 25 topics was selected. For 3-level hierarchical clustering situation is more complicated.

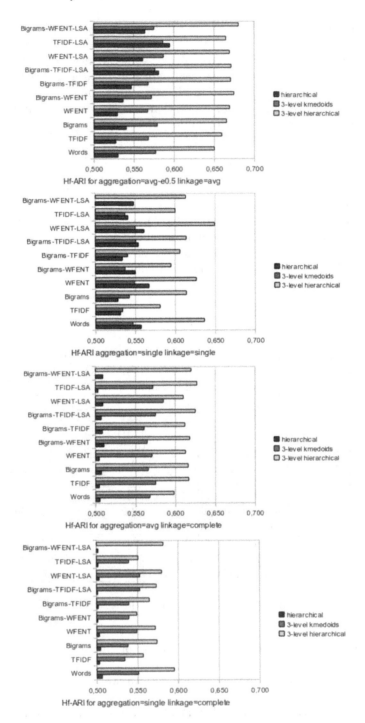

Fig. 9. Comparison of clustering methods

For aggregation=avg-e0.5 and linkage=avg high numbers were chosen. For aggregation=single and linkage=single low values are preferred.

5.8 The Best Hierarchy

In our experiments we tested over 2000 configurations (different representations, weighting schemes, similarity methods, clustering methods, number of LSA topics). Table 8 presents top 10 experiments' configurations with the highest Hf-ARI values.

Table 8. Configurations with the highest Hf-ARI value

Leaves	Linkage	Representation	Clusters	Hf-ARI
avg-e0.5	avg	Bigrams-WFENT-LSA300	290 / 40	0.680
avg-s0.5	avg	Bigrams-WFENT	320 / 40	0.677
avg-s0.75	avg	WFENT-LSA150	250 / 40	0.677
avg	avg	Bigrams-WFENT-LSA400	300 / 40	0.676
avg-s0.75	single	Bigrams-WFENT-LSA500	220 / 30	0.676
avg	avg	Bigrams-WFENT-LSA350	280 / 40	0.675
avg-s0.5	single	Bigrams-WFENT-LSA250	320 / 30	0.675
avg-e0.5	avg	Bigrams-WFENT	310 / 30	0.674
avg-e0.5	avg	Bigrams-WFENT-LSA200	210 / 40	0.674
avg-e0.5	single	Bigrams-WFENT-LSA350	200 / 30	0.674

All the results were obtained for 3-level hierarchical clustering.
Hf-ARI for comparing MSC tree to itself is equal to 0.759.

First column stands for similarity aggregation method (section 3.1). Second describes linkage type in clustering. Three typical methods were considered: average, single and complete linkage. All the best results were obtained for 3-level hierarchical clustering (for details see section 5.6). In the third column document representations are shown. For LSA we tested different weighting schemes and number of topics between 25 and 500. To calculate similarity we used either Tversky (for words and bigrams) or Cosine-like method (in other cases). The fourth column stands for number of nodes (clusters) at different levels of obtained hierarchy. First number stands for number of nodes at M level and second at H level.

The best obtained hierarchy is for bigrams with $WF \times ENT$ weighting and LSA applied with $avg - e0.5$ similarity aggregation method. This result is consistent with the best result obtained in the section 5.2 what suggest that both measurements give similar results.

In the table 9 fragments of the best obtained tree are shown. In the second column identified problems are described. Simplified visualisation of the tree in comparison to the original MSC tree is shown in the figure 10.

Table 9. Problems in the best obtained tree

Tree fragments	Problems
...((62G05) (62G07) (62G10) (62M05))...	some leaves are merged too late
...((20F36) (20D10 20D30 20E15 20F05 20K20)))...	some leaves are merged too fast
...((17B37) (22E40) (32M05) (32M15) (37D40) (43A80) (43A85) (53C15) (53C20) (53C21) ... (53C50) (53C55) (53D50) (58E20) (58J20) (58J35) (58J50) (58J60) (49Q05 53A10) (17B10 17B20 17B35 22E30 22E45 22E46 22E47))...	big groups of leaves of different kind are merged
... ((60K25 90B22))...	small clusters out of totally different codes are created
... ((91B14)) ((91B28)) ...	leaves that should be merged are left separated
...(65M15 65M60 65N15 65N25 65N30 65N55 74S05 76M10)...	single leaves are glued to groups of different type

Original hierarchy   Reconstructed hierarchy

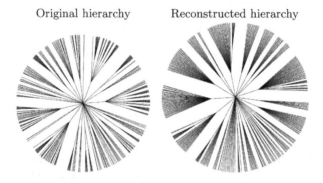

Fig. 10. Simplified visualisations of trees

5.9 Interesting Results

We reviewed our best hierarchy. We checked all clusters of size 2 and 3 at M level. From such clusters we extracted those pairs of MSC leaves that were in the same cluster but had no common prefix (e.g. 60K25 and 90B22). The list of extracted pairs is shown in the table 10.

An analysis of the table 10 leads to interesting conclusions. Our reconstruction process glued leaves that were strongly linked despite the fact that they were in different branches of the hierarchy. For example 60K25 ('Queueing theory') was merged with 90B22 ('Queues and service'). It is clear that this two leaves must be very similar but the first was placed in 60 ('Probability theory and

Table 10. 'Wrong' pairs in clusters of size 2 and 3

Leaf	Explanation	Leaf	Explanation
60K25	Queueing theory	90B22	Queues and service
35P25	Scattering theory for PDE	47A40	Scattering theory
49Q05	Minimal surfaces	53A10	Minimal surfaces, surfaces with prescribed mean curvature
32S65	Singularities of holomorphic vector fields and foliations	37F75	Holomorphic foliations and vector fields
34C25	Periodic solutions	37J45	Periodic, homoclinic and heteroclinic orbits; ...
11F70	Minimal surfaces, surfaces with prescribed mean curvature	22E50	Representations of Lie and linear algebraic groups over local fields
35Q30	Stokes and Navier-Stokes equations	76D05	Navier-Stokes equations
35Q30	Stokes and Navier-Stokes equations	76N10	Existence, uniqueness, and regularity theory

stochastic processes') and the second in 90 ('Operations research, mathematical programming').

Very similar analysis is shown in the table 11. We considered all pairs of MSC leaves and extracted those that were in the same cluster but had no common prefix. In the next step, we casted leaves to the highest level of the hierarchy (extracted prefixes of length 2) and counted pairs.

An analysis of the table 11 reveals similar conclusions. Some of the groups of leaves are placed in the same cluster even though they are in different branches. For example 'Numerical analysis' very often co-occurred with 'Fluid mechanics'. This is connected to the fact that numerical methods are widely used in applications of fluid mechanics.

6   Summary and Conclusions

In this paper we studied the problem of recreating the hierarchy of codes of a subject classification system. Our goal was to find a method of constructing, based on metadata of mathematical publications, a tree that would be as close to the original MSC 2000 tree as possible.

In order for the goal to be meaningful, we first had to decide what it means that two given trees are similar and to quantify that similarity. To this end we studied and developed novel methods of assessing tree similarity. After a series of experiments we have chosen Hf-ARI measure (cf. Section 2) as it had the best properties among all the evaluated candidates.

Next, we have selected a method of quantifying document similarity and devised optimization in computing similarity matrices. Finally, we have performed

Table 11. The most often 'wrong' pairs

Count	MSC leaf	Description	MSC leaf	Description
24	17*	Nonassociative rings and algebras	22*	Topological groups, Lie groups For transformation groups
12	65*	Numerical analysis	74*	Mechanics of deformable solids
12	65*	Numerical analysis	76*	Fluid mechanics For general continuum mechanics
4	35*	Partial differential equations	76*	Fluid mechanics For general continuum mechanics
2	90*	Operations research, mathematical programming	60*	Probability theory and stochastic processes For additional applications
2	76*	Fluid mechanics For general continuum mechanics	74*	Mechanics of deformable solids
2	37*	Dynamical systems and ergodic theory	32*	Several complex variables and analytic spaces For infinite-dimensional holomorphy
2	34*	Ordinary differential equations	37*	Dynamical systems and ergodic theory
2	22*	Topological groups, Lie groups For transformation groups	11*	Number theory
2	35*	Partial differential equations	47*	Operator theory
2	49*	Calculus of variations and optimal control; optimization	53*	Differential geometry For differential topology

a series of experiments aimed at calibrating our solution. Each experimental result was accompanied by analysis and conclusions (cf. Section 5).

6.1 Future Work

During our research we have identified several problems occurring in reconstruction process e.g. some leaves are glued too early whereas other too late. The problems decrease quality of obtained results. We believe that their influence can be reduced by modifications in representation and clustering algorithms. Incorporating new features (e.g. link-based) can also help.

Another direction of our works would be to examine more deeply correlation between nodes in MSC hierarchy. This research could lead to the proposal of a modified structure. The structure could be more efficient in some applications such as automatic classification.

Acknowledgements. This work is supported by the National Centre for Research and Development (NCBiR) under Grant No. SP/I/1/77065/10 by the Strategic scientific research and experimental development program: "Interdisciplinary System for Interactive Scientific and Scientic-Technical Information."

References

1. ICML 2010 Tutorial on Metric Learning (2010),
   `http://www.eecs.berkeley.edu/~kulis/icml2010_tutorial.htm`
2. Ahmad, S.B., Cakmak, A., Ozsoyoglu, G., Hamdani, A.A.: Evaluating Publication Similarity Measures. IEEE Data Eng. Bull. 28(4), 21–28 (2005),
   `http://sites.computer.org/debull/A05dec/bani-ahmad.pdf`
3. Brouwer, R.: Extending the rand, adjusted rand and jaccard indices to fuzzy partitions. Journal of Intelligent Information Systems 32, 213–235 (2009),
   `http://dx.doi.org/10.1007/s10844-008-0054-7`, 10.1007/s10844-008-0054-7
4. Denäud, L., Guãnoche, A.: Comparison of distance indices between partitions. In: Batagelj, V., Bock, H.H., Ferligoj, A., Aiberna, A. (eds.) Studies in Classification, Data Analysis, and Knowledge Organization, Data Science and Classification, pp. 21–28. Springer, Heidelberg (2006),
   `http://dx.doi.org/10.1007/3-540-34416-0_3`
5. Fowlkes, E.B., Mallows, C.L.: A method for comparing two hierarchical clusterings. Journal of the American Statistical Association 78(383), 553–569 (1983),
   `http://dx.doi.org/10.2307/2288117`
6. Fuller, M., Zobel, J.: Conflation-based comparison of stemming algorithms. In: Proceedings of the Third Australian Document Computing Symposium, pp. 8–13 (1998)
7. Hãllermeier, E., Rifqi, M., Henzgen, S., Senge, R.: Comparing fuzzy partitions: A generalization of the rand index and related measures. IEEE Transactions on Fuzzy Systems (to appear)
8. Jiang, D., Tang, C., Zhang, A.: Cluster analysis for gene expression data: a survey. IEEE Transactions on Knowledge and Data Engineering 16(11), 1370–1386 (2004),
   `http://dx.doi.org/10.1109/TKDE.2004.68`
9. Sathiyakumari, K., Preamsudha, V., Manimekalai, G.: A Survey on Document Clustering Using Different Techniques. International Journal of Computer Applications in Technology 2(5), 1534–1539 (2011)
10. Lee, M., Pincombe, B., Welsh, M., Bara, B.: An empirical evaluation of models of text document similarity. Lawrence Erlbaum Associates, Chicago (2005)
11. Lin, Z.: Link-based Similarity Measurement Techniques and Applications. Ph.D. thesis, The Chinese University of Hong Kong (2011)
12. Manning, C.D., Raghavan, P., Schtze, H.: Introduction to Information Retrieval. Cambridge University Press, New York (2008)
13. Nakov, P., Popova, A., Mateev, P.: Weight functions impact on lsa performance. In: EuroConference RANLP 2001 (Recent Advances in NLP), pp. 187–193 (2001)

14. Porter, M.F.: An algorithm for suffix stripping. In: Readings in Information Retrieval, pp. 313–316. Morgan Kaufmann Publishers Inc., San Francisco (1997), http://dl.acm.org/citation.cfm?id=275537.275705
15. Rand, W.: Objective criteria for the evaluation of clustering methods. Journal of the American Statistical Association 66(336), 846–850 (1971)
16. Řehůřek, R., Sojka, P.: Software Framework for Topic Modelling with Large Corpora. In: Proceedings of the LREC 2010 Workshop on New Challenges for NLP Frameworks, pp. 45–50. ELRA, Valletta (2010), http://is.muni.cz/publication/884893/en
17. Salton, G., Wong, A., Yang, C.S.: A vector space model for automatic indexing. Commun. ACM 18(11), 613–620 (1975), http://doi.acm.org/10.1145/361219.361220
18. Shepard, R.: The analysis of proximities: Multidimensional scaling with an unknown distance function. i. Psychometrika 27, 125–140 (1962), http://dx.doi.org/10.1007/BF02289630, 10.1007/BF02289630
19. Tversky, A.: Features of similarity. Psychological Review 84(4), 327–352 (1977)

Semantic Evaluation
of Search Result Clustering Methods[*]

Sinh Hoa Nguyen[1,2], Wojciech Świeboda[1], and Grzegorz Jaśkiewicz[1]

[1] Institute of Mathematics, Warsaw University
Banacha 2, 02-097 Warsaw, Poland
[2] Polish-Japanese Institute of Information Technology
Koszykowa 86, 02008, Warszawa, Poland
{hoa,wswieb,jaskiewicz}@mimuw.edu.pl

Abstract. This paper is a continuation of the research on designing
and developing a dialog-based semantic search engine for SONCA system[1] which is a part of the SYNAT project[2]. In previous papers we proposed some extensions of Tolerance Rough Set Model (TRSM), which
can be used to improve the search result clustering algorithms. In this
paper, we investigate the problem of quality analysis of presented solutions. We propose some semantic evaluation measures to estimate the
quality of the proposed search clustering methods. We illustrate the proposed evaluation method on the base of the Medical Subject Headings
(MeSH) thesaurus and compare different clustering techniques over the
commonly accessed biomedical research articles from PubMed Central
(PMC) portal. The experimental results are showing the advantages of
our new clustering methods which are implemented in SONCA system.

Keywords: Search results clustering, semantic evaluation, MeSH,
PubMed.

1 Introduction

It is a good practice in each software project to implement quality assurance
and quality control (QA/QC) procedures. Such QA/QC procedures are useful
to improve the transparency, consistency, comparability, completeness, and confidence in every component of the investigated project. This paper presents a

---

[*] The authors are partially supported by the grant N N516 077837 from the Ministry
of Science and Higher Education of the Republic of Poland and by the National Centre for Research and Development (NCBiR) under Grant No. SP/I/1/77065/10 by
the strategic scientific research and experimental development program: "Interdisciplinary System for Interactive Scientific and Scientific-Technical Information". The
research of the first author has been partially supported by the statutory research
grant founded by Polish-Japanese Institute of Information Technology.
[1] Search based on ONtologies and Compound Analytics.
[2] 'Interdisciplinary System for Interactive Scientific and Scientific-Technical Information' (www.synat.pl).

R. Bembenik et al. (Eds.): *Intell. Tools for Building a Scientific Information*, SCI 467, pp. 393–414.
DOI: 10.1007/978-3-642-35647-6_24   © Springer-Verlag Berlin Heidelberg 2013

quality control method for search results clustering algorithms, which are the most important tools in all semantic search engines.

This paper is a part of the ongoing project, called SONCA (Search based on ONtologies and Compound Analytics), realized at the Faculty of Mathematics, Informatics and Mechanics of the University of Warsaw. It is part of the project 'Interdisciplinary System for Interactive Scientific and Scientific-Technical Information' (www.synat.pl). SONCA is an application based on a hybrid database framework, wherein scientific articles are stored and processed in various forms. SONCA is expected to provide interfaces for intelligent algorithms identifying relations between various types of objects. It extends typical functionality of scientific search engines by more accurate identification of relevant documents and more advanced synthesis of information. To achieve this, concurrent processing of documents needs to be coupled with the ability to produce collections of new objects using queries specific for analytic database technologies.

Ultimately, SONCA should be capable of answering the user query by listing and presenting the resources (documents, Web pages, etc.) that correspond to it *semantically*. In other words, the system should have some *understanding* of the intention of the query and of the contents of documents stored in the repository as well as the ability to retrieve relevant information with high efficacy. The system should be able to use various knowledge sources related to the investigated areas of science. It should also allow for independent sources of information about the analyzed objects, such as, e.g., information about scientists who may be identified as the stored articles' authors.

In previous paper [13], we presented a generalized schema for the problem of semantic extension for snippets. We investigated two levels of extension. The first one was related to extending a concept space for data representation and the second one was related to associating terms in the concept space with semantically related concepts. The first extension level is performed by adopting semantic information extracted from a document content such as citations or from document meta-data like authors, conferences. The second one is achieved by application of Tolerance Rough Set Model [18][11] and domain ontology, in order to approximate concepts existing in documents description and to enrich the vector representation.

This paper is an extension of [13]. We perform a deep analysis of the proposed approach on the database of articles on biomedical topics PubMed. Our investigation focuses on developing semantic evaluation methods of clustering results. We focus strictly on external validation criteria of clustering results, which aim to compare a clustering with expert tags provided by experts. In most previous approaches to this problem, the tagging provided by experts determined a partitioning of documents. We propose extensions of some of these measures so that they compare a clustering with a multi-label expert tagging (i.e., several tags can be assigned to a single article). The biomedical concepts assigned to articles in PubMed Central data by experts are adopted to the evaluation process as the source of validation information. In the paper we present the evaluation methods and show the quality of semantic extension models obtained by presented methods.

The paper is organized as follows. In Section 2 we recall the general approach to enriching of snippet representation using *Generalized* TRSM. Section 3 presents the setting for experiments and validation methods based on expert's tags. Section 4 is devoted to experiments, followed by conclusions in section 5.

2 Extended Snippet Representation

In this section we recall the main extended representation models of document (or snippet) by some additional semantic information including citations, authors and affiliations of cited papers, publishers and publication years of the cited papers and semantic concepts related to the document. Our approach is based on some new applications of Tolerance Rough Set Model, which is a special case of a generalized approximation space [18], in Semantic Search. Three Tolerance Rough Set models will be discussed in this section: standard TRSM, extended TRSM by citations and extended TRSM by semantic concepts. We also present, how to apply these models to enrich documents' representation.

2.1 Tolerance Rough Set Model

Tolerance Rough Set Model (TRSM) was developed in [6,4] as a basis to model documents and terms in Information Retrieval, Text Mining, etc. With its ability to deal with vagueness and fuzziness, Tolerance Rough Set Model seems to be a promising tool to model relations between terms and documents. In many Information Retrieval problems, especially in document clustering, defining the relation (i.e. similarity or distance) between document-document, term-term or term-document is essential. It has been noticed [4] that a single document is usually represented by relatively few terms[3]. This results in zero-valued similarities which decreases quality of clustering. The application of TRSM in document clustering was proposed as a way to enrich document and cluster representation with the hope of increasing clustering performance.

The idea is to capture conceptually related index terms into classes. For this purpose, the tolerance relation R is determined as the co-occurrence of index terms in all documents from D. The choice of co-occurrence of index terms to define tolerance relation is motivated by its meaningful interpretation of the semantic relation in the context of IR and its relatively simple and efficient computation.

Let $D = \{d_1, \ldots, d_N\}$ be a set of documents and $T = \{t_1, \ldots, t_M\}$ set of *index terms* for D. With the adoption of Vector Space Model [10], each document d_i is represented by a weight vector $[w_{i1}, \ldots, w_{iM}]$ where w_{ij} denotes the weight of term t_j in document d_i. TRSM is an approximation space $\mathcal{R} = (T, I_\theta, \nu, P)$ determined over the set of terms T as follows:

---

[3] In other words, the number of non-zero values in document's vector is much smaller than vector's dimension – the number of all index terms.

– **Uncertainty function:** The parametrized uncertainty function I_θ is defined as

$$I_\theta(t_i) = \{t_j \mid f_D(t_i, t_j) \geq \theta\} \cup \{t_i\}$$

where $f_D(t_i, t_j)$ denotes the number of documents in D that contain both terms t_i and t_j and θ is a parameter set by an expert. The parameter θ is called an *tolerance threshold*. The set $I_\theta(t_i)$ is called the *tolerance class* of term t_i. One can observe the higher the tolerance threshold the smaller tolerance class and vice versa.

– **Vague inclusion function:** To measure degree of inclusion of one set in another, the vague inclusion function is defined as is defined as

$$\nu(X, Y) = \frac{|X \cap Y|}{|X|}$$

It is clear that this function is monotone with respect to the second argument.

– **Structural function:** All tolerance classes of terms are considered as structural subsets: $P(I_\theta(t_i)) = 1$ for all $t_i \in T$.

The membership function μ for $t_i \in T$, $X \subseteq T$ is then defined as $\mu(t_i, X) = \nu(I_\theta(t_i), X) = \frac{|I_\theta(t_i) \cap X|}{|I_\theta(t_i)|}$ and the lower and upper approximations of any subset $X \subseteq T$ can be determined – with the obtained tolerance $\mathcal{R} = (T, I, \nu, P)$ – in the standard way

$$\mathbf{L}_\mathcal{R}(X) = \{t_i \in T \mid \nu(I_\theta(t_i), X) = 1\}$$
$$\mathbf{U}_\mathcal{R}(X) = \{t_i \in T \mid \nu(I_\theta(t_i), X) > 0\}$$

2.2 Standard TRSM

Let $D = \{d_1, \ldots, d_N\}$ be a set of documents and $T = \{t_1, \ldots, t_M\}$ the set of *index terms* for D. Tolerance Rough Set Model for term space was described in the previous section.

$$\mathcal{R}_0 = (T, I_\theta, \nu, P)$$

In this model a document, which is associated with a *bag of words/terms*, is represented by its upper approximation, i.e. the document $d_i \in D$ is represented by

$$\mathbf{U}_\mathcal{R}(d_i) = \{t_i \in T \mid \nu(I_\theta(t_i), d_i) > 0\}$$

The extended weighting scheme is inherited from the standard *tf-idf* by:

$$w_{ij}^* = \begin{cases} (1 + \log f_{d_i}(t_j)) \log \frac{N}{f_D(t_j)} & \text{if } t_j \in d_i \\ 0 & \text{if } t_j \notin \mathbf{U}_\mathcal{R}(d_i) \\ \min_{t_k \in d_i \wedge t_j \in I_\theta(t_k)} w_{ik} \frac{\log \frac{N}{f_D(t_j)}}{1 + \log \frac{N}{f_D(t_j)}} & \text{otherwise} \end{cases}$$

In the formula above $f_d(t_j)$ and $f_D(t_j)$ denote a frequency of term t_j in document d and in corpus D, respectively. The extension ensures that each term occurring in the upper approximation of d_i but not in d_i itself has a weight smaller than the weight of any terms in d_i. Normalization by vector's length is then applied to all document vectors:

$$w_{ij}^{new} = \frac{w_{ij}^*}{\sqrt{\sum_{t_k \in d_i} (w_{ij}^*)^2}}$$

The example of standard TRSM is presented in Table 1.

Table 1. Example snippet and its two vector representations in standard TRSM

Title: EconPapers: Rough sets bankruptcy prediction models versus auditor			
Description: Rough sets bankruptcy prediction models versus auditor signalling rates. Journal of Forecasting, 2003, vol. 22, issue 8, pages 569-586. Thomas E. McKee. ...			
Original vector		**Enriched vector**	
Term	Weight	Term	Weight
auditor	0.567	auditor	0.564
bankruptcy	0.4218	bankruptcy	0.4196
signalling	0.2835	signalling	0.282
EconPapers	0.2835	EconPapers	0.282
rates	0.2835	rates	0.282
versus	0.223	versus	0.2218
issue	0.223	issue	0.2218
Journal	0.223	Journal	0.2218
MODEL	0.223	MODEL	0.2218
prediction	0.1772	prediction	0.1762
Vol	0.1709	Vol	0.1699
		applications	0.0809
		Computing	0.0643

2.3 Extended TRSM by Citations

Let $D = \{d_1, \ldots, d_N\}$ be a set of documents and $T = \{t_1, \ldots, t_M\}$ the set of *index terms* for D. Let $B^{in} = \{b_1^{in}, \ldots, b_K^{in}\}$ be a set of bibliography items that cite the documents from D (*inbound citation*) and $B^{out} = \{b_1^{out}, \ldots, b_L^{out}\}$ be a set of items that are cited by documents from D (*outbound citation*)
The extended tolerance rough set model for terms and citations is a triple:

$$\mathcal{R}_1 = (\mathcal{R}_T, \mathcal{R}_B^{in}, \mathcal{R}_B^{out})$$

where $\mathcal{R}_T = (T, I_{\theta_T}, \nu, P)$ and $\mathcal{R}_B^{in} = (B^{in}, I_{\theta_B^{in}}, \nu, P)$, $\mathcal{R}_B^{out} = (B_{out}, I_{\theta_B^{out}}, \nu, P)$ are TRSM defined for term space T and bibliography item spaces B^{in}, B^{out}, respectively.

In the extended model, each document $d_i \in D$ is associated with a triple $(T_i, B_i^{in}, B_i^{out})$, where T_i is the set of terms that occur in d_i, B_i^{in} and B_i^{out} are the sets of inbound and outbound citations into and from d_i. Therefore each document can be represented by a triple of upper approximations, i.e.,

$$d_i \dashrightarrow (\mathbf{U}_{\mathcal{R}_T}(d_i), \mathbf{U}_{\mathcal{R}_B^{in}}(d_i), \mathbf{U}_{\mathcal{R}_B^{out}}(d_i))$$

where

$$\mathbf{U}_{\mathcal{R}_T}(d_i) = \{t_i \in T \mid \nu(I_{\theta_T}(t_i), d_i) > 0\}$$
$$\mathbf{U}_{\mathcal{R}_B^{in}}(d_i) = \{b_i \in B^{in} \mid \nu(I_{\theta_B^{in}}(b_i), B^{in}|d_i) > 0\}$$
$$\mathbf{U}_{\mathcal{R}_B^{out}}(d_i) = \{b_i \in B^{out} \mid \nu(I_{\theta_B^{out}}(b_i), B^{out}|d_i) > 0\}$$

where $B^{in}|d_i$ and $B^{out}|d_i$ denotes the set of inbound and outbound citation for d_i, respectively.

Once can observe some properties of the extended model.

- if $B^{in} = B^{out} = \emptyset$ and $\theta_T = card(D) + 1$, documents in D are represented by their original text without any extension.
- if $B^{in} = B^{out} = \emptyset$, documents in D are represented by extended text (according to TRSM).
- if $\theta_T = \theta_B^{in} = \theta_B^{out} = card(D) + 1$, document in D are represented by an original text and original inbound and outbound citations d_i.
- if $B^{in} \neq \emptyset$ and $B^{out} \neq \emptyset$ and $\theta_B^{in} = \theta_B^{out} = card(D) + 1$, documents in D are represented by the extended text (according to TRSM) and original inbound and outbound citations.

The influence of such representation models to the quality of search result clustering will be discussed in the last sections.

2.4   Extended TRSM by Citations and Semantic Concepts

Let $D = \{d_1, \ldots, d_N\}$ be a set of documents and $T = \{t_1, \ldots, t_M\}$ a set of *index terms* for D. Let $B^{in} = \{b_1^{in}, \ldots, b_K^{in}\}$ and $B^{out} = \{b_1^{out}, \ldots, b_L^{out}\}$ be the sets of inbound and outbound citation for D . Let C be the set of concepts from a given domain knowledge (e.g. the concepts from MeSH).

The extended tolerance rough set model based on both citations and semantic concepts is a tuple:

$$\mathcal{R}_2 = (\mathcal{R}_T, \mathcal{R}_B^{in}, \mathcal{R}_B^{out}, \mathcal{R}_C, \alpha_n)$$

where $\mathcal{R}_T$, $\mathcal{R}_{Bin}$, $\mathcal{R}_{Bout}$ and $\mathcal{R}_C$ are tolerance spaces determined over the set of terms T, the sets of bibliography items B^{in}, B^{out} and the set of concepts in the knowledge domain C, respectively. The function $\alpha_n : \mathcal{P}(T) \longrightarrow \mathcal{P}(C)$ is called the *semantic association* for terms. For any $T_i \subset T$, $\alpha_n(T_i)$ is the set of n concepts most associated with T_i, see [19]. In this model, each document $d_i \in D$ associated with a triple $(T_i, B_i^{in}, B_i^{out})$ is represented by a tuple:

$$d_i \dashrightarrow (\mathbf{U}_{\mathcal{R}_T}(d_i), \mathbf{U}_{\mathcal{R}_B^{in}}(d_i), \mathbf{U}_{\mathcal{R}_B^{in}}(d_i), \alpha_n(T_i))\}$$

In Figure 1 we can see an example of a biomedical article and the list of concepts associated with the article.

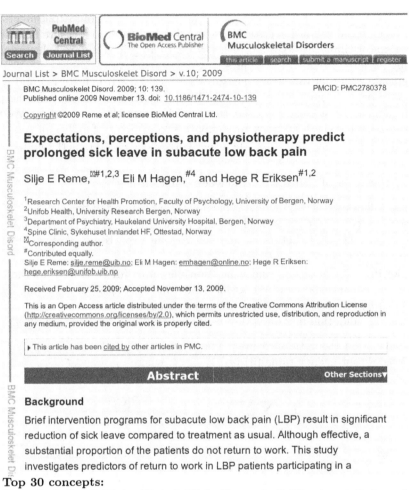

Fig. 1. An example of a document and a list of top 30 concepts assigned to its snippet by the semantic tagging algorithm

3   Semantic Evaluation of Clustering

This section is an extension of our earlier work in [12,13]. In our experiments, we focus on external validation criteria. External criteria aim to measure the degree of overlap (or similarity) of a clustering w.r.t. to a "gold standard" clustering or tagging, usually provided by experts.

Typically external validation measures are used for hard clustering and when experts provide a partition of objects. These criteria aim to compare the partition induced by a clustering and the partition provided by experts. We will shortly extend their definitions to compare two arbitrary families of sets (one corresponding to the clustering, the other corresponding to expert tags).

3.1   Measures of Similarity between Partitions

In what follows we briefly review typical external validation measures.

Let $C = \{C_1, \ldots, C_k\}$ be a clustering of X ($|X| = N$), $\bigcup_{i=1}^{k} C_i = X$ and $C_i \cap C_j = \emptyset$ ($i \neq j$). Let $E = \{E_1, \ldots, E_l\}$ be a partition of X defined by experts, $\bigcup_{i=1}^{k} E_i = X$, $E_i \cap E_j = \emptyset$ ($i \neq j$).

Following [14], most external validation measures (or broader – methods of comparing clusterings) can be classified into three groups: *pair-counting based*, *set-matching based* and *information theoretic* measures.

External validation measures that we discuss fall into two of these categories.

– Information theoretic measures. These measures can be expressed as functions of contingency table (3) $(a_{i,j})_{j=1,\ldots,l}^{i=1,\ldots,k}$, where $a_{i,j} = |C_i \cap E_j|$.
  • Mutual information

$$I = \sum_i \sum_j \tilde{P}(C_i \cap E_j) log_2 \frac{\tilde{P}(C_i \cap E_j)}{\tilde{P}(C_i)\tilde{P}(E_j)} \quad (1)$$

  where $\tilde{P}$ denotes empirical probability (relative frequency), i.e.:

$$\tilde{P}(C_i \cap C_j) = p_{i,j} \propto a_{i,j}$$

  • Normalized mutual information[10]

$$NMI = \frac{I}{(H(C) + H(E))/2} \quad (2)$$

  (also called *symmetric uncertainty* [21]), where $H(C)$ and $H(E)$ denote entropies of marginal empirical distributions of C and E respectively. There are numerous other ways of normalizing (and adjusting for chance) I, see e.g. [14].

– Pair-counting based measures. A *pair* of objects (o_1, o_2) is a True Positive (TP) when it belongs to the same expert class and to the same cluster. Similarly we define FP (False Positive), TN (True Negative) and FN (False Negative). Below we briefly mention typical measures defined in terms of confusion matrix of pairs of objects (2):

- Rand Index[16]

$$RI = \frac{TP + TN}{TP + FP + FN + TN} \tag{3}$$

  is an analogue to accuracy in classification. See also [5],[14] for discussion of normalized variants.
- Jaccard coefficient[20]

$$J = \frac{TP}{TP + FP + FN}. \tag{4}$$

- Typical Information Retrieval measures [10]: Precision, Recall, F_β.

3.2 Measures of Similarity between Arbitrary Families of Sets

Since there are usually several expert tags assigned to a document and clusters themselves may be overlapping (e.g. in fuzzy clustering or clusters output by LINGO algorithm) we will now propose extensions of measures from previous sections to measures that compare two arbitrary families of sets or two set covers C and E.

Families C and E no longer have to be partitions. A set E_i will now correspond to the subset of documents tagged by i-th expert tag (rather than to the i-th partition provided by expert). Previous works by other authors on this problem include [1] (Fuzzy Clustering Mutual Information) and [7] (for comparing set covers). We consider two approaches: extending the pair-counting measures and extending the probabilistic model in which information theory based measures are defined.

Pair-Counting Measures. Pair-counting measures only require definitions of TP, FP, TN, FN for pairs of objects (2). These, in turn, only require definitions of cluster-similarity and expert-similarity relations between pairs of objects. We define similarity of documents with respect to experts and similarity of documents with respect to a clustering as follows: We will say document A with the set of assigned expert tags H_A and document B with the set of assigned expert tags H_B are (expert) θ-similar if $\frac{|H_A \cap H_B|}{|H_A \cup H_B|} > \theta$ (Another similarity relation which we considered is $d_1 \sim d_2 \iff |H_A \cap H_B| > \theta$). We will say that document A and document B are cluster-similar if they belong to at least one cluster. These definitions determine the confusion matrix and hence pair-counting measures.

Information Theoretic Measures. Let us now focus on Mutual Information (this discussion is also valid for the remaining information theoretic measures). Consider a table T consisting of triples (d, i, j), where d denotes a document, i denotes a clustering and j denotes an expert tag, which lists all such triples such that $d \in C_i \cap E_j$. If both the clustering and the "gold standard" (expert tagging) induce partitions, then each document corresponds to exactly one row in table T. Mutual information in this case measures the amount of information between

Table 2. Measures defined on confusion matrix count the number of *pairs* of objects (whether they are similar or dissimilar)

		Cluster-similar	
		Yes	No
θ-similar	Yes	True Positive	False Negative
	No	False Positive	True Negative

random variables $c(X)$ and $e(X)$ where X denotes a randomly (and uniformly) drawn document, and $c(d)$, $e(d)$ are functions mapping (deterministically) the document d to a cluster and an expert tag.

If a document may belong to several clusters (or no clusters) and several (or no) tags may be assigned to a document, the simplest extension is to assume uniform distribution over all rows (triples) in matrix T. This leads to well defined measures (I, NMI, conditional and marginal entropies, etc.) of two arbitrary families of sets. Moreover, these measures are defined using exactly the same formulas.

In our experiments, we wished to preserve an equal contribution of each document into the measure (i.e. uniformity over documents, not over triples), which leads to another model: We draw (uniformly) a document X at random. Independently, we draw (uniformly) a cluster $c(X)$ and an expert tag $e(X)$ out of clusters (and tags) assigned to the document. It defines $c(X)$ and $e(X)$. We define I as the mutual information between random variables $c(X)$ and $e(X)$. This approach leads to a well defined measure when C and E are covers of X (otherwise, $c(X)$ or $e(X)$ might be left undefined). Information theoretic measures are now functions of table (3) $(p_{i,j})_{i \in 1...k}^{j \in 1...l}$ defined as follows: $p_{i,j} = N^{-1} \sum_{d \in C_i \cap E_j} n_C(d)^{-1} n_E(d)^{-1}$, where $n_C(d)$ is the number of clusters that contain document d and $n_E(d)$ is the number of expert tags assigned to document d.

However, please note that if we extend Mutual Information and marginal entropies as above, Normalized Mutual Information (normalized by the average of marginal entropies) may no longer attain 1 as the maximal value.

3.3   Semantic Evaluation of Clustering: Example

In practice, the available expert data only provides a tagging (rather than a partition) of objects, hence we will use measures of similarity between set covers.

Table 3. In order to calculate information theoretic measures (e.g. Mutual Information) it suffices to know empirical probabilities $p_{i,j}$

		Expert tags				
		E_1	E_2	E_3	...	E_l
	C_1	$p_{1,1}$	$p_{1,2}$	$p_{1,3}$	...	$p_{1,l}$
	C_2	$p_{2,1}$	$p_{2,2}$	$p_{2,3}$	...	$p_{2,l}$
Clusters	C_3	$p_{3,1}$	$p_{3,2}$	$p_{3,3}$	...	$p_{3,l}$
	...	...	...	...	...	...
	C_k	$p_{k,1}$	$p_{k,2}$	$p_{k,3}$	...	$p_{k,l}$

Let us consider an artificial example with a set of three documents $X = \{d_1, d_2, d_3\}$ and two clusters $C_1 = \{d_1, d_2\}$, $C_2 = \{d_3\}$. Experts assigned labels to documents as follows:

$$d_1 \mapsto \{Cow, Milk, Dairy\ animals, Food\}$$
$$d_2 \mapsto \{Milk, Calcium, Magnesium, Food\}$$
$$d_3 \mapsto \{Chemistry, Calcium, Alkaline, Metal\}$$

In other words, experts define $E = \{E_1, E_2, \ldots, E_9\}$, which correspond to concepts (*Alkaline, Calcium, Chemistry, Cow, Dairy Animals, Food, Magnesium, Metal, Milk*) such that $E_{Alkaline} = E_1 = \{d_3\}, E_{Calcium} = E_2 = \{d_2, d_3\}, \ldots$

Let us first consider all pairs of documents. We will calculate which documents are 0.2-similar to each other.

pair	$\dfrac{\lvert H_A \cap H_B \rvert}{\lvert H_A \cup H_B \rvert}$	θ − similar?
(d_1, d_2)	$\frac{2}{6}$	yes
(d_1, d_3)	$\frac{0}{8}$	no
(d_2, d_3)	$\frac{1}{7}$	no

Confusion matrix is as follows:

	θ-similar	θ-dissimilar
same cluster	TP=1	FP=0
different clusters	FN=0	TN=2

Hence e.g. $RI = \dfrac{TP + TN}{TP + TN + FP + FN} = 1.$

Table $(a_{i,j})$ used for calculation of I as well as marginals used for entropy calculations are presented in Table 4.

Mutual information $I \approx 0.75$, while marginal entropies are $H(C) \approx 0.92$ and $H(E) \approx 3.08$. Hence $NMI \approx 0.38$.

Table 4. Marginal empirical distributions used for entropy calculation

Terms	Frequency in Cluster 1	Frequency in Cluster 2	Frequency total
Milk	$\frac{2}{12}$	0	$\frac{2}{12}$
Cow	$\frac{1}{12}$	0	$\frac{1}{12}$
Dairy animals	$\frac{1}{12}$	0	$\frac{1}{12}$
Food	$\frac{2}{12}$	0	$\frac{2}{12}$
Calcium	$\frac{1}{12}$	$\frac{1}{12}$	$\frac{2}{12}$
Magnesium	$\frac{1}{12}$	0	$\frac{1}{12}$
Chemistry	0	$\frac{1}{12}$	$\frac{1}{12}$
Alkaline	0	$\frac{1}{12}$	$\frac{1}{12}$
Metal	0	$\frac{1}{12}$	$\frac{1}{12}$
total	$\frac{2}{3}$	$\frac{1}{3}$	

4 Experiment Set-Up

The aim of our experiments is to explore clusterings induced by different snippet extensions:

- (S) baseline – i.e. a snippet which consists of 20 words surrounding a query term, with no additional information,
- (T) lexical extension via TRSM,
- (M) semantic extension using induced MeSH-terms,
- (IO) reference-based extensions (I stands for "inbound" and O stands for "outbound" edges in the citation graph).

In experimental section that follows we will refer to these extensions using symbols S, T, M, IO as denoted in brackets above.

The diagram of our experiments is shown in Figure 2. A path (from a query to search result clustering) consists of three stages:

- Search and filter documents matching a query. Search result is a list of *snippets* and corresponding document identifiers.
- Extend representations of snippets and citations data by *TRSM-derived terms* and *semantically similar concepts* from MeSH ontology (terms used for clustering were automatically assigned by an algorithm[19], terms used for validation were assigned by human experts).
- Cluster document search results.
- Calculate external validation measures using tags provided by MEDLINE indexers.

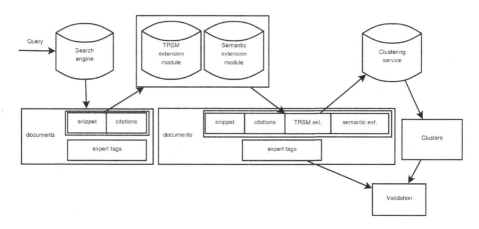

Fig. 2. Experiment diagram

4.1 Text Corpus

The text corpus in our study is PubMed Central Open Access Subset database, which consists of ~ 200000 documents (approximately a 10% of PubMed Central database), which are free and available under licences similar to Creative Commons license. We further limit our experiments to the subset of documents for which MeSH terms (see below) are available.

4.2 Expert Tags

Medical Subject Headings (MeSH) is a controlled vocabulary created (and continuously being updated) by the United States National Library of Medicine. It is used by MEDLINE and PubMed databases of citations for indexing documents by human experts. A large part of these citations contain links to full text articles in PubMed Central database. MeSH vocabulary defines subject headings (descriptions) and their (optionally) accompanying subheadings (qualifiers). A single document is typically tagged by 10 to 18 heading-subheading pairs.

For example, document (paper) "PubMed Central: The GenBank of the published literature." by Roberts R. J. ([17]) is assigned the following tags:

subject heading	subheading
Internet	
MEDLINE	economics
Periodicals as Topic	economics
Publishing	economics

There are approximately 25000 subject headings and 83 subheadings in the 2009 version of MeSH. Furthermore, there is a rich information accompanying MeSH: descriptor hierarchy, additional information regarding subject headings (e.g. scope notes, annotations, allowable qualifiers, historical notes, etc.), qualifier hierarchy and Supplementary Concept Records. Currently we only use the hierarchy of qualifiers[4] by exchanging a given qualifier by its (at most two) topmost ancestors or roots in some of our experiments.

We focus on Medical Subject Headings and subheadings as the source of validation information because we aim to discover semantic relationships between documents. Our choice of expert tags (headings, subheadings or subheading roots) determines how precisely we wish to interpret expert opinion.

source	possible tags	expert tags assigned to example document
subject headings	$\sim$ 25000 (MeSH 2009)[5]	Internet, MEDLINE, Periodicals as Topic, Publishing
subheadings	83	economics
subheading roots	23	organization & administration

4.3   Example Queries

In order to perform validation one needs a set of search queries that reflect actual user usage patterns. Our source of typical user queries is the daily log previously investigated by Herskovic et al. in [3] and by Lu et al. in [8].

For experiments described in this paper, we extracted a subset of $\sim$ 100 out of 1000 most frequent one-term queries for which our search engine returned at least 200 documents. For each query, the search engine further restricted the set of returned documents to top 200 relevant documents.

4.4   Snippet Results and Extended Snippets

A snippet from the article and citation information provides the raw data for clustering (see table 5). This data will be further extended by MeSH terms inferred from snippets using ESA algorithm and additional terms provided by TRSM (see table 6). MeSH terms provided by experts (in MEDLINE database) provide the basis for validation and are not used in clustering.

The goal of our experiments is to cluster snippets (or extended snippets) rather than documents, whereas MeSH tags were assigned to whole documents. A snippet may be semantically related to a different set of concepts than the whole document. While we understand the limitation of our approach, we stress the fact that snippets may be also interpreted as (very short) document summaries (within the context of a given query).

4.5   Clustering Methods

This section is a brief recap of a more elaborate description in [12]. We performed experiments regarding extensions using three clustering algorithms: *KMeans*[9],

---

[4] http://www.nlm.nih.gov/mesh/subhierarchy.html

Table 5. We limit the ranked list of documents d_i matching a query to top 200 documents. Each of these documents is assigned a list of pairs of subject headings h and subheadings s (some of which may be blank). These are used for validation. On the other hand, terms from a snippet S and citations (either inbound I or outbound O) are used as the raw data to clustering algorithm. All of these lists are of varying lengths. The actual range of index i is omitted in this example for clarity.

document	MeSH	raw data
d_1	$(h,s)_{1,i}$	$(S)_{1,i}$, $(I)_{1,i}$, $(O)_{1,i}$
d_2	$(h,s)_{2,i}$	$(S)_{2,i}$, $(I)_{2,i}$, $(O)_{2,i}$
...	...	...
d_{200}	$(h,s)_{200,i}$	$(S)_{200,i}$, $(I)_{200,i}$, $(O)_{200,i}$

Table 6. Snippets are further used to derive additional extensions – MeSH terms M (assigned by the algorithm) and terms from TRSM extension T

document	MeSH	raw data	semantic extension	TRSM extension
d_1	$(h,s)_{1,i}$	$(S)_{1,i}$, $(I)_{1,i}$, $(O)_{1,i}$	$(M)_{1,i}$	$(T)_{1,i}$
d_2	$(h,s)_{2,i}$	$(S)_{2,i}$, $(I)_{2,i}$, $(O)_{2,i}$	$(M)_{2,i}$	$(T)_{2,i}$
...	...	...	...	...
d_{200}	$(h,s)_{200,i}$	$(S)_{200,i}$, $(I)_{200,i}$, $(O)_{200,i}$	$(M)_{200,i}$	$(T)_{200,i}$

Hierarchical Clustering[2], and *Lingo*[15]. Lingo and KMeans operate on feature space (e.g. vector space) representation of objects, while Hierarchical Clustering only requires object-object distance function. While objects in our experiments correspond to snippets (rather than whole documents), we will nevertheless use the word "document" in this section to point out the similarity with documents represented in terms of term-document matrix. We decided to base all of our experiments on a vector space model representation by augmenting the term-document matrix with additional matrices: inbound.citation-document matrix, outbound.citation-document matrix and concept-document matrix. While Lingo by the design operates on a single matrix (it is based on SVD decomposition), for the remaining two algorithms we experimented with distances defined on subspaces corresponding to different representations: our experiments assessed whether a semimetric derived from cosine similarity, an Euclidean distance or a combined distance should be used for extensions. For Hierarchical Clustering, we assessed which linkage type should be used.

While label generation is remarkably important, we have not addressed this issue in our current experiments. Label generation is intrinsically linked with cluster calculation in Lingo and we currently left the result intact. For the remaining two algorithms we assigned three distinct most frequent words present in snippets of documents in a given cluster as the label of the cluster.

5   Results of Experiments

The aim of this section is to evaluate the semantic enrichment models. The objective of our investigation is

1. to examine whether the sematic extensions improve the clustering quality and to verify if it is robust against varying evaluation criterions.
2. to find the proper clustering algorithm for every type of semantic extensions
3. to investigate the time performance of clustering algorithms and to check if the sematic extensions reduce a speed of clustering algorithms.

Our experiments are carried out on three clustering algorithms: (1) K-Means clustering, (2) Hierarchical clustering and (3) Singular Value Decomposition (SVD) based clustering (Lingo algorithm). To evaluate clustering results the Rand Index measure is applied.

5.1   Empirical Assessment of Enrichment Models

The goal of our experiments is to examine if the semantic extension improve the clustering quality and if it is robust against different evaluation criterions. As a basic model we use snippets (**S**) to document representation.

All extension models can be divided into four groups:

1. *Standard TRSM*:
   - **S**nippet enriched by **T**erms on tolerance classes - (**ST**)
2. *Standard TRSM extended by citations*:
   - **S**nippet enriched by **I**nbound and **O**utbound citations and **T**erms in tolerance classes (**SIOT**)
3. *Standard TRSM extended by semantic concepts*:
   - **S**nippet enriched by semantic concepts adopted from MeSH ontology and **T**erms in tolerance classes (**SMT**)
4. *Standard TRSM extended by citations and semantic concepts*:
   - **S**nippet enriched by **I**nbound and **O**utbound citations, semantic MeSH concepts and **T**erms in tolerance classes (**SIOMT**)

Altogether we have five models for document representation: one basic and four extended models.

The first step in our experiment is to determine the optimal value of tolerance threshold θ in the TRSM. Results show that the optimal values of θ parameter varies upon evaluation measures. For *RandIndex* the clustering quality is best for high θ ($\theta_{RI} = 15$). It means, the best result is achieved when tolerance classes are small. Howerver in the cases of $F - measure$ or $Purity$ measure, the best performance occurs when tolerance threshold is large ($\theta_F = 2$, $\theta_{Pur} = 4$). The relationship between θ values and clustering quality is shown in Picture 3 (the result is obtained from the Lingo clustering algorithm).

Next to compare the basic representation model and the extended ones, the search result clustering (for about 70 most frequently asked questions from a

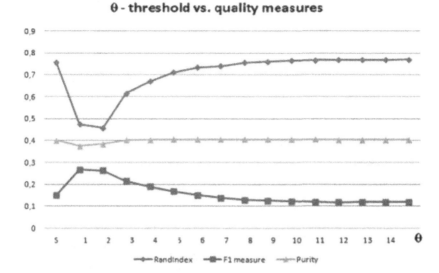

Fig. 3. The optimal tolerance threshold varies with quality measures

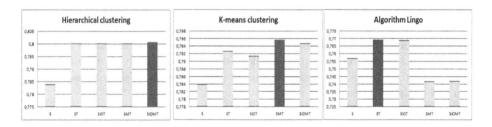

Fig. 4. Comparing the quality of basic presentation model and four enrichment models. An evaluation is based on 23 subheadings.

biomedical field) based on five models are achieved. The representation model is better if the quality of clusters is higher. In Figure 4, one can compare the quality of the basic representation (Snippet) and another extension models. In all cases the extended representations show to be better (the higher Rand index). When hierarchical clustering algorithm was applied, the full extension (SIOMT) was slightly better then another extensions. But when K-means was used the representation enriched by semantic concepts from MeSH ontology shows to be the best. In the case of Lingo algorithm, the standard TRSM is most appropriated.

To examine the robustness of enrichment models against varying evaluation criterions, we investigate the quality of these models over two another evaluation methods: based on *83 subheadings* and based on about 25000 *headings* assigned by experts to the documents.

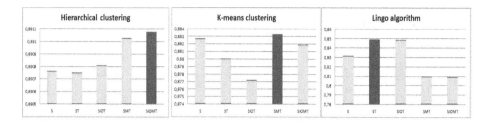

Fig. 5. Comparing the quality of standard presentation model and three enrichment models. An evaluation is based on 83 subheadings.

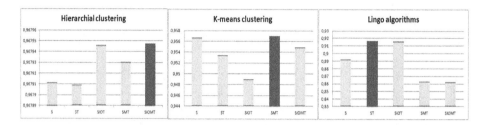

Fig. 6. Comparing the quality of standard presentation model and three enrichment models. An evaluation is based on headings.

Figure 5 and Figure 6 present the quality of different enrichment methods comparing to the standard representation model. One can observe, in the both cases and for all evaluation measures the semantic extensions show an advantage over the basic representation.

5.2   Quality of Clustering Algorithms

The aim of this section is to compare the quality of clustering algorithms impact with the extension types and to select the best algorithm for every kind of extension. We use the set of 23 subheadings to defined a similarity of documents. In Figure 7, the quality of considered clustering algorithms are presented. In all cases the hierarchical algorithm show to be the best (according to the Rand Index measure). K-means algorithm is a little worse then the first one. The Lingo is in the third place.

5.3   Time Performance of Clustering Algorithms

This set of experiments investigates the time required for each clustering algorithm. For every type of extension we calculate the average performance time of K-Means, hierarchical i Lingo algorithm. The results are presented in Figure 8. One can observe that all three algorithms are effective, all cluster a set of 200 search results in less than 20 seconds. This time is feasible for an online

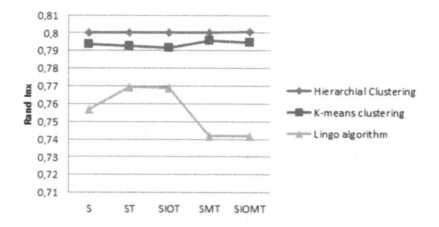

Fig. 7. The quality of clustering algorithms

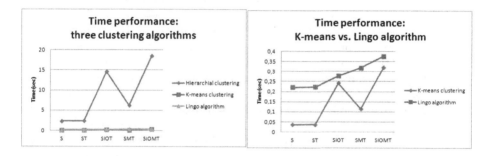

Fig. 8. The time performance of clustering algorithms

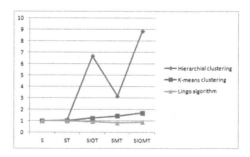

Fig. 9. The clustering time vs. extension types

clustering system. However the K-Means and Lingo algorithm are about 30 time quicker than the hierarchical one. The average time of last algorithms is lower then 0,4 sec for one query.

The last experiment focuses on analyzing, if the extended representations prolong the clustering time. For every clustering algorithm (hierachical, k-means or Lingo algorithm), we calculate the ratio of clustering time running on an extended model to that one, which is achieved on the basic model. One can observe that in the case of K-means and Lingo algorithm, the extended models are comparable with the basic one in the terms of computation time. However the hierarchical algorithm achieved on a full extended model (SIOMT) is several time slower then such one, which is running in the basic model, see Figure 9.

6   Conclusions and Further Work

Rapidly increasing size of scientific article meta-data and text repositories, such as MEDLINE or PubMed Central (PMC), emphasizes the growing need for accurate and scalable methods for automatic tagging and classification of textual data. For example, medical doctors often search through biomedical documents for information regarding diagnostics, drugs dosage and effect or possible complications resulting from specific treatments. In the queries, they use highly sophisticated terminology, that can be properly interpreted only with a use of a domain ontology, such as Medical Subject Headings (MeSH). In order to facilitate the searching process, documents in a database should be indexed with concepts from the ontology. Additionally, the search results could be grouped into clusters of documents, that correspond to meaningful topics matching different information needs. Such clusters should not necessarily be disjoint since one document may contain information related to several topics.

We had proposed a generalized Tolerance Rough Set Model for text data. The proposed method has been applied to extend the poor representation of snippets in the search result clustering problem. This functionality has been implemented within the SONCA system and tested on the biomedical documents, which are freely available from the highly specialized biomedical paper repository called PubMed Central (PMC).

The experimental results suggest the following facts:

- The proposed enrichment methods seem to be accurate approach to the considered problem.
- Varying richness of document representation displays a trade-off between important structural properties of resulting clusters, namely – the specificity of groups and the number of unassigned documents. Hence, results that confirm closer to a postulated balance may be found if we prepare appropriate document representations.

Our future plans are briefly outlined as follows:

- extend the experiments with different semantic indexing methods that are currently designed for SONCA system,
- analyze label quality of clusters resulting from different document representations,

- conduct experiments using other extensions (e.g. citations along with their context; information about authors, institutions, fields of knowledge or time),
- clustering of other types of objects (e.g. authors).
- experiments and theoretical study of measures proposed in this article.

References

1. Cao, T., Do, H., Hong, D., Quan, T.: Fuzzy named entity-based document clustering. In: Proc. of the 17th IEEE International Conference on Fuzzy Systems (FUZZ-IEEE 2008), pp. 2028–2034 (2008)
2. Hastie, T., Tibshirani, R., Friedman, J.: The Elements of Statistical Learning. Springer Series in Statistics. Springer New York Inc. (2001)
3. Herskovic, J.R., Tanaka, L.Y., Hersh, W., Bernstam, E.V.: A day in the life of pubmed: analysis of a typical day's query log. Journal of the American Medical Informatics Association, 212–220 (2007)
4. Ho, T.B., Nguyen, N.B.: Nonhierarchical document clustering based on a tolerance rough set model. International Journal of Intelligent Systems 17(2), 199–212 (2002)
5. Hubert, L., Arabie, P.: Comparing partitions. Journal of Classification 2, 193–218 (1985)
6. Kawasaki, S., Nguyen, N.B., Ho, T.-B.: Hierarchical Document Clustering Based on Tolerance Rough Set Model. In: Zighed, D.A., Komorowski, J., Żytkow, J. (eds.) PKDD 2000. LNCS (LNAI), vol. 1910, pp. 458–463. Springer, Heidelberg (2000)
7. Lancichinetti, A., Fortunato, S., Kertász, J.: Detecting the overlapping and hierarchical community structure of complex networks. New Journal of Physics 11, 033015 (2009)
8. Lu, Z., Kim, W., Wilbur, W.: Evaluation of query expansion using MeSH in PubMed. Information Retrieval 12(1), 69–80 (2009)
9. MacQueen, J.B.: Some methods for classification and analysis of multivariate observations. In: Cam, L.M.L., Neyman, J. (eds.) Proceedings of the Fifth Berkeley Symposium on Mathematical Statistics and Probability, vol. 1, pp. 281–297. University of California Press (1967)
10. Manning, C.D., Raghavan, P., Schtze, H.: Introduction to Information Retrieval. Cambridge University Press, New York (2008)
11. Nguyen, H.S., Ho, T.B.: Rough document clustering and the internet. In: Pedrycz, W., Skowron, A., Kreinovich, V. (eds.) Handbook of Granular Computing, pp. 987–1004. Wiley & Sons (2008)
12. Nguyen, S., Świeboda, W., Jaśkiewicz, G., Nguyen, H.: Enhancing search result clustering with semantic indexing. In: 3rd International Symposium on Information and Communication Technology (SoICT 2012), pp. 71–80. ACM (2012)
13. Nguyen, S.H., Świeboda, W., Jaśkiewicz, G.: Extended Document Representation for Search Result Clustering. In: Bembenik, R., Skonieczny, L., Rybiński, H., Niezgodka, M. (eds.) Intelligent Tools for Building a Scient. Info. Plat. SCI, vol. 390, pp. 77–95. Springer, Heidelberg (2012)
14. Nguyen, X.V., Epps, J., Bailey, J.: Information theoretic measures for clusterings comparison: Variants, properties, normalization and correction for chance. Journal of Machine Learning Research 11, 2837–2854 (2010)
15. Osiński, S.: An algorithm for clustering of web search result. Master's thesis, Poznan University of Technology, Poland (June 2003)

16. Rand, W.M.: Objective criteria for the evaluation of clustering methods. Journal of the American Statistical association 66(336), 846–850 (1971)
17. Roberts, R.J.: PubMed Central: The Gen. Bank of the published literature. Proceedings of the National Academy of Sciences of the United States of America 98(2), 381–382 (2001)
18. Skowron, A., Stepaniuk, J.: Tolerance approximation spaces. Fundamenta Informaticae 27(2-3), 245–253 (1996)
19. Szczuka, M., Janusz, A., Herba, K.: Semantic Clustering of Scientific Articles with Use of DBpedia Knowledge Base. In: Bembenik, R., Skonieczny, Ł., Rybiński, H., Niezgódka, M. (eds.) Intelligent Tools for Building a Scient. Info. Plat. SCI, vol. 390, pp. 61–76. Springer, Heidelberg (2012)
20. Tan, P.-N., Steinbach, M., Kumar, V.: Introduction to Data Mining, 1st edn. Addison-Wesley Longman Publishing Co., Inc., Boston (2005)
21. Witten, I.H., Frank, E.: Data Mining: Practical Machine Learning Tools and Techniques, vol. 2. Morgan Kaufmann (2005)

Automatic Extraction of Profiles from Web Pages

Piotr Andruszkiewicz and Beata Nachyła

Institute of Computer Science,
Warsaw University of Technology,
Warsaw, Poland
{P.Andruszkiewicz,B.Nachyla}@ii.pw.edu.pl

Abstract. Web pages are usually unstructured and Information Extraction from them is not trivial. In the paper we describe the process of Information Extraction on the example of researchers' home pages. For this reason we applied SVM, CRF, and MLN models. Performed analysis concerns texts in English language only.

Keywords: Information Extraction, probabilistic graphical models, SVM, CRF, MLN, resercher profiling.

1 Introduction

Nowadays, when information can be found on web pages, it is important to collect available information from such a source. We consider the system storing information about science, e.g., universities, researchers, publications, conferences, journals. Hence, we aim to enrich collected data with information from unstructured web pages of, e.g., researchers, conferences. We want to extract information we need and store it in a structured form because extracted contents will be converted to instances of appropriate concepts and integrated in a knowledge base. A source of the concepts definitions will be the Ontology of Science [1].

For specific types of objects, very specific web resources should be analyzed. For acquiring information about researchers, the home pages of researchers can be used as information sources. We focus on this type of information and check different methods of information extraction. Thus, we extract from home pages researchers' profiles, that is, properties (e.g., position, affiliation of a person, university which granted, e.g., Ph.D. degree to a researcher) that characterize the objects.

There are various methods for information extraction, thus, each property can be extracted using different methods with different accuracy, starting from simple regular expressions, through such classifiers as a decision tree [2], SVM [3], Naïve Bayes [4], to Conditional Random Fields [5] or Markov Logic Networks [6].

R. Bembenik et al. (Eds.): *Intell. Tools for Building a Scientific Information*, SCI 467, pp. 415–431.
DOI: 10.1007/978-3-642-35647-6_25   © Springer-Verlag Berlin Heidelberg 2013

2   Related Work

Profile extraction from web pages has been discussed and studied several times. The Artequakt system extracts automatically information from web pages and generates biographies of artists using a rule based extraction [7]. The ArnetMiner system [8] harvests information about researchers, publications, and conferences from structuralized sources. It also extracts information about researchers from their home pages [9].

In information extraction many models have been utilized: Support Vector Machine (SVM) [3], Hidden Markov Models [10], Conditional Random Fields [5], Markov Logic Networks [6,11].

Support vector machines (SVM) algorithms represent a group of supervised learning methods that can be applied to classification or regression. Support vector machines are an extension to nonlinear models of the generalized portrait algorithm developed by V. Vapnik [3]. Support Vector Machines have been applied in various applications, in particular in information extraction.

Conditional Random Field (CRF) is a conditional probability of a sequence of labels given a sequence of observations [5]. The linear-chain CRFs model only the linear-dependencies between observations and are not able to model hierarchical dependencies [12]. In order to take into account hierarchies, more general Conditional Random Model, e.g., Tree-structured Conditional Random Fields, needs to be applied. Both types of CRFs, linear-chain and tree-structured, have been utilized in information extraction from researchers' home pages [13].

A Markov Logic Networks model was succesfully applied to solving great variety of tasks known in artificial intelligence, machine learning, data mining and text exploration. As was presented in [14], the approach can be considered as generalization of well–known techniques of intelligent and statistical analysis like Binomial Distribution, Multinomial Distribution, Logistic Regression and Hidden Markov Models. The huge power of expression of the MLN let to achieve promising results in experiments with documents classification, entity resolution [15], extracting semantic networks [16] and also in information extraction from text. The information extraction using MLN was presented in [17].

3   Processing of Scientific Information by Means of CRF and MLN

In automatic information extraction, Conditional Random Fields (CRF) and Markov Logic Networks (MLN) can be used to perform the researchers' profiling, which is the process of obtaining the values associated with the different properties that constitute a person model. One example of such a property can be a current university of a researcher. This and other properties are likely to be found on researchers' home pages. In the researchers' profiling, automatic extraction should be used because manually annotating the profile for a large set of researchers is obviously tedious and time consuming. To this end, a unified approach to researchers' profiling based on the textual data tagging proposed in

[13] can be adopted. In this approach, the problem is treated as assigning tags to input texts, with each tag representing one profile property, for instance, position, affiliation, etc. As proposed in [13], we assume the usage of the following tags: *position, affiliation, bsuniv* (a name of university, where a bachelor degree was obtained)*, bsdate* (date of receiving bachelor degree)*, bsdegree* (a bachelor degree symbol, like *B.Sc.*)*, bsmajor* (a branch of science bachelor thesis concerned)*, msdegree, msmajor, msuniv, msdate* (the four preceding tags regard master degree and their explanation is analogous to the bachelor degree concerning tags)*, phddegree, phdmajor, phduniv, phddate* (the four tags regarded with an information about a doctorate)*, interests, address, email, phone, fax, publication, homepage,* and *resactivity* (research activity). In the sequel, we will use terms tagging and annotating interchangeably.

3.1 Conditional Random Fields

Conditional Random Fields (CRF) are often used as a tagging model. In this application, the CRF model is a representation of conditional probability distribution of a sequence Y of tags given a sequence X of observations, i.e., $P(Y|X)$. Linear-chain CRF is a basic type of a CRF in which dependencies are assumed to occur only between consecutive observations in a sequence. During a learning phase, a CRF model is built based on a tagged training dataset by means of an iterative algorithm using the Maximum Likelihood Estimation approach. The learnt CRF model can be used now to tag new sequences of observations by assigning respective tags with the highest likelihood. In a more general type of a CRF relationships are represented by means of a graph rather than a chain. In particular, CRF with a tree structure can be used to model a hierarchical dependency.

3.2 Data Preparation for Profiling Task

The goal of a profiling task, given a web page of a researcher, is to assign each phrase occurring on this page one tag from a given set of possible tags such as affiliation, received degrees, etc. To this end, we use tagging models built from a training set consisting of manually annotated home pages or web pages introducing researches.

Both, the learning phase and the annotation phase were preceded by the following data preparation activities:

- text tokenization,
- assignment of token types,
- assignment of possible types of tags,
- creation of features functions.

In the text tokenization task, the first step of preparing data for extraction of useful information from the identified web pages, we parsed a HTML content of web pages to identify tokens. After separating text into tokens, we assigned a type of token to each token. As in the ArnetMiner [8], we indicated five types of tokens: standard word, special word, image token, term, and punctuation mark. Special words are tokens with content corresponding to one of five types: email address (type EMAIL), IP address (IP), URL (URL), date (DATE), number (NUMBER), percentage (OTHER), words containing special symbols e.g., 'MSc', 'Ph.D.' (type OTHER), unnecessary tokens e.g., '− − − −' and '###'(type OTHER) , '< image >' tags in a HTML file (type IMAGE). Terms are base noun phrases, for instance, University of California, National Institute of Health, extracted from the web pages. Punctuation marks include period, question, and exclamation mark. To process web pages, we employed GATE and JAPE (Java Annotation Patterns Engine). When possible, we used annotations provided by GATE, and for cases harder to extract we wrote special rules in JAPE, for instance, in the case of web pages addresses.

The separated tokens formed the basic units and the pages formed the sequences of units in our tagging problem. In tagging, given a sequence of units, the most likely corresponding sequence of tags is determined by using trained tagging model. Each tag corresponds to a property defined in the schema of researcher profile.

During the third step of data preparation we assigned a tag to each token based on its corresponding type. For tokens classified as standard word, we used all possible tags. For special word there were nine possible tags: position, affiliation, email, address, phone, fax, bsdate, msdate, phddate. For image token only two tags were considerd: photo and email. For term token, position, affiliation, address, bsmajor, msmajor, phdmajor, bsuniv, msuniv, and phduniv tags were assigned. Thus, in this approach each token could be assigned several possible tags.

In order to train a CRF model, in the last step of data preparation we created three types of features, which were based on features functions presented in [9]. These features can be easily incorporated in the model by defining Boolean-valued feature functions, for instance, a Boolean-valued function that returns 1 when a token is capitalized and 0 otherwise.

For a standard word, the features included:

- Word features. Whether the current token is a standard word.
- Morphological features. The morphology of the current token: part of speech, whether the token is capitalized.

For special tokens:

- Special tokens. Whether the current token is a special word, type of the special token, for example, e-mail, phone number, date.

For term tokens:

- Type of the term provided by GATE: organization, job title, location, person, other.

For each token:

- Positive words. Whether the current token contains positive words for a category. Positive words were chosen semi-automatically; that is, we chose words which the most frequently co-occur in a sentence with words from a given category and occur less frequently in a sentence with words from the rest of categories. Positive words for some categories needed to be manually adjusted.
- Image tokens were treated as special words with IMAGE type.

3.3 Test Linear–Chain Conditional Random Fields Model Implementation

In order to test the approach to researchers profile extraction presented above, we used data provided by ArnetMiner team [18]. The set contains of 898 manually annotated home pages or introducing pages of researches. 600 web pages were used as a training set, the remaining web pages constituted a test set. Both sets were prepared in a way presented in the previous section. In order to estimate the linear CRF model, we used the implementation available at [19], which is the implementation of linear–chain Conditional Random Fields algorithm described in [5].

In the first test, the following features were used: part of speech, whether the token is capitalized for standard word, type of the special token, type of the term token. During training 64858 features functions were generated. The process of training the model took about 140 seconds. The overall F1–measure (for all tags anotated in the data set) was 50.72. In order to improve the results, we used all described features. Unfortunately, the provided implementation crashed due to segmentation fault during training, hence we checked two other CRF implementations and presented the obtained results in the following subsection.

3.4 Test of External CRF Libraries

Within our project, implementations of the CRF model from two open–source CRF libraries; namely CRFall and CRF++, were tested due to the failure of implementation mentioned in Section 3.3. Both libraries support only a linear dependency structure. The CRFall library was written for MATLAB environment, while CRF++ was implemented in C++. In the experiments, we used all features previously described; that is, 72 binary features. As previously, manually tagged dataset from ArnetMiner resources was selected as a source of training and testing data. The following results on sample texts were accomplished by using CRFall library. The task of two categories recognition, position and affiliation, was considered:

```
As [position] corporate vice president [/position] of the [affiliation]
Natural Interactive Services Division [/affiliation] NISD at [
affiliation] Microsoft Corp. [/affiliation] Kai- Fu Lee was responsible
for the development of the technologies and services for making the user
```

interface simpler and more natural. NISD includes technologies and
products for speech natural language advanced search and help and
authoring and learning technologies . The mission of NISD is to make
these technologies usable and useful for Microsoft 's customers . Lee
joined Microsoft in 1998 and was the founder of Microsoft Research Asia
which has since become one of the best research laboratories in the
world with a prolific publication and product transfer record . Before
coming to Microsoft he was the president of Cosmo Software the Silicon
Graphics Inc . SGI multimedia software business unit . Before that he
was [position] vice president [/position] and [position] general manager
[/position] of [affiliation] Silicon Graphics [/
affiliation] Web products division responsible for several product lines
and the company 's corporate Web strategy . Before joining SGI Lee spent
six years at Apple most recently as vice president of the company 's
interactive media group which developed QuickTime QuickDraw D QuickTime
VR and PlainTalk speech technologies . Prior to his position at Apple he
was an [position] assistant professor [/position] [affiliation] at
Carnegie Mellon University [/affiliation] where he developed the world '
s first [position] speaker-independent speaker [/position] [affiliation]
independent [/affiliation] continuous speech-recognition system . While
at [affiliation] Carnegie Mellon [/affiliation] Lee also developed the
world-champion [position] champion [/position]computer program that
plays the game Othello and that defeated the human world champion in
1988 . Lee holds a doctorate in computer science from [affiliation]
Carnegie Mellon University [/affiliation] and a bachelor 's in computer
science with highest honors
from [affiliation] Columbia University [/affiliation]. Lee is a Fellow of
the IEEE .

There are two types of dependencies modeled by CRF. First is a dependency
of current state value ($Y_t = y_t$) on observations ($X_t = x_t$), expressed by so
called state feature functions: $f_k(y_t \mid x_t)$. Subscript k is a sequential number
of given feature function used for indexing purpose. We are assuming that we
previously selected K state feature functions indexed by k=1,..., K. In our
case of tagging text tokens with information labels, current state is a label of
currently considered token. Observations, that form vector x_t, are the values of
features of current token determined in data preparation step. The latter type
of dependency is a correlation between labels of sequential tokens, modeled by
transition feature functions: $g_l(y_t|y_(t-1))$. We assume, that we have L transition
feature functions indexed by l=1,..., L. Each feature function has its own weight:
λ_k and μ_l for state and transition components, respectively. That weights are
parameters of the model being optimized during training process. Assuming
above conditions, joint conditional distribution of current token label being equal
to value y_t, can be expressed as:

$$p(Y_t = y_t \mid X_t = x_t, Y_{t-1} = y_{t-1}) = e^{\left(\sum_{k=1}^{K} \lambda_k f_k(y_t|x_t) + \sum_{l=1}^{L} \mu_l g_l(y_t|y_{t-1}) \right)} \quad (1)$$

The main drawback of CRFall implementation is lack of ability to train transition
dependency coefficients μ_l. They can only take constant value, predefined by

user. Moreover learning procedure is not well–optimized. The amount of training documents could not exceed a few dozen due to long training time. To overcome these problems, another CRF software was investigated. CRF++ library was selected. Same two–class experiment, for position and affiliation labels, with new implementation was performed. This time we were able to use over 6 hundreds of tagged web pages instead of about 60 of them taken for training with CRFall. Despite of increasing size of training set tenfold we observed similar learning time as previously. Results on the sample piece of test document are presented below:

```
As corporate [position] vice president [/position] of the [affiliation]
Natural Interactive Services [/affiliation] Division NISD at Microsoft
Corp. Kai- Fu Lee was responsible for the development of the
technologies and services for making the user interface simpler and more
natural . NISD includes technologies and products for speech natural
language advanced search and help and authoring and learning
technologies . The mission of NISD is to make these technologies usable
and useful for Microsoft 's customers . Lee joined Microsoft in 1998 and
was the [position] founder [/position] of [ affiliation] Microsoft
Research Asia [/affiliation] which has since become one of the [position
] best [/position] research laboratories in the world with a prolific
publication and product transfer record . Before coming to Microsoft he
was the [position] president [/position] of Cosmo Software the [
affiliation] Silicon Graphics Inc [/affiliation]. SGI multimedia
software business unit . Before that he was [position] vice
president [/position] and general manager of Silicon Graphics Web
products division responsible for several product lines and the company
's corporate Web strategy . Before joining SGI Lee spent six years at
Apple most recently as vice president of the company 's interactive
media group which developed QuickTime QuickDraw D QuickTime VR and
PlainTalk speech technologies . Prior to his position at Apple he was an
 [position] assistant professor [/position] at [affiliation] Carnegie
Mellon University [/affiliation] where he developed the world 's first
speaker speaker-independent independent continuous speech-recognition
system . While at Carnegie Mellon [affiliation] Lee [/affiliation] also
developed the world-champion champion computer program that plays the
game Othello and that defeated the human world champion in 1988 . Lee
holds a doctorate in computer science [affiliation] from Carnegie Mellon
 University and [/affiliation] a bachelor 's in computer science with
highest honors [affiliation] from Columbia
University [/affiliation]. Lee is a Fellow of the IEEE .
```

The second task was to simultaneously extract information about receiving Ph.D. and B.Sc. diplomas. In the experiment following eight information labels were used: bsdegree, bsmajor, bsuniv, bsdate, phddegree, phdmajor, phduniv, phddate. In this case the difficulty lies in similar contents of some pairs of tags like phddate and bsdate, phduniv and bsuniv, phdmajor and bsmajor. Due to this, they can be often confused. All of this data appears also less frequently in the training set than the position or affiliation tags. With the CRF++ library we have received following labels on a sample piece of text:

```
Steve received his [bsdegree] B.S [/bsdegree]. in [phdmajor] [bsmajor]
 Computer Science [/phdmajor] [/bsmajor] from [phduniv] [bsuniv] National
  Taiwan University [/phduniv] [/bsuniv] in [bsdate] 1991 [/bsdate] and
 his M.S . and [phddegree] Ph .D. [/phddegree] in [phdmajor] Electrical [
 /phdmajor] [bsmajor] Engineering [/bsmajor] from [bsuniv] Stanford
 University [/bsuniv] in 1995 and [phddate] 2000 [/phddate] respectively.
```

There were a few hundreds of pages used for learning. For comparison, results for training dataset consisting of only 40 documents, were much less precise:

```
Steve received his B.S [phdmajor]. in Computer Science [phddate][phduniv]
 from [/phddate] National Taiwan University [/phdmajor] [/phduniv] in [
 phddate] [phdmajor] [phduniv] 1991 [/phddate] [/phdmajor] [/phduniv] and
 his M.S . and [phddegree] Ph . D. [/phddegree] in Electrical [phdmajor]
 Engineering [phduniv] from Stanford University [/phdmajor] [/phduniv]
 in [phddate] [phdmajor] [phduniv] 1995 [/phddate] [/phdmajor] [/phduniv]
 and 2000 respectively .
```

In the last experiment emails, phone and fax numbers were retrieved. Content of such data is relatively well–structured and usually easy to express by regular language. A special set of the JAPE rules was prepared to extract tokens of non–standard type with email addresses and phone/fax numbers. JAPE is a kind of regular language, provided by GATE library, allowing to edit and change set of annotations of tokens, prescribed to them by GATE text–processing procedure, at any step of text–processing. Positive words for phone category (like 'Phone:', 'Tel:') and fax category (like 'Fax:'), preceding often demanded numbers, were also very useful in distinguishing between the tokens of that classes. Following lists contain results of evaluation previously trained CRF model on test dataset. There are true addresses or numbers in the left columns (expected to be labeled) and contents of appropriate tags, assigned by CRF model in the right ones:

– email:

```
gvj@site.uottawa.ca            wade@csd.uwo.ca
jzalamea@ac.upc.es             klastrup AT itu . dk
a . j . winter at bris . ac . uk   sima@cs.umd.edu
a . j . winter at bris . ac . uk   zhangfan at cs dot ust dot hk
                                  fanzhangfd at gmail dot com
 .                                a s h v i n e e c g . t o r o n t
devetak@usc.edu                   o . e d u
segal@cs.bgu.ac.il             stephane . grumbach AT inria . fr
ytan@u.washington.edu          aejnioui@mail.ucf.edu
xvg@berkeley.edu xvg1@grupovera.   Ravi_Sankar
 com                           <IMAGE_src="email.gif"_alt=""/>
cidon@ee.technion.ac.il        k.chao@coventry.ac.uk
erafalin at cs . tufts . edu   hkondo@cs.takushoku-u.ac.jp
stox@cg.ac.yu                  rhayward@mail.pse.umass.edu
wade@csd.uwo.ca
```

memik@ece.northwestern.edu
j-ku@northwestern.edu
dlharrison@mchsi.com
delis@cti.gr
treytnar@edacentrum.de
suri at cs . ucsb . edu
sandeep@cs.albany.edu
gvj@site.uottawa.ca
font
jzalamea@ac.upc.es
winter at bris . ac
winter at bris . ac .
segal@cs.bgu.ac.il
gdas@cse.uta.edu[/email]<b>Phone
with
cs . tufts .
stox@cg.ac.yu
wade@csd.uwo.ca
wade@csd.uwo.ca

klastrup AT itu . dk
sima@cs.umd.edu
zhangfan at cs dot ust dot hk
fanzhangfd
gmail
grumbach AT inria .
School_of_Mathematical
k.chao@coventry.ac.uk
sakama sys . wakayama-u . ac . jp
rhayward@mail.pse.umass.edu
memik@ece.northwestern.edu
j-ku@northwestern.edu
delis@cti.gr
treytnar@edacentrum.de
suri at cs . ucsb .
co . uk .
.

j.kelsey@ucl.ac.uk

– phone

ext._6686
6979
617_627_5128
x86994
7218 5029
301-405-0688
2358-6992
8261 4462
6264 3087
407.823.4757
8123
413-577-1317
+49_511_762_5030

ext._6686
6979
617_627_5128
7218 5029
301-405-0688
2358-6992
407.823.4757 Dept
8123
413-577-1317
+49_511_762_5030

– fax

7055
2358-1477
6264 7458
407.823.5835
7570
413-545-0082
417-889-3704
+49_511_762_5051

7055
2358-1477
6264 7458
407.823.5835
7570
413-545-0082
417-889-3704
+49_511_762_5051

3.5   Hierarchical CRF

Factorial CRF is a more general approach than Linear–chain Conditional Random Fields. In particular, Factorial CRF can model dependencies across hierarchically laid–out information, i.e., parent–child dependencies. As the properties of the researchers' profile can be ordered in such a hierarchy, as the next step of CRF application in information extraction we have used Factorial CRF.

To model Factorial CRF we used GRaphical Models in the Mallet (GRMM) library [20]. GRMM, a toolkit for inference and learning in graphical models, was created as an add–on package to MALLET [21], a Java package for statistical natural language processing, document classification, clustering, topic modeling, information extraction, and other machine learning applications to text.

In order to conduct the profiling task, we have implemented two–level factorial CRF in GRMM. To train the model, again we have used data provided by the ArnetMiner team [18]. We have based the parent–child dependencies on the structure of the training data [9]. The parent–child dependencies we applied are summarized in Table 1.

Table 1. Definition of parent–child structure for Factorial CRF

Block Type	Profile property
Photo	Person photo
Basic information	Position, Affiliation
Contact information	Fax, Phone, Address, E-mail
Educational history	Phddate, Phduniv, Phdmajor, Msdate, Msuniv, Msmajor, Bsdate, Bsuniv, Bsmajor

3.6   Tree–structured Conditional Random Fields

Tree–structured Conditional Random Fields (TCRF) as a more general approach than Linear–chain Conditional Random Fields, can model dependencies across hierarchically laid–out information; that is, parent–child and sibling dependencies [9]. Researches profile properties can be ordered in such a hierarchy and model with TCRF improving the results.

In researchers' profile extraction and TCRF model, the home page or introducing page is the observation. The tree is built by converting the web page into a DOM tree. The root node denotes the web page, each leaf node represents the word token, and the inner node denotes the information block (for instance, a block containing educational history). Thus, the label of the inner node of the tree represents one type of the information block. The label of the leaf represents one of the profile properties, for instance, a date when a researcher received Ph.D. The profile properties, the information blocks, and the relationships between them are the same as for hierarchical CRF and are shown in Table 1.

Tree–structured Conditional Random Fields can model hierarchical relationships between profile properties, which better reflects the structure of information presented on home pages or introducing web pages of researches. Moreover, the unified approach (Linear–chain CRF and Tree–structured CRF), in contrast to the separated method, where the annotations of different properties are interdependent, can take advantage of the interdependencies between the subtasks of profiling and achieve better performance in researcher profiling than the separated methods.

3.7 Markov Logic Networks

Experiments performed with Conditional Random Fields indicate some weak points of this model. Especially, extracting Ph.D. and B.Sc. degrees information showed that additional constrains could improve localization of some categories. We could also want to incorporate into model knowledge from other sources, like system databases, ontologies and web searches in form of hints improving profile extraction performance. To obtain this, an unified way of expressing constrains generated from heterogeneous data sources is needed. A model which can capture these requirements is Markov Logic Networks (MLN). In context of the discussion led in the previous sections, we can consider MLN as a generalization of the Conditional Random Fields method. It is a relational model with dependencies described by set of first order logic formulas. Term network means arbitrary structure of undirected dependency graph. Definitions of dependency relations are provided by user as set of first order logic formulas (clauses). Each of them consists of functions and predicates, that can be connected with logical operators and preceded by quantifiers. Functions and predicates can take variable or constant arguments. Concrete objects in data (like tokens or identifiers of tokens, values of feature functions applied to concrete instances of tokens) are represent by constants. Variables represent types of constants, for instance variable named t can represent arbitrary token. Logical formulas consisting of only one predicate are called the atoms (or atomic formulas). Set of logical formulas (clauses) forms knowledge based constraints. Every logical constraint owns a weight, which denotes probability of its accomplishment. Sample in data, which does not meet provided constraints, decreases their probability levels. It is also allowed to include hard formulas, that has to be respected by any data (it is equivalent to set infinite weights). In case of using Markov Logic Network model an additional step of data preparation is required. This is a transformation of dataset in the latter form of feature functions' values to a dataset consisting of predicates and logic functions' instances with values of their parameters set to constants. The constants represent values of that parameters induced from data, for example values of feature functions. Output tags of tokens are also expressed as predicates. The predicates usually have names of possible labels and take one argument indicating a token. For example, in task of recognizing position and affiliation tags for each token w in the dataset there are three predicates modeling possible tags of w: POSITION(w), AFFILIATION(w) and NONE(w). In general, we can split set of formulas into two parts. First of them includes all formulas, that contains at least one predicate using to modeling output tags and is called a query. The latter, namely an evidence, holds all remaining formulas. Datasets consists of predicates in ground form, that is having no variable arguments, being replaced by appropriate constants representing data. In case of training set, query predicates for each token are known from data. The dataset contains one entry for each token with predicate indicating its true tag. In testing or working datasets query predicates are not provided. The aim of learning process is to find value of weights of all formulas from query part. Selected weights should lead to maximal probability of existence of data from training set. As in CRF model, entropy maximization rule is applied. Since according to Hammersley–Clifford theorem, each Markov Network is equivalent to some Gibbs distribution, there is an immediate formula describing world state to be optimized [22]:

$$P(Y = y \mid X = x) = \frac{1}{Z}e^{\left(w_i n_i(x,y)\right)}, \tag{2}$$

where Y denotes query part of formulas set and X evidence part. w_i is a weight of i-*th* formula and n_i is number of times it is satisfied in data. The most likely state is tried to be found:

$$w'_Y = \underset{w_Y}{\arg\max}\, P(y \mid x), \tag{3}$$

where w_Y is a vector of weights corresponding to query formulas. Monte Carlo version of Boolean satisfability local search algorithm, namely WalkSAT, is most often used to solve this task. The MaxWalkSAT is able to find solution for weighted SAT problem. Further details about MLN and methods of its training and inference are introduced in [11]. In order to proof selection of described method, a task of Ph.D. and B.Sc. data was taken into consideration. Following additional rules were assumed to be helpful: A token having organization name property appearing as one of the six sequential tokens immediately after phddegree (bsdegree) labeled word is assumed to be phduniv (bsuniv respectively). A token having date property appearing as one of the six sequential tokens immediately after phddegree (bsdegree) labeled word is assumed to be phddate (bsdate respectively). Except of them, only six other rules were used:

- A token with DATE property might be a phddate/bsdate
- A token containing positive word of phddegree/bsdegree category might belong to phddegree/ bsdegree
- A token containing positive word of phduniv/bsuniv category might belong to phduniv/ bsuniv

Training set consisted of 40 pages. Using Tuffy 0.3, very efficient MLN library [23], following output was obtained:

```
Steve received his [bsdegree] B.S . [/bsdegree] in Computer Science from
  [phduniv] National [/phduniv] [bsuniv] Taiwan University [/bsuniv] in [
  bsdate] 1991 [/bsdate] and his M.S . and [phddegree] Ph [/phddegree] . D
  [phddegree] . [/phddegree] in Electrical Engineering from [phduniv]
  Stanford University [/phduniv] in [phddate] 1995 [/phddate] and [phddate
  ] 2000 [/phddate] respectively.
```

It is also possible to build unsupervised Markov logics performing clustering task. We are going to test this approach for profile extraction and name disambiguation tasks in near future. It could let us avoid tedious preparation of training data for Polish language.

4   Experimental Evaluation

Plots in Figure 1 present test results of Markov Logic Network and Conditional Random Fields models used to extraction phddate, bsuniv and affiliation information. We used CRF with linear structure. Due to performing affiliation category test, we prepared models for an extraction of two tags: position and affiliation. In phdate and bsuniv categories tests, we trained the models for an extraction of 8 tags concerning PH.D, and B.Sc. degrees: phddegree, bsdegree, phddate, bsdate, phduniv, bsuniv, phdmajor and bsmajor. In the two latter tests we applied additional rules for building Markov Logic Network described in section 3.7. The results are in form of ROC lines prepared separately for each of three tests and for each of two types of models. Every of MLN models was trained with 40 documents and every of CRF chains was learned with 400 documents. Table 3 shows the results for Factorial CRF. We used GRMM library to train FCRF on 248 documents. We present the results only for one point (combination of true positive and false positive rates) instead of ROC curve because GRMM returned only assigned values of tokens, not the probabilities.

Considering ROC curves for CRF and MLN models, we can observe significant improvement in case of extraction of degree information by means of MLN in the comparition to the CRF chain. In case of affiliation label the results of CRF model are better.

The test showed that the most difficult to extract was the bsuniv tag. That class usually contains names of specific organization that are very often not correctly recognized as a special word of ORGANIZATION type by the GATE library. It seems to be very common that only part of the names are correctly tagged with the bsuniv label that significantly influences presented outcomes.

Comparing the results for CRF, FCRF, and MLN (at the given point, determined by FCRF) we can observe that the best results are obtained by means of different methods. For the phddate, we obtained the best results for MLN. The results for CRF and FCRF are almost the same. However, for bsuniv, FCRF achieved slightly better results than CRF and significantly better than MLN.

The results would be improved by preparing training data set with homepages that contains more instances of the concerned tags. The Table 2 shows the amounts of tokens in training data set being part of each of the tags. The contribution of training examples for each class is very low comparing to size of whole data set. Selection of documents that contains the tags could not be enough in this case. A typical homepage contains information about each of received degree only once, and at the same time plenty of tokens ascribed to other classes. It results in extremely imbalanced data sets. A data set balancing techniques could bring significant improvement. One could try to remove randomly part of tokens

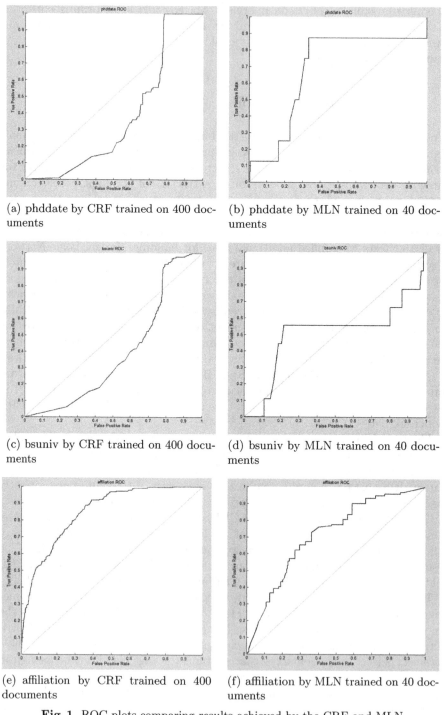

(a) phddate by CRF trained on 400 documents

(b) phddate by MLN trained on 40 documents

(c) bsuniv by CRF trained on 400 documents

(d) bsuniv by MLN trained on 40 documents

(e) affiliation by CRF trained on 400 documents

(f) affiliation by MLN trained on 40 documents

Fig. 1. ROC plots comparing results achieved by the CRF and MLN

Table 2. Number of training tokens and their percentage contribution in whole training set for three selected categories

	affiliation	phddate	bsuniv
MLN train set	314	13	17
	1.56%	0.06%	0.08%
CRF train set	3298	218	358
	0.15%	0.14%	0.23%
FCRF train set	1717	63	100
	2.31%	0.09%	0.14%

Table 3. The results of Factorial CRF

Tag	True positive rate	False positive rate	Precision	Recall	F
phddate	0.6667	0.7460	0.6667	0.2540	0.3678
bsuniv	0.8125	0.7400	0.8125	0.2600	0.3939
affiliation	0.6613	0.7851	0.6613	0.2149	0.3244

using a stratification sampling. The other possibility is to use segmentation of homepage, that would enable us to select paragraphs containing information of interests and build a data set only from their content.

5 Conclusions and Future Work

Extraction of specific information from text is a difficult and complicated task. Parts of text that carrying important contents are related to each other. That relations are often very complex and irregular. It is difficult to catch all of them with a single model or even with a single type of modeling. We have shown that the best results can be obtained by means of different models. So the first hint for one who builds information extraction system is to split different types of information into groups of similar content or structure. We split information appearing on homepages into groups regarding contact data (tags: `phone`,`email`, `fax`), obtaining scientific degree (tags, like for instance: `phddegree` or `bsuniv`) and an employment information (`position` and `affiliation` classes). That approach lets one save a lot of time that we could spend building more general and complex models. Some type of information are in most cases very simple to be recognized by trivial approaches like regular expressions or short dictionaries. It would be inappropriate to try to retrieve contact data with relational techniques like Markov Logic Networks. It would lead to overfitted models that often results in poor performance. On the other hand, highly related semantically data like degree and employment tags are well suited to be caught with relational models. The data that are commonly integrated with a structure of the document should be described by structure dependent algorithms like Factorial or Tree–Structured CRF.

The great improvement brings an usage of especially selected tools of natural language processing like named noun phrases and good tokenization that works well with domain specific abbreviations. In the current version our feature extraction implementation uses default tools provided by GATE library. We suppose that changing some of them should make better obtained outcomes, what is the subject for future work.

Last but not least is training data preparation. Selecting a proper set of documents and selecting appropriate size of training data is crucial for time of computations and the quality of final solution. Balancing categories seems to have important meaning in information extraction, where great majority of instances in data sets are useless. We observed that relational models have better results in cases where classes have similar number of representatives. We will try to prove experimentally our observation in the future.

References

1. Synat system ontology, http://wizzar.ii.pw.edu.pl/passim-ontology/:
2. Quinlan, J.R.: Induction of decision trees. Machine Learning 1(1), 81–106 (1986)
3. Cortes, C., Vapnik, V.: Support-vector networks. Machine Learning 20(3), 273–297 (1995)
4. Manning, C.D., Raghavan, P., Schütze, H.: Introduction to information retrieval. Cambridge University Press (2008)
5. Lafferty, J.D., McCallum, A., Pereira, F.C.N.: Conditional random fields: Probabilistic models for segmenting and labeling sequence data. In: Brodley, C.E., Danyluk, A.P. (eds.) ICML, pp. 282–289. Morgan Kaufmann (2001)
6. Domingos, P.: Real-World Learning with Markov Logic Networks. In: Boulicaut, J.-F., Esposito, F., Giannotti, F., Pedreschi, D. (eds.) PKDD 2004. LNCS (LNAI), vol. 3202, pp. 17–17. Springer, Heidelberg (2004)
7. Kim, S., Alani, H., Hall, W., Lewis, P.H., Millard, D.E., Shadbolt, N.R., Weal, M.J.: Artequakt: Generating tailored biographies with automatically annotated fragments from the web. Presented at the Semantic Authoring, Annotation and Knowledge Markup (SAAKM) 2002 Workshop at the 15th European Conference on Artificial Intelligence (ECAI 2002), pp. 1–6 (2002)
8. Tang, J., Zhang, J., Yao, L., Li, J., Zhang, L., Su, Z.: Arnetminer: extraction and mining of academic social networks. In: Li, Y., Liu, B., Sarawagi, S. (eds.) KDD, pp. 990–998. ACM (2008)
9. Tang, J., Yao, L., Zhang, D., Zhang, J.: A combination approach to web user profiling. TKDD 5(1), 2 (2010)
10. Ghahramani, Z., Jordan, M.I.: Factorial hidden markov models. Machine Learning 29(2-3), 245–273 (1997)
11. Richardson, M., Domingos, P.: Markov logic networks. Machine Learning 62(1-2), 107–136 (2006)
12. Zhu, J., Nie, Z., Wen, J.R., Zhang, B., Ma, W.Y.: Simultaneous record detection and attribute labeling in web data extraction. In: Eliassi-Rad, T., Ungar, L.H., Craven, M., Gunopulos, D. (eds.) KDD, pp. 494–503. ACM (2006)
13. Yao, L., Tang, J., Li, J.Z.: A unified approach to researcher profiling. In: Web Intelligence, pp. 359–366. IEEE Computer Society (2007)

14. Domingos, P., Richardson, M.: Markov logic: A unifying framework for statistical relational learning. In: Proceedings of the ICML 2004 Workshop on Statistical Relational Learning and its Connections to Other Fields, pp. 49–54 (2004)
15. Singla, P., Domingos, P.: Entity resolution with markov logic. In: ICDM 2006 Proceedings of the Sixth International Conference on Data Mining, pp. 572–582. IEEE Computer Society Press (2006)
16. Kok, S., Domingos, P.: Extracting semantic networks from text via relational clustering (2008)
17. Poon, H., Domingos, P.: Joint inference in information extraction. In: Proceedings of the 22nd National Conference on Artificial Intelligence, pp. 913–918. AAAI Press (2007)
18. http://arnetminer.org/labdatasets/profiling
19. http://keg.cs.tsinghua.edu.cn/persons/tj/software/KEG_CRF/
20. http://mallet.cs.umass.edu/grmm/index.php
21. http://mallet.cs.umass.edu/
22. Kok, S., Domingos, P.: Learning the structure of markov logic networks. In: Raedt, L.D., Wrobel, S. (eds.) ICML. ACM International Conference Proceeding Series, vol. 119, pp. 441–448. ACM (2005)
23. http://research.cs.wisc.edu/hazy/tuffy/

Chapter VI
Multimedia Data Processing

A Distributed Architecture for Multimedia File Storage, Analysis and Processing*

Andrzej Dziech[1], Andrzej Glowacz[1], Jacek Wszolek[1],
Sebastian Ernst[2], and Michał Pawłowski[2]

[1] AGH University of Science and Technology, Department of Telecommunications,
al. Mickiewicza 30, 30-059 Kraków, Poland
`dziech@kt.agh.edu.pl`, {`aglowacz,jacek.wszolek`}`@agh.edu.pl`
[2] AGH University of Science and Technology, ACC Cyfronet AGH,
ul. Nawojki 11, 30-950 Kraków, Poland
`ernst@agh.edu.pl, bajdala@gmail.com`

Abstract. The paper presents a prototype of a distributed storage, analysis and processing system for multimedia (video, audio, image) content based on the Apache Hadoop platform. The core of the system is a distributed file system with advanced topology, replication and balancing capabilities. Metadata for stored media objects is being managed in a separate database in conformance with well-known metadata models and standards. In addition, the system allows implementation of efficient multimedia data analysis (e.g. face recognition) and processing (e.g. transcoding, digital watermarking) algorithms. Interoperability with external systems, such as the currently-developed Infona is provided by means of an OAI-PMH-compliant interface for metadata harvesting. More in-depth integration is possible by making storage, analysis and processing services available to external entities using the SYNAT integration platform. The article evaluates several important aspects of the proposed distributed repository, such as security, performance, extensibility and interoperability.

Keywords: distributed storage, distributed processing, MapReduce, multimedia, Hadoop, HDFS, performance evaluation, scalability, high availability.

1 Introduction

The SYNAT project [2] aims at establishment of an universal and open repository platform with hosting and transmission features for network knowledge resources. The system is intended to become the widest publicly available repository of open knowledge in Poland, interlinking various existing digital libraries as well as future data sets. The system is going to be used by scientific, educational and cultural societies. The Project is managed by a consortium of top Polish universities and national libraries.

* Work financed by the National Centre for Research and Development (NCBiR) within project no. SP/I/1/77065/10.

R. Bembenik et al. (Eds.): *Intell. Tools for Building a Scientific Information*, SCI 467, pp. 435–452.
DOI: 10.1007/978-3-642-35647-6_26   © Springer-Verlag Berlin Heidelberg 2013

In the scope of managing and processing visual content within the system, one of the key objectives of the project is development of a distributed repository of visual data and related innovative services.

The first research goal is creation of an open database system. It will take advantage of ontologies for design of the database structure and should allow for convenient integration with multimedia processing tools. These tools include content-based multimedia indexing, digital watermarking, video quality assessment and transcoding-based optimisation. Innovative elements also include detection and prevention of distributed attacks against database system. The solution includes distribution of network and processing resources, advanced control of network connections along with adequate security policy. However, that feature is beyond the main scope of this paper.

The second goal is related to development of distributed metadata repositories and methods of advanced search in federation of repositories containing visual data. Access interfaces will be based on common standards such as SQI [12] and protocols standardised by the Open Archives Initiative[1]. Furthermore, interlinking various databases requires unification of metadata description of stored multimedia objects. In addition, management of user profiles will enable complex search functionalities. Collaborative, Content-based or Demographic Filtering strategies can be adopted to assist users in limiting the number of search results. Collaborative Filtering is expected to be a useful solution in such scenario; however, some scientific users may prefer to have it implemented as an option. That topic is the main subject of another paper.

The final objectives of the project are strictly related to multimedia content and comprised of content-based multimedia indexing, digital watermarking technologies and methods of high-quality content distribution.

Content-based indexing methods provide metadata information about real objects occurring in multimedia content. For example, video sequences stored in the system can be processed in order to find actor appearances in a movie. This can be achieved using automatic face recognition based on advanced image processing and a database of known actors.

Digital watermarking is a set of techniques that operate directly on multimedia content and provide advanced functionality in the area of data integrity, metadata verification, identification of distribution channel or protection of sensitive data. New watermarking techniques are also under research in this task.

The metadata system will store additional information about content quality. The content distribution system is focused on providing multimedia data at a given quality level to the end user. It will implement adaptation and visual data transcoding functions as well as fragmentary object delivery (e.g. delivery of part of a scanned manuscript or just the audio stream from video recording).

The overall architecture of the repository and its modules are presented in Figure 1.

---

[1] http://www.openarchives.org

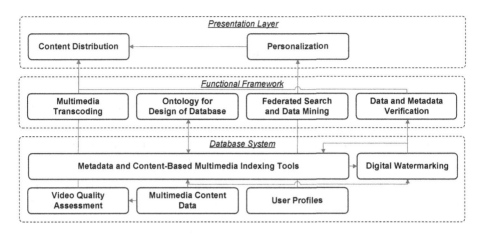

Fig. 1. Overall architecture of the repository of visual data

2   Motivation and State of the Art

One of the main objectives of the SYNAT project is creation of a distributed visual data repository. Repository users can be divided into two categories: data providers and data consumers (end users). Different functionality is offered for each category:

- Data providers can store multimedia data in a safe manner (replication, horizontal scalability and load balancing is provided by the Apache Hadoop[2] framework). All multimedia content must be described with an appropriate metadata information (Dublin Core). Additionally, before the data is available online, it can be watermarked and processed in different ways, mainly for legal (copyright protection) and safety-related issues (e.g. face recognition for privacy protection). After making the visual data accessible, it's automatically made searchable and accessible for end user and other repositories (OAI-PMH[3]).
- End users can browse and discover available resources through a rich GUI and a search interface. Also, automatic transcoding is used to enable delivery of multimedia objects attuned to specific devices (PC, tablet, mobile) and adapt it to network bandwidth capabilities.

3   Hadoop: A Brief Description

Apache Hadoop is an open-source platform which allows for implementation of systems capable of storing and processing data in a distributed, scalable and reliable manner. It was conceived as an open-source implementation of Google's

---

[2] http://hadoop.apache.org/
[3] http://www.openarchives.org/pmh

concepts such as the Google File System [7] and the MapReduce distributed programming model [6].

Hadoop consists of three base modules:

- *Hadoop Common*, which consists of the base utilities used to manage the two main functional modules,
- HDFS (*Hadoop Distributed File System*), allowing for high-performance, redundant access to data (further described in Section 3.1),
- *Hadoop MapReduce*, which provides relatively easy means of implementing distributed data processing programs (further described in Section 3.2).

Hadoop was chosen as the base solution for various reasons described later in the paper, including features of the HDFS file system which make it well-suited to multimedia applications, described in Section 3.1.

Numerous additional tools and software packages have been developed to facilitate the application of Hadoop in different fields. These include Cassandra[4] (a column-based NoSQL database), HBase[5] (Apache's implementation of the Google BigTable Concept [5]), Hive (a distributed data warehouse system) and Mahout (a machine learning and data mining library). Availability of such rich tools will allow for future extension of the platform with new functionality.

3.1 HDFS

HDFS is a distributed file system, aimed primarily at fault tolerance and scalability. It is a crucial element of the Hadoop platform and has been designed to be run on large clusters of relatively low-cost hardware. HDFS is aimed primarily at providing high-throughput access to large datasets, and copes best with large files.

According to [4], HDFS is based on the following assumptions, which have been analyzed with regard to applicability for the multimedia data repository:

1. **Tolerance of hardware failures.** Data replication, fault detection and quick recovery help maintain availability even when some components of a cluster are non-functional.
2. **Support for large data sets.** Files in HDFS are expected to be large (GBs or TBs in size). Efficient support for such volumes is achieved via scalability to hundreds of machines in a single cluster. This makes HDFS ideal for storage of large files, which is often the case with multimedia data. Support for small files (such as metadata, documents or single images) can be achieved through a dedicated database on top of HDFS (see Section 4.2) or by utilizing sequence files (see Sections 3.2 and 4.1).
3. **Streaming access model.** HDFS is not fully POSIX-compliant. As the main focus is on large throughput rather than low latency, it cannot be used as a general-purpose file system. This poses no problem with regard

---

[4] http://cassandra.apache.org/
[5] http://hbase.apache.org/

to retrieval or processing of multimedia files; however, implementation of a real-time streaming server may require additional buffering machines or utilization of techniques outlined in point 2.

4. **Write once, read many.** Files in HDFS are intended to be stored once and retrieved numerous times, without the need for modification or appending of data. Again, this fits well with the intended application.

5. **Near-local file processing.** As processing nodes are usually located on the same machines as data storage nodes, processing of data can take place very near the data itself.

6. **Portability.** As Hadoop (and HDFS) is entirely Java-based, nodes running different platforms may co-exist in the same cluster.

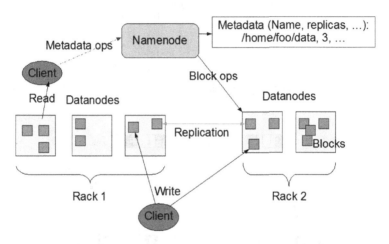

Fig. 2. Architecture of a HDFS cluster [4]

Files in HDFS are split into blocks; the default block size is 64MB, but it can be adjusted centrally and on a per-file basis. The architecture of a HDFS cluster has been shown in Figure 2. It consists of the following entities:

- *data nodes:* the main storage facilities, they store the data blocks,
- *name node:* one per cluster, it serves as the central point, keeps track which blocks are kept on which *data nodes*.

Clients first connect to the *name node* and, upon authorization, are directed to individual data nodes in order to retrieve or store data blocks.

Redundancy and fault tolerance is implemented by means of *block replication*. The default replication factor for the entire cluster can be configured and it can be set individually for selected files or directories. Replica placement policies can be adjusted to fit individual needs; the main goal is to minimize bandwidth consumption and read latency. In order to avoid data corruption, all blocks are checksummed when they are stored. Even distribution of data among *data nodes* is ensured by the *balancer*; if free space runs low, it can be reclaimed by either adding *data nodes*, emptying the *trash* folder, or decreasing the replication factor [4].

3.2   MapReduce

The MapReduce programming model [6] has its roots in functional programming and relates to its two operations: *map* and *fold*. The *map* operation is used to described an operation conducted on parts of the input dataset (which can be executed in a parallel manner), whereas *fold* is used to aggregate the results (which first have to be sorted by the key).

In MapReduce, the map phase corresponds to the *map* operation, and the reduce phase roughly corresponds to the *fold* function. Procedures executed in both phases use operate on key/value pairs; they can therefore be described as [10]:

$$\text{map} : (k_1, v_1) \rightarrow [(k_2, v_2)]$$
$$\text{reduce} : (k_2, [v_2]) \rightarrow [(k_3, v_3)]$$

where (k_1, v_1) represents the input data elements, and $[(k_3, v_3)]$ – the resulting dataset.

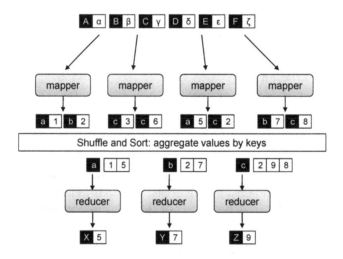

Fig. 3. A simplified presentation of a MapReduce program

A simplified view of MapReduce has been presented in Figure 3. All input key/value pairs are transformed by mappers into intermediate key/value pairs which are then shuffled and sorted according to the key. Then, the intermediate pairs are fed into the reducers, which aggregate the values from intermediate pairs sharing the same key.

The architecture of Hadoop's MapReduce is similar to HDFS in that it fits the master/slave model. The main controlling node, the *job tracker*, keeps a record of all jobs executed in the cluster. It also keeps track of all available processing nodes and assigns tasks dynamically. Every "slave" node runs a *task tracker*, which receives tasks from the *job tracker* and executes the necessary procedures.

Hadoop's MapReduce processing flow extends the general MapReduce model with the *combiner*, which is usually identical to the reducer and is used to aggregate intermediate key/value pairs from a single mapper.

In Hadoop, there are several ways to implement MapReduce algorithms. The most commonly-used native method involves implementing map and reduce procedures in Java, in classes which extend the `MapReduceBase` abstract class, and implement the `Mapper` or `Reducer` interfaces, respectively [3]. The keys and values should implement the `WritableComparable` interface; depending on the input and output file formats, the key and value can either be text (`Text` class), boolean (`BooleanWritable`), numeric (`IntWritable`, `DoubleWritable`, `FloatWritable`) or binary (`BytesWritable`).

Input files are either text files (which are fed line by line into mappers), *sequence files* or *HAR files* (Hadoop ARchive). A sequence file is a flat file consisting of binary key/value pairs [14]; this format can be used by MapReduce as input or output files, and is always used to store intermediate key/value pairs. HAR files are slightly different, as they maintain an index of all the files; they are therefore more costly to create and process, and are best used purely for archival purposes [15]. Intuitively, these two formats could be likened to a tarball, commonly used in UNIX systems.

If the map and reduce functions need to use external, non-Java libraries (such as the OpenCV image processing library [1]), they can either use JNI-based bindings (such as JavaCV) and keep the Java code, or use one of the two additional techniques which can be used to integrate non-Java software with Hadoop's MapReduce.

The first method, *Hadoop Streaming*, is actually a utility which allows users to create and run jobs with any executables (e.g. shell utilities) as the mapper and/or the reducer. This provides enormous flexibility, but also comes with two drawbacks. First, Hadoop Streaming introduces a slight overhead; however, as shown in Sections 5.2 and 5.3, this overhead can be compensated by faster execution of native procedures. The second drawback is support for file formats – Hadoop Streaming works well with text files, but sequence files need to be converted to `Text` objects (using `SequenceFileAsTextInputFormat`). This drawback limits the applicability of Hadoop Streaming, but it can also be circumvented, as shown in Section 4.1.

The other method for non-native MapReduce algorithm implementation is *Hadoop Pipes*. It allows C++ code to use HDFS and MapReduce. According to [3]: „The primary approach is to split the C++ code into a separate process that does the application specific code. In many ways, the approach will be similar to Hadoop streaming, but using Writable serialisation to convert the types into bytes that are sent to the process via a socket." While Hadoop Pipes can be useful for some applications in the multimedia file repository, this method has not yet been investigated.

MapReduce has been used for multimedia data processing before. [8] presents extensive results for bulk conversion of image files with MapReduce, and [9] provides a general analysis of the applicability of MapReduce for video data conversion and estimates the cost of such operations with regard to Amazon's S3 and EC3 services. Another approach is presented in [13], which describes HIPI (the Hadoop Image Processing Interface).

The approach presented in this paper is different in that it describes an architecture for a fully-featured multimedia file repository, which stores multimedia objects (which include but are not limited to images) along with their metadata, provides multimedia processing capabilities and which can be integrated with external systems, both with regard to metadata harvesting (i.e. in a federation of repositories) and as a remote data processing service.

4   Repository Overview and Architecture Description

This section presents the proposed architecture of the multimedia file repository. In Section 4.1, issues related to storing multimedia files in HDFS are described. Section 4.2 discusses both the technical and standards-related aspects of storing metadata. Section 4.3 presents three methods for implementing multimedia file processing algorithms in Hadoop MapReduce. Finally, Section 4.4 describes possible integration of the multimedia file repository with external systems.

4.1   Multimedia File Storage and Retrieval

HDFS is well-suited to storing multimedia files. However, certain precautions need to be taken into account. First, while Hadoop easily deals with a reasonable number of very large files, dealing with very large numbers of small files is a real problem [15]. The block size (default: 64MB) can be adjusted to the expected average file size, but this will not reduce the number of files. Every object (file, directory, block) takes up about 150 bytes of the *name node's* memory. Therefore, storing large numbers (e.g. billions) of files is not feasible with current hardware.

Therefore, if the stored files are in general smaller than the block size, it is best to place them in a single, larger archive. As described in Section 3.2, Hadoop provides two means of accomplishing this, namely *sequence files* and *HAR archives*. By comparison, the aforementioned HIPI Interface [13] utilises its own format (HIB – HIPI Image Bundle), which largely resembles the two former approaches.

For application in the multimedia file repository, sequence files seem to be the more appropriate solution. Therefore, experiments have been conducted using both raw files stored directly in HDFS and sequence files. While generation of sequence files can be time-consuming, it can be performed in parallel and therefore scales rather well (as long as the size of sequence files is kept within reasonable limits). This can be done directly from raw HDFS files, or from intermediate TAR archives [11].

From the purely storage-oriented point of view, direct storage of files in HDFS is the most efficient method, since it does not require the creation of sequence files/HAR archives or any file retrieval interfaces (files can be retrieved directly). The choice of the storage method does however have some impact on processing performance, as described in Section 4.3.

For the repository prototype, two storage models have been chosen:

- direct file storage in HDFS,
- sequence files.

Files can be uploaded directly to HDFS using command-line tools or Hadoop's Java API. Since the repository will have numerous purposes, it has been decided that file storage and retrieval services can be implemented as needs for interoperability arise. More details on interoperability have been presented in Section 4.4.

For the purpose of experiments (see Section 5), sequence files are generated using a MapReduce job in a distributed, parallel manner. First, files are split into batches (in a way to best fit the configured block size), and the HDFS URIs for files in each batch are placed in index files. Then, mapper-only jobs are started with each batch index file as input. These jobs use the `SequenceFileOutputFormat` class for output, and generate the individual output sequence files.

4.2 Metadata and Structured Data Storage

Metadata is data which describes other data. It provides context, content, structure or some other additional information about data. Traditional definition of metadata is "data about data". In digital repositories, all kinds of different metadata are used, including descriptive, structural and administrative metadata. Although experiments described in further sections do not use metadata in any way (their processing due to small size is insignificant), we have decided to give them some attention to provide the reader with a full overview of the proposed solution. Metadata are an inseparable part of every repository solution and need to be taken into account from an early design stage (especially their storage method and format). Firstly, we will focus on descriptive metadata standards, which are used to describe stored objects and their content for fast and efficient searching, browsing and discovery, and then we will describe some possibilities of storing metadata information in solutions based on Hadoop.

The most popular and frequently used metadata schemes and standards for knowledge and science repositories are:

- Dublin Core[6] – a metadata standard used for description of web resource for the purpose of discovery. It's one of the most frequently used metadata standards in digital libraries. For example, it is used by Europeana[7] and the biggest Polish digital library solution dLibra[8].

---

[6] http://dublincore.org/
[7] http://www.europeana.eu/
[8] http://dlibra.psnc.pl/

- MARC 21[9] – MARC (Machine-Readable Cataloging) is an international standard, developed by the Library of Congress for description of bibliographic items. MARC 21 is a family of standards for representation and transmission of bibliographic and related information in machine-readable form. It is mainly used in online cataloging solutions. In Poland it's used by NUKat[10].
- IEEE LOM[11] – IEEE Learning Object Metadata is a data model used to describe learning digital resources, usually in learning management systems.

All of the aforementioned metadata standards only allow for textual description of multimedia objects, which at first may not seem to be the best solution for a multimedia repository. Nonetheless, we decided to use Dublin Core as a primary and mandatory metadata standard for the description of stored digital resources. This choice is based on fact that Dublin Core is used in OAI-PMH, which is one of the most frequently used protocols for metadata exchange between repositories and one of the main possible solutions for interoperability with external systems (for example, it is ued to connect and integrate with Infona[12] – a web portal created as a central access point to polish science and knowledge repositories). Dublin Core is also the preferred metadata model in Federacja Bibliotek Cyfrowych[13] and Europeana. Usage of textual metadata will allow for fast and efficient discovery of stored content. Also, additional metadata schemes strictly dedicated for multimedia content like MPEG-7 descriptors are both allowed and inevitable for advanced functionalities such as query by example. Metadata describing multimedia objects can be stored directly in HDFS or in databases like Hadoop and Casandra, which have been described below.

Hadoop Distributed File System (HDFS) is the basic storage system designed to be use by Hadoop applications. However, Hadoop can also cooperate with database architecture instead of file system. This can be particularly useful in the case of storing small files. HBase (Hadoop Database) and Cassandra are two main databases systems which can be integrated with Hadoop and MapReduce.

HBase (Hadoop Database) an open-source implementation of a distributed database BigTable, presented in [5] by Google. HBase is a NoSQL (non-relational) database and has been optimized for operations typical of MapReduce.

Cassandra is also an open-source distributed database management system. It was originally developed by Facebook and now it is (as is HBase) an Apache Software Foundation project. Similarly, is a non-relational database and since 2011 it has Hadoop integration with MapReduce support.

Despite many similarities, HBase and Cassandra are separate, independent database systems. HBase integration with MapReduce is more complete mostly due to a fact that it is built on HDFS and Hadoop. Cassandra's MapReduce support is limited only to the Java API because it does not support Hadoop

---

[9] http://www.loc.gov/marc/

[10] http://www.nukat.edu.pl/

[11] http://ltsc.ieee.org/wg12/files/IEEE_1484_12_03_d8_submitted.pdf

[12] http://www.infona.pl/

[13] http://fbc.pionier.net.pl/

Streaming. On the other hand, HBase depends on Hadoop's name node, and in case of a name node failure the entire database can (and actually will) become unreachable.

4.3   Multimedia Data Processing

The two storage models outlined in Section 4.1 are being used for image processing experiments. In the first case (with files being stored directly in HDFS), additional text files with HDFS URIs of each file are being generated; they serve as the input for MapReduce jobs.

For the experiment, three processing modes have been selected:

- direct HDFS storage, index files, native Java processing,
- direct HDFS storage, index files, Hadoop Streaming-based processing,
- sequence file storage, native Java processing.

Sequence files have not been tested with Hadoop Streaming, due to limitations described in Section 3.2.

Selected Multimedia Processing Algorithms. Face recognition has been chosen as the primary task for multimedia processing experiments. Automatic face recognition is a method which enables identification of individuals in visual content, e.g. marking actor names in archived video sequences. It requires two interlinked steps: face detection and identification.

The face detection first locates face candidates in an image for further processing. Such methods can be subdivided into two classes. First group uses skin detection algorithms with colour filters and additional shape information. Another methods operate on characteristic face regions such as position of eyes, mouths, nose, cheeks or ears. These features can be extracted in different ways.

The latest version of OpenCV [1] provides face detectors based on a cascaded detection engine which utilises the AdaBoost algorithm. The result of detection is a location of face in image.

Face identification uses information from detection step to build a map of similarities between extracted and known objects (faces). Since objects are described by vectors with possibly correlated values in a multi-dimensional space, different mathematical methods can be applied, e.g. Principal Component Analysis. Eventually some measure is applied to find similarities, e.g. euclidean distance.

Whereas face identification requires database of known faces, face detection is a pure visual processing algorithm and can be efficiently distributed across multiple processing nodes. Details and obtained results are presented in following sections.

Other Multimedia Processing Tools. In a typical scenario, multimedia library customers may require content with various quality depending on the end user platform. To provide the content with various quality, either real-time encoding/decoding must be performed on the server side, or the library must include already encoded/decoded content at an appropriate quality level. The second scenario is usually implemented when the processing power is not suffi-cien. By using the created platform, it is also possible to integrate multimedia library processing with a Hadoop cluster. Depending on the needs, the library policy and the available processing power, both scenarios can be implemented.

4.4   Platform Interoperability

One of the most important goals of the SYNAT project is to create an open and interoperable platform. To ensure that open standards and software solutions can be easily utilised, the main focus was put on technologies based on the Java platform. Java-related interoperability was one more argument to use Hadoop. Additionally, the usage of the Dublin Core metadata model and Open Archives Initiative protocols should allow for quick integration with the platform created by ICM SYNAT team, as wee as future it integration with existing repositories such as Federacja Bibliotek Cyfrowych or Europeana.

5   Conducted Experiments

This section presents the experiments conducted on the repository prototype and discusses the obtained results.

5.1   Experiment Setup

Two types of machines participated in the experiment: the *master* and the *slave*. The *master* is a virtual server created on the VMware ESXi platform, which was also used to prepare the system image for the and the *slave* machines. The Ubuntu Server 12.04 LTS operating system was installed on all machines involved in the experiment.

The system image, prepared on the ESXi *slave* virtual machine, was dis-tributed and copied into hard disk drives of 24 identical computers. Each com-puter had 2 GB RAM, one Intel Core 2 Duo E6550 CPU and a 120 GB Hard Drive. All 24 PCs were directly connected to a Gigabit Ethernet switch; they received IP address and from local DHCP server, hostnames were taken from the DNS. The system image cloned on all 24 PCs allowed for dynamic allocation of the desired number of nodes (PCs) to create the Hadoop cluster. The master node was located on ESXi and was also directly connected to all worker nodes.

Hadoop 1.0.2 was installed on all machines, with appropriate master/slave configurations. The master machine ran the *name node* and the *job tracker*, and each of the slave machines ran a *data node* along with a *task tracker*.

Multimedia processing experiments were based around face recognition on still JPEG files. Datasets of various sizes have been prepared and uploaded to the repository. Sequence files for the datasets have been generated as described in Section 4.3.

Face detection mapper jobs have been prepared using OpenCV 2.4 in two variants:

- native Java jobs, using the JavaCV[14] JNI library,
- C++ programs, using OpenCV natively, launched using Hadoop Streaming.

Due to limitations of the OpenCV library, data passed to sequence file-based mapper jobs had to be saved to a disk and loaded back into memory, but due to the complexity of the face detection procedure, the file save and load time was negligible.

5.2   Basic Scalability Tests

The basic scalability tests served as a proof of the repository concept and were aimed at verifying whether the selected technology and approach can be applied to multimedia processing. The tests involved face recognition on 10 high-resolution (12-megapixel) JPEG photos.

The processing times for each configuration have been shown in Table 1. A plot of the results is shown in Figure 4.

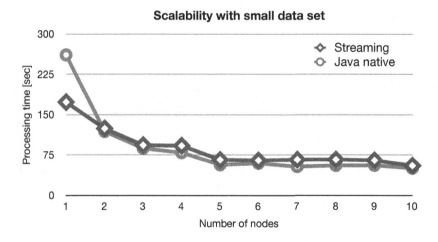

Fig. 4. Processing times for small data set with 1–10 processing nodes

A certain almost constant startup and finish time is present and rises very slowly with the number of processing nodes; it has been estimated to start at around 40 seconds for the Hadoop Streaming version, and 30 seconds for the Java native version. Since the presented results represent the entire job startup, processing and finish time, it must be stated that the infrastructures scales rather well.

---

[14] http://code.google.com/p/javacv/

Table 1. Processing times for small data set with 1–10 processing nodes

Number of nodes	Streaming [s]	Java native [s]
1	172.85	170.23
2	123.77	118.35
3	92.80	87.01
4	91.90	78.97
5	65.66	56.29
6	64.65	59.15
7	65.81	52.86
8	65.99	54.92
9	64.68	55.21
10	54.72	49.95

To further clarify those characteristics, the total CPU load and memory utilisation plots have been presented in Figure 5. Since the number of input files was 10, it is clearly visible that the initial 8 files were being processed by 4 nodes in parallel, and the remaining two were being processed by two nodes, while the other two were idle.

Since sequence files would not make any sense with such a small data set, results for these experiments are not significant and have not been presented.

5.3   Large-Volume Processing

The large-volume processing scenario involved performing face detection on a set of almost 900 high-resolution JPEG files using 10 and 20 nodes, respectively. Since the size of the input files was around 4–7 MB each, they have been split into groups of 20 to obtain sequence files which would fit into two 64 MB HDFS blocks.

These sequence files were then supplied to identical mapper-only jobs using all three processing models (direct storage + index files + Java, direct storage + index files + Hadoop Streaming, sequence files + Java); for details see Section 4.3. The results of the experiments have been presented in Figure 6 and Table 2.

5.4   Discussion of Results

Both the basic scalability and the large-volume processing experiments have shown that the prototype already scales well and is well suited to multimedia data storage and processing. The load distribution among nodes was even, and the processing could take full potential of the available resources.

Since the job startup and finish times are almost constant, they can play a significant role with very small batches; they can even exceed the processing time itself.

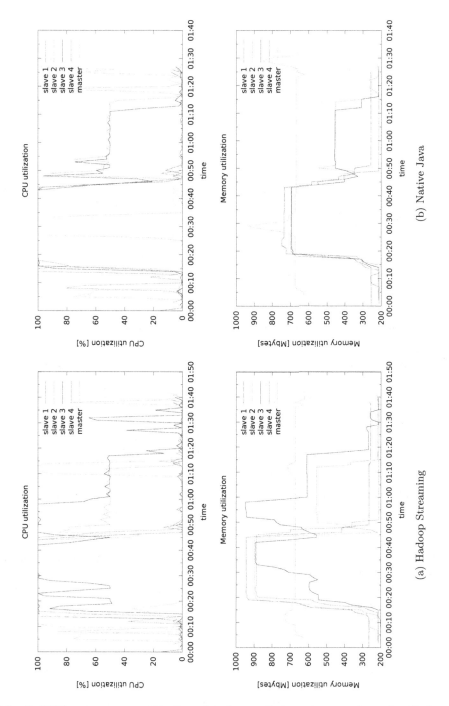

Fig. 5. CPU and memory utilisation plots for small dataset processing using Hadoop Streaming and native Java on 4 nodes

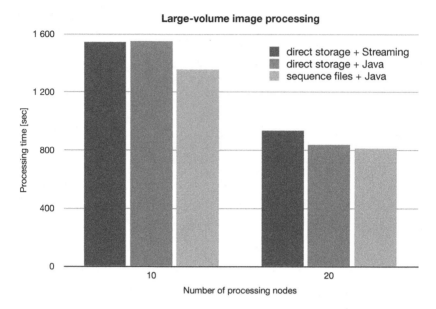

Fig. 6. Processing times for large data set with 10 and 20 processing nodes

Table 2. Processing times for large data set with 10 and 20 processing nodes

Processing mode	Time [s]	
	10 nodes	20 nodes
direct storage + Streaming	1 544.50	934.96
direct storage + Java	1 550.13	838.85
sequence files + Java	1 356.86	812.53

Therefore, the proposed architecture is more useful for high-latency, large-volume processing rather than ad-hoc processing requests for small amounts of data. Such requests should be handled by utilities which retrieve a file from the repository and process it locally.

In addition, large-volume processing experiments (see Section 5.3) have shown that, as predicted, sequence files have a slight performance edge over index file-based approaches. This is primarily because processing tasks can be allocated to nodes which already have given blocks, thus reducing network usage as well as the overhead due to data transfers. However, generation of sequence files also bears a measurable cost, which does not let us choose this approach as the single best option.

Moreover, due to limitations of certain multimedia processing software libraries (such as OpenCV), data needs to be retrieved from HDFS (or saved, if delivered as sequence file) to the processing node's local hard drive before it can be loaded for processing. However, this overhead seems insignificant for computationally-intensive processing jobs.

6   Conclusions and Future Work

The results obtained using a prototype based on the presented approach look promising. Apache Hadoop is very well-suited to storage of large data files, and even rough integration with multimedia processing software has shown good scalability and performance. Moreover, the "large file problem" can be alleviated in several ways.

However, the presented architecture is the first step and will be subjected to further refinement. Firstly, it should be possible to integrate the data processing jobs more tightly with the external libraries to avoid unnecessary storage media writes and reads. Secondly, more extensive tests, including stress tests, with even larger data sets and higher numbers of nodes can be executed to further examine the characteristics of the repository. Finally, certain data processing routines can be integrated more tightly with MapReduce, i.e. they can be broken down into *map* and *reduce* functions rather than being executed as an external process.

References

1. Open CV Open Source Computer Vision (2012),
   `http://opencv.willowgarage.com`
2. SYNAT Project (2012), `http://www.synat.pl`
3. Apache Software foundation: Hadoop 1.0.2 API Documentation (2012),
   `http://hadoop.apache.org/common/docs/r1.0.2/api/`
4. Borthakur, D.: HDFS Architecture Guide, Version 1.0.2 (2012),
   `http://hadoop.apache.org/common/docs/r1.0.2/hdfs_design.html`
5. Chang, F., Dean, J., Ghemawat, S., Hsieh, W.C., Wallach, D.A., Burrows, M., Chandra, T., Fikes, A., Gruber, R.E.: Bigtable: A Distributed Storage System for Structured Data. In: OSDI 2006: Seventh Symposium on Operating System Design and Implementation (2006)
6. Dean, J., Ghemawat, S.: Map Reduce: Simplified Data Processing on Large Clusters. Communications of the ACM 51(1), 1–13 (2008),
   `http://www.usenix.org/events/osdi04/tech/full_papers/dean/dean_html/`
7. Ghemawat, S., Gobioff, H., Leung, S.T.: The Google file system. ACM SIGOPS Operating Systems Review 37(5), 29 (2003),
   `http://portal.acm.org/citation.cfm?doid=1165389.945450`
8. Kim, M., Lee, H., Cui, Y.: Performance Evaluation of Image Conversion Module Based on MapReduce for Transcoding and Transmoding in SMCCSE. In: 2011 IEEE Ninth International Conference on Dependable, Autonomic and Secure Computing, pp. 396–403 (2011),
   `http://ieeexplore.ieee.org/lpdocs/epic03/wrapper.htm?arnumber=6118756`
9. Kiran, G.V.R.: Online Video Conversion Using Map Reduce. Tech. rep., International Institute of Information Technology, Hyderabad
10. Lin, J., Dyer, C.: Data-Intensive Text Processing with MapReduce, 1st edn. Morgan & Claypool Publishers (January 2010)
11. Sierra, S.: A Million Little Files (2008),
    `http://stuartsierra.com/2008/04/24/a-million-little-files`

12. Simon, B., Massart, D., van Assche, F., Ternier, S., Duval, E., Brantner, S., Olmedilla, D., Miklós, Z.: A simple query interface for interoperable learning repositories. In: Proceedings of the 1st Workshop on Interoperability of Web-Based Educational Systems, pp. 11–18 (2005)
13. Sweeney, C., Liu, L., Arietta, S., Lawrence, J.: HIPI: A Hadoop Image Processing Interface for Image-based MapReduce Tasks. Master's thesis, University of Virginia (2011)
14. White, T.: Hadoop: The Definitive Guide. O'Reilly Series, vol. 54. O'Reilly Media (2009)
15. White, T.: The Small Files Problem (2009), http://www.cloudera.com/blog/2009/02/the-small-files-problem/

Low-Level Music Feature Vectors Embedded as Watermarks

Janusz Cichowski, Piotr Czyżyk, Bożena Kostek, and Andrzej Czyżewski

Gdańsk University of Technology, Multimedia Systems Department, Gdańsk,
80-233, Poland
ksm@sound.eti.pg.gda.pl

Abstract. In this paper a method consisting in embedding low-level music feature vectors as watermarks into a musical signal is proposed. First, a review of some recent watermarking techniques and the main goals of development of digital watermarking research are provided. Then, a short overview of parameterization employed in the area of Music Information Retrieval is given. A methodology of non-blind watermarking applied to music-content description is presented. The system architecture for the embedding and recovery of the watermarks, along with the algorithms implemented, are described. The robustness of the watermark implemented is tested against audio file processing, such as re-sampling, filtration, time warping, cropping and lossy compression. Procedures for simulating musical signal alteration are explained with a focus on the influence of lossy compression on the degradation of the embedded watermark. The advantages and disadvantages of the proposed approach are discussed. An outline of future applications of the methodology introduced is also included.

Keywords: Music Parameterization, Feature Vectors, Watermarking Techniques, Music Information Retrieval, Lossy Compression.

1 Introduction

Multimedia services, which provide common access to digital entertainment including games and exchange of multimedia content via Internet communities such as YouTube or Facebook, are becoming increasingly popular and, as a consequence, there are a growing number of scientific researchers involved in the area. These are web services that specialize in multimedia which allow uploading, sharing, managing and downloading data for each user that is logged in. This kind of data access is popular because it is free-of-charge, its distribution is unregulated and the data flow is unrestricted. One of the most common types of data exchange is related to audio content. More and more services provide audio streams instead of audio files. There are also applications which need to operate on audio files, such as the web services developed within the SYNAT project [9][25].

One of the proposed services within the SYNAT project, developed at the Multimedia Systems Department, is the restoration of degraded audio recordings. A number of novel algorithms widely described in the literature [10][11][32], have been

R. Bembenik et al. (Eds.): *Intell. Tools for Building a Scientific Information*, SCI 467, pp. 453–473.
DOI: 10.1007/978-3-642-35647-6_27   © Springer-Verlag Berlin Heidelberg 2013

implemented to enhance the audio quality by automatic sound restoration [8]. One of the goals of our project is to design an open multimedia repository for science, education and an open knowledge community. In general, it is controversial to protect the intellectual property in a repository of open type, however some particular law regulations and legislation may enforce the copyright management procedures. Content stored in such a repository may have different license agreements, thus for example previewing a file may be permissible but its distribution is forbidden. The large amount of multimedia data, in forms such as text, audio or video, requires appropriate management and intellectual property protection. Moreover, much of the content has to take the form of multimedia files instead of online streams.

The created service enables the logged-in user to become acquainted with the practical applications of digital-signal-processing algorithms to reconstruction of audio archives. In order to facilitate the reconstruction procedure, it is necessary to design algorithms for automatic audio-video content quality assessment and restoration, to manage efficiently the archive repository and to automatically search the stored audio files. By management one should understand a watermarked service to protect the website user's audio entries and as well as the restored audio files. On the other hand, search and retrieval process, based on audio parameterization, should make possible to identify various restored versions of the same audio degraded source.

Recent research carried out in the Multimedia Systems Department of the Gdansk University of Technology has been focused on an evaluation of a methodology for copyright protection which would prevent the illegal distribution and counterfeiting of audio data. Providing the data as files requires increased efforts to enforce the copyrights and protect against piracy. The method that is commonly used and is able to improve the security of information in digital multimedia is digital audio watermarking. The idea of watermarking has been adopted in commercial DRM (digital rights management) systems such as Varance [39] or Cinavia [7]. Previous research described in the literature [12] led to the implementation of a watermarking algorithm suitable for controlling illegal distribution and tracking piracy sources. An analysis of the watermarking methods was performed according to pre-existing algorithms based on spread spectrum [5], echo hiding [6], DCT (Discrete Cosine Transform) [13], DWT (Discrete Wavelet Transform) [31] and approaches employing DWT together with SVD (Singular Value Decomposition) [2][3][4][26][40]. The balance between algorithm efficiency, processing time and watermark audibility was achieved using a DWT domain.

The experience gained by the authors in previous research stages provided the motivation to engage the developed algorithms in a different way. The large amount of audio data causes problems with efficient management and retrieval. The more sophisticated web services try to provide upgrades that allow users to browse the audio repository using novel approaches, such as QBE (Query-by-Example) [16], QBH (Query-by-Humming) [30] or QBM (Query-by-Mood) [35]. MIR (music information retrieval) systems transform the time or frequency domain available in audio tracks into semantic fuzzy terms. The repository may also be explored using subjective expressions instead of strictly defined titles, authors, albums or other ID3 tags. The set of available phrases is usually finite, but searching is conducted using semantic keywords. The appropriate categorization allows the assignment of audio tracks to adequate membership groups, which are linked with specific sets of phrases.

The manual classification of audio tracks is time consuming and error prone. In addition, because of high redundancy characterizing audio signals, parameterization is applied to automate classification procedures. Different types of parameters are extracted from audio files and crucial features are organized in a vector. Instead of listening to the music and tagging its content or performing waveform analysis, the feature vector enables the classification of the audio track by employing decision systems or statistical computation.

Therefore, the main objective of the proposed approach is to embed the feature vector as a watermark inside the audio signal. The extracted feature vector allows the retrieval of audio content – only the encapsulation of a watermark is necessary. In general, any modification to a signal causes quality degradation. The presented research, however, focuses on such signal alterations that avoid or minimize unfavorable influence on a signal features. The results obtained during subjective tests are presented in the last section.

The remainder of the paper is organized as follows. An audio parameterization basis developed within the SYNAT project is described in Section 2 [25]. An audio watermarking algorithm in the wavelet transform domain is presented in Section 3. The principles of the non-blind feature vector extraction employing DWT are presented in Section 4. Attacks simulation methodology, comparison of different formats of audio lossy compression and an explanation of possible threats are included in Section 5. The results obtained from the experiments performed are described in Section 6. The final Section contains conclusions regarding future system upgrades to reduce existing disadvantages.

2   Parameterization

Parameterization is a very broad subject, it includes various approaches based on signal processing, but very often related to the specific type of signal (e.g. speech, audio, music, medical signal) to be parameterized. Even within the same type of signal different methods [17][19][21][22][23][24][36][37][38][41] may be employed regardless the fact that for the purpose of efficiently performing parameterization MPEG 7 standard was proposed in 2000 [18][29].

Within the SYNAT project, an attempt has been made to objectively assess whether it is possible to specify a feature vector which would be appropriate for various experimental music databases of different sizes. Firstly, some chosen parameters were extracted from audio signals from two databases. Then, feature vectors, which included these parameters, were created and a thorough statistical analysis was carried out.

The experience gained during the stage of music genre classification may be very useful to transfer this knowledge to the audio restoration archive repository. Especially as, in the current study an idea occurred to contain the feature vector in the watermark included in a music excerpt. This may serve two purposes, a music excerpt will be watermarked for file sharing security and at the same time it will be possible to automatically search using the feature vector extracted from the watermark.

Below, the background of the searching for optimum feature vectors is presented. For this purpose, as mentioned before, two databases were constructed. The first database was prepared for the ISMIS '2011 conference [20], and the second one was constructed with the "help" of a Music Robot, which collected the available music files from the Internet [25]. The second database is practically limited only to physical resources such as the capacity of disks. However, it has been restricted for practical reasons.

The ISMIS database created in the Department of Multimedia Systems contains musical pieces of 60 composers/artists [25]. From the albums within the same style, 15-20 musical pieces of each musician/composer have been chosen. Musical genre categories in the created database were limited to basic styles, such as classical music, jazz, blues, rock, heavy metal and pop. For each piece 25-second-long-excerpts were extracted and parameterized. The excerpts were evenly distributed in time. The parameterization involved extraction of the following parameters: Mel-Frequency Cepstral Coefficients (MFCC - 20 descriptors), 127 descriptors of the MPEG-7 standard, in addition time-related 'dedicated' parameters (24 descriptors). The parameter vector includes 171 descriptors and the total number of parameter vectors in the whole database amounts to 21,680. In addition, it must be added that the parameters were normalized to be in the range of (-1,+1). Since MPEG-7 features and Mel-Frequency Cepstral-Coefficients are widely presented in a rich literature related to this subject there they will only be listed here [42]:

- parameter 1: Temporal Centroid,
- parameter 2: Spectral Centroid average value,
- parameter 3: Spectral Centroid variance,
- parameters 4 - 37: Audio Spectrum Envelope (ASE) average values in 34 frequency bands,
- parameter 38: ASE average value (averaged for all frequency bands),
- parameters 39 – 72: ASE variance values in 34 frequency bands,
- parameter 73 – averaged ASE variance,
- parameters 74,75 – Audio Spectrum Centroid – average and variance values,
- parameters 76,77 – Audio Spectrum Spread – average and variance values,
- parameters 78 – 101 – Spectral Flatness Measure (SFM) average values for 24 frequency bands,
- parameter 102 – SFM average value (averaged for all frequency bands),
- parameters 103 – 126 – Spectral Flatness Measure (SFM) variance values for 24 frequency bands,
- parameter 127 – averaged SFM variance parameters,
- parameters 128 – 147 – 20 first MFCC (mean values),
- parameters 148 – 171 – dedicated parameters in the time domain based on the analysis of the envelope distribution in relation to the RMS value, explained further on.

The dedicated parameters are related to the time domain. They are based on the analysis of the distribution of sound sample values in relation to the root mean square values of the signal (RMS). For this purpose three reference levels were defined: r_1, r_2 and r_3 – equal to namely 1, 2, 3 RMS values of the samples in the analyzed signal frame. The first three parameters are linked to the number of samples that are exceeding the levels: r_1, r_2 and r_3 [42].

$$p_n = \frac{count(samples\_exceeding\_r_n)}{length(x(k))} \tag{1}$$

where $n=1,2,3$ and $x(k)$ is the signal frame analyzed.

The initial analysis of the values of parameters p_n showed to be difficult, because the RMS level in the analyzed excerpts sometimes significantly varies within the frame. In order to deal with this problem another approach was introduced. Each 5-second frame was divided into 10 smaller segments. In each of these segments the p_n parameters (Eq. 2) were calculated. As a result a P_n sequence was obtained:

$$P_n = \{p_n^1, p_n^2, p_n^3, \ldots, p_n^{10}\} \tag{2}$$

where p_n^k, $k= 1 \ldots 10$ and $n=1,2,3$ as defined in Eq. 1.

In this way, 6 new features were defined on the basis of the P_n sequences. New features were defined as the mean (q_n) and variance (v_n) of P_n, $n=1,2,3$. Index n is related to the different reference values of r_1, r_2 and r_3.

$$q_n = \frac{\sum_{k=1}^{10} p_n^k}{10} \tag{3}$$

$$v_n = \mathrm{var}(P_n) \tag{4}$$

In order to supplement the feature description, additional three parameters were defined. They are calculated as the 'peak to RMS' ratio, but in three different ways described below:

- parameter k_1 calculated for a 5-second frame,
- parameter k_2 calculated as the mean value of the ratio calculated in 10 sub-frames,
- parameter k_3 calculated as the variance value of the ratio calculated in 10 sub-frames.

The last group of the dedicated parameters is related to the observation of the threshold crossing rate value. This solution may be compared to a more general idea based on a classical zero crossing rate parameterization (ZCR). The ZCR parameterization is widely used in many fields of automatic sound recognition. The extension of this approach is a definition of a threshold crossing rate value (TCR) calculated analogically as ZCR, but by counting the number of signal crossings in relation not only to zero, but also to the r_1, r_2 and r_3 values. These values (similarly as in the case of the other previously presented parameters) are defined in three different ways: for an entire 5-second frame and as the mean and variance values of the TCR calculated for 10 sub-frames. This gives 12 additional parameters to the feature set [25][42].

The entire set of dedicated parameters consists of 24 parameters that supplement 147 parameters calculated based on MPEG-7 and mel-cepstral parameterization

[25][42]. Parameters r1, r2, r3 indirectly pass such important information as: audio rhythm, tempo and bit per minute (BPM) which describe time–related nature of audio data. The parameters are necessary for music retrieval engines to narrow searchable data set when identifying musical genres.

In the second database, called SYNAT, realized by Gdansk University of Technology (GUT), music was acquired from amazon.com (http://www.amazon.com) by the constructed music robot. Below in Table 1 the effects of the music robot working for ca. 65.5 hours are shown. At that time the database included 30 thousand files (the size of music files was 8991 MB), while now it stores 50 thousand pieces of music.

Table 1. SYNAT music database

No.	Music genre	Number of artists	Number of music excerpts
1	Pop	1498	3920
2	Rock	1008	3159
3	Country	496	1829
4	R&B	648	1788
5	Classical	864	1725
6	Soundtracks	929	1711
7	Rap & Hip-Hop	602	1707
8	Christian & Gospel	603	1634
9	Latin Music	614	1490
10	Children's Music	592	1439
11	Jazz	597	1434
12	Alternative Rock	635	1410
13	Dance & DJ	802	1374
14	International	654	1225
15	New Age	496	1204
16	Folk	417	964
17	Miscellaneous	364	891
18	Opera & Vocal	378	858
19	Blues	337	834
20	Hard Rock & Metal	366	691
21	Broadway & Vocalists	255	378
22	Classic Rock	75	203
	BROADWAY*	2	2
	Sum	13232	31870
	* it seems that two files have been incorrectly assigned the genre (it is BROADWAY should be Broadway & Vocalists)		

The SYNAT database is a set of parameters obtained from the analysis of the excerpts of mp3-quality recordings. The database stores 191-feature vectors, which are in majority the same as those in the ISMIS database. However, their positions within a vector are different. The differences between the databases result from two things. The SYNAT database lacks certain parameters that ISMIS database includes, which is a consequence of the analysis band limited to 8kHz in SYNAT. On the other hand SYNAT has been supplemented with a series of statistical parameters which ISMIS database does not comprise. These statistical parameters are variances of mel-cepstral coefficients. Thus the optimized feature vector served as a basis for further experiments.

3   Feature Vector Watermarking

A vector of parameters, of the length of 191 feature values, 8 bits per value, is loaded and translated into a binary sequence. The feature vector resolution was reduced; a usually used floating point with 32 bits per value was limited to 8 bits per value. The developed watermarking methodology provides low data capacity at the expense of high imperceptibility and robustness. Additional binary markers, at the beginning and at the end of the vector, are generated. The marker function is to separate the repetitive vectors. The length of the watermark binary sequence is 1544 bits in total (1528 bits of information plus 8 bits of start sequence and 8 bits of end sequence). Each feature vector is associated with a specific audio track located in the reference database.

The watermarking process is held in the DWT domain. The original input signal is divided into non-overlapping frames. The size of a frame influences the watermark bitrate. The relationship between bitrate and sample rate is given in Table 2.

Table 2. The dependencies of audio sample rate, frame width and watermark bitrate

	256	512	1024	2048
8 kHz	31.2 bps	15.6 bps	7.8 bps	3.9 bps
11.025 kHz	43.1 bps	21.5 bps	10.8 bps	5.4 bps
16 kHz	62.5 bps	31.2 bps	15.6 bps	7.8 bps
22.05 kHz	86.1 bps	43.1 bps	21.5 bps	10.8 bps
32 kHz	125 bps	62.5 bps	31.2 bps	15.6 bps
44.1 kHz	172.3 bps	86.1 bps	43.1 bps	21.5 bps
48 kHz	187.5 bps	93.7 bps	46.8 bps	23.4 bps
96 kHz	375 bps	187.5 bps	93.7 bps	46.8 bps
192 kHz	750 bps	375 bps	187.5 bps	93.7 bps

When the frame size decreases the bitrate increases. However, an increased concentration of watermarked frames negatively influences the audible quality of recordings. The wider the frame, the fewer artifacts are generated during the watermark-embedding process, but at the cost of a smaller watermark bitrate (also described as the payload of a watermark). The frame size of 1024 samples for a 44.1 kHz sample rate is a compromise between high bitrate and quality degradation. The assumptions allow for achieving a bitrate equal to 43.1 bps, which is highlighted in Table 1. Among the audio signal samples each frame is able to cover one bit of the embedded watermark.

A simplified schema for embedding a watermark in an audio signal is shown in Fig. 1.

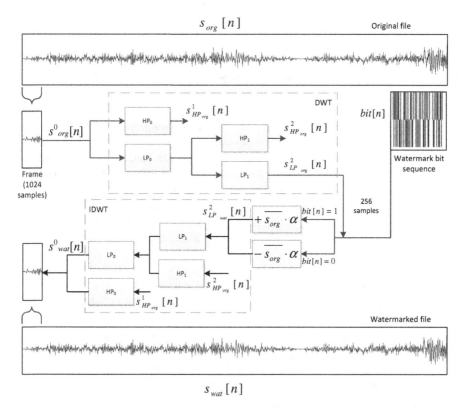

Fig. 1. The process of file watermarking

The embedding of the watermark is realized in the second level low-frequency-DWT-decomposition of the signal. The width of an analysis frame from the time domain is decreased in the DWT domain: with a frame length of 1024 the analysis frame is 256 samples long. The wavelet decomposition is executed using a pair of forward high-pass HP and low-pass LP filters. After signal filtering, both filtered signals are down-sampled. The second level of DWT decomposition is executed only on the down-sampled low-pass signal. The second level of decomposition provides frames containing 256 samples. The values of samples in analysis frames are modified according to the proposed relation (Eq. (5)):

$$bit\ 1\ \Rightarrow\ s^2_{LP_{wat}}[n] = s^2_{LP_{org}}[n] + \overline{s_{org}} \cdot \alpha$$
$$bit\ 0\ \Rightarrow\ s^2_{LP_{wat}}[n] = s^2_{LP_{org}}[n] - \overline{s_{org}} \cdot \alpha$$

(5)

where:

$\overline{s_{org}}$ - average sample value of the entire file

$s^2_{LP_{wat}}[n]$ - output sample value of the DWT second level low-pass component

$s^2_{LP_{org}}$ - original sample value from the current DWT second level low-pass
              component

α - parameter of watermarking strength

The α parameter determines watermarking strength and is inversely proportional to the fidelity of the watermarked file (understood in terms of the identity of the content of the watermarked file in comparison to the source file). The watermark binary sequence is embedded continuously, that is, if the binary sequence is finished it is sequentially repeated to the end of an audio file. In the case of a stereo file, the watermark is embedded in both channels in turn: an odd bit is embedded in the left channel and an even bit in the right channel. If the number of samples at the end of the channel is smaller than the frame size these samples are not watermarked. After the second-level modification, a frame is transformed back into the time domain using two levels of IDWT (Inverse Discrete Wavelet Transform), including recursive LP and HP filters. Then, the filtered signal is up-sampled and added together, thus the resulted signal contains the watermark embedded in the DWT domain.

The last stage of the process is not directly related to the watermarking algorithm but allows the watermark decoder to find the first audio frame of the watermark. The importance of the synchronization procedure demanded the embedding of the additional time domain signature, which makes it possible to find the beginning of a watermark in an audio signal. Audio files are watermarked from the beginning to the end so that the beginning of the file should be the beginning of the watermark. Unfortunately, different modifications (described in detail in Section 5) generate artifacts and produce redundant samples at the ends and beginnings of files. To avoid

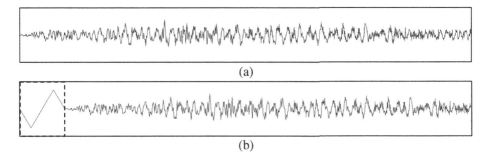

(a)

(b)

Fig. 2. Signal waveforms: a) original, b) original with start marker

blinding the decoder by simple modifications, an additional marker is inserted before the first audio sample. The waveform of the synchronization descriptor is the same for both encoder and decoder, so using signal correlation or mean square error it is possible to define the marker localization among the audio samples. The position of the signature determines the first sample of the watermarked signal. Example waveforms without (a) and with (b) a synchronization signature are presented in Fig. 2.

4   Extraction of Features from a Watermark

The most important condition for the proposed system is that the embedded watermark can be extracted with no errors. If the BER (bit error rate) is too high the feature vector will change significantly and thus losing its main function. The non-blind extraction mechanism [5] requires the availability of the original and the watermarked signals to extract the information hidden in the watermark. The idea of the extraction procedure is presented in Fig. 3.

The non-blind method was chosen to ensure both high imperceptibility and high capacity of the watermark. The watermarking strength and capacity can be maintained independently, but at the expense of audio quality.

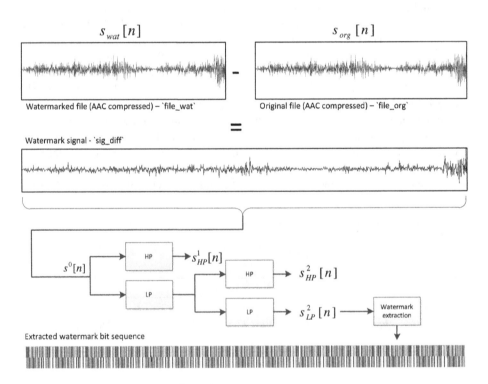

Fig. 3. Watermark extraction process

Watermark extraction is similar to the embedding process. The schema for extraction is shown in Fig. 3. For the extraction of the watermark two files are required: the original uncompressed one (*'file_org'*) and the watermarked file (*'file_wat'*). If the watermarked file is compressed into a different format, then some pre-processing must be done. Moreover, if the files have a different quality a downgrading procedure is required for quality-rate compensation. The *'file_org'* must be compressed to the exact format of *'file_wat'*.

The difference between the low frequency parts of the second level DWT of two files (*'sig_diff'*) represents the watermark signal. The *'file_wat'* is searched for the marker in the time domain, defining the beginning of an actual watermark signal. This ensures that synchronization is maintained even after lossy-compression codecs have added irrelevant data to the beginning of the signal. The *'sig_diff'* is divided into non-overlapping frames containing 256 (because of second-level DWT decomposition) samples each. Two watermark bits are extracted from each frame (one for the left and one for the right channel). If the mean frame value of the *'sig_diff'* is greater than zero the extracted bit equals 1. Otherwise, the extracted bit equals 0. Finally, the binary sequence is saved in a text file for comparison with the binary sequence that was originally embedded.

5   Simulation of Attacks

Many attacks that affect the robustness of the watermark can be performed on the user's side, since the audio file may be processed in many different ways, such as by resampling, filtration, time warping, cropping and lossy compression. All types of unexpected processing can pose a threat to the extraction of a bit sequence hidden in the embedded watermark. The treatment of simple modifications of audio signals as potential watermark attacks has been described in the literature [28][34]. This potential processing is not always aimed at watermark deletion or corruption. Nevertheless, modifications may be dangerous whether they are intentional or unintentional and can lead to watermark degradation.

It is not feasible to embed a watermark that will withstand all types of attacks. Lossy audio compression is the most likely of the attack types in the digital data flow. The lossy compression formats that are commonly used, such as AAC (Advanced Audio Coding) [14], MP3 (MPEG-1/MPEG-2 Audio Layer 3)[27] and VORBIS (Xiph.org Foundation)[33] allow users to modify data size at the expense of data quality. As is shown in Fig. 4, encoding a file to another format influences the watermark. The waveform in Fig. 4 (a) presents a fragment of an extracted watermark signal which represents the difference between the original and the watermarked file in the second- level low-frequency DWT domain [12]. The same procedures of embedding and extracting the watermark were performed for the signal shown in Fig. 4 (b), after applying MP3 lossy compression with a CBR (constant bitrate) equal to 320 kbps. An audio sample containing classical music was used in the experiment.

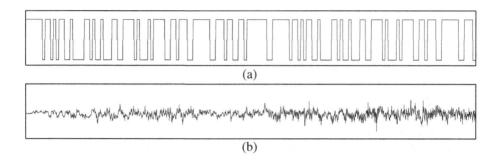

Fig. 4. A watermark signal extracted from: a) raw audio data, b) compressed audio data

The waveforms presented in Fig. 4 are significantly different, but both contain the same fragment of the watermark. The implemented extraction algorithm is able to extract the binary watermark sequence from (a) and (b) waveforms. The raw audio data were obtained from the PCM file; the compressed file was generated from the raw data using MP3 format with 320 kbps bitrate. However, the algorithms had to be tested for different musical genres, lossy compression codecs and for different compression bitrates to ensure the correctness of the watermark extraction.

To simulate realistic situations a large test set is required. The set employed for simulations was generated automatically and included six musical genres (Beat, Blues, Classical, Jazz, Reggae, Rock), three lossy compression formats (AAC, MP3, VORBIS) and ten quality levels for each codec. For each genre there were 9 fragments of music of a duration of 30 seconds each. The recompression procedure was executed using open source C++ libraries, such as Lame [27] for MP3, Faac [14] for AAC and OggVorbis [33] for VORBIS.

Finally, the set of 3240 audio files, including 1620 reference audio files (original files, without watermark, compressed with specific codec and bitrate) and 1620 watermarked audio files, was generated. The large amount of files generated enabled the simulation of potential attacks that is described in the following section.

Resampling attacks change the sampling frequency of the whole audio file and so also affect the embedded watermarks. The frames containing watermark bits are shrunk during the down-sampling and are expanded during up-sampling. The main problem during the extraction of the watermark is de-synchronization and change in frame length. The implemented time-based synchronization method only shows the beginning of the first frame and is insufficient to determine the frame length. For this reason the proposed method cannot withstand resampling attacks.

Time-domain attacks, such as time-warping and cropping, affect some of the watermark frames. The algorithm itself can withstand these kinds of attacks because of the synchronization method implemented. The feature vector used in this study was 1528 bits long, and the maximum capacity of a 30 second watermark is 2582 bits. Because of this only one full watermark sequence was embedded in the files used for the research, and the second iteration contained only 60 % of the sequence. The redundancy would have to be much greater for the watermark to resist time-domain attacks. In the next stage of research, the embedding of a smaller feature vector will be tested against these kinds of attacks. Changing the frame length from 1024 to 512, which would double the data payload and so increase redundancy, will also be tested. The performed experiments were oriented on the audio lossy compression influence on the watermarked data. The system in its current form is not robust in terms of time domain attacks. At present, a functionality of the attack type detection is not implemented in the described system. The detection of the type of an attack that allows reconstructing the watermarking is envisioned for the upgraded version of the systems.

6   Results

There was a large set of test data, including six musical genres with nine example tracks for each. The three different music codecs were employed at ten different output bitrates. The final test set contained 1620 watermarked files together with the same number of reference audio tracks. The large number of audio tracks analyzed meant that comprehensive research could be done into the influence of the audio lossy compression on watermarked features vectors.

These measures were adopted to compare audio signals: PSNR (peak signal to noise ratio) and MSE (mean square error) [1]. In the case of the binary comparison of the original and extracted watermark sequences the most suitable measure is BER (bit error rate). BER shows how many bits have been extracted erroneously. The measure presented is used to analyze the quality of the watermark extraction [15]. It shows the efficiency of the watermarking system and its ability to recover embedded information. The lower the value of BER the less errors there are in the recovered watermark.

The simulations compared watermarked and reference audio files. After two-level DWT, the audio signals were subtracted from each file and the watermark was recovered in the differences. There is little guidance from earlier research [12], which has concerned fingerprinting and user-signature embedding. This requires embedding a watermarked sequence containing five times fewer bits than is the case when watermarking features vectors. The watermarked sequence is able to multiply 1.6 times inside 30 seconds of audio recording. When a watermarked sequence is characterized by low redundancy there is a greater chance of errors occurring.

The most important aspect of the first part of the experiment was to measure the BER for different compression formats with various bitrates for several musical genres. These were: Beat, Blues, Classical, Jazz, Reggae and Rock. The simulations took into consideration various codec bitrates: 64, 80, 96, 112, 128, 169, 192, 224, 256 and 320 expressed in kbps (kilobits per second). The set of nine representative samples was measured and the results obtained were averaged for each genre.

The results obtained during simulations are presented in Fig. 5. The x axes are related to the output-file bitrate. The y axes present the BER value. Where BER is equal to 0, no errors were encountered.

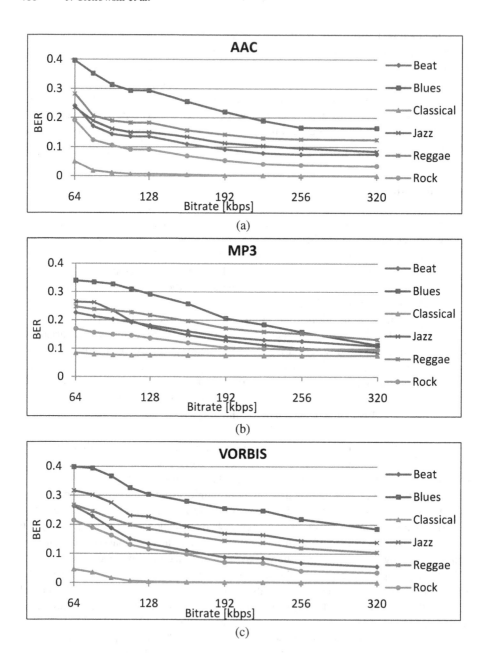

Fig. 5. Bit error rate according to different audio bitrates: a) AAC, b) MP3, c) VORBIS

The first part of the results obtained was expected, the BER decreases together with increasing audio bitrate (audio quality). For each compression standard the lowest BER was observed for classical music and the highest for blues music. The other genres were in the middle. AAC (Fig. 5 (a)) and VORBIS (Fig. 5 (c)) had the widest dynamic range of BER: for 64 kbps BER equals ca. 0.4 for blues and ca. 0.05 for classical, while for 320 kbps BER equals ca. 0.18 for blues and ca. 0.001 for classical. The MP3 format has a narrow dynamic for the BER parameter: for 64 kbps BER equals ca. 0.33 for blues and ca. 0.09 for classical, while for 320 kbps BER equals ca. 0.13 for blues and ca. 0.09 for classical. The BER for classical is constant independent of the bitrate. The BER characteristics adopt a linear trend at a bitrate of 128 kbps for each codec. Optimal BER characteristics without significant fluctuations are observed above 128 kbps, which is an optimal bitrate value. Where the influence of genre on watermark robustness is concerned, some genres, such as classical and rock, are very suitable for watermarking. There are also genres that cause error-prone watermark extraction, such as blues and reggae. The weakest results are obtained employing MPEG-2 Layer 3 format, where no BER value is below 0.075 for classical music, though for other formats for similar bitrate and genre BER is equal to ca. 0.001.

The comprehensive research was carried out according to the different codecs, genres and bitrates. The results presented in Fig. 5 allowed the most suitable codec for audio compression after watermarking to be selected. Desirable musical genres for watermarking were identified. Once the analysis of extraction quality had been completed there was a need to attend to a quantity analysis. The number of vectors extracted with the specific BER are organized in bar charts. The comparison involves a BER equal to 0, which means errorless extraction. Three other values were taken into consideration, including BER less than or equal to 0.1, which means that less than 10 % of bits were extracted erroneously. Watermark sequences extracted with errors of 25 % or less were calculated and, finally, watermarks extracted with BER greater than or equal to 0.4 were analyzed. The bitrates were first analyzed based on the conditions described. The results are shown in Fig. 6.

The results show how many vectors were recovered as binary sequences with an appropriate error rate. The analysis of the number of vectors extracted without any errors (Fig. 6 (a)) was surprising: there was no error-free extraction for the MP3 format even for the best audio quality. The number of properly extracted vectors increases proportionally to bitrate, so that a rate of 320 kbps produces proper extraction of 15 vectors for the VORBIS codec and of 13 for the AAC codec. The error-free extraction rate was slightly better for VORBIS than for AAC independent of bitrate. In summary, the number of properly extracted watermarks was 72 for AAC and 89 for VORBIS. The specific features of the MP3 codec mean that it is not possible to extract watermarked data without errors. The charts for BER ≤ 0.1 (Fig. 6 (b)) and BER ≤ 0.25 (Fig. 6 (c)) are similar for more cases, with AAC recording the best performance and VORBIS the worst. There are several exceptions. At rates of 112 kbps and 128 kbps VORBIS gives better results, while for the highest bitrate, 320 kbps, the MP3 format is the best. The number of extracted vectors with less than 10 % of errors was 262 for AAC, 247 for MP3 and 245 for VORBIS. For BER less than or equal to 0.25 there were 428 vectors for AAC, 417 for MP3 and 370 for VORBIS. The AAC codec allows the greatest number of vectors to be extracted in the middle range of BER. Turning to Fig. 6 (d), there were also vectors extracted with BER greater than or equal to 0.4, while only 13 vectors were extracted for AAC, 22 for

MP3 and 38 for VORBIS. As we have already stated, AAC delivers the best results, MP3 is in the middle and VORBIS is the poorest performer. Meanwhile, at BER equal to 0 VORBIS was the best, but for other conditions it was worse than AAC and MP3.

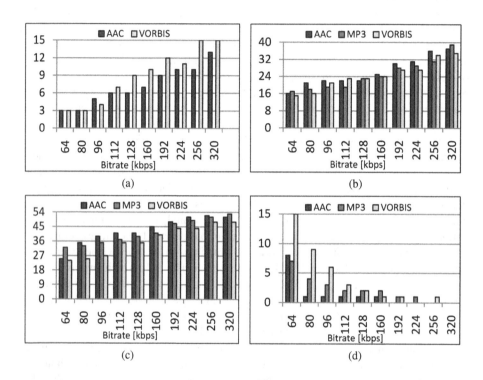

Fig. 6. Number of extracted vectors according to variety of bitrates with specific error rate conditions: a) BER = 0.0, b) BER ≤ 0.1, c) BER ≤ 0.25, d) BER ≥ 0.4

A quantity analysis regarding the different musical genres was performed. The number of vectors extracted with specific BER conditions inside specific musical genres was calculated. The results of the analysis are presented in Fig. 7.

Fig. 7 (a) presents the numbers of vectors extracted without any errors from files representing various musical genres. The highest number of vectors were extracted properly from audio files with classical music: for AAC there were 48 and for VORBIS 57. Turning to the other genres, rock permitted the extraction of 13 vectors for AAC and of 12 vectors for VORBIS, while from the reggae genre there were 11 for AAC and 13 for VORBIS. There were also 7 vectors extracted from the beat genre after VORBIS compression. There was no error-free vector extraction for the jazz and blues genres. The results for BER ≤ 0.1 (Fig. 7 (b)) and BER ≤ 0.25 (Fig. 7 (c)) were similar: the highest number of vectors was extracted from audio files representing the classical genre, followed by rock, beat, reggae, jazz and blues. The number of vectors

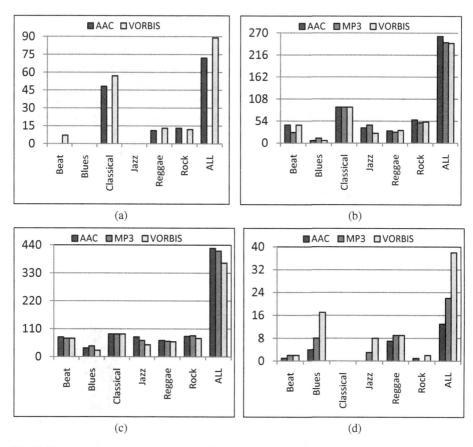

Fig. 7. Number of extracted vectors according to various musical genres with specific error rate conditions: a) BER = 0.0, b) BER ≤ 0.1, c) BER ≤ 0.25, d) BER ≥ 0.4

extracted with BER greater than or equal to 0.4 (Fig. 7 (d)) was 0 for the classical genre in each test case. For each compression format the highest number of extracted vectors with BER ≥ 0.4 was for blues, reggae and jazz. There was also a small number of erroneously extracted vectors from the beat and rock genres.

The results allowed us to identify a lossy audio compression format that does not cause watermark degradation and is therefore the most suitable. It was AAC. The analysis of the musical genres allowed us to specify the features that are necessary during watermark embedding for each of them. The audio files representing the classical or rock genres can be watermarked employing a small α factor, which determines the watermarking strength. When watermarking genres such as blues, reggae, jazz and beat a higher watermarking strength is required. It is necessary to balance the α parameter and the audibility of the watermark.

Signal degradation caused by the watermarking procedure was tested subjectively. A group of experts had to choose a better audio recording from a pair of audio samples, i.e. the original and the watermarked track. There were 15 experts, and two test series were performed to enhance a reliability of the performed experiment. Each series contained six pairs of audio samples corresponding to specific music genres. The obtained results are presented in Fig. 8. If the watermarked file was assessed as better than the original one, then one point was added to the pool, otherwise one point was subtracted. The final result for specific genre shows differences between the compared audio tracks. A positive value means that the watermarked file was assessed better than the original file, while a negative value proves that the watermark was audible and detected by the expert, a value close to zero means that the watermark was not detected and the audio files were comparable.

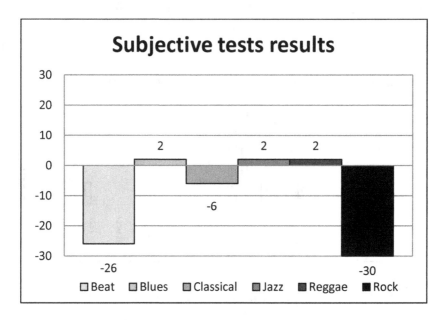

Fig. 8. Results obtained during subjective audio tests

The obtained results demonstrate that watermarking generates distortions. They differ, however, for music genres. Degradation of quality caused by the watermarking procedure was the highest for Rock, then for Beat and finally for Classical genres. Generally, experts have not observed any degradations in Blues, Jazz and Reggae. These differences depend on genre and the dynamic range of a track. Audible distortions were caused by watermarking because this type of modification produces the overdrive of the audio signal.

7   Conclusions

The proposed watermarking approach has been developed as a solution for multimedia content in the exploration, classification and categorization of digital repositories. The proposed approach is focused on audio-content processing, but a similar approach may be employed for other types of content.

The watermarking algorithm is able to embed a watermark in audio domains, but we should remember that watermark extraction is as necessary as watermark embedding. Information covered as a watermark allows for the recovery of feature vectors. The quality of watermark extraction influences the proper categorization of audio content. A large number of errors leads to wrong categorization, which is then of little use. It is necessary to optimize the watermarking parameters for the transmission channels. If the lossy audio compression format can be predicted and other factors, such as distortion, accounted for, then the negative influence of the transmission channels on the watermarked data can be reduced.

It is necessary to undertake comprehensive research into different types of modification, such as filtration, resampling, down-mixing, multiple recompression and D/A → A/D conversion. The test set was nevertheless sufficiently large to estimate the behavior of real life systems. There were a representative number of musical genres and several audio samples were included for each of them. Subjective tests show that watermarking generates distortions, however they differ for music genres. Generally, experts have not observed any degradations in Blues, Jazz and Reggae. Though detectable for other genres, they didn't disturb much the retrieving of watermarking.

Further improvements should also be implemented. In the existing system the binary sequence has no correction codes to avoid the occurrence of errors. The non-blind extraction is also inefficient: the reference audio track has to be found and, in using it, the watermark is extracted. The upgrades and modification are aimed at blind extraction.

Acknowledgements. This research was conducted and partially founded within project No. SP/I/1/77065/10, 'The creation of a universal, open, repository platform for the hosting and communication of networked knowledge resources for science, education and an open knowledge society', which is part of the Strategic Research Programme, 'Interdisciplinary systems of interactive scientific and technical information' supported by the National Centre for Research and Development (NCBiR) in Poland.

References

1. Al-Haj, A.M.: Advanced Techniques in Multimedia Watermarking: Image, Video and Audio Applications, New York (2010)
2. Al-Haj, A., Mohammad, A.: Digital Audio Watermarking Based on the Discrete Wavelets Transform and Singular Value Decomposition. European Journal of Scientific Research 39(1), 6–21 (2010)
3. Lei, B.Y., Soon, I.Y., Li, Z.: Blind and Robust Audio Watermarking Scheme Based on SVD–DCT. Signal Processing 91(8), 1973–1984 (2011)
4. Bhat, K.V., Sengupta, I., Das, A.: A New Audio Watermarking Scheme Based on Singular Value Decomposition and Quantization. Multimedia Tools and Applications 52(2-3), 369–383 (2011)

5. Bloom, J.A., Cox, I.J., Fridrich, J., Kalker, T., Miller, M.L.: Digital Watermarking and Steganography, Boston (2008)
6. Ciarkowski, A., Czyżewski, A.: Performance of Watermarking-Based DTD Algorithm under Time-Varying Echo Path Conditions. Intelligent Interactive Multimedia Systems and Services 6, 69–78 (2010)
7. Cinavia website, http://www.cinavia.com/
8. Czyżewski, A., Kostek, B., Kupryjanow, A.: Automatic Sound Restoration System - Concepts and Design. In: International Conference on Signal Processing and Multimedia Applications, pp. 1–5 (July 2011)
9. Czyżewski, A., Kupryjanow, A., Kostek, B.z.: Online Sound Restoration for Digital Library Applications. In: Bembenik, R., Skonieczny, L., Rybiński, H., Niezgodka, M. (eds.) Intelligent Tools for Building a Scient. Info. Plat. SCI, vol. 390, pp. 227–242. Springer, Heidelberg (2012)
10. Czyżewski, A., Kostek, B., Kupryjanow, A.: Online Sound Restoration for Digital Library Applications. In: SYNAT Workshop - Post-Conference Event by the International Symposium on Methodologies for Intelligent Systems (ISMIS 2011), Warsaw, Poland (July 2011)
11. Czyżewski, A., Maziewski, P., Kupryjanow, A.: Reduction of parasitic pitch variations in archival musical recordings. Signal Processing 90, 981–990 (2010)
12. Czyżyk, P., Cichowski, J., Czyżewski, A., Kostek, B.z.: Analysis of Impact of Lossy Audio Compression on the Robustness of Watermark Embedded in the DWT Domain for Non-blind Copyright Protection. In: Dziech, A., Czyżewski, A. (eds.) MCSS 2012. CCIS, vol. 287, pp. 36–46. Springer, Heidelberg (2012)
13. Dutta, M.K., Gupta, P., Pathak, V.K.: Perceptible Audio Watermarking for Digital Rights Management Control. In: 7th International Conference on Information, Communications and Signal Processing, vol. 1, pp. 55–59 (2009)
14. Faac/Faad (MPEG 2/4 AAC encoder/decoder library) project website, http://www.audiocoding.com/
15. Furht, B., Kirovski, D.: Multimedia Encryption and Authentication Techniques and Applications, Florida (2006)
16. Helen, M., Lahti, T.: Query by Example Methods for Audio Signals. In: 7th Nordic Signal Processing Symposium, pp. 302–305 (June 2006)
17. Holzapfel, A., Stylianou, Y.: Musical genre classification using nonnegative matrix factorization-based features. IEEE Transactions on Audio, Speech, and Language Processing 16(2), 424–434 (2008)
18. Hyoung-Gook, K., Moreau, N., Sikora, T.: MPEG-7Audio and Beyond: Audio Content Indexing and Retrieval. Wiley & Sons (2005)
19. Jang, D., Jin, M., Yoo, C.D.: Music genre classification using novel features and a weighted voting method. In: ICME, pp. 1377–1380 (2008)
20. Kostek, B.z.: Music Information Retrieval in Music Repositories. In: Skowron, A., Suraj, Z. (eds.) Rough Sets and Intelligent Systems - Professor Zdzisław Pawlak in Memoriam. ISRL, vol. 42, pp. 463–489. Springer, Heidelberg (2013)
21. Kostek, B.: Perception-Based Data Processing in Acoustics. In: Applications to Music Information Retrieval and Psychophysiology of Hearing. SCT. Springer, Heidelberg (2005)
22. Kostek, B.: Soft Computing in Acoustics, Applications of Neural Networks. In: Fuzzy Logic and Rough Sets to Musical Acoustics. STUDFUZZ, Physica Verlag, Heildelberg (1999)

23. Kostek, B., Czyzewski, A.: Representing Musical Instrument Sounds for their Automatic Classification. Journal Audio Engineering Society 49, 768–785 (2001)
24. Kostek, B., Kania, Ł.: Music information analysis and retrieval techniques. Archives of Acoustics 33(4), 483–496 (2008)
25. Kostek, B., Kupryjanow, A., Zwan, P., Jiang, W., Raś, Z.W., Wojnarski, M., Swietlicka, J.: Report of the ISMIS 2011 Contest: Music Information Retrieval. In: Kryszkiewicz, M., Rybinski, H., Skowron, A., Raś, Z.W. (eds.) ISMIS 2011. LNCS(LNAI), vol. 6804, pp. 715–724. Springer, Heidelberg (2011)
26. Lalitha, N.V., Suresh, G., Sailaja, V.: Improved Audio Watermarking Using DWT-SVD. International Journal of Scientific and Engineering Research 2(6), 1–7 (2011)
27. Lame (MPEG Audio Layer 3 encoder/decoder library) project website, http://lame.sourceforge.net/
28. Lang, A., Dittmann, J., Spring, R., Vielhauer, C.: Audio Watermark Attacks: from single to profile attacks. In: Proceedings of the 7th Workshop on Multimedia and Security, New York (2005)
29. Lindsay, A., Herre, J.: MPEG-7 and MPEG-7 Audio - An Overview, vol. 49(7/8), pp. 589–594 (2001)
30. Lu, L., Saide, F.: Mobile ringtone search through query by humming. In: IEEE International Conference on Acoustics, Speech and Signal Processing, pp. 2157–2160 (March/April 2008)
31. Maha, C., Maher, E., Chokri, B.A.: A blind audio watermarking scheme based on Neural Network and Psychoacoustic Model with Error correcting code in Wavelet Domain. In: Third International Symposium on Communications, Control and Signal Processing, pp. 1138–1143 (2008)
32. Maziewski, P., Kupryjanow, A., Czyżewski, A.: Drift, Wow and Flutter Measurement and Reduction in Shrunken Movie Soundtracks. In: 124th Audio Engineering Society, Amsterdam, Holland (May 2008)
33. OggVorbis (VORBIS encoder/decoder library) project website, http://www.vorbis.com/
34. Petitcolas, F.A.P., Anderson, R.J., Kuhn, M.G.: Attacks on Copyright Marking Systems. In: Aucsmith, D. (ed.) IH 1998. LNCS, vol. 1525, pp. 218–238. Springer, Heidelberg (1998)
35. Plewa, M., Kostek, B.: Creating Mood Dictionary Associated with Music. In: 132nd Audio Engineering Society, Budapest, Hungary (April 2012)
36. Schierz, A., Budka, M.: High–Performance Music Information Retrieval System for Song Genre Classification. In: Kryszkiewicz, M., Rybinski, H., Skowron, A., Raś, Z.W. (eds.) ISMIS 2011. LNCS(LNAI), vol. 6804, pp. 725–733. Springer, Heidelberg (2011)
37. The International Society for Music Information Retrieval / International Conference on Music Information Retrieval, http://www.ismir.net/
38. Tzanetakis, G., Essl, G., Cook, P.: Automatic musical genre classification of audio signals. In: Proc. Int. Symp. Music Information Retrieval, ISMIR (2001)
39. Varence website, http://www.verance.com/
40. Vongpraphip, S., Ketcham, M.: An Intelligence Audio Watermarking Based on DWT-SVD Using ATS. In: Global Congress on Intelligent Systems, GCIS 2009, pp. 150–154 (May 2009)
41. Zwan, P., Kostek, B.: System for Automatic Singing Voice Recognition. Journal Audio Engineering Society 56(9), 710–723 (2008)
42. Zwan, P., Kostek, B., Kupryjanow, A.: Automatic Classification of Musical Audio Signals Employing Machine Learning Approach. In: 130th Audio Engineering Society, London, May 13-16 (2011)

Intelligent Control of Spectral Subtraction Algorithm for Noise Removal from Audio

Andrzej Czyżewski

Gdansk University of Technology, Multimedia Systems Department
ul. Narutowicza 11/12, 80-233 Gdansk, Poland
ac@pg.gda.pl

Abstract. In this paper 'soft computing' algorithms for audio signal restoration are considered in regard to a practical digital sound library application. The methods presented are designed to reduce empty channel noise, being applicable to the restoration of noisy audio recordings. The audio signal is processed iteratively by a noise-reduction algorithm based on an intelligent comparator, improving the signal-to-noise ratio slightly at each iteration. At each time step, a fuzzy reasoning algorithm processes two values representing spectral power density estimates considered as linguistic variables. We describe a comparator module based on a neural network which approximates the distribution representing a non-linear function of spectral power density estimates. We demonstrate experimentally that the methods examined may produce meaningful noise reduction results without degrading the original sound fidelity. They have been applied to a practical Internet-based sound library (http://www.youarchive.net).

1 Introduction

The YouArchive.net service is a technology demonstrator employing archival sound restoration methods developed within the SYNAT project [2]. The demonstrator implements many algorithms to restore the quality of audio recorded on older media, including several conceived by the author [3][4][5][6][10][11]. This paper specifically addresses the spectral subtraction method applied to noise removal from old audio recordings.

There is much research in the area of sound restoration and noise reduction. For example, Godsill & Rayner [7] or Yektaerian & Amirfattahi [13] and Read & Meyer [12] investigated the problem in the context of old records, archival tapes and motion film restoration. Typically the problem of noise estimation in archival recordings is approached by employing advanced statistical methods [1][7]. However, benefits related to applying computational intelligence to this task can also be exploited [9][11].

An application was developed to simplify the process of acquiring and uploading audio material for processing – setting aside server-side components. The application, called YouArchive Recorder, is portable (written in Java programming language) and seamlessly usable from the user's web browser by means of Java WebStart

R. Bembenik et al. (Eds.): *Intell. Tools for Building a Scientific Information*, SCI 467, pp. 475–488.
DOI: 10.1007/978-3-642-35647-6_28 © Springer-Verlag Berlin Heidelberg 2013

technology. The Online Sound Restoration System concept and principles were described in an earlier paper [2], so details of it will not be discussed here.

This system implements various algorithms for archival audio restoration. Nevertheless, it is worth mentioning that it features a typical "wizard" user interface which guides the user through the less obvious steps, starting with the creation of the recording profile which describes the source of the recording, e.g., turntable, cassette player, reel player, pre-recorded audio file (Fig. 1); and the audio interface used to acquire the sampled data (Fig. 2).

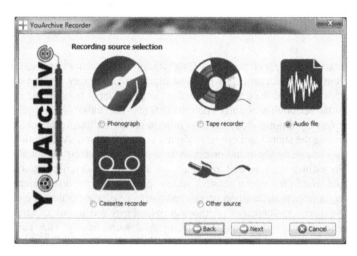

Fig. 1. Archival sound library system - recording source selection

The recording profile is saved in the application settings, stored locally on the user's computer or remotely in the user's private folder on the YouArchive.net website. The next step is the identification of the audio material being recorded. The user may provide the name of the artist, album title and the titles of the tracks. These metadata may be enriched by performing a search of an online music database similar to the CDDB system present in some media players [14]. The number and duration of the tracks is used to aid the recording, post-processing and segmentation of the tracks.

Depending on the choice of the signal source defined in the recording profile, the user is presented with an option to record the audio material using a familiar tape-recorder-like interface with Record, Pause and Stop buttons, or to select a pre-recorded audio file. After the recording is done or the file selected, its post-processing is performed, including the track segmentation. The metadata entered by the user or obtained from an online database are attached to the tracks, which are now ready to be uploaded to the YouArchive.net website. In order to accomplish this, the user is expected to set the privacy level of the recorded tracks and enter his credentials (which will be stored in encrypted form in application settings). The credentials are verified over a secure connection prior to upload by the server-side components. After the tracks are uploaded, the server-side component indexes them, adding references to the audio files and associated metadata to a database, and the automatic restoration process is started. Upon its completion, the user receives an e-mail notification with a link to the reconstructed tracks. This concludes the wizard operation, as the uploaded tracks are now visible within the YouArchive.net service and may be managed through the browser-based interface. In Fig. 3 the processes invoked by the user are depicted together with the software tools applied to satisfy their demands.

The application constitutes a practical sound archive system to which many algorithms can be applied in order to remove noise and distortions. A selection of algorithms that apply soft computing to the removal of broadband noise are now discussed. Although signal reconstruction already has quite a long history, the use of soft computing and algorithms from the domain of artificial intelligence can still induce innovative methods with interesting results.

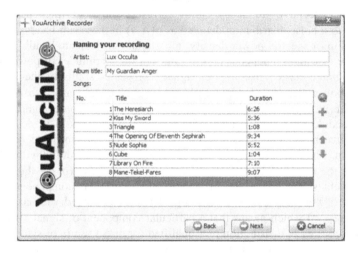

Fig. 2. Part of the archival recordings library system - recording identification

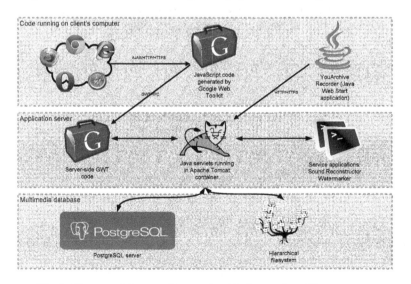

Fig. 3. Diagram of front-end layer of archival recordings library system

2   Intelligent Control of the Spectral Subtraction Algorithm

A general schema of the elaborated noise reduction method is presented in Fig. 4. The method uses spectral subtraction controlled by a learning algorithm. The algorithm being discussed is an extension to these methods through application of intelligent methods [9].

Two kinds of signal appear at the output of the system: $\tilde{s}$ - empty channel noise, and noise-affected output signal $y = x + s$, where x is the original signal unaffected by noise and s is the noise affecting the signal recorded on an old medium. It is assumed that signals x and s are correlated to some extent. An extracted noise pattern is analysed in consecutive frequency bands corresponding to critical hearing bands which leads to a noise distribution denoted as $\rho(A, t)$ in these sub-bands. In fact, it is the distribution of spectral power density in particular critical sub-bands. For a given sub-band, the distribution depends on the values of the amplitudes of power spectrum A, at time moment t. The noise-affected signal is processed similarly; the distributions of spectral power density of a noise-affected signal $\sigma(A, t)$ are also evaluated in consecutive critical sub-bands. Then, distributions ρ and σ are compared in the reasoning module acting as an intelligent comparator. Its task is to detect implicit relations between the two distributions, i.e., to define the measure of similarity between them. The comparator outputs a vector of coefficients defining the degree of similarity of the two distributions. The vector is denoted as $p(\rho, \sigma)$; the greater the coefficient values, the higher the probability that particular components of signal $y = x + s$ are of noise type. That is why signal y is then filtered with compliance to distribution $p(\rho, \sigma)$ which means that a particular component of signal $s + n$ is either decreased according to distribution p or qualified for direct use in re-synthesis (based on distribution $1 - p$). Consequently, the desired signal (the estimate of a non-affected signal) appears at the output of the system.

In the method described in this paper, an audio signal may be processed repeatedly by the noise-reduction algorithm, improving the signal-to-noise ratio with each iteration.

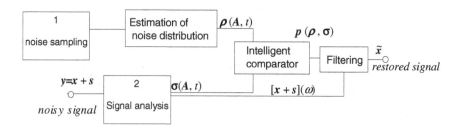

Fig. 4. General block schema of noise reduction system

Denotations: y – noisy signal, $\tilde{x}$ - original signal estimate, $[x+s](\omega)$ – spectrum of noisy signal composed of original signal x and noise s, $\rho(A,t)$ - distribution of spectral power density of noise, $\sigma(A,t)$ - distribution of spectral power density of noise affected-signal, $p(\rho,\sigma)$ — distribution defining signals similarity

2.1   Method Description

Let us assume, that a useful signal $x(k)$ is distorted by additive noise $s(k)$. Thus, the resulting noise-affected signal may be described in the time domain as:

$$y(k) = x(k) + s(k) \tag{1}$$

After the transform to the domain of frequency is done using the DFT algorithm (Discrete Fourier Transform), the above relation adopts the following form:

$$Y(\omega) = X(\omega) + S(\omega) \tag{2}$$

In order to obtain an expression defining the useful signal $X(\omega)$, a formula may be written as:

$$X(\omega) = Y(\omega) - S(\omega) = \left(1 - \frac{S(\omega)}{Y(\omega)}\right) \cdot Y(\omega) \tag{3}$$

However, in practice, it is very difficult to obtain an exact description of the distorting signal $S(\omega)$ in an explicit form. That is why estimate $\tilde{S}(\omega)$ of noise $S(\omega)$ must be used. Then, the only resulting signal is an estimate $\tilde{X}(\omega)$ of the original useful signal $X(\omega)$, i.e.:

$$S(\omega) \rightarrow \tilde{S}(\omega) \Rightarrow X(\omega) \rightarrow \tilde{X}(\omega) \tag{4}$$

Thus, the estimated useful signal $\tilde{X}(\omega)$ may be expressed as:

$$\tilde{X}(\omega) = \left(1 - \frac{\tilde{S}(\omega)}{Y(\omega)}\right) \cdot Y(\omega) = H(\omega) \cdot Y(\omega) \tag{5}$$

where: $H(\omega)$ — transfer function of noise reduction system,
    $Y(\omega)$ — spectrum of useful signal x affected by noise s

As a result of filtration, subsequent spectrum lines of a signal being reconstructed may be decreased by a defined value, or left unchanged. The subtraction is carried out in compliance with the distribution $p(\tilde{S}(\omega), Y(\omega))$, to some extent reflecting the degree of similarity between signals x and s, which can be expressed as:

$$H(\omega) = f(\tilde{S}(\omega), Y(\omega)) = \begin{cases} p_i & \text{according to distribution}: \ p(\tilde{S}(\omega), Y(\omega)) \\ 0 & \text{according to distribution}: \ 1 - p(\tilde{S}(\omega), Y(\omega)) \end{cases} \tag{6}$$

where: p_i — is the value subtracted from the amplitude of the i-th spectrum line that was obtained through distribution $p(\tilde{S}(\omega), Y(\omega))$.

The signal $\tilde{S}(\omega)$ is only an estimate of the distorting signal, thus it is very difficult to explicitly define the distribution $p(\tilde{S}(\omega), Y(\omega))$, especially in cases when noise $S(\omega)$ is non-stationary. In order to evaluate the distribution $p(\tilde{S}(\omega), Y(\omega))$ without making a *priori* assumptions about signal or noise distributions, reasoning based on learning algorithms is applied.

2.2   Noise Distribution Estimation

A measure of noise content in the mth band may be expressed by ratio ρ of the number of power spectrum lines representing the empty channel noise to the total number of power spectrum lines in this band. At the time moment t_j, the discussed ratio may be calculated with the following formula:

$$\rho(A,t) = \frac{n(A,t)}{X}\bigg|_{t=t_j} \tag{7}$$

where: $n(A,t)$ — the number of noise components of power A,

X — total number of spectrum lines in a given critical band

In the case of digital signals, however, the values of power and time are discrete. Thus, in the kth moment of time the following expressions are the equivalents of formula (7):

- for non-stationary noise:

$$\rho(\Delta_i,t_j) = \frac{n(\Delta_i,t_j)}{X}\bigg|_{j=k}; \quad i=1,...,I \tag{8}$$

- for stationary noise:

$$\rho(\Delta_i) = \frac{\overline{n}(\Delta_i)}{X}; \quad i=1,...,I \tag{9}$$

where: Δ_i – is set of power values for lines from subband: $[(i-1)\cdot\Delta, (i+1)\cdot\Delta)]$,

Δ - is the width of the elementary subband,

$n(\Delta_i,t_j)$ - number of noise-affected spectrum lines which have power from i-th subband Δ_i,

$\overline{n}(\Delta_i)$ - average number of noise-affected spectrum lines which have power from the i-th subband Δ_i,

I - number of subbands Δ_i, the whole power density band is divided into. For the maximum line value MAX in the m-th critical band we may define the following relation:

$$\lceil MAX \rceil = \Delta \cdot I \tag{10}$$

where: $\lceil \cdot \rceil$ - *entier* operation

In experiments, value Δ falling within the subband from $\Delta=2$ to $\Delta=16$ were taken.

Distributions of the power density of noise expressed by formula (7) presented in Fig. 5 satisfy the following relation:

$$\sum_{i=1}^{I} \rho(\Delta_i, t_j)\bigg|_{j=k} \le 1; \quad j, k = 1, 2, \ldots \tag{11}$$

The estimation of the current signal distribution is carried out analogously. Thus, for a discrete form of a noise-affected signal, the following relation is used:

$$\sigma(\Delta_i, t_j) = \frac{n(\Delta_i, t_j)}{X}\bigg|_{j=k}; \quad i = 1, 2, \ldots, I; \quad j, k = 1, 2, \ldots \tag{12}$$

where: $n(\Delta_i, t_j)$ — the number of spectrum lines of a noise-affected signal with power from the i-th subband Δ_i.

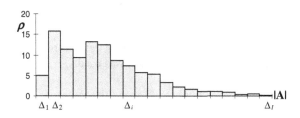

Fig. 5. Distribution of spectrum power density ρ in a single critical band within a defined time interval

2.3 Application of Intelligent Algorithms

As already mentioned, intelligent algorithms were applied to define the current distribution $p(\tilde{S}(\omega), Y(\omega))$. The estimation of this distribution was done by an intelligent comparator (as in Fig. 4). Signals ρ and σ were supplied to its input, and the comparator output distribution $p(\rho, \sigma)$. The implementation of the intelligent comparator whose operation is based on fuzzy reasoning and neural processing is briefly discussed below.

Let us consider the mth critical band and the ith sub-band of spectral power density Δ_i in this sub-band. Assume the following denotations:

a) ρ – spectral power density of noise estimated according to formula (7),
b) σ – spectral power density of noise estimated according to formula (12),
c) p – measure defining the degree of similarity between signals supplied to the comparator at its two inputs.

2.3.1 Application of Fuzzy Reasoning

At any moment of time, two values are supplied to the inputs of intelligent comparator, and it generates one value at its output. The values may be considered as linguistic variables. Thus, the input linguistic variables are: ρ and σ, and the output linguistic variable is p. Fuzzy sets of attributes for particular variables are in all cases denoted as (*Low*, *Medium*, *High*) and defined by membership functions $\mu(\rho)$, $\mu(\text{current})$, $\mu(p)$. The shapes of these membership functions are presented in Fig. 6. The comparator operates on a set of heuristic rules, presented in Table 1.

Table 1. Set of rules of a reasoning module based on fuzzy logic

1. **IF** ρ **is**	*Low*	**AND** σ **is**	*Low*	**THEN** p	**is**	*High*
2. **IF** ρ **is**	*Low*	**AND** σ **is**	*Medium*	**THEN** p	**is**	*Medium*
3. **IF** ρ **is**	*Low*	**AND** σ **is**	*High*	**THEN** p	**is**	*Low*
4. **IF** ρ **is**	*Medium*	**AND** σ **is**	*Low*	**THEN** p	**is**	*Medium*
5. **IF** ρ **is**	*Medium*	**AND** σ **is**	*Medium*	**THEN** p	**is**	*High*
6. **IF** ρ **is**	*Medium*	**AND** σ **is**	*High*	**THEN** p	**is**	*High*
7. **IF** ρ **is**	*High*	**AND** σ **is**	*Low*	**THEN** p	**is**	*Low*
8. **IF** ρ **is**	*High*	**AND** σ **is**	*Medium*	**THEN** p	**is**	*Medium*
9. **IF** ρ **is**	*High*	**AND** σ **is**	*High*	**THEN** p	**is**	*High*

To map the output fuzzy set P to an exactly defined real value p_i properly using defuzzification, the centre of gravity method was used:

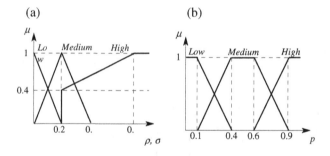

(a)   (b)

Fig. 6. Experimentally evaluated shapes of membership functions for: (a) input variables ρ and σ, (b) output variable p

$$p_i\left(\tilde{S}(\omega), Y(\omega)\right) = \frac{\int\limits_{P} \mu_p\left(p, \tilde{S}(\omega), Y(\omega)\right) \cdot p \, dp}{\int\limits_{P} \mu_p\left(p, \tilde{S}(\omega), Y(\omega)\right) dp} \tag{13}$$

that is:

$$p_o\left(\tilde{S}(\omega), Y(\omega)\right) = \frac{\int\limits_{P} \mu_p(p, \rho, \sigma) \cdot p \, dp}{\int\limits_{P} \mu_p(p, \rho, \sigma) \cdot p \, dp} \tag{14}$$

where: p_i - real value of input linguistic variable p,

$\tilde{S}(\omega)$, $Y(\omega)$ - noise estimate and noise-affected signal spectrum

P - the domain of fuzzy set and variable p.

2.3.2 Application of Neural Networks

The comparator module may alternatively be based on a neural network. The network's task is to non-linearly approximate distribution $p(\widetilde{S}(\omega), Y(\omega))$ which is a certain non-linear function of values ρ, σ. Hence, the network should have at least two input neurons representing these variables and one output neuron representing the degree of similarity between input signals p. The degree is expressed numerically and it designates the possibility of removing lines that have power from sub-band Δ_i of the mth critical band of hearing. The appropriate structure in this case is a unidirectional multilayer network which may be extended by additional functional links [15]. In the presented implementation, in both cases, a network with a single hidden layer was used. The number of neurons in this layer was between 10 to 20. The structure for networks of this type is shown in Fig. 7.

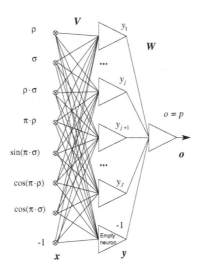

Fig. 7. Structure of applied neural network - functional link neural network

With matrix operators Γ defined in formula (15), the approximation $p(\rho,\sigma)$ may be obtained through the subsequent relation (16):

$$\Gamma[q] = \begin{bmatrix} f_1(q_1) & 0 & ... & 0 \\ 0 & f_2(q_2) & ... & 0 \\ ... & ... & ... & ... \\ 0 & 0 & ... & f_Q(q_Q) \end{bmatrix} \tag{15}$$

where: $f_1, f_2, f_1{'}, f_2{'}, ...$ — neuron activation functions with their derivatives

$$p_o = p(\widetilde{S}(\omega), Y(\omega)) = p(\rho, \sigma) = \Gamma[W \cdot \Gamma[Vx]] \tag{16}$$

where: V and W — weight matrixes for subsequent network layers

x — input vector which for both: a classic unidirectional network and a network with functional links may be defined by the following relation:

$$x^T = \begin{cases} [\rho, \sigma \cdot t]^T \\ [\rho, \sigma \cdot t, \rho \cdot \sigma, \sin(\rho \cdot \pi), \sin(\sigma \cdot t \cdot \pi), \cos(\rho \cdot \pi), \cos(\omega \cdot \pi)]^T \end{cases} \quad (17)$$

The nature of the problem enforces application of the supervised training method. Data are supplied by an expert, who defines a desired output value for each subsequent combination of input values ρ and σ achieved from the analysis of sound samples. Because a gradient training algorithm with error backpropagation is used, it is necessary to employ neuron units of a continuous activation function.

2.4   Examples of Noisy Audio Signal Restoration

An obtained example of an audio signal reconstruction by means of the elaborated algorithm is graphically presented in Fig. 8. In the example a male voice is affected by stationary noise of a subjectively high level. Time distribution of the spectral power density of a noise-affected signal is illustrated in Fig. 8a, and the corresponding analysis of the signal representing the reconstruction, subjectively perceived as the best, is shown in Fig. 8b. In this case, an intelligent comparator was implemented based on fuzzy logic. Spectrum analysis of the noise and useful signals employed a Fourier transform. The length of the frame was assumed to be 1024 samples, and the length of overlaps equalled 512 samples. The sampling frequency was 44.1kHz, and the windowing function was the Hamming function. Various signals affected by stationary noise were reconstructed with the help of the described method. In these reconstructions, the noise pattern was taken from the nearest silence passage occurring before the signal fragment being reconstructed. To achieve this, a procedure of speech signal detection which examines the characteristics of speech resonance (autocorrelation analysis of signal periodicity) was introduced into the algorithm. Because speech passages were distinguished from silence passages, it was possible to obtain patterns of ρ, which might have changed along with the noise affecting the useful signal. In the process of system optimization, we used the subjective preferences of an expert who could influence the final effects of the system operation by the choice of membership functions and fuzzy rules.

Results obtained with functional link neural networks (see Fig. 7) are demonstrated in Fig. 9. In Fig. 9a the spectral analysis of a 5 second-long utterance spoken by the male is presented. It was recorded using 8 kHz sampling frequency. Noise in the recorded excerpt was detected automatically using the energy-based voice activity detector. Based on the noise profile the whole signal was whitened [2] and then neural-network controlled spectral subtraction was performed. In Fig. 9b the spectral analysis of the reconstructed signal is shown, revealing a strong noise reduction effect.

(a)

(b)

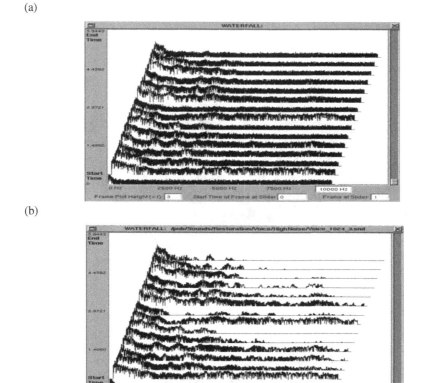

Fig. 8. Spectrum analysis of speech signal samples affected by strong noise of non-stationary character: (a) before restoration; (b) after restoration (results obtained earlier)

More experiments with the comparator module based on the neural network were conducted recently employing speech and musical excerpts and their results were assessed with listening tests (see Table 2). Distortions were removed from digitized 5 second-long audio excerpts sampled with 44.1 kSa/s. The developed on-line sound restoration service [2] makes an efficient tool for acquiring audio material for the

Table 2. Results of subjective evaluation of audio restoration – average MOS subjective grades

Sample / MOS Grade increase	Fuzzy reasoning	Neural network
1 - speech	2 ->4	2 ->4
2 – musical quartet	1 ->3	1 ->3
3 – symphonic music	2 ->5	2 ->4
4 – pop music	3 ->5	3 ->5
5 – folk music	2 ->4	2 ->4

(a)

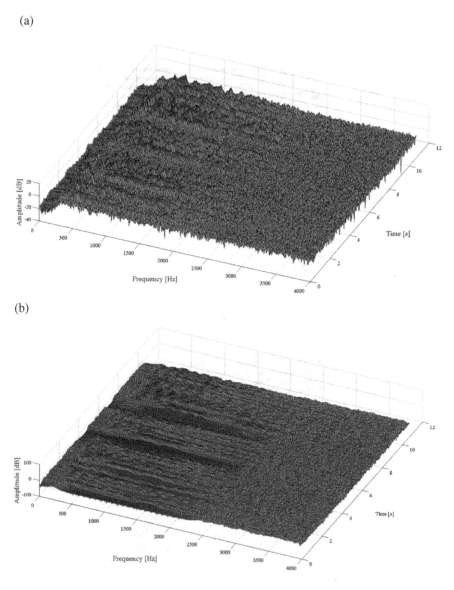

(b)

Fig. 9. Spectral analyses of speech excerpt- original noisy signal (a); signal denoised employing functional link neural network as noise estimates comparator (b)

restoration. Several restoration attempts were made employing both noise reduction methods described above. Results were assessed subjectively by the author and his co-workers. Recently some more formal subjective test sessions were organized [16]. The 5-grade Mean Opinion Score (MOS) scale was used for expressing opinions concerning the effect of restoration, where 1 - denotes the lowest quality and

5 – means entirely satisfying result. Averaged results reflecting the performance of elaborated sound restoration algorithms applied to the processing of various kinds of audio signals were gathered in Tab. 2. Overall results achieved are positive, however no universal rules were found reflecting the dependency of the kind of the audio content (speech, music) on the algorithm that is optimal for its restoring.

3   Summary

The engineered system for non-stationary noise reduction has been described very briefly and the Internet sound restoration system was also mentioned (www.youarchive.net). Two soft computing approaches were implemented in regard to noise suppression, namely the fuzzy logic control of spectral subtraction and the neural network controller acting as an intelligent comparator of noise estimates. A number of experiments and case surveys were carried out. The results obtained show that the soft computing approach to noise reduction is a robust tool for the improvement of quality of archival audio recordings. A drawback of this solution is serious computational complexity related to training of the neuronal algorithms. Nevertheless, as proved by listening tests, the soft-computing approach is fully applicable to non-stationary noise suppression in practice.

Acknowledgments. The research was partially supported within the project No. SP/I/1/77065/10 entitled: "Creation of universal, open, repository platform for hosting and communication of networked resources of knowledge for science, education and open society of knowledge", being a part of Strategic Research Programme "Interdisciplinary system of interactive scientific and technical information" funded by the National Centre for Research and Development (NCBR, Poland).

References

1. Cabras, G., Canazza, S., Montessoro, P.L., Rinaldo, R.: Restoration of Audio Documents with Low SNR: a NMF Parameter Estimation and Perceptually Motivated Bayesian Suppression Rule. In: Proc. of Sound and Music Computing Conference, Barcelona, pp. 314–321 (July 2010)
2. Czyżewski, A., Kupryjanow, A., Kostek, B.z.: Online Sound Restoration for Digital Library Applications. In: Bembenik, R., Skonieczny, L., Rybiński, H., Niezgodka, M. (eds.) Intelligent Tools for Building a Scient. Info. Plat. SCI, vol. 390, pp. 227–242. Springer, Heidelberg (2012)
3. Czyzewski, A.: Digital sound. Academic Press Exit, Warsaw (2001)
4. Czyżewski, A., Ciarkowski, A., Kaczmarek, A., Kotus, J., Kulesza, M., Maziewski, P.: DSP Techniques for Determining "Wow" Distortion. Journal of the Audio Engineering Society 55(4), 266–284 (2007)
5. Czyżewski, A., Królikowski, R.: Noise Reduction in Audio Signals Based on the Perceptual Coding Approach. In: Proceedings of the IEEE Workshop on Applications of Signal Processing to Audio and Acoustics, New Paltz, NY, USA, October 17-20, pp. 147–150 (1999)

6. Czyżewski, A., Maziewski, P., Kupryjanow, A.: Reduction of Parasitic Pitch Variations in Archival Musical Recordings. Signal Processing, Special Issue of Signal Processing: Ethnic Music Restoration 90(4), 981–990 (2010)
7. Godsill, S.J., Rayner, P.: Digital Audio Restoration - A Statistical Model-Based Approach, ch. 8, pp. 171–190. Springer, London (1998)
8. Kamath, S.D., Loizou, P.C.: A Multi-Band Spectral Subtraction Method for Enhancing Speech Corrupted by Colored Noise. In: Proc. of ICASSP 2002, Orlando, FL (May 2002)
9. Kostek, B.: Applying computational intelligence to musical acoustics. Archives of Acoustics 32(3), 617–629 (2007)
10. Kulesza, M., Czyżewski, A.: Tonality Estimation and Frequency Tracking of Modulated Tonal Components. J. Audio Eng. Soc. 57(4), 221–236 (2009)
11. Królikowski, R., Czyżewski, A.: Noise Reduction in Telecommunication Channels Using Rough Sets and Neural Networks. In: Zhong, N., Skowron, A., Ohsuga, S. (eds.) Proceedings of the 7th International Workshop on Rough Sets, Fuzzy Sets, Data Mining, and Granular-Soft Computing, Ube, Yamaguchi, Japan, pp. 100–108. Springer, Berlin (1999)
12. Read, P., Meyer, M.P.: Restoration of Motion Picture Film. Butterworth Heinemann, Oxford (2000)
13. Yektaeian, M., Amirfattahi, R.: Comparison of Spectral Subtraction Methods used in Noise Suppression Algorithms. In: ICICS (2007)
14. CDDB – Compact Disc Database, http://en.wikipedia.org/wiki/CDDB
15. Nanda, S.K., Tripathy, D.B.: Application of Functional Link Artificial Neural Network for Prediction of Machinery Noise in Opencast Mines. Advances in Fuzzy Systems 2011, Article ID 831261 (2011)
16. Przyłucka, K., Kostek, B., Czyżewski, A.: Testing audio restoration algorithms. In: VDT Conf. Cologne, November 22-25 (in print, 2012)

Automatic Transcription of Polish Radio and Television Broadcast Audio

Danijel Koržinek, Krzysztof Marasek, and Łukasz Brocki

Polish-Japanese Institute of Information Technology, Warsaw, Poland

Abstract. This paper describes a Large-Vocabulary Continuous Speech Recognition (LVCSR) system for the transcription of television and radio broadcast audio in Polish. This work is one of the first attempts of speech recognition of broadcast audio in Polish. The system uses a hybrid, connectionist recognizer based on a recurrent neural network architecture. The training is based on an extensive set of manually transcribed and verified recordings of television and radio shows. This is further boosted by a large collection of textual data available from online sources, mostly up-to-date news articles. The paper describes and evaluates some of the key components of the architecture. The system is also compared to a conventional HMM-based architecture. An application of the described system in indexing and search of terms within audio and video transcripts is also described.

Keywords: automatic speech recognition, broadcast news transcription, speech indexing, statistical language models, speech corpora, machine learning.

1 Introduction

Transcription of broadcast audio speeches is one of the staples of Automatic Speech Recognition (ASR) technology [1,2,3,4]. A substantial increase of interest in this domain has been noticed, particularly in the recent years. This is both thanks to the increased performance of such systems and greater demand in the commercial environment.

The performance of ASR in the broadcast news task has always been a challenge for various reasons:

- varying quality of recordings (studio, outside, phone calls, etc.)
- the channel is mixed with several sources of sound (background music, sound effects)
- the manner of speech is fast and natural (compared to dictation)
- cocktail-party effect (many people talking simultaneously)
- rapid speaker change (making it difficult to adapt)

Nevertheless, the interest in solving this problem was always high, because of its applicability in the radio and television business. Among other uses, it is been primarily applied to captioning, both as an aid for the deaf or hard of hearing

R. Bembenik et al. (Eds.): Intell. Tools for Building a Scientific Information, SCI 467, pp. 489–497.
DOI: 10.1007/978-3-642-35647-6_29   © Springer-Verlag Berlin Heidelberg 2013

and for translation. This work will however mostly discuss its use in indexing and search.

There is a growing trend of presenting information in video or at least audio form instead of text. This is thanks to the ease of creation and distribution of such information in the recent years as well as greater acceptance of such material by consumers. Whatever the reason for this trend, however, it puts us at a problematic situation where it is getting increasingly hard to maintain and index large quantities of streaming audio and video. A common solution consists of attaching various kinds of meta-data to the streams, but this has very limited use and demands constant and manual administration. The real solution lies in the application of audiomining techniques and technology.

When dealing with broadcast radio and television, the main interest is in recognizing speech. To do this we have several options: we can apply either keyword detection or full audio transcription. Keyword detection [12] is useful mostly when we want to find words from a very limited vocabulary and we can predict what the user may look for. Furthermore, it does not have much use in indexing, since we would first need to know what to look for and that would require the transcription anyway. It is therefore much more convenient to use full audio transcription for indexing and search. Another benefit is that transcription has many more uses than keyword detection and is thus a more studied and familiar field of science with lots of tools and techniques available.

The state of speech recognition in Polish is still very weak compared to other languages, even though it is improving at a fast pace and should not be considered an under-resourced language for very long [25,22]. Several research projects have emerged in the last couple of years dealing with the topics of automation in the telephony environment [5], transcription of legal documents [8] and recently speech-to-speech translation in different settings [9]. Commercially, there have been a few local startups and a few attempts by world market leaders, but none have yet achieved real adaptation of LVCSR in the field, with the exception of a few (Google and Apple) which include ASR as a free service with their other products. This makes it quite difficult for companies and organizations which might seek to use such technology and find that it either does not fulfil their needs or is too expensive for them to invest into. It has been our goal for many years now to contribute in the research of Polish ASR and help bring it to such a level to be both useful and attainable by the larger country audience.

2   System Architecture

The system used in the paper is an evolution of the authors' system used in previous projects [6]. The acoustic layer is a hybrid Recurrent Neural-Network based model used to recognize phonemes. The feature front-end is the standard MFCC and energy feature set with delta and acceleration coefficients. The decoder is a Viterbi-like decoder modified to utilize the acoustic model outputs properly. In phoneme recognition on the TIMIT corpus, the system achieves comparable results to other well known equivalent systems [10].

The system was previously used with a grammar-based language layer used to successfully recognize domain-constrained tasks with up to several thousands of concepts per grammar state achieving recognition rates in excess of 90%. This was also true when the system was used in noisy telephony environments [7].

A preprocessing subsytem was independently trained to perform preliminary sound level normalization and speech detection. The normalization was performed in batch to achieve unit variance and zero mean on all files individually. The speech detection was then trained in a similar fashion to the acoustic model to recognize speech, non-speech and silence events.

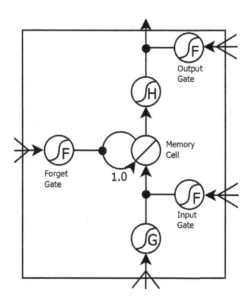

Fig. 1. A block diagram of an LSTM memory block

An Artificial Neural Network (ANN), known as Long-Short Term Memory (LSTM) was used for the acoustic modelling [10]. LSTM neural nets are a special type of Recurrent Neural Networks (RNN), which can be trained using the gradient descent method. The characteristic topology of these networks guarantees that the backward propagated error remains at a constant level. It is proven that such networks can learn dependencies 1000 time steps apart. This is a feat unrealizable by any standard RNNs. An LSTM network consists of the so-called memory blocks (fig. 1). Each block contains three gates which control a special memory cell used for storing information. The gates are actually simple perceptrons with sigmoidal activation functions. The values returned by these functions are limited to the <0,1> range. The input of the memory cell is first multiplied by the output of the input gate. Therefore, if the input gate returns a value close to zero this will stop the signal trying to reach the memory cell. If the value of the input gate is close to one, the signal will reach the memory cell almost unchanged. The output gate works almost the same way. The only difference is that it affects the output of the network. There is one more gate inside the memory block used for resetting of the memory cell. If the value of the forget gate is close to zero the value saved in the memory cell will be erased, but if it's close to one, it will stay the same. A very good analogy to the memory block is an electronic chip which performs read, store and reset memory operations. The memory block of the LSTM is actually a differentiable version of such a chip.

For the purposes of the current system, a language model was developed to aid the process of LVCSR. Since Polish language is highly inflected, a straightforward n-gram approach does not work too well. The inflection causes the vocabulary size to grow several times compared to languages like English. Larger vocabularies usually require more training data and they considerably reduce the speed and accuracy of the decoding process. On the other hand, the different inflected forms of a particular word stem play a consistent role in the sentence structure. That is why a combination of three language models responsible for words, word stems and grammar classes was used. The models were linearly interpolated and their interpolation weights were determined using a machine learning approach minimizing the perplexity on a validation set (see section 4 for details).

3   Speech Database

To train the acoustic models a database of TV and radio shows was recorded using DTV/radio tuner hardware and downloaded from online streams. It turned out it was considerably easier to obtain the data online since several streams were available at once. To achieve the same thing on a tuner, we would need a separate device for each stream. An initial collection of around 100-150 hours of various shows was saved in 16 kHz. 16-bit uncompressed audio format. The quality of Internet streams was sufficient for our experiments and allowed us to make a more robust system.

The database was then manually transcribed as obtaining the transcriptions automatically turned out to be impossible. The transcription took a group of 4-8 people several months to complete. The first database consisted of 78 hours of data. The transcription was done in such a way to remove any non-speech events. To save time, all incomplete and inaudible words (as judged by the transcribers) were also removed. This greatly helped in acquiring a completely correct dataset useful for training of a clean acoustic model. To deal with real-world data, where incomplete words, interruptions and non-speech events are plentiful, other means could be used, such as garbage modelling [13]. This could be tuned on a much smaller, more accurately transcribed dataset.

To achieve acceptable results, much more data needed to be used than what was available in the core data set. For the acoustic model training several acoustic corpora was used for initial training of the models. The main was the Polish Speecon corpus [14,15], which consists of recordings from 550 speakers, about 1 hour (session time) each, from various regions of the country. The acoustic environment varies greatly but much of it is meeting room style. Apart from that, a few in-house corpora, mostly recorded in the studio were also used.

For language modelling, the transcriptions of the training set were primarily used, but even the largest acoustic corpora make for weak language corpora. That is why other sources of available data were seeked out. Polish language corpora do exists (e.g. IPIPAN [16], Rzeczpospolita [17]). This was combined with the usual up-to-date data available online (newspapers, blogs, encyclopediae).

The final model was built from 3 different sub-models: a model of grammar features, word lemmas and words themselves combined together using linear interpolation. A $(\mu + \lambda)$ Evolution Strategy [23,24] that minimized the perplexity of the final model on the development set was used to find the optimal weights of this interpolation process. The weights of the different corpora were also optimized in this process. This type of an Evolutionary Algorithm is especially suited for optimizing continuous valued variables, where the $(\mu + \lambda)$ optimization helps with speeding up the convergence.

4   Evaluation Results

The core dataset was split into 90% for training and development and 10% for testing (single-fold). As mentioned in the previous chapter, the initial models were first trained on other corpora and then adapted to the core data of this particular task. Initial training was necessary to avoid over-training to the training set. Also, when dealing with context-based models (e.g. triphones) using a small training set would make the system react poorly when faced with unseen contexts. The files in the test set were reasonably long: up to 1 minute of only speech.

The language models were initially trained on a manually prepared language corpus where they achieved a perplexity (described at the end of this section) of 376. They were then used to annotate and disambiguate the texts of the core training and test sets. After retraining the models with the new data the perplexity on the core test set was 246. The combination of word, stem and grammar models was considerably better than the simple trigram word model trained using SRI LM toolkit [18] that achieved values of perplexity 293. The vocabulary was set to around 40k words, which gave around 14k word stems and about 600 grammar classes.

The experiment results are summarized in the table below. Most common errors were related to Out-Of-Vocabulary words in a limited vocabulary language model. The size of the vocabulary was chosen to 40 thousand words because of the design issues and memory requirements of the current system implementation. Another problem would often arise when speakers changed. It seems that these rapid changes would interfere with the acoustic model contexts, not to mention that some speakers were just badly recognized compared to others. Finally, background noise and multiple people speaking at the same time (also known as the "cocktail party effect") would cause greatest errors. Even though the dataset was cleaned to remove all the non-speech and inaudible speech, it

Table 1. Experiment results comparing our system to the Julius baseline using models from IRSTLM on a 30k and 60k vocabulary. (I-insertions, S-substitutions, S-deletions, H-correct).

Model	Accuracy	Correctness	LP perplexity	I	S	D	H
Our system	**65**	67	246	168	1814	417	4637
IRSTLM 30k	**44**	52	289	611	2424	832	3605
IRSTLM 60k	**47**	56	276	595	2512	544	3812

was decided to leave background noise in the recordings as long as the person transcribing the data could understand the spoken word without too much effort.

A standard HMM system based on GMM acoustic models was also trained in HTK toolkit [19] and tested in the Julius [21] decoder using a language model created by IRSTLM toolkit [20]. This system performed considerably worse compared to our own. It was not especially tuned and this result may be due to the difficult acoustic quality of the data.

The results presented in table 1 are a standard measure of ASR performance based on the Levenshtein distance between the output of the system and the correct reference. Correctness is the number of correctly recognized tokens (H) divided by the number of tokens in reference, normalized to the range of $< 1,100 >$. Accuracy also subtracts the insertion errors (I) from the correct tokens (H) to produce an even more accurate depiction of performance. It's worth noting that accuracy measures can produce results smaller than zero.

Perplexity is a measure of quality of the language model given a specific sequence of words as reference. It is a measure of fit of the particular model to the reference and is not an objective measure by itself, however it is highly correlated with the accuracy of recognition. Perplexity is calculated as the exponent of the entropy of the particular probability language model [11]:

$$PP = 2^{-\sum_x p(x)\log_2 p(x)} \tag{1}$$

5   Search and Visualisation Demonstration

A prototype of a search engine and visualisation tool was developed to demonstrate the usability of the described technology (see fig. 2). After the user looks for a keyword he is presented with a list of video clips whose transcripts contain the looked-for term. After choosing a specific item, the user is presented with the particular clip which immediately skips to the place where the particular keyword is used.

The clip visualization tool contains two useful features. Captioning is used to display the transcription of the audio overlaid on the clip itself. This is done using HTML5 video and audio streaming capabilities combined with custom Javascript code to display the subtitles. This will most likely be superseded by the HTML5 <track> tag [26] that was not supported by any browsers at the time of writing this paper. Finally, the bottom of the page contains the whole

Subtitle player

zostało nam pół minuty to może wróćmy sobie do jednego z bohaterów
internetu pan nazywa się dawid chełmecki jest kandydatem na radnego
jaworzna na śląsku został zaproszony w ostatniej chwili bo kolega nie mógł
dojechać na debatę w lokalnej telewizji która się nazywa C T V jaworzno no i
tak to wyglądało nie wiedział że na żywo sport to jest coś czym należy

Fig. 2. Screen grab of a website displaying the video player and captioning page. Current subtitle is displayed overlaid on the top of the video and it is also highlighted in the text below. A user can also click on any fragment to jump immediately there as shown by the border around the words underneath the cursor.

transcription of the clip. It allows the user to quickly glance over the whole discussion and skip to any part of the clip using a single click.

6   Conclusion

The paper described a system for automatic transcription of broadcast radio and news audio. The technology is still at its early stages when it comes to the Polish language, but it's usefulness is definitely helping move things forward. The example described in the paper is just one of many its uses and with laws requiring more shows contain subtitles for the deaf and hard of hearing, such solutions may just become a necessity.

Our initial experiments show several improvements are possible and required to make our system feasible for real world use. First, improvements on the front-end and signal processing, specifically when it comes to speech detection and removal of unwanted noise is required. Adaptation through robust speaker change and identification could prove to be very useful in improving the quality of recognition. Finally, larger and more accurate language models should help with out-of-vocabulary situations, as well as methods of acquiring and injecting new and trending words into the language model.

Acknowledgments. The work was sponsored by the National Centre of Research and Development (NCBiR) under grant No. SP/I/1/77065/10 by the strategic scientific research and experimental development program: "Interdisciplinary System for Interactive Scientific and Scientific-Technical Information".

References

1. Pallett, D.S., Fiscus, J.G., Garofolo, J.S., Martin, A., Przybocki, M.: Broadcast News Benchmark Test Results: English and Non-English Word Error Rate Performance Measures. In: Proceedings of 1999 DARPA Broadcast News Workshop (1998)
2. Neto, J., Meinedo, H., Viveiros, M., Cassaca, R., Martins, C., Caseiro, D.: Broadcast news subtitling system in portuguese. In: Proc. ICASSP 2008, Las Vegas, USA (2008)
3. Meinedo, H., Abad, A., Pellegrini, T., Trancoso, I., Neto, J.: The L2F Broadcast News Speech Recognition System. In: Proc. Fala 2010, Vigo, Spain (2010)
4. Dharanipragada, S., Franz, M., Roukos, S.: Audio-Indexing For Broadcast News. In: Proceedings of TREC6 (1997)
5. Marasek, K., Brocki, Ł., Koržinek, D., Szklanny, K., Gubrynowicz, R.: User-Centered Design for a Voice Portal. In: Marciniak, M., Mykowiecka, A. (eds.) Aspects of Natural Language Processing. LNCS (LNAI), vol. 5070, pp. 273–293. Springer, Heidelberg (2009)
6. Koržinek, D., Brocki, Ł.: Grammar based automatic speech recognition system for the Polish language. In: Recent Advances in Mechatronics (2007)
7. Brocki, Ł., Koržinek, D., Marasek, K.: Voice Portal for Public City Transportation, Interfejs użytkownika - Kansei w praktyce (2009)
8. Szymański, M., Klessa, K., Lange, M., Rapp, B., Grocholewski, S., Demenko, G.: Opracowanie modeli akustycznych na potrzeby systemu rozpoznawania mowy ciągłej z zastosowaniem dużych leksykalnych baz danych, Best Practices - Nauka w obliczu społeczeństwa Cyfrowego, Poznań, pp. 280–289 (2010) ISBN 978-83-7712-032-3
9. EU-Bridge - Bridges Across the Language Divide, European Union project in the 7th Framework Programme, under grant agreement n°287658, http://www.eu-bridge.eu
10. Graves, A., Fernández, S., Gomez, F., Schmidhuber, J.: Connectionist temporal classification: labelling unsegmented sequence data with recurrent neural networks. In: Proceedings of the 23rd International Conference on Machine Learning, Pittsburgh, Pennsylvania (2006)
11. Jelinek, F.: Statistical Methods for Speech Recognition. MIT Press, Cambridge (1998)
12. Wöllmer, M., Eyben, F., Graves, A., Schuller, B., Rigoll, G.: Bidirectional LSTM Networks for Context-Sensitive Keyword Detection in a Cognitive Virtual Agent Framework. Cognitive Computation (2010)
13. Kemp, T., Jusek, A.: Modelling unknown words in spontaneous speech. In: ICASSP Conference Proceedings, Acoustics, Speech, and Signal Processing (1996)
14. Marasek, K., Gubrynowicz, R.: Multi-level Annotation in SpeeCon Polish Speech Database. In: Bolc, L., Michalewicz, Z., Nishida, T. (eds.) IMTCI 2004. LNCS (LNAI), vol. 3490, pp. 58–67. Springer, Heidelberg (2005)

15. Marasek, K., Gubrynowicz, R.: Design and Data Collection for Spoken Polish Dialogs Database. In: Language Resources and Evaluation Conference 2008, Marrakech Morroco (2008)
16. Przepiórkowski, A.: Korpus IPI PAN. Wersja wstępna, The IPI PAN Corpus: Preliminary version. IPI PAN, Warszawa (2004)
17. Korpus Rzeczpospolitej, http://www.cs.put.poznan.pl/dweiss/rzeczpospolita
18. Stolcke, A.: SRILM – An Extensible Language Modeling Toolkit. In: Proc. Intl. Conf. on Spoken Language Processing, vol. 2, pp. 901–904. Denver (2002)
19. Young, S.J., Young, S.J.: The HTK Hidden Markov Model Toolkit: Design and Philosophy. Entropic Cambridge Research Laboratory, Ltd. (1994)
20. Federico, M., Bertoldi, N., Cettolo, M.: IRSTLM: an Open Source Toolkit for Handling Large Scale Language Models. In: Proceedings of Interspeech, Brisbane, Australia (2008)
21. Lee, A., Kawahara, T., Shikano, K.: Julius — an open source real-time large vocabulary recognition engine. In: Proc. European Conference on Speech Communication and Technology (EUROSPEECH), pp. 1691–1694 (2001)
22. Lööf, J., Gollan, C., Ney, H.: Cross-Language Bootstrapping for Unsupervised Acoustic Model Training: Rapid Development of a Polish Speech Recognition System. In: INTERSPEECH 2009, Brighton, United Kingdom, September 6-10 (2009)
23. Michalewicz, Z.: Genetic algorithms + Data Structures = Evolution Programs. Springer (1994)
24. Michalewicz, Z., Fogel, D.B.: How to Solve It: Modern Heuristics. Springer (1999)
25. Marasek, K.: Large vocabulary continuous speech recognition system for Polish. Archives of Acoustics 28(4), 293–304 (2003)
26. W3C, HTML 5 Working Draft March 29, 2012, revision 1.5612, section 4.8.9 The track element (2012)

Chapter VII

Advanced Software Engineering

Exploring the Space of System Monitoring

Janusz Sosnowski, Piotr Gawkowski, and Krzysztof Cabaj

Institute of Computer Science, Warsaw University of Technology
Nowowiejska 15/19, 00-665 Warsaw, Poland
{jss,p.gawkowski,k.cabaj}@ii.pw.edu.pl

Abstract. During system exploitation and maintenance an important issue is to evaluate its operational profile and detect occurring anomalies or situations which may lead to such anomalies (anomaly prediction issue). To resolve these problems we have studied the capabilities of standard event and performance logs which are available in computer systems. In particular we have concentrated on checking the morphology and information contents of various event logs (system, application levels) as well as the correlation of performance logs with operational profiles. This analysis has been supported with some tools and special scripts.

Keywords: system monitoring, event and performance logs, dependability.

1   Introduction

Monitoring event and performance logs is gaining more and more interest both during the system development and exploitation. This technique delivers traces of system operation and behavior. They are very useful in detecting hardware, software, administration and user problems such as errors, performance bottlenecks, configuration flaws, inconsistencies with the workload profiles (especially if they change in time). Most research on monitoring is targeted at evaluating system availability and dependability (e.g. based on downtimes) [5, 6, 8, 14, 16]. Recently we observe more papers on anomaly predictions [2, 4, 15]. The presented results usually relate to some specific systems observed for long time period e.g. telecommunication, cluster systems [5, 8, 14, 20]. These systems have been the subject of heavy and complex workloads. Monitoring systems in the Institute, in particular the developed data repository (grant no. SP/I/1/77065/10), we have observed low CPU usage and a few hardware errors (which were reported as frequent in the literature). Nevertheless, many system configuration and administrative problems occurred, which are neglected in the literature. Hence, it is not easy to profit the gained experience dealing with other systems. Moreover, analyzing logs it is needed to get some knowledge of occurring events, formats of their registration as well as their categorization and information contents. Experience in this area is not published. Hence we have decided to gain some experience in this respect within our computer environments. In particular we are interested in identifying log characteristics related to normal and abnormal system operation. This experience is targeted at establishing monitoring schemes for the developed data

R. Bembenik et al. (Eds.): *Intell. Tools for Building a Scientific Information*, SCI 467, pp. 501–517.
DOI: 10.1007/978-3-642-35647-6_30   © Springer-Verlag Berlin Heidelberg 2013

repository. The repository is still in progress, so the collected logs relate to relatively short period of observations and in fact are still not sufficient to get broader experience and closed to the future real operation. Hence, we decided to monitor additionally other computers to get a wider view on the considered problems.

Having collected many logs from various computers we tried to characterize them and define problems for advanced analysis. This is described in section 2. Section 3 deals in more detail with event log problems. Specific problems related to monitoring disc storage are presented in section 4. Section 5 presents some results of analyzing CPU usage of the developed repository. Conclusions are summarized in section 6.

2   Features of Event and Performance Logs

The computer system operation can be in a large extent assessed by tracing event and performance logs. In general there are many sources of event logs which provide various data describing the status of system components, operational changes related to initiation or termination of services, program updates, configuration modifications, execution errors, etc. Depending upon the system some classes of logs (e.g. related to different sources) are stored in a common file or in independent files (e.g. security events specifying authorization problems, user login and logout events, events generated by applications, etc.).

In Unix systems we have many event logs generated by syslog or other programs. We concentrate on logs generated by the system e.g.: auth.log (reports on correct or erroneous user loggings, connection crashes, etc.), daemon.log (gives some information related to cooperation with DNS server), last.log (in particular it provides information on login and logouts of the users) and application logs (related to various used programs).

In Windows systems we can distinguish 3 classes of logs: system, security and application logs. The system log comprises events registered by the operating system related to problems with loading IO drivers, system restarts and crashes, events related to some standard applications, etc. Security logs comprise events reporting correct or incorrect user logins, accesses to system resources (creating, opening, deleting files, etc.). Application logs relate to individual programs and they include information on correct licensing, authorization, appearing problems during the execution, etc. It is worth noting that newer system versions comprise more separate application log files. Moreover, the generated logs usually provide better description of the system normal or abnormal behavior.

Typically events in logs comprise date and time, some information on event source (computer name, program, etc.), category (informative, warning, critical, etc.) and a text message describing the event (usually loose unformatted text). Unix and Linux logs are characterized by low level of structuralization. Windows logs seem to be more systematic with well-defined fields in particular they comprise event ID, which together with event source and event type define event class. The formats and information contents of system generated logs may differ upon the system and the configured environment. In the case of application related logs we observed higher diversity and application specific log messages. Depending upon the logs we encounter text, csv, xml or other formats. Looking at various event logs we can find a large number of registered events each day, of course this number depends upon the system usage. Manual analysis of events is practically impossible. Hence it has to be supported with appropriate tools and analytical methods. This needs syntactic and semantic analysis of logs. The syntactic analysis allows identifying appropriate fields, variables, specific objects (e.g. email addresses, www links). Semantic analysis allows better classification of logs (this involves in a large extent detailed analysis of text messages). This process is more complex for Unix logs than Windows. Just for an illustration we give an example of two events related to these systems:

— Linux:

Jul 1 12:01:44 g108 ntpd[2659]: synchronized to 192.168.19.201, stratum 2
    After the date and time we have node name g108, process name ntpd, its PID and
    the message field comprising the address to the synchronization server

— Windows:

<date and time>.NET runtime 1022 information, RestartingIIS Admin. For more information see help and support center at http://go.microsoft.com/fwlink/events.asp
    After the date and time field (presented in a symbolic way) we have log source
    name, event ID, event type and the message which comprises a link to the help
    web page.

The presented event for Windows refers to some web page where additional comment is available. These simple event records give some impression on incomprehensiveness of logs especially if they appear in a large quantity. Moreover their interpretation needs some experience. Hence an important issue is to study their morphology and meaning basing on a large benchmark of logs. For this purpose some tools are needed (outlined in the next section). The extracted knowledge will be helpful to define schemes of identifying normal and abnormal behaviors.

Classifying event logs we use regular expressions derived on the gained experience [11], in particular they allow us to de-parameterize event records. This process is enhanced with some semantic analysis of event messages. To give some view on this problem we give a sample statistics of an application log in the first author workstation. This log comprised over 44000 event records with included text messages based on about 1500 different words (total over 700 000 words). The semantic analysis is targeted at correlating these messages with some typical problems e.g. errors, performance, program updates, inaccessible files.

In most computer systems various data on performance can be collected in appropriate counters [9, 10, 11]. These counters are targeted at performance objects such as processor, physical memory, cache, physical or logical discs, network interfaces, server or service programs (e.g. web services), I/O devices. Many counters (variables) can be attributed to an object so as to characterise its operational state, usage level, performance properties, etc. The designers can also add counters characterizing application properties (e.g. transaction delays). These mechanisms are useful for evaluating system dependability, predicting threats to undertake appropriate corrective actions, etc. The list of counters, which can be configured, is very long. In Windows systems we can program counters for hundreds of variables [19]. In case of Unix or Linux systems standard performance parameters are more restricted, but more detailed parameters can be added. Collecting performance data involves some CPU overhead. Hence an important issue is to select appropriate measures (variables). Analysing the collected performance measures we can identify resource bottlenecks, workload profiles and anomalous behaviour. Many practical problems are caused by system resource exhaustion, moreover some problem sources can be hidden by other coupled problems. The most important is monitoring the operation of the system storage and CPU. Some experience with this problem we present in sections 4 and 5.

It is worth noting that in practice log and performance monitoring can be considered as complementary. Moreover, some anomaly in event logs can trigger deeper analysis of selected performance parameters or vice versa. In the case of a stable application (with fixed workflow) we can model correlations between performance metrics or events and detect anomalies if these correlations are broken. In the case of workflow fluctuation it is important to identify characteristic operational phases or trends and adapt analysis to these phases. This can be done basing on some general approach or enhanced with domain specific knowledge. It is worth noting that various operational problems quite often result from system runtime factors (e.g. access sequences, concurrency level, resource usage) which cause contextual anomalies. Hence, an important issue is also to extract normal operation characteristics.

3   Exploring Event Logs

Usually event logs comprise many registered records describing events which at first glance look a little bit discouraging and difficult to interpret. The included text messages comprise natural language words, numbers and various symbolic notations. Events are registered during normal system operation (correct) as well as during some anomalous or erroneous situations. Moreover, for some anomalies we have no direct trace event in the log. Analyzing event logs we can try to identify characteristics of normal operation and symptoms of anomalies. In this process it is important to define some features (measures) which can be helpful. We can distinguish here semantic and time features. Within the semantic features we take into account event source, type, severity, information significance, etc. Depending upon the system some of these features can be attributed automatically and specified in appropriate well defined fields or hidden (need to be searched). Having collected and analyzed event logs from many computers we have found that the standard classification is not precise and coarse grained. Moreover, it can be misleading by qualifying an event as a simple information type in the case of its real criticality and vice versa.

(a

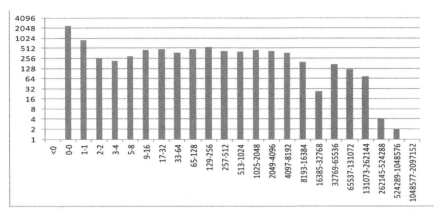

b)

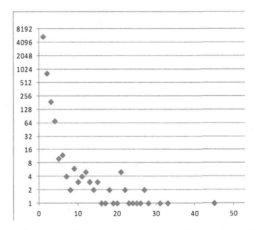

Fig. 1. Time characteristics of application log events: a) distribution of time between subsequent events, b) distribution of event cardinality within 1 second time slots

An important feature is distribution of registered events in time. In particular we can generate time plots (for all events or selected event types) with different granularity (minutes, hours, days, etc.) and in different time perspectives. Here we can observe various regularities e.g. periodically higher rates for working days' hours, long term trends. Some of such regularities are visible in aggregated plots presenting distribution of events in correlation with day hours, week days, and months. In particular they characterize workloads and workflow. Potential anomalies may result in changes of typical images of observed features. Such characteristics can trigger more detailed analysis e.g. distribution of time between events, distribution of event frequency which are shown in Fig.1 for a typical workstation (result of monitoring 5 months of the system operation). Fig. 1a shows the number of events (y-axis) which succeeded another event after the specified time period (x-axis) in seconds. Both axis are logarithmic. The statistics relate only to application logs (total 8979 events: 68 warn-ings, 79 errors). The bars on the right hand of the plot (longer time distance between events) relate to switching off the computer each day late afternoon and switching on in the morning, the longest periods relate to weekends or vacation time. More interest-ing is detecting bursts of events. Fig. 1b gives a distribution of 1 second time periods comprising the specified number of events. The biggest number of events within 1 sec period was 45 (occurred only once). For 5472 periods we have got only one event, 1083 cases with 2-4 events and 67 cases with 5 or more events. In practice we can use different ranges of time slots (e.g. 10s or 1 minute), this can be set after gaining more experience. Slots with bigger number of events can be considered as suspicious, however for some activities they can be accepted e.g., in our case such bursts were generated by McAfee (antivirus) in relevance to blocking access to some files.

Another important issue is analyzing event correlations. In particular we can try to find characteristic event sequences. In particular we can identify pairs of events <ev1, ev2>, defined in such a way that ev1 precedes ev2. In practice such analysis is rea-sonable at the level of event classes (at least with excluded time stamps). We did such analysis for Windows systems taking into account event classes defined by event source and ID, we call this an aggregated event $E_i = \{$event source$_i$, event ID$_i\}$. So we analyze pairs of the form $<E_i, E_j>$. In practice we have a lot of events in logs however the cardinality of event sources and IDs is usually quite limited. For example in the considered workstation we have got 12759 events within syslog (for 5 months). They were generated by 14 event sources and related to 31 different event IDs. This re-sulted in 31 aggregated events (the same number of IDs) and 88 different aggregated event pairs. In the case of random distribution of events we could receive about 961 different pairs $<E_i, E_j>$. Hence, the low number of detected pair classes (88) confirms some regularities and polarization within the sequence of all events (total number of event pairs -12758). Deeper analysis gave us quite interesting statistics. The most frequent aggregated pairs are generated by Service Control Manager (SCM):

<{SCM, 7035}, {SCM, 7036}>   – 3388 (support), confidence 0.85
<{SCM, 7036}, {SCM, 7035}>   – 2937 (support), confidence 0.51
<{SCM, 7036}, {SCM, 7036}>   – 2320 (support), confidence 0.40
<{SCM, 7035}, {SCM, 7035}>   – 570 (support), confidence 0.12

The number of different aggregated pairs with support (number of detected pairs) higher than 300 was 11. The number of pairs with confidence (support/number of all pairs) exceeding 0.5 was 32, however only 13 comprise support exceeding 10. This statistics confirms significant event correlation. Similar properties have been identi-fied for other logs and computers. Having analyzed these pairs we have found that most of them related to normal behavior. For example they related to program update requests and confirmations of correct update. Sometimes negative confirmation appeared. There are also longer unique event sequences related to program license authorization. Such event sequence analysis is useful to filter out events related to normal operation which results in facilitating the search of anomalies. Moreover, it can be useful in identifying workflow properties. More formal analysis can be based of profile graphs with nodes corresponding to aggregated events and edges to the corresponding pairs. In the discussed example we have 31 nodes and 88 edges. In most nodes the fan-out is in the range 1-4, The maximal fan-out was 19 for one node {SCM, 7036} the second highest fan-out was 6 for {SCM, 7035}. Similarly the node fan-in was typically in the range 1-4, the highest values were 12 for {SCM, 7035}, 9 for {SCM, 7036}, and 5 for {EventLog, 6009}.

In general event categorization (abstracting [13]) is a quite complex problem and some data mining techniques can be helpful here. This is confirmed by some preliminary semantic analysis of event log messages (compare section 2).

In most cases events relate to normal operation however some of them can be considered as critical, others can be qualified as normal or abnormal in the context of the neighboring events or more complex context conditions. For example system restart is abnormal if it is not preceded by event of system closing or some program update requests. Moreover, some spurious repetition of restarts can also be attributed to some problems. In standard logs we may have specified event types as errors or warnings, however in practice this categorization is misleading, and quite often events specified as critical in fact are not critical, on the other hand those with info specification may be critical. Another problem relates to abnormal situations with no direct image in event logs, sometimes it is signaled as a console message (visible to the user) but not registered, moreover they can be misleading. Sometimes only the user or the administrator observes some anomaly. We have observed quite a few of such situations. Hence, it is recommended to create a special log for collecting such information. It is worth noting that many problems result from various system (hardware, software) inconsistencies and configuration flows. As an illustration we present an anomaly related to the mail server used in our Institute. This anomaly is visible in the plot of Fig. 2 which shows the mailing queue. Most of the time the queue is at the level 1-4 (close to the x-axis). In June the queue was constant at a level 300 (due to some disturbance), more critical situation (queue exceeding 3000) appeared on October in relevance to the Passim project (within Synat). Some configuration inconsistency at the user side caused generation of infinite loop of emails in response to a received multicast email. Both situations have been eliminated by some configuration refinements.

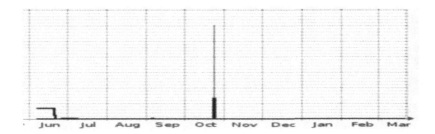

Fig. 2. Queue length of emails in the mail server (y-axis range [0,3500])

This critical situation was not reported in standard event logs. So this confirms the need for checking other internal system state features including performance measures. Some of them are discussed in the sequel.

4   Storage Monitoring

One of the main IT subsystems is storage. In advanced configurations it is built up on arrays (matrixes) of hard disks. The redundancy (i.e. with different RAID levels) assures data availability. However, many dependability and manageability aspects have to be considered while designing the storage. For instance, many advanced RAID controllers has a whitelist of HDDs that are allowed to operate. So, users are obligated to buy new disks from the manufacture's distribution channel only. Additionally a failure of a disk in an array may require buying exactly the same disk model in order to fulfill the controller's requirements – that might be a serious problem after a few years. From dependability perspective, the time to obtain a replacement disk may require to shut down the array to prevent array failure due to its degraded state (i.e. any other disk failure may cause the whole array failure). That means that the data on the array might be unavailable if no spare disk is owned. Another issue is related to the RAID controller operation. It is worth to note that advanced RAID controllers (standalone or in disk matrixes) do not offer the possibility to directly monitor the disks states (e.g. with SMART technology discussed later) limiting the accessibility to the internal health data. Instead, the controller manages the disks state in an array by itself. Even in many popular RAID controllers (e.g. built-in at motherboards) any sector relocation made internally by a disk is considered as a disk failure. It can be considered as a correct behaviour as the sector reallocation is a common predicate of a drive failure [22]. During the research two such hard disks were identified – both with different number of sector reallocations (Fig. 3). However, both are working well (out of the RAID as standalone drives) for almost three years till now from the first reallocation instant. These disks pass the manufacturer's diagnostic software tests – so, they are obviously healthy. The history (unfortunately not the whole) of their reallocations is presented in Fig. 3.

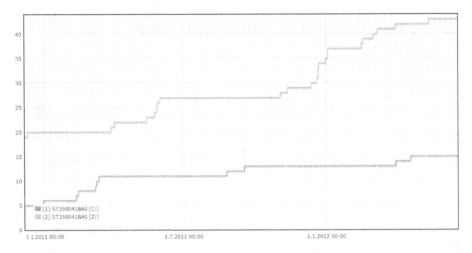

Fig. 3. The Reallocated Sectors Count observed for different Seagate's HDDs (grey vertical stripes marks weekends)

Despite of the failure probability, the storage system should be also monitored to easy manageability (e.g. profiling the utilization, trace the performance bottlenecks, manage maintenance plans and maintain the budget for further spending). Regardless to the developed redundancy levels, the monitoring of the whole storage as well as single drive's state is a key issue to be addressed. One of the rarely investigated sources of information about the storage state is SMART (*Self-Monitoring, Analysis and Reporting Technology*) data which is provided by modern hard disks. The scope of its possible exploration along with storage performance monitoring at the system and application layers are very attractive for further storage management and dependability modeling as the administrators could be warn before the actual storage failure.

Due to the complexity of possible hard disk drives defects, the early failure prediction or even some error detection is complicated. However, it can be based on monitoring techniques. Manufactures of the disk drives noticed the necessity of early failure prediction at the very low level – at the hard disk itself. The first idea of disk self-monitoring was introduced by IBM with their reliability prediction technology called *Predictive Failure Analysis* (PFA) in 1992. It evaluated and been redefined by the SFF Committee and finally taken under the T13 committee and specified as a part of the ATA standard [21]. Now, the SMART is available in all contemporary disks.

The SMART specification says that the intention is "... to protect user data and minimize the likelihood of unscheduled system downtime that may be caused by predictable degradation and/or fault of the device" [21]. The word *predictable* is very crucial as the practice shows that the SMART efficiency is limited [23]. It is actually not in contradiction to the SMART goals as it was created to cover "predictable degradation and/or fault of device". As the disk failure can have many different origins (faulty electronic components, mechanical defects, buggy firmware, etc.), SMART monitors (internally at the HDD electronics level) quite rich set of attributes describing disk operation and health (related to environmental conditions, performance, internal error contamination ratios, physical aspects of servo mechanisms, etc.). Unfortunately, even though the SMART is standardized, the interpretation of many attributes is not publically available and is a proprietary secret of the disk manufacture. Moreover, the set of attributes and its interpretation can differ between models of the given manufacture. Some differences can also be present within the same disk model but supplied with different firmware versions. Possibly, it is one of the reasons for still very low level of popularity and practical usage. The other factor is probably mentioned earlier limitations of accessibility to the SMART data in many advanced storage systems (i.e. inaccessible for the third party software). Nevertheless, such controllers use the SMART to predict and diagnose failing disks.

Despite the problems described above, the SMART delivers some interesting data hard to get other ways (e.g. internal temperature, number of sector reallocations, data transmission errors, hardware error correction efficiency). Moreover, the SMART data can be correlated with other sources of monitoring sources, especially the performance counters and system logs (discussed later on). Figure 4 presents the temperature fluctuations of three monitored disks in one of Institute's workstations. The first disk has additional cooling fan installed. The spike of the disk temperature to over 46°C took place as the fan failure. Other interesting observation is the correlation of the small temperature rises of two disks during the weekends (greyed vertical stripes) which denotes the weekly disk-to-disk backup operations (disks 1 and 2 – the temperature of disk 3 was not changing so much). This was correlated with the system/application logs. Monitoring the storage temperature can detect a hardware failure or can be used as a reference for proper setting of the cooling system (e.g. server rooms are often undercooled) saving the power costs.

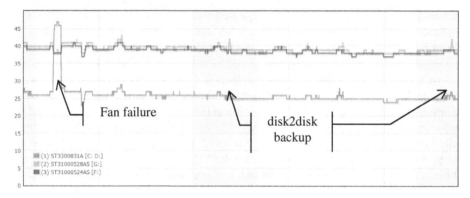

Fig. 4. Identification of specific events with hard disks temperature

In order to analyze the SMART data the *Disk Monitoring and State Acquisition System* (DiMAS) was developed in the Institute. It collects SMART data from the monitored systems (i.e. nodes) and stores that data in central database for further analysis. Figure 5 presents DiMAS architecture.

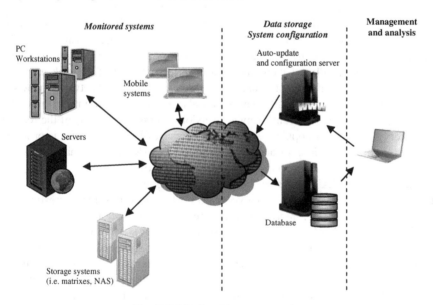

Fig. 5. DiMAS architecture overview

At each monitoring node the agent capable of SMART reading is to be installed. The agent should be operating all the time of the system runtime - even if no user is currently logged on. Moreover, it should automatically be turned on during the target node startup. So, the agent is a system service in Windows and it should be a daemon in case of Linux (not implemented yet). So far, the Windows agent is implemented using the *CrystalDiskInfo* [22] application's core to read the SMART attributes from the disks. Other agents (e.g. for Linux, AIX), supporting also the integration with the performance monitoring, are planned to be implemented in the future.

The configuration of all agents is maintained through the central configuration file distributed through the web page. It simplifies the system management as any configuration changes will automatically propagate to all nodes (each node periodically downloads the up to date configuration). One of the most important attribute of the configuration file is the encrypted connection string to be used by the nodes for the database. Thanks to that, the relocation or reconfiguration of the central database can be easily made. Similarly to the configuration file, the automatic updates of the agents are planned to be implemented. The other important feature of the node agent is local buffer for collecting SMART data during unavailability of the central database. It is especially important for laptops - they locally collect the data and push it to the database as it is be available in the network. That also simplifies the system management (e.g. the database can be maintained offline or relocated) while assuring uninterruptable data collection.

The collected data can also be extended towards storage performance counters/logs. In particular, the Windows performance counters include details on each physical disk like current and average values of, for example, number of bytes read/written per second and the read/write queue lengths. The poor performance of the particular disk should also be considered as one of the major indicators of its health (e.g. long latencies due to mechanical or platter problems).

5   Performance Oriented Analysis of Developed Repository

Preliminary experiments concerning detection of normal and abnormal computer system behavior have been conducted in the environment, which is provided by the Institute of Computer Science (Fig. 6). The monitored machine hosts a small Passim repository, which mimics the future environment. For the monitoring purposes standard *snmpd* program is used. It enables remote collecting of crucial performance counters via SNMP (Simple Network Management Protocol) [3]. This program runs on the monitored machine (see Fig. 6). All other activities associated with periodic data acquisition and analyses are performed on other machines. This approach minimizes time overhead and adaptation changes within the monitored machine - preliminary experiments proved low impact on the operation of that system.

Selected system performance data are collected remotely by SNMP protocol and stored locally at the monitoring machine in separate text files. For this purpose a special, custom made script, is periodically executed (once per five minutes) by standard Unix *cron* system utility. The implemented custom shell script utilizes SNMP tool *snmpwalk*, which allows selecting only interesting parts (in SNMP usually called *branches*) from SNMP message information base (MIB) [3]. All these actions are performed within the dedicated monitoring machine. In the developed experiments only data concerning processes activities on the monitored machine are used. For the experiments purpose three branches, identified by particular OID (i.e. object identifiers and associated names) are used: given process CPU utilization (.1.3.6.1.2.1.25.5.1.1.1, HOST-RESOURCES-MIB::hrSWRunPerfCPU), process name (.1.3.6.1.2.1.25.4.2.1.2, HOST-RESOURCES-MIB::hrSWRunName) and execution parameters (.1.3.6.1.2. 1.25.4.2.1.5, HOST-RESOURCES-MIB::hrSWRun Parameters). The process identification number (pid), used for unique identification of particular process, is taken directly from the received OIDs.

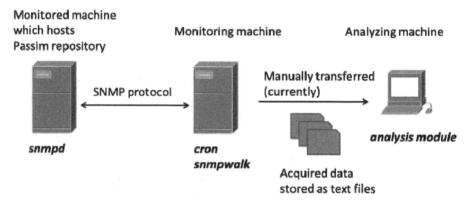

Fig. 6. Topology of testbed installation associated with PASSIM project

During the preprocessing phase, the data described above (concerning each process) is transformed into the item set where a particular item represents interesting features like, for example, PID, process name or CPU usage. Additionally, aggregated item sets (describing all the processes from a few subsequent measurements) are examined in the further analysis. At this stage, the *frequent sets* discovery technique can be used [1]. The main advantage of this approach is ability to detect a repeatable pattern. For example, we can detect a particular process with a unique PID or a set of processes (e.g. of the same application) which usually utilize CPU in very similar way (e.g. intensively). Moreover, the technique can automatically find associations of objects on different attributes (e.g. correlate different processes by names not only PIDs). The discovery of frequent sets is performed on static data and gives only static pattern description. However, *jumping emerging patterns* which are used during this research enable detection of changes in the analyzed data [7]. Jumping emerging pattern (JEP) is discovered between two data sets. It is represented by a frequent set that appears in one data set but is not observed in the second one. In conducted experiments, the detected JEPs are associated with an important process that was accidentally shut down or with the execution of CPU intensive task, etc.. JEPs are the basis for further characterization of normal system behavior and detection of anomalies.

5.1   Detection of Normal and Abnormal Behavior

Detection of the abnormal activity bases on finding the normal state of the computer system and the identification of state changes. For this purpose some data mining methods are used - *frequent sets* for detection of normal computer system state and *jumping emerging patterns* for detection of state changes. The detailed description of abnormal behavior detection process is described in the following section. Additionally, Figure 7 contains an outline of the detection process.

As mentioned, the performance counters of processes at the monitored system are periodically collected using SNMP. The measurement takes place each 5 minutes. In Fig. 7, such measurement is presented as a vertical line at the x-axis. Every two adjacent measurements are compared; and, in effect, all the processes, which utilize CPU or have just been created during this period, are determined. For each process, an item set is constructed which contains process PID, identifier of the process name and identifier of the class describing the amount of used CPU time. The item set is added to the analysis window. In the next step, the frequent sets are discovered in this window. For example, *window 1* in the Fig. 7 contains information concerning processes that are running in the time range from t_1 to t_n. During preliminary experiments various windows lengths and corresponding minimal support values were examined. Basing on these results, window size of the 12 measurements (corresponding to 1 hour) and minimal support of 12 were used during further experiments to mitigate irrelevant data fluctuations while keeping the reasonable responsiveness of the system. Frequent sets detected in such a way represent processes that are always active at the machine. Conducted experiments indicate that this method allows for an automatic detection of particular processes connected to machine role (e.g. database, HTTP or application server). The processes are described by detected frequent sets, containing items corresponding to the PID or identifier of a process name. Additionally, abnormal activity that lasts longer than 1 hour can be easily detected. The sliding window technique allows easy recalculation of frequent sets that describe the last hour as soon as a new measurement is collected. This is presented in the Fig. 7, when a new measurement is taken at the time t_{n+1}. Item sets describing processes at the time instant t_1 are removed and new item sets concerning process at t_{n+1} are added. In effect, *window 2* contains information from time range $<t_2, t_{n+1}>$. Thanks to jumping emerging patterns (JEPs), changes in the monitored machine state can be revealed: they are described by disappeared frequent sets and some new frequent sets appearing in adjacent windows. Each pattern can be easily interpreted as the addition of the new process or the termination of the important one (potentially abnormal case). Further section presents some empirical results showing the applicability of this technique.

5.2   Experimental Analysis Module

Currently the analysis module is implemented as an independent program that can be used for manual analysis of data acquired by the monitoring machine. There are plans to integrate program as real-time monitoring tool installed at the monitoring system. In

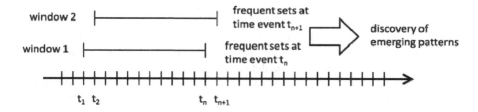

Fig. 7. Two consequent sliding windows in which frequent sets are discovered and later compared for detection of jumping emerging patterns

such solution, information concerning all detected anomalies can be sent directly to the administrator of the monitored machine. In the first step, the implemented analytical module preprocesses the collected data describing all executed processes, initially stored as text files. At the end of this process, the data have the form suitable for data mining frequent sets method. Finally, multiple items sets, each describing one process, are produced. In the second step all item sets are added to the sliding window in which frequent sets are discovered. When window is moved to the next time instant, changes in processes are discovered using jumping emerging patterns. The performed experiments show that even in normal operation of modern operating system many processes responsible for various maintaining tasks are periodically executed. Due to this fact presenting information on each process which changes its state is impractical. For this purpose some filtering rules are implemented, for example, only coincident changes of a few processes in the same instant are presented. Finally, only the most important events are directly presented to the user and for the further analysis all events are stored in logs. Additionally, generation of normal behavior model can be correlated with some other factors for its better understanding. First experiments that correlated frequent processes with CPU high usage are promising. Listing 1 shows exemplary output generated by implemented analytical module. It relates to the situation with high CPU utilization due to the backup tasks which were performed few times per month. Thanks to the frequent sets analysis, a set of different processes used by a backup procedure (unrelated at first glance) have been successfully identified.

```
CPU High utilization period starts at 01:55 and lasts 11
consecutive windows
pid 1998 (java), 1998, Process name "java",  - support 11
pid 30125 (Agent), 30125, Process name "Agent",  - sup-
port 11
Process name "backup",  - support 11
100009, Process name "cpio",  - support 10
100009, Process name "ssh",  - support 11
100009, Process name "gzip",  - support 11
```

Listing 1. Sample output from the monitoring module showing detected frequent sets in the interesting period when high CPU utilization was observed

The most important information is at the beginning and at the end of the sample listing. In the first line information concerning time is presented. Discovered high CPU utilization period starts at 1:55 and lasts for 11 subsequent windows (each 5 minutes long). The last four lines explain the root cause for high CPU utilization. All names associated with detected processes can be easily traced to a backup procedure. The remaining lines, associated with process named *java* and *Agent*, describe processes that were also active during automatically detected high CPU period. Further analysis proved that these processes should always be working on the monitored machine.

Preliminary experiments, conducted in the presented environment using implemented prototype of the analysis module, prove that this methodology can be used for detecting various anomalies. During the analysis of normal operation of the small Passim repository for more than two months, a few anomalies described in the sequel have been detected.

As presented, the proposed method can automatically detect a set of processes that works on the monitored machine almost all the time. In the most cases such processes are critical for service, which the machine provides to the users. Additional investigation shows that automatically detected process with *java* name is associated with *JBoss* application server hosting logic for Passim repository. During the conducted experiments this process was accidentally shut down two times. In effect, users lost access to data provided by the repository. What should be emphasized in one case, system administrator was informed by our team before other users realized any problems with this service, despite of real-time automatic reporting feature.

Problems with appearing and disappearing services are observed in a few other cases. In one case, the problem was associated with improper startup configuration of the monitored machine. After the reboot one of the analytical processes did not start automatically as was expected. In the second case, an improper firewall configuration denied all access for SNMP - in effect the implemented software detects two days long machine shutdown. Some detected issues are not revealed till now and require additional detailed analysis. One issue is associated with appearance of new process after machine reboot – called *hstat_col*. The second one is connected with sudden rise of time needed for backup activity. Due to high CPU usage this procedure was periodically detected each Wednesday and Friday. At the beginning of analyzed period backup procedure lasts one hour. At the end of analyzed period this time rise more than two times. Additionally to those, minor detected events are associated with detection of the monitored machine shutdowns or some temporary network outages. This information can be used for statistic researches concerning services or network availability to the end users.

6   Conclusion

This research proved the potential of event and performance logs as a source of detecting system anomalies, finding workload properties, operational profiles, etc. A key issue is collecting appropriate knowledge based on real data. For this purpose we have developed some supplementary tools (or scripts) targeted at collecting and processing various logs e.g. extraction of some statistical and other properties, log categorization (de-parameterization and normalization), identifying anomaly signatures. The gained experience showed the need of creating additional logs filled in by system administrators and users. Their content can be correlated with event and performance logs to diagnose the anomaly source. This approach was helpful in resolving some occurring problems in our computing environment. Another important issue is to avoid detecting anomalies in case of workload changes (they have to be identified by the monitoring processes).

In further research we plan to develop more advanced log analysis algorithms in particular correlating features of different logs (multidimensional approach). This will provide indications to domain experts helping distinction between normal and anomalous system behavior. This approach will be verified in the developed data repository system.

Acknowledgement. This work is supported by the National Centre for Research and Development (NCBiR) under Grant No. SP/I/1/77065/10.

References

1. Agrawal, R., Srikant, R.: Fast algorithm for mining association rules. In: Bocca, J.B., Jarke, M., Zaniolo, C. (eds.) Proceedings 20th International Conference on Very Large Databases, pp. 487–499. Morgan Kaufmann Publishers Inc., San Francisco (1994)
2. Brandt, J., et al.: Using probabilistic characterization to reduce runtime faults in HPC systems. In: 8th IEEE Int. Symp. on Cluster Computing and the Grid, pp. 759–764 (2008)
3. Case, J., Fedor, M., Schoffstall, M., Davin, J.: A Simple Network Management Protocol, SNMP (1990)
4. Chandola, V., Baerjee, A., Kumar, V.: Anomaly detection, a survey. ACM Computing Surveys 41(3) (2009)
5. Chen, C., Singh, N., Yajnik, M.: Log analytics for dependable enterprise telephony. In: Proc. of 9th European Dependable Computing Conference, pp. 94–101. IEEE Comp. Soc. (2012)
6. Cinque, M., et al.: A logging approach for effective dependability evaluation of computer systems. In: Proc. of 2nd IEEE Int. Conf. on Dependability, pp. 105–110 (2009)
7. Dong, G., Li, J.: Efficient mining of Emerging Patterns: Discovering Trends and Differences. In: Proceedings of the 5th International Conference on Knowledge Discovery and Data Mining (SIGKDD 1999), San Diego, USA, pp. 43–52 (1999)
8. Gmach, D., et al.: Workload analysis and demand prediction of enterprise data canter applications. In: 10th IEEE Int. Symposium on ISSWC, pp. 171–180 (2007)
9. John, L.K., Eeckhout, L.: Performance evaluation and benchmarking. CRC Press (2006)
10. Król, M., Sosnowski, J.: Multidimensional monitoring of computer systems. In: Proc. of IEEE Symp. and Workshops on Ubiquitous, Autonomic and Trusted Computing, pp. 68–74 (2009)
11. Kubacki, M., Sosnowski, J.: Analysing Linux log profiles in Linux systems. In: Borzemski, L., et al. (eds.) Information Systems Engineering, pp. 135–143 (2011) ISBN 978-83-7493-630-9
12. Magalhaes, J.P., Silva, L.M.: Adaptive profiling for root-cause analysis of performance anomalies in Web based applications. In: Proc. of IEEE Int. Symp. on Network Computing and Applications, pp. 171–178 (2011)

13. Naggapan, M., Vouk, M.A.: Abstracting Log Lines to Log Event Types for Mining Software System Logs. In: Proceedings of Mining Software Repositories (Co-Located with ICSE 2010), pp. 114–117 (2010)
14. Oliner, A., Stearley, J.: What Supercomputers Say: A Study of Five System Logs. In: Proc. of the IEEE/IFIP International Conference on Dependable Systems and Networks, pp. 575–584 (2007)
15. Salfiner, F., Lenk, M., Malek, M.: A survey of failure prediction methods. ACM Computing Surveys 42(3), 10.1–10.42 (2010)
16. Simache, C., Kaaniche, M.: Availability assessment of SunOS/Solaris Unix systems based on syslog and wtmpx log files; a case study. In: Proc. of the 11th Pacific Rim International Symposium on Dependable Computing (2005)
17. Sosnowski, J., Poleszak, M.: On-line monitoring of computer systems. In: Proc. of IEEE DELTA Workshop, pp. 327–331 (2006)
18. Sosnowski, J., Gawkowski, P., Cabaj, K.: Event and performance logs in system management and evaluation. In: Information Systems in Management XIV, Security and Effectiveness of ICT Systems, pp. 83–93. WULS Press, Warsaw (2011)
19. Ye, N.: Secure Computer and Network Systems. John Wiley& Sons, Ltd., Chichester (2008) ISBN 978-0-470-02324-2
20. Li, Y., Zheng, Z., Lan, Z.: Practical Online Failure Prediction for Blue Gene/P: Period-based vs. Event-driven. In: Proc. of the IEEE/IFIP International Conference on Dependable Systems and Networks Workshops, pp. 259–264 (2011)
21. Information technology - AT Attachment 8 - ATA/ATAPI Command Set (ATA8-ACS), T13/1699-D (2006)
22. CrystalDiskInfo homepage, http://crystalmark.info/
23. Pinheiro, E., Weber, W.D., Barroso, L.A.: Failure Trends in a Large Disk Drive Population. In: Proc. of the 5th USENIX Conference on File and Storage Technologies (2007)

Handling XML and Relational Data Sources with a Uniform Object-Oriented Query Language

Emil Wcisło, Piotr Habela, and Kazimierz Subieta

Polish-Japanese Institute of Information Technology, Warsaw, Poland
{ewcislo,habela,subieta}@pjwstk.edu.pl

Abstract. Where data integration is considered, a natural choice is establishing a canonical data model which needs to fulfill two essential requirements. Firstly, it has to be suitable and expressive enough from the consuming applications point of view. Secondly, it should include constructs allowing for a possibly straightforward mapping of the data structures of the resources being integrated. The object data model based on classes seems to be an intuitive and sufficiently expressive choice for this role. However, the other data models used in the resources being integrated result in a mismatch that needs to be handled in order to avoid the productivity and maintainability penalty. This paper presents the data store model used by the SBQL query and programming language and discusses the already developed and future features of the language aimed at uniform handling of relational and XML data.

1 Introduction

The integration of external, independently created and managed data sources can be very challenging even in case of technologically and lexically homogeneous environment. The differences of semantic or temporal nature, content languages, units and defaults or the autonomy and security requirements, to mention those few exemplary categories, may require non-trivial behavioral elements for that data mapping. Considering the amount this inherent complexity, it is of primary importance to mitigate the accidental complexity caused by the differences between the data models of resources participating in the integration. Hence it is essential to select and establish a canonical data model and make the mappings to and from it possibly straightforward.

For the object oriented database and integration environment being developed by us [1] and the programming/ query language SBQL used with it, an advanced variant of class-based object model has been chosen. In comparison to the model known from popular object oriented programming languages, it directly supports composition (allowing to nest complex objects) and is open for other features like dynamic inheritance based on roles. As a relatively rich data model, this has been chosen to serve as the canonical model to which the contributing data sources are mapped in order to be consumed by SQBL application. In this paper the issues of achieving such data model alignment around an object data model, including the updateability of data sources, are discussed. Although the problem of integration outlined here is of unidirectional nature (that is, we consider handling other kind of a data source from our object environment

R. Bembenik et al. (Eds.): *Intell. Tools for Building a Scientific Information*, SCI 467, pp. 519–528.
DOI: 10.1007/978-3-642-35647-6_31 © Springer-Verlag Berlin Heidelberg 2013

and not vice versa), it is still relatively difficult due to substantial differences between the data models considered. While the native solutions for particular source data models are known and realize their goals, our challenge is handling at least their essential features in our model in a way that would not obscure our object-oriented environment and keep it as clear and intuitive as possible. As two prominent kinds of the source data environments we are going to consider the integration with relational database management systems and with XML documents.

This paper is organized as follows. Section 2 presents the starting point of our design, which is an abstract data storage model for our object-oriented database environment. Section 3 confronts it with the requirements of handling relational data. In section 4 the characteristics of XML data and the limitations of the object model resulted from them are outlined. Section 5 provides a definition of the extended object store model and outlines the semantics associated with it. In section 6 a brief description of the approach to design the data source adapters is provided. Section 7 concludes.

2   Initial Object Store Model

The foundational element of a language environment and its underlying data model is the abstract data store definition. It determines the amount of information included in the data structure and provides the basis for a precise definition of the language semantics. For the semantics definition we follow the Stack Based Archtecture (SBA) approach [1]. The definition of such structure is supplemented by the expression results data model, which is usually richer due to additional structures the derived data can constitute. Our reference object store model considered in this work as the starting point is called AS1 [1] and is defined as a five-tuple <S, C, R, CC, SC> consisting of the following:

- S - set of objects,
- C - set of classes (special kind of objects),
- R - set of identifiers of the root objects (visible in a global environment as the starting points for the navigation through the object store graph),
- CC - relation $i_c \times i_c$ of respective classes, denoting the class generalization and inheritance hierarchy,
- SC - relation $i_s \times i_c$, denoting the respective objects being instances of particular classes.

Each object in our data store, apart from its value, has its own name (not unique) and identifier (unique). This leads to defining the object as a triple. The following kinds of objects have been defined[1]:

---

[1] An attentive reader would note important differences of this model from e.g. Java object model. The first one, clearly inspired by the database systems field, is making the name a property of each object (rather than just allowing associating named references with it). Another one is providing an explicit object identifier also to primitive objects. From the conceptual point of view it is essential for updateability. It does not necessarily imply supporting associations leading to primitive objects.

- $<i, n, v>$ - representing primitive objects having identifier i, external name n and atomic value v,
- $<i1, n, i2>$ - a pointer object, where $i2$ is a reference to another object,
- $<i, n, T>$ - a complex object, where T represents a set of objects. The recursive nature of this definition allows complex object to include not only primitive and pointer objects, but also nested complex objects with no conceptual limit on the depth of the nesting.

On the schema definition aspect, classes support declaring, among others, the attributes with their multiplicities (similarly as the in the UML [2] class diagram).

Regarding the query result model, additional constructs need to be supported. Apart from object identifiers and primitive values, this includes the following building blocks that may constitute complex values:

- Binder - is a pair N(x) used to provide a result with auxiliary name. N is an external name, while x is any result of a query,
- Bag - resulting from the queries that may return non-unique values or repeating identifiers,
- Sequence - resulting from the use of ordering clause in queries,
- Struct - a structure (tuple) constructor, providing a value consisting of one or more, usually named, elements being arbitrary query results.

Note that with the absence of the requirement for the struct to have all its elements named (that is - being binders in our terminology), the distinction between it and a bag or a sequence blurs. However, having a separate value constructor is justified by the clarity of semantics (e.g. consider the roles the bag and struct play during the typical evaluation of the Cartesian product operator). Moreover, the semantics of our query operators assumes "flattening" of direct nesting between collections and between structs, hence resorting to the struct is needed to achieve collection nesting. From the practical point of view this is anyway necessary to assure a meaningful (consumable) result, and, except for syntactic differences, works similar to other object-oriented expression languages like e.g. OCL [3].

Another foundational element of the query / programming language semantics is the Environment Stack (ES). To provide it with all the names occurring in the environment of a given evaluation step, it is populated with binders. During expression evaluation ES is scanned top-down, which means - from the most local towards more global environments represented by particular sections. The notion of section is essential for collection processing as it allows the expression to bind a number of elements. While it is recommended to refer to the systematic description of SBA, let us briefly explain here the interrelation between the store model and the query processing, using a very simple example. Assuming that there are root objects named Emp of class EmployeeClass that defines the fields FirstName, LastName, Salary, Job, Subordinate (the last of multiplicity 0 or more), the processing of a query Emp where Job = 'programmer' would (conceptually) proceed as follows.

- The initial state of ES is populated with the names of objects found in the set R of root objects. More precisely the binders $N(i_x)$ appear in the section of the ES.

Among them there are a certain number of binders Emp(i_x) representing each instance of Emp in the store.

- Subexpression Emp becomes evaluated and, in effect, the bag of identifiers of all the Emp objects is placed on the Result Stack (*RS*).
- For each identifier of the Emp object collected from the *RS* the subquery Job = 'programmer' is evaluated. To perform it, the *ES* must be first updated with the section of binders representing the methods of the class EmployeeClass and with the binders representing subobjects of the currently evaluated instance - in our case the binders FirstName, LastName, Salary, Job and zero or more binders Subordinate. The availability of binder Job(i_j) will allow to determine the identifier of a respective primitive object Job, its retrieval from the store and extraction of its value (so-called dereferencing) to perform the comparison.
- After completing all the iterations, the identifiers of the Emp objects for which the predicate evaluated to true, would be placed, instead of the original Emp query result, on the *RS*.

The availability of object identifiers allows to perform actions against those objects like updating, removal or value retrieval (those actions, parameterized with object identifier, are called *update*, *delete* and *read* respectively). Their explicit definition is essential from the point of view of the virtual updateable view mechanism, in which all those actions can be overridden with explicit behavior specification defined against the virtual (derived) data. The *read* action against a complex object should not be confused with opening the local environment (called operation *nested* in the SBA) for accessing subobjects (e.g. as performed by the where operator in the above example). The normal behavior of the *read* action in that case would be returning a complex structure of binders containing the respective subobject values.

Note that the overall pattern of the SBA is very flexible and hence capable of supporting much more sophisticated data models. The necessary adjustments would include extension of the store model (the core aspect as it determines the amount of information built in into the data model) and the respective query result model extensions and new operator definitions.

The shape of the data model presented above can be described as an attempt to combine the mainstream concepts of object oriented programming languages (and - to some extent - certain notions of conceptual modeling as known from UML) with the characteristics of database management systems. This data model seems fully suitable as long as we deal with regular, structured data and the evolution of objects over their lifespan is limited.

3   Adapting Relational Data

Relational data model uses highly regular structures (with slight irregularities coming from the use of null values only) that can be easily mapped onto our object model. However, several semantic details need to be resolved and the usability and performance factors have to be taken into account. The relational data adapter can be realized as a set of updateable object views, offering a set of complex objects for each relational table contributed for integration [4]. The simplicity of the relational model makes it possible to reflect the core conceptual information encapsulated in the relational schema design, into the schema definition describing those views.

Following the semantic foundations outlined in the previous section, we now need to review our object store model and then to outline the mapping onto it. Clearly, our complex object definition is suitable for representing a table content, with table name mapped onto the object's name and the column names reflected in its subobject names. Identifiers need to be determined. For simplicity, we assume here requiring the presence of a single-column primary key in each table. In this case our identifier can be constructed as the tuple $<m, t, i>$, where m is the identifier of the database module generated for the data source and assigned during the import and containing the adapter views definition, t is the name of the table and i is the value of the primary key.

The fact it was not necessary to extend our reference store model removes the necessity of revising further layers of the language environment. The basic variant that assumes having all the columns represented explicitly with their values as virtual object attributes, as implemented in the current version of our prototypes, may be however refined so as to better exploit the expressiveness of the object model. Namely, we suggest removing the attributes representing foreign keys and, in place of them, adding bidirectional associations (as the pairs of virtual pointers) between the respective objects. The primary key derived attributes on the other hand would be left available as the actual values in case of some keys may bear business meaning.

The last aspect that should be outlined here is the optimization of queries. Contrary to other, more ad-hoc data solutions, relational database usually contain their native, efficient optimization means. Hence the challenge for the adapter design is to attempt reusing them rather than depending on "naïve" fine-grained interface implying performing all the query evaluation on the integration environment side.

4   XML Data Model Characteristics

A superficial look at the XML data model, especially when examining some simpler and more regular cases (resulting e.g. from dumping programming language objects or relational database tables), could lead to the conclusion that this is a hierarchical structure an object data model with composition can absorb easily. Indeed the repetitive and optional information resultant from including multiple subelements or skipping subelements, can be easily accommodated with appropriate multiplicities of respective object attributes. Also the fact our data model assumes object names to be inherent properties of each object, nicely matches with the nature of XML element tags that impose name for each element.

The notion of XML attributes, as the construct distinct from subelements poses some complication, but this can be approached with a notational convention (e.g. preceding attribute names with @ character, as in the XPath language [5]). Also the XML namespaces, although requiring special handling in data storage and expression syntax, do not constitute a conceptual problem from the point of view of our data model.

However, there are several features of XML data model that differ radically from the intuitions grown in the database management field. Firstly, the notions of tuples and collections are not distinguished at the XML side. In both cases they are represented as sequences of subelements. Secondly, the order of subelements is assumed to bear information in the XML data model. While this could be perceived as an unnecessary overhead from the database point of view (rows of a relational database are not ordered; similarly, objects in object databases are often kept in unordered containers), this becomes natural when considering e.g. the Web markup documents (i.e. when processing a document consisting of elements that represent its chapters we may expect their order remain undistorted). For that matter XML guarantees maintaining the order of all the elements. Hence, e.g. if a document chapter consists of heading, section and side-note elements, it is not sufficient to record the order of headings, sections and sidenotes separately for each of those element kinds. Hence, speaking in UML terms, defining a class `Article`, having attributes `heading`, `section` and `sidenote`, each declared as multiplicity 0..* and having a flag `isOrdered`, would not suffice to maintain the document structure information in such a schema.

Finally, the most problematic feature of mapping XML data is the so-called mixed content. This allows element's having both regular text and subelements, with the subelements interwoven in arbitrary positions within the text. The presence of subelements prevents mapping such element as a simple string-type object attribute, which would be possible for a purely textual content elements. A similar, although less difficult issue occurs in the case of textual element having (XML) attributes.

Comparing our database environment design with the XML expression and query language state of the art reveals also some "cultural" differences. One of them is the XML ability to represent highly irregular data, uncommon to database field. The irregularity of that data can be also reflected with a very general or even missing document schema definition, which undermines strong type checking. The optionality of identifiers in XML document breaks the principle of internal identification [6,1], thus making the document update problematic.

5   Extended Object Store Model and Processing Routines

As can be observed from the previous section, the data model differences between XML and the canonical object-oriented model considered by us are substantial. If the main assumptions of such a data environment (e.g. object orientedness, strong type checking, precise schema definition) are to be maintained, some compromise with respect to directly supported features needs to made.

Clearly, our goal is to take advantage of the intuitiveness of representing XML attributes and some elements as well, with attributes of primitive type. Hence, to reach the "low hanging fruits", we are going to assume that each XML element would produce an object which:

- records its position with respect to its sibling objects, but only those of the same name and (assumed) type,
- is capable of providing some primitive value if requested so.

The first feature requires augmenting our store model with collection ordering capability. Hence we need to add another, optional complex object type:

- <i, n, Q> - where Q represents a sequence of objects.

Let us note that this store model feature could allow a complete ordering of all subobjects of a given object (that is, also those representing different attributes in the schema). However, in our approach we are going to provide that capability only within particular object attributes. Hence we would be able to formulate for example a query `Article.Chapter[2]`, but not `Article.*[5]` (which could mean any feature of the Article that occupies fifth position in the sequence of its children objects). Although the latter would satisfy the requirements towards dedicated XML query languages [5,7], it is not compatible with the strong type checking feature of our environment.

Regarding the semantic details of processing, the mutual ordering of such multi-valued attribute instances requires providing their binders on the *ES* with additional numbering so as to assure their sequence position information.

The second of the abovementioned features is more sophisticated. It is the place of making the compromise to provide simple mapping of primitive element, but at the same time to offer some consistent and possibly useful functionality in case of the mixed content. Our idea here is to introduce a kind of "hybrid" objects by differentiating the kinds of complex content as "visible" by operation *nested* and action *read*. Namely, any XML element having if only one attribute or a one nested element, would produce a complex object in our mapping. The *nested* operation, if invoked would then locate all the object attributes as usual, whereas the *read* action would always return just a raw content of the source XML element (that is, everything located between its opening and closing tag) as a string.

Let us consider an element `<link url= "http://www.example.com">Sample</link>`. For the object derived from it, an expression `link` would return just a string value "`Sample`". However, a programmer may also treat this as a complex object and access its feature with expression `link.@url` , which would return a string value "`http://www.example.com`".

A less natural, but consistent and still somewhat useful, is the result of applying this approach to mixed content element. Assuming there is e.g. an element `chapter` containing text, inside which elements like `person`, `place` and `link` occur, in our approach a programmer could retrieve, update or remove e.g. the third element `link` contained in that object. However, by accessing that complex object features, they would not be able to determine the exact location of that subelement with respect to other kind of subelements and the enclosing text content. Those details can be determined, if needed, by retrieving the complete raw content of the element as a single string value and parsing it. Also, since the abovementioned hybrid objects need to be distinguished in the type system anyway - their base class may provide generic tools known from typical XML programming interfaces like e.g. Document Object Model [8].

Updating the XML data source also faces some limitations. Firstly, according to our abstract store model, we need to assign each object (which in this mapping means - each XML element and attribute) a unique identifier. We could construct such as the tuple of values <m, x, a>, where m is a module identifier assigned during importing a particular document, x is the position of the element in the XML document (making it possible to retrieve using the (//*) [x] XPath expression), whereas a only applies to attributes and contains the attribute name. Note that such identification is not reliable enough for creating association links to such virtual objects in case the document is updateable. It might be desirable to consider a limitation for such linking that allows them only for elements that posses their XML identifiers and hence can be identified with a tuple <m, i, a> instead, where i is the XML-side identifier value.

Regarding the updates it can be noted that our object environment side strong type checking may not be capable of enforcing all the fine details of the XML document schema definition. In such case an additional validation run against the XML Schema definition (as known from e.g. JAXB interface [9]) should be considered to assure the document consistency.

6   Adapter Design

In the preceding sections we have outlined the mapping issues conceptually, focusing on the matching the constructs around the abstract store model. This is the most foundational layer of the problem as it reduces the contributed data sources to the constructs of the query evaluation semantics used in our canonical environment. However, this leaves open the actual practical scenario of bringing the data and the architecture of incorporating the optimizations. The discussion included in this section focuses on the XML, but most of the issues are applicable to relational data sources as well.

Regarding the overall scenario of integration, there are two main options. The mapping outlined earlier can be used so as to simply retrieve the data source and populate our object store with its copies. This is the simplest scenario which additionally removes most of the updateability concerns mentioned earlier. It is shown in the Fig. 1 in its bottom left part. This way the XML document wrapper prototype has been implemented in our ODRA environment. However, such an approach can be in many cases inacceptable due to the contributed data source lifecycle shape (e.g. its subsequent outside updates etc.).

Hence, in this paper we mainly consider the other option, which assumes working on the original data rather than on its copies, using the updateable virtual views mechanism. This is represented in the bottom right part of Fig.1.

Note however the straightforward application of this approach would assume shifting the whole query evaluation onto the side of our object-oriented environment. Considering the simple exemplary query presented earlier, this would mean we need to retrieve on the object-oriented side not just the selected objects Emp, but also the values of the Job subelements for all the Emp elements in the document (with all the respective virtual identifier construction and mappings involved). The analogous problem would occur for other kinds of data sources imported, most notably the relational databases. The drawback in this case is the excessive number of calls and unnecessary layer traversals. A more optimum approach is detecting during the static analysis and encapsulating those expression abstract syntax tree fragments that depend solely on the respective external data source and replacing them with the call containing the generated query code native to a given data source [10]. This step can be performed together with other optimizations performed against the syntax tree (like e.g. removing dead subqueries, avoiding unnecessary iterations or using indices) and has similar nature. Hence the presence of the *query API* element in the figure above. The call of such query performing API operation would result in placing on the *RS* the set of virtual identifiers of the objects representing the respective query result. Due to the view transparency, the further processing of that result may proceed exactly like in case of regular, concrete objects stored in our environment.

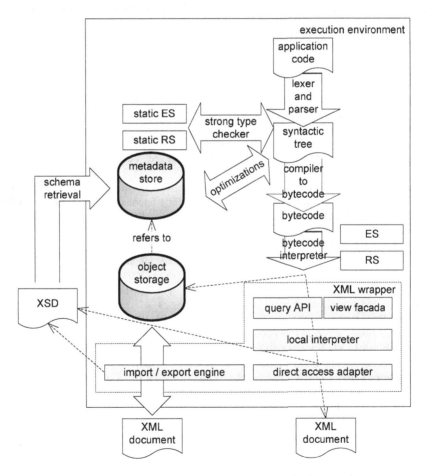

Fig. 1. XML wrapper as an element of the integrated database environment

7   Conclusions and Future Work

In this paper the issues of integrating external data sources under a common, canonical object-oriented data model have been presented. Among the two kinds of data sources described here, the XML structures are much more demanding with respect of their adequate and straightforward mapping. In order to balance the ease of use with the completeness of support, a compromise solution has been proposed. This includes an extension of the abstract storage model and introducing a specific, "hybrid" behavior of XML-derived complex objects, inspired by the updateable object views concept. Considering the performance issues, the approach of encapsulating and delegating subqueries to the data source has been suggested. Our existing prototype implementations include the schema aware XML importer and the basic variant of the relational database wrapper in the ODRA system. The current work is focused on the development of similar functionality for an object query language SBQL for Java [11] and the extension of this integration capabilities towards the virtual view based wrapper approach.

References

1. Stack-Based Architecture (SBA) and Stack-Based Query Language (SBQL) website, Polish-Japanese Institute of Information Technology (2011), http://www.sbql.pl
2. Object Management Group. OMG Unified Modeling Language (OMG UML), Superstructure. Version 2.2. formal/2009-02-02, http://www.omg.org
3. Object Management Group. Object Constraint Language. Version 2.2.formal/2010-02-01, http://www.omg.org
4. Adamus, R., Kaczmarski, K., Stencel, K., Subieta, K.: SBQL Object Views - Unlimited Mapping and Updatability. In: Proceedings of the First International Conference on Object Databases, ICOODB 2008, Berlin, March 13-14, pp. 119–140 (2008) ISBN 078-7399-412-9
5. World Wide Web Consortium. XML Path Language (XPath) 2.0. W3C Recommendation (December 14, 2010), http://www.w3.org/TR/xpath20/
6. Subieta, K.: Theory and construction of object query languages. Editors of the Polish-Japanese Institute of Information Technology, 522 pages (2005) (in Polish)
7. World Wide Web Consortium. XQuery 1.0: An XML Query Language. W3C Recommendation (December 14, 2010), http://www.w3.org/TR/xquery/
8. World Wide Web Consortium. Document Object Model (DOM), http://www.w3.org/TR/DOM/
9. Laun, W.: A JAXB Tutorial (2012), http://jaxb.java.net/tutorial/
10. Wiślicki, J., et al.: Implementation and Testing of SBQL Object-Relational Wrapper Supporting Query Optimisation. In: Proceedings of the 1st Intl. Conf. on Object Databases, ICOODB 2008, Berlin, March 13-14, pp. 39–56 (2008)
11. SBQL4J - Stack-Based Query Language for Java website, http://code.google.com/p/sbql4j/

Architecture for Aggregation, Processing and Provisioning of Data from Heterogeneous Scientific Information Services

Cezary Mazurek, Marcin Mielnicki, Aleksandra Nowak,
Maciej Stroiński, Marcin Werla, and Jan Węglarz

Poznań Supercomputing and Networking Center
ul. Z. Noskowskiego 12/14, 61-704 Poznań, Poland
{mazurek,marcinm,anowak,stroins,mwerla,weglarz}@man.poznan.pl

Abstract. One of the tasks undertaken in PSNC in the frame of the SYNAT project is the design of architecture for the scientific information system allowing to integrate heterogeneous distributed services necessary to build knowledge base and innovative applications based on it. This paper contains overall description of the designed architecture with the analysis of its most important technical aspects, features and possible weak points. It also presents initial results from the test deployment of the first prototype of several architecture components, including initial aggregation and processing of several millions of metadata records from several tenths of network services.

Keywords: data aggregation, REST architecture, distributed systems, metadata processing.

1 Introduction

One of the first milestones in the history of development of Polish digital libraries is the deployment of the Wielkopolska Digital Library (http://www.wbc.poznan.pl/). This digital library was officially opened in October 2002, as an outcome of cooperation between PSNC and Poznań Foundation of Scientific Libraries [1]. It was very important step for the future of Polish digital libraries for two reasons. Firstly, it was created to be a shared hardware and software infrastructure for several scientific and cultural institutions from the region. Secondly, it was based on dLibra Digital Library Framework, which was developed in PSNC as a configurable and reusable software package designed for wide use. The success of Wielkopolska Digital Library became a motivation for other institutions and regions of Poland to adopt the same approach and create dLibra-based regional digital library [2]. The significant increase of the number of Polish digital libraries observed since 2005 (6 such libraries were available at that time), resulted in the network of 91 digital libraries available in the middle of 2012, giving access to more than a million of object from collections of several hundred memory institutions (see http://fbc.pionier.net.pl/owoc/list-libs for up-to-date list).

R. Bembenik et al. (Eds.): *Intell. Tools for Building a Scientific Information*, SCI 467, pp. 529–546.
DOI: 10.1007/978-3-642-35647-6_32   © Springer-Verlag Berlin Heidelberg 2013

In 2006, with over a dozen of digital libraries existing in the PIONIER network, PSNC decided to start works on the development of a new central service, responsible for the aggregation of digital objects' metadata from distributed digital libraries in order to provide single point of access to the resources of these libraries [3]. The new service, named PIONIER Network Digital Libraries Federation, is briefly described in the second chapter of this paper, with a focus on the evolution of its architecture.

Constant increase of the amount, complexity and heterogeneity of aggregated data, increase of the number of covered sources, as well as rising interest in the federation service caused a need to rethink the idea of this service and redesign its architecture. Therefore the general aim of designing an architecture for aggregation, processing and provisioning of data from heterogeneous scientific information services was set as one of the key aims of PSNC works in the SYNAT national research project. It was assumed that such architecture should allow:

— automated discovery of services available in the system and dynamic introduction of new services,
— unified means of access to functionality offered by particular types of services,
— the possibility to use services of the designed system on various levels of architecture from lowest system layers, up to integration of services implemented in the end-user's web browser,
— large scaling possibilities both in terms of data storage and processing, with the possible use of cloud technologies.

Beside, one of the novel things was the requirement to aggregate data from wide range of scientific information systems (i.e. not only from digital libraries, but also from library catalogues, museum inventory systems, etc.) and to provide all the data in a way which would be useful for creation of knowledge base on top of such data.

First stage of works was devoted to the analysis of data and functionality which is made available by services potentially useful in the knowledge base construction process. For this purpose, as a thematic scope for the knowledge base the cultural heritage domain was chosen. The analysis also covered the (passive or active) way in which a service is able to interact with the designed data aggregation system. On this basis an architecture for open data aggregation, processing and provisioning system based on REST [4] services was designed. It is described in the third chapter of this paper.

The new architecture was evaluated with ATAM method [4], which is shortly described in the fourth chapter, and deployed in a test environment. The test deployment and run of data aggregation process, together with the deployment architecture and example application written on top of that architecture, is the subject of chapter five. The paper ends with a summary.

2   PIONIER Network Digital Libraries Federation

2.1   Overview

As described in the introduction, the increasing number of digital libraries in Poland caused the need for a new service allowing to explore resources made available by these digital libraries via a single access point. Such new service was desired not only by end-users but also by third-party data aggregation services interested in Polish cultural heritage resources available on-line. The mission of the new service was expressed in the following way (http://fbc.pionier.net.pl/owoc/about?id=about-fbc):

"The Digital Libraries Federation mission is to:

— facilitate the use of resources from Polish digital libraries,
— increase the visibility and popularity of resources from Polish digital libraries in the Internet,
— enable new advanced network services based on the resources from Polish digital libraries to Internet users and digital libraries creators."

The initial functionality of the federation was searching in the metadata of digital objects published by cooperating digital libraries (both simple and advanced). Beside of aggregation of metadata of already available objects, data describing objects planned for the digitization in the nearest future was also gathered. This allowed to provide automated reporting about digitization plans of Polish digital libraries and introduce innovative service for detection and prevention of duplicated digitization. Additional services provided by the federation provide the functionality of OAI identifiers resolution, reuse of a single user profile in many digital libraries and statistical reports about the aggregated metadata. Later on searching functionality was extended by the browsing possibilities, providing easy and attractive access to recently added, most frequently and most recently read objects. These features required also the extensions of the scope of aggregated information to include thumbnails of available objects.

After being well received by the end-users, the service attracted the attention of large scale international metadata aggregators like Europeana, DART-Europe and OCLC WorldCat and soon started to play the role of fully automated intermediary, facilitating the access to data about distributed digital collections of Polish memory institutions [6].

Figure 1 shows the growth of the number of metadata records aggregated by the federation during five years period of its operational activity. It can be noticed that the initial number of less than 100 000 metadata records increased tenfold during this time, reaching 1 million in the middle of June 2012. The number of data sources also increased, from less than twenty to over ninety.

Both the development of the federation functionality and the increase of scale and complexity of data aggregation operations, resulted in the evolution of the federation architecture, which is the subjects of the next section.

2.2   Evolution of the Federation Architecture

Initial architecture of the Digital Libraries Federation was three-tiered, consisting of web application layer, services layer and data layer (see Figure 2). The web application layer provided all the DLF portal functionality, basing on the services layer in which the data aggregation and processing was implemented [7]. The services layer was divided into two sets of services responsible respectively for metadata of available digital objects and metadata of objects planned for digitization. Initially each services set consisted of one core service implementing majority of business logic and three supporting services: authorization and authentication service, services registry and notification service used for asynchronous event passing between services. These services were using data layer – consisting of relational database (initially PostgreSQL, later Oracle) and Apache Lucene – for data storage.

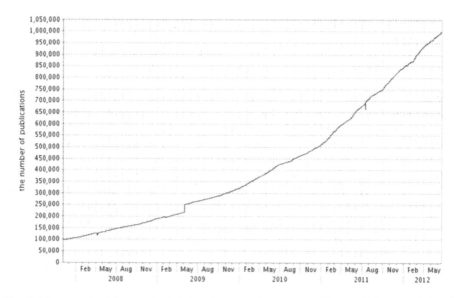

Fig. 1. The growth of the number of metadata records aggregated by the PIONIER Network Digital Libraries Federation (source: http://fbc.pionier.net.pl/)

Development of functionality provided by the federation portal and increased amount of data led to the division of the core service into the following services:

— Metadata aggregation and provisioning service – responsible for metadata aggregation, storage and access to the metadata (incl. basic validation and transformation operations).
— Metadata indexing service – responsible for the creation of full-text index of aggregated metadata used by searching service.
— Metadata searching service – responsible for providing the searching functionality, based on the full-text metadata index built by indexing service.

Such division allowed to separate on different physical machines the three most expensive operations: data aggregation and access, data indexing and data searching. This obviously allowed to increase   the overall performance of the federation portal. It also significantly improved the stability as the index creation process was separated from the index read requests, which were very intense during the peak traffic periods. Each night these two services used the lowest traffic time (around 4 AM) to synchronize indexes.

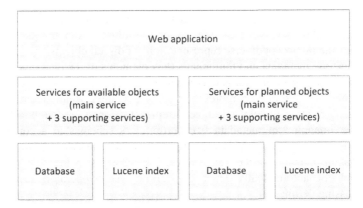

Fig. 2. Overview of the initial Digital Libraries Federation architecture

Later on, the development of digital objects' thumbnails aggregation functionality resulted in the creation of one more service responsible for downloading of the thumbnails or, in the case of objects without thumbnails, downloading of entire objects and automated generation of thumbnails.

The final architecture was in general very similar to the initial one, but the number of core services increased from one to four (plus three supporting services). Lessons learned from several years of development and production maintenance of the Digital Libraries Federation Service were taken into account while designing the new architecture in the frame of SYNAT project. This architecture together with its rationale is described in the next chapter.

3   New Design of Architecture for Data Aggregation, Processing and Provisioning

3.1   Desired Features of the Architecture

The desired features of the architecture for data aggregation, processing and provisioning are outcomes of two types of requirements – functional and non-functional – which are shortly described in the following paragraphs.

Functional Requirements. Analysis of data and functionality provided by selected services being in scope of the designed information system for humanities shown that very often the data sources are proprietary systems and do not have standardized data access interface (and there is no real chance to change this situation). The approach chosen during the design of Digital Libraries Federation was that each data source must be compatible with the OAI-PMH protocol [8] and must expose metadata in the Dublin Core schema [9]. For several data sources of the Federation it was very problematic to fulfill such requirements. Especially the data mapping from proprietary schema to the Dublin Core schema done on the data source side was causing troubles in production environment and very often was not updated when the proprietary schema was changed, resulting in loss of data. Therefore it was assumed that the new architecture should be extensible enough to support multiple, often unique data access interfaces and should accept diverse representations of digital objects.

For this purpose metadata and thumbnails (key elements of the data model of Digital Libraries Federation) were defined as possible kinds of representations of a digital object. Another example of such representation may be the textual content of the object. The transformation of selected objects' representations to a common form (e.g. the mentioned mapping of metadata from proprietary schemas to a common one) should be done for purpose of particular applications, as a process performed by the designed system and not by the data source. The changes in the source metadata schema should be automatically detected to allow system operators to review the related processing definitions.

Additionally there should be possibility to define simple transformations of representations and to join them into more complex chains (pipes).

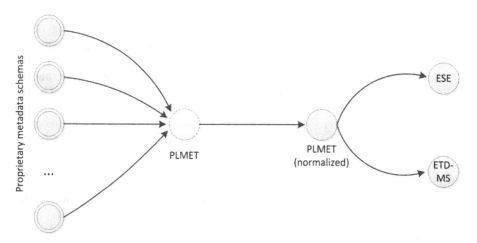

Fig. 3. Basic example of metadata transformation workflow

Example scenario for such requirements (depicted in Figure 3) is the two step transformation of heterogeneous metadata from digital libraries to several commonly used schemas:

1. Metadata expressed in proprietary schemas should be transformed to PLMET schema [10], with additional operations like values normalization (date, language, type, etc.) and cleaning. Metadata from various sources expressed in a common schema is a good base for building applications like Digital Libraries Federation portal.
2. Metadata expressed in PLMET schema should be transformed into Europeana Semantic Elements [11] schema for the purpose of cooperation with Europeana and in parallel into ETD-MS [12] metadata schema for the purpose of cooperation with services like NDLTD.

In the depicted scenario metadata expressed in proprietary metadata schemas is transformed to the normalized form of PLMET via an intermediary form of PLMET (dashed circle in Figure 3). Then, the normalized PLMET is the basis for further processing into ESE and ETD-MS metadata schemas. The non-normalized PLMET form does not have to be (and should not) be stored in the system, but it gives the possibility to reuse the component implementing PLMET -> PLMET (normalized) in several transformation workflows from proprietary metadata schemas to PLMET (normalized). The PLMET (normalized), ESE and ETD-MS are in such case schemas exposed to external systems.

Non-functional Requirements. One of the key non-functional aspects of architecture in large production quality systems are the scalability and robustness. Experiences from the Digital Libraries Federation shown that the most serious scalability issues are related to the increase of the amount of aggregated data or the complexity of data processing workflow. The scalability issues related to services provided to the end-user via web portal are significantly less problematic. Faults appearing in the Digital Libraries Federation operations are in majority of cases caused by technical problems or unexpected behavior on the data providers side. Real-life examples can be: hang-outs during data transmission, very long data provider response times or a situation when a digital library holding 100 thousand objects, which usually adds or updates up to 300 objects daily, suddenly reports that 90% of all objects were updated during single working day. The last case matches quite well the "burst compute" pattern [13] for cloud computing and at the moment administrator's intervention is necessary to deal with it.

Another important non-functional requirement which should be mention is interoperability. As written in the functional requirements section, the designed architecture should allow to aggregate data from heterogeneous sources, which expresses the need of interoperability in contact with data providers. The interoperability should be also assured on the level of interfaces which will be used by external systems willing to access the aggregated data, offering access in a technology neutral manner.

The metadata aggregation, processing and provisioning system should be also ready for several years long operation period possible without major architecture re-designs, therefore maintainability and extensibility, as well as operability, availability and reliability are additional important non-functional requirements. It should be possible to easily extend the system features in the areas of aggregation, processing and provisioning. The data aggregation and processing should be separated from data provisioning processes, to make possible to access the data even when data aggregation and processing services are not functioning for some reason or are overloaded with unexpected amount of data. On the other hand the increased number of end-users requests should not influence data aggregation operations. The system should also minimize the involvement of administrator in daily system routines, automatically performing all regular operations and alerting only when human intervention is really necessary.

Finally the security and privacy aspects should be mentioned. The legal context in which the designed system will operate, makes these aspects less demanding. In case of Digital Libraries Federation, the aggregated data is already available on the Internet, therefore the authorization is necessary only for create/update/delete operations and read operation may be initially done without special authorization. Although this may change in the future, therefore the system should be extensible enough to introduce full authentication and authorization mechanisms later on.

3.2   Description of the Proposed Architecture

The schema of proposed architecture for data aggregation, processing and provisioning is presented in Figure 4. It consists of five main layers described in the following paragraphs (in bottom-up order).

Source Data Layer. Consists of heterogeneous distributed services giving access to metadata and optionally also to data described by the metadata. Examples of such services are digital libraries (metadata and data) or library catalogues (just metadata, describing physical objects from library collections).

Data Aggregation and Processing Layer. In this layer three groups of components (services) can be distinguished. First group (on the right) is responsible for the aggregation of data. It consists of several agents – services which are able to extract metadata from different types of data sources. It was assumed that for each type of data source an agent will be developed capable to communicate with that data source. Of course one type of agent can also be compatible with several types of data sources (for example joint OAI-PMH and OAI-ORE [14] agent). Depending on the needs there can be several instances of agents of the same type deployed, resulting in a cloud of agent services, which are managed by component named agents manager. This component utilizes information from data sources manager and schema manager in order to effectively assign agents to data sources. It is also responsible for agents monitoring and collection of statistics from agents. New data source is added by registering it in data sources manager. Schema manager is responsible for information about data schemas used by particular sources and it enables the automated detection of changes in schema definitions.

Agents store aggregated source data in the component named objects database, which together with message broker forms second group of components in this layer. Objects database is a component responsible for storing data in the data storage layer (described below). It also sends notifications to the message broker about each modification performed on the aggregated data (create/update/delete operations). Other services can subscribe in the broker to receive particular types of notifications.

Such asynchronous notification is the basis for efficient processing of aggregated data, which is done by third group of components in this layer (on the left). Notifications about new or updated data is received by conversion manager. This service allows to define processing paths as shown in Figure 3. If the notification is related to creation or update of a metadata record which type is in the beginning of one of defined processing paths, such path is applied to this record. In order to do so, the

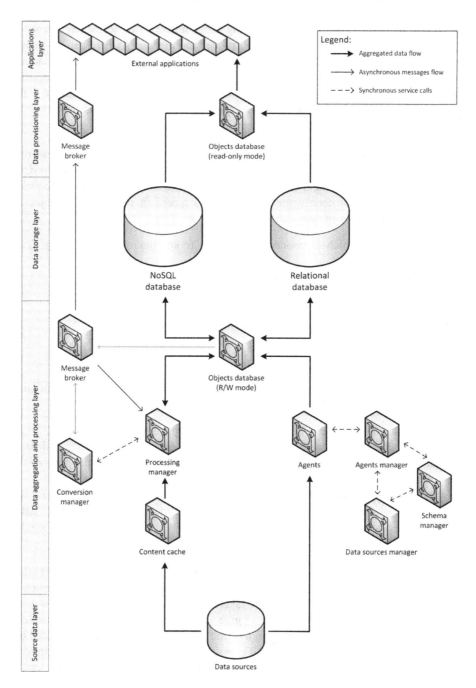

Fig. 4. Schema of proposed architecture for data aggregation, processing and provisioning

conversion manager sends proper processing order message to the message broker. Such messages are received by one of the processing managers (there can be many instances of processing managers) which are responsible for getting the metadata record from objects database, processing it according to chosen path and storing the outcome back to the objects database. If in some cases the processing requires download of additional digital content from the data source (e.g. in case of thumbnail generation), then optional content cache component may be used.

Data Storage Layer. The objects database in the previously described layer stores data in the data storage layer. This layer consists of two types of databases: NoSQL database and relational database. The relational database is designed to store only basic information about each aggregated metadata record or other kind of object's representation. This information includes:

— the id of object which is connected with the representation (provided by data source),
— the unique ID of the representation in the data storage layer,
— date of its first appearance in the storage layer,
— date of last modification of the representation,
— date of deletion (if the object was deleted and is no longer available),
— current checksum,
— data source,
— data schema,
— information if the particular representation is in its source form, or if it is an outcome of one of defined processing paths.

The NoSQL database should be used to store the representations of objects. Such mixed approach was chosen to allow easy selective queries on the aggregated data without the need of duplication of the data in the NoSQL database for the purpose of every possible selection criteria.

In the basic variant instead of NoSQL database, a distributed file system could be used, but the database gives significantly better possibilities for extension of the data model in the future (for example when authorization will be required on the single object's representation level).

Data Provisioning Layer. Access to aggregated and processed data is offered via the data provisioning layer. This layer contains duplicated instances of two components from the data aggregation and processing layer: objects database and message broker. Objects database uses data storage layer to give access to the data (in read-only mode). The second message broker copies from the first one messages about new, updated or deleted entries in the objects database and exposes them for interested external applications. Notifications generated by the conversion manager (and potentially other internal system notifications) are ignored by the data provisioning message broker.

Applications Layer. Applications layer is a place for external services which want to access resources aggregated and processed by the described system. Such applications can use services from the data provisioning layer. Examples of such applications are described in section 5 of this paper.

Selected Non-functional Aspects. In order to assure interoperability it was decided that REST protocols will be used to expose functionality provided by all system components, especially the objects database API. The multi-tiered architecture is constructed in a way which allows data provisioning layer and data aggregation and processing layer to work independently. For both these layers the data storage layer is necessary, but modern relational and NoSQL database system can be configured and deployed in a way which ensures very high availability (including database clusters with data synchronization between nodes).

Introduction of several parallel processing managers and data aggregation agents makes the system more immune to unexpected behavior of data sources, or data anomalies in the aggregated data. Beside it allows to achieve better scalability of the system, also because the workload distribution related to data processing is handled by the asynchronous messaging system, in which a single queue of messages (representing data processing tasks) can be serviced by a group of processing managers. The size of such group can be easily scaled up or down as needed without system reconfiguration – it is just a matter of subscription of particular processing manager in the message broker.

Additional instances of message broker and objects database make the security aspects easier to achieve, as the instances located in the data provisioning layer work in the read-only mode. Because of the use of REST protocols, such read-only mode can be achieved as easily as by proper firewall configuration (e.g. by not allowing HTTP requests using other methods than GET). The use of NoSQL database gives the possibility to introduce more complex authorization information in the future, but of course in such case adaptation of the objects database will be also needed.

Agents manager and conversion manager services makes the entire services infrastructure more maintainable and operable, as they were designed to monitor the data aggregation and processing components, provide overview of the current system state and alert administrators when needed.

4   Evaluation of the Designed Architecture with ATAM

The architecture described in the previous chapter was evaluated with Architecture Tradeoff Analysis Method (ATAM) by a group of PSNC employees involved in the SYNAT project and experienced in digital libraries domain. This chapter provides an overview of the evaluation process and notes the most important findings.

ATAM is a method for evaluating software architecture relative to quality attribute goals (chosen for example according to ISO/IEC 9126 standard [15]). Evaluation process consists of nine steps, which are divided into four phases:

— Presentation

1. Present the ATAM
2. Present business drivers
3. Present architecture

— Analysis

4. Identify architectural approaches
5. Generate quality attribute tree
6. Analyze architectural approaches

— Testing

7. Brainstorm and prioritize scenarios
8. Analyze architectural approaches

— Reporting

9. Present results

Presentation Phase. At the beginning of the ATAM process the evaluation leader presented the evaluation method to the gathered stakeholders and answer questions about the process. In the second step the project manager presented business drivers for the system. He showed currently working service (Digital Libraries Federation) and pointed out its disadvantages and limitations (most of them was mentioned in sections 2.2 and 3.1 of this paper). Then he described the concept of the new design and expected quality criteria, including interoperability, extendibility, performance and tolerance of incorrect or unexpected behavior of sources (see section 3.1 of this paper for more details). The only technical limitation imposed on the new architecture was the programming language – Java. In the third step the architect presented the architecture (described in section 3.2) using Kruchten's 4+1 model view [16]. This model consists of four views (logical, development, process and physical) and a set of use case scenarios for system components. These scenarios included:

— regular nightly new or updated data aggregation,
— addition of new type of data source,
— addition of new data source,
— addition of new type of data processing component.

Analysis Phase. In the next step architectural approaches were identified by the architect and afterwards the quality attribute utility tree was generated. A utility tree defines quality factors for non-functional requirements annotated with level of priority and difficulty and scenario for each quality factor. Selected elements from the utility tree are presented in Table 1 (grouped according to ISO/IEC 9126 characteristics). The most significant elements (i.e. those with the highest "importance x complexity" result) are underlined.

Analysis phase ended with step during which the architectural approaches were analyzed and evaluated against each of prepared scenarios. Key finding was that the usage of a NoSQL database system and limitation which it introduces on the size of representations of digital objects introduces a risk factor in the context of potential change of requirements in the future.

Table 1. Selected elements from the utility tree prepared during the ATAM session

Impor tance	Com plexity	Tree elements
		Functionality
		Interoperability
3	2	Aggregation of data from heterogeneous systems
3	2	Provisioning of metadata records converted to PLMET schema
2	2	Provisioning of thumbnails of digital objects
3	2	Provisioning of textual contents of objects
		Security
2	1	Authorization of data aggregation agents in the objects database
		Reliability
		Fault tolerance
2	1	Aggregation of data in case when some of the data source are not available
2	1	Aggregation of data from unstable sources
2	2	Resistance to invalid source data
		Efficiency
		Time behavior
3	2	All new data should be aggregated and processed in a daily cycle (during 24h period)
		Maintainability
		Changeability
3	1	Possibility to add new data source
2	2	Possibility to add new type of data source
2	2	Possibility to add new data schema

Testing Phase. Phase of testing involved the larger stakeholder group – all members of the PSNC digital libraries team. In the seventh step stakeholders expanded set of scenarios by adding 20 new ones and selected the most important of them in voting process. In the next step the architecture was evaluated against those chosen scenarios. The highly ranked scenarios elicited from participants are presented below with short statements about the outcome of their use for architecture evaluation:

— Failure of the relational database, resulting in inconsistency between databases (relational and NoSQL) - very serious, but unlikely event. Both databases (i.e. the data storage layer) play crucial role in the system and should be deployed with high availability in mind. Additionally the objects database component can provide the functionality of synchronizing the state of both databases. The contents of the relational database will be in such case the more significant source of information.

— Adding new data source with a large number of already available objects – typical "burst compute" scenario, which can be handled by the architecture because of the possibility to dynamically create dedicated data aggregation agent for such data source and to add new processing managers in order to make the processing faster. If there will be no possibilities to do so (e.g. because of lack of hardware resources), other option is the reconfiguration of the message broker to separate messages related to the new data source from messages related to other data sources. Such separate channel should allow to avoid starvation-like effect, where data from regular nightly updates waits for several days in the processing queue because of the large number of tasks generated earlier by new data source.

— A client wants to get all records from the system – the REST interface of the objects database was designed to deal with such cases, which appear for example in the data delivery to Europeana service. In such case the response will be partitioned and the client will be asked to submit a number of subsequent requests to get consecutive parts of the response.

— Large number of external clients sending requests at the same time (DDoS) – such scenario is very hard to handle in majority of cases. The suggested solution in this case is based on the assumption that the designed system will not be directly available to end-users, but only to trusted services deployed on the applications layer. If such trusted service will be under DDoS attack, the objects database located in the data provisioning layer may decide that the service generates too much traffic and have to be blocked for some time. To enable this additional traffic monitoring mechanism will have to be introduces in this layer. Fortunately, again because of the use of REST protocols as the API for objects database, external DDoS protection solutions designed for HTTP-based applications may be reused [17]. In the worst scenario the data provisioning layer may be disabled, which will not influence the data aggregation and processing activities realized in the lower layers.

— A source by mistake always reports that all records are modified – this case is handled by the objects database, as it stores checksums and modification timestamps for all stored representations. Particular data aggregation agent can use this information to initially validate the records obtained from the data source. Even if this is not done by the agent, such checking procedure is done also by the objects database each time the data update request is received. This allows to avoid further processing of data which is for example updated according to the modification timestamp, but has the same checksum as its previously submitted version.

— An unexpectedly large PDF file with high resolution image is used to generate thumbnail of some digital object, which may cause out of memory errors during data processing – this is a case, where data obtained by the client causes serious error in the processing manager. First of all good coding practices impose proper handling of such situation, so it should not cause the termination of the processing manager service, but rather a warning message in the system log. Such problematic piece of data should be of course skipped to allow processing of other tasks. Data stored in the relational database in connection with processing paths defined in the conversion manager allow to detect such skipped objects representations even if the warning message will not be handled by the system administrator for some reason. Such detection of unprocessed data could be added as a periodic task allowing to determine one of the parameters of the system health status.

Presentation Phase. At the end information collected in the ATAM process are was presented to the stakeholders. Several tips were noted for the components implementation stage.

The evaluation of the architecture with ATAM provided very valuable feedback for the system designers and implementers. The identified scenarios can be used as test cases on the stage of automated functional tests. During the evaluation several remarks were collected pointing issues which should be properly handled at the implementation stage of particular components. The evaluation confirmed that the architecture is well designed, as all the scenarios could be handled by it, with proper implementation of components. No architecture change requests were made.

It is also worth to notice, that some of the analyzed scenarios if occurred, would cause serious troubles in the present Digital Libraries Federation system, leading in the worst case to limited availability of this service or bottlenecks in data processing. This is a proof that the newly designed architecture has higher quality and better matches the requirements than the old one, which was designed several years earlier with significantly smaller amount of data in mind.

5   Test Deployment

To practically evaluate the proposed design, the described architecture was implemented and deployed in test environment. The following paragraphs shortly describe the deployment configuration, the test data aggregation process and two example applications deployed in the applications layer on top of developed prototype services.

Deployment Environment Configuration. The prototype implementation of particular components of the described architecture was deployed on a cluster of servers in virtualized environment. For particular layers of the architecture the following configuration was used (compare with Figure 4 for better overview):

— Source data layer – tenths of servers distributed all over the country – as data sources for the test deployment all the data sources of the Digital Libraries Federation and additionally NUKAT (Polish National Union Catalogue) bibliographic database [18] were used.
— Data aggregation and processing layer – two virtual servers:
  • first one for all data aggregation components, objects database and the message broker;
  • second one for all data processing components.

— Data storage layer – two virtual servers:
  • first one for relational database instance (PostgreSQL);
  • second one for NoSQL database instance (Apache Cassandra [19])

— Data provisioning layer – for the purposes of test applications described here objects database and message broker from the data aggregation and processing layer were reused.

Data Aggregation and Processing. The test environment was initially deployed in the middle of 2011 and after a year of operational activity, including nightly data updates and processing the objects, in June 2012 the database contained more than 8.5 million records aggregated from 75 sources and expressed in 77 different data schemas. Around 1.5 million of records were the outcome of processing of the data aggregated from Polish digital libraries to PLMET and ESE schemas (according to the workflow shown in Figure 3). The number of records which were marked as deleted during this period was around 5 thousand. After some initial adjustments in the hardware configuration, no serious performance problems were noticed during the data aggregation and processing operations in all this period.

Data Provisioning and Test Applications. Two applications were used to test the data provisioning facilities of the deployed prototype. Firstly, it was used by the knowledge base building services developed in PSNC in the A10 stage of the SYNAT project. Initial task of the periodically repeated knowledge base construction process done by these services is to download all metadata records aggregated from Polish digital libraries and all bibliographic data taken from the NUKAT catalogue and to process the data to the form of RDF triples [20]. Such mass download and processing operation lasts around 1 hour for one million of NUKAT bibliographic records and 1 hour and 54 minutes for 1.6 million digital libraries metadata records. Half of the downloaded digital libraries metadata records are expressed in PLMET schema and are downloaded in batches, in the same way as the NUKAT records. The second half are records expressed in the ESE schema, which are downloaded one by one (because of the present implementation of the knowledge base construction process). This is the reason why the total pace of digital libraries metadata records download and processing is around 14 000 records per minute, and the pace for NUKAT bibliographic records is 16 000 records per minute. This experiment shows that the designed architecture is capable of delivering large amount of data to real applications and that the various download possibilities offered by the objects database API (batch and single records download) are necessary.

The second developed application was aiming to provide the functionality of the full text metadata index using Apache Solr framework [21]. It is a popular open source enterprise search platform and a part of the Apache Lucene project. Core Solr features of full-text indexing and searching are provided by Lucene library, and the added value are REST-like HTTP/XML and JSON APIs which make Solr very easy to use. In the described application Apache Solr was used to store and search through metadata describing objects from digital libraries.

A special application, the indexing module, has been developed. This module retrieves XML documents from objects database (as a reaction to notification received from the message broker) and sends them to Solr server for indexing. The indexing module also reacts to the modification or deletion of an object in the objects database, to maintain consistency between full text indexes and the data storage layer. As mentioned above, metadata records from digital libraries are expressed in PLMET and ESE schemas, and encoded in XML. The transformation between XML data and Lucene full text index documents is made by Apache Tika toolkit. This toolkit was extended to be able to properly transform PLMET and ESE records.

The aim of the second application was to verify whether the long term data synchronization between external application (Solr in this case) and the data storage layer can be achieved on the basis of notifications sent out by the message broker. Practical implementation and deployment of application working according to this scenario confirmed this assumption.

6   Conclusions

This paper presented the new architecture for aggregation, processing and provisioning of data from heterogeneous scientific information services which was designed as a part of the SYNAT project. Rationale for this architecture features were described and strongly motivated by previous PSNC's experiences with the PIONIER Network Digital Libraries Federation. The architecture was verified theoretically, during ATAM session, and practically by prototype implementation and deployment. Additionally two applications were built on top of the deployed architecture, showing its usefulness and capability to deal with large amounts of data.

Future works include further development of the prototype components and deployment of new services on top of these services, this time in a production environment. The possible optimizations which can be also done in the future are related to more automated and intelligent management of aggregation agents and processing managers, with a focus on workload balancing, including automated deployment of new instances of such services in the cloud environment when needed.

Acknowledgements. Works described in this paper are financed as a part of the SYNAT project, which is a national research project aimed at the creation of universal open repository platform for hosting and communication of networked resources of knowledge for science, education, and open society of knowledge. It is funded by the Polish National Center for Research and Development (grant no SP/I/1/77065/10).

References

1. Górny, M., Gruszczyński, P., Mazurek, C., Nikisch, J.A., Stroiński, M., Swędrzyński, A.: Zastosowanie oprogramowania dLibra do budowy Wielkopolskiej Biblioteki Cyfrowej. In: Zeszyty Naukowe Wydziału ETI Politechniki Gdańskiej. Technologie Informacyjne, May 18-21. I Krajowa Konferencja Technologie Informacyjne, pp. 109–117. Wydawnictwo Politechniki Gdańskiej, Gdańsk (2003)
2. Mazurek, C., Stroiński, M., Werla, M.: Wdrażanie regionalnych bibliotek cyfrowych w sieci PIONIER w oparciu o środowisko dLibra. In: INFOBAZY 2005 – Bazy Danych dla Nauki, Gdańsk, September 25-27, pp. 58–64. Centrum Informatyczne TASK, Gdańsk (2005) ISBN 83-908112-3-5

3. Lewandowska, A., Mazurek, C., Werla, M.: Enrichment of European Digital Resources by Federating Regional Digital Libraries in Poland. In: Christensen-Dalsgaard, B., Castelli, D., Ammitzbøll Jurik, B., Lippincott, J. (eds.) ECDL 2008. LNCS, vol. 5173, pp. 256–259. Springer, Heidelberg (2008)
4. Dudczak, A., Mazurek, C., Werla, M.: RESTful Atomic Services for Distributed Digital Libraries. In: 1st International Conference on Information Technology, Gdańsk, May 18-21, pp. 267–270 (2008) ISBN 978-1-4244-244-9
5. Kazman, R., et al.: The architecture tradeoff analysis method. In: Proceedings of Fourth IEEE International Conference on Engineering of Complex Computer Systems, ICECCS 1998, pp. 68–78 (1998) ISBN 0-8186-8597-2
6. Mazurek, C., Mielnicki, M., Werla, M.: Selective harvesting of regional digital libraries and national metadata aggregators. In: The Proceedings of 9th ACM/IEEE-CS Joint Conference on Digital Libraries, Austin, TX, USA, June 15-19 (2009) ISBN 978-1-60558-322-8
7. Mazurek, C., Parkoła, T., Werla, M.: Building Federation of Digital Libraries Basing on Concept of Atomic Services. In: ACM/IEEE Joint Conference on Digital Libraries, JCDL 2008, Pittsburgh, PA, USA, June 16-20. ACM (2008) ISBN 978-1-59593-998-2
8. Lagoze, C., et al.: The Open Archives Initiative Protocol for Metadata Harvesting (2002), http://www.openarchives.org/OAI/openarchivesprotocol.html
9. Dublin Core Metadata Terms, http://dublincore.org/documents/dcmi-terms/
10. PLMET Metadata schema, http://dl.psnc.pl/community/display/FBCMETGUIDE
11. Europeana Semantic Elements Specification (ver. 3.4) (2011), http://version1.europeana.eu/c/document_library/get_file? uuid=77376831-67cf-4cff-a7a2-7718388eec1d&groupId=10128
12. ETD-MS: an Interoperability Metadata Standard for Electronic Theses and Dissertations version 1.00, rev. 2, http://www.ndltd.org/standards/metadata/etd-ms-v1.00-rev2.html
13. Rosenberg, J., Mateos, A.: The Cloud at Your Service. The when, how, and why of enterprise cloud computing (November 2010) ISBN 9781935182528
14. Open Archives Initiative - Object Reuse and Exchange. ORE Specifications and User Guides - Table of Contents, http://www.openarchives.org/ore/1.0/toc.html
15. ISO/IEC 9126 Standard: Software engineering – Product quality – Part 1: Quality model, http://www.iso.org/iso/iso_catalogue/catalogue_tc/catalogue_detail.htm?csnumber=22749
16. Kruchten, P., Lago, P., van Vliet, H.: Building Up and Reasoning About Architectural Knowledge. In: Hofmeister, C., Crnković, I., Reussner, R. (eds.) QoSA 2006. LNCS, vol. 4214, pp. 43–58. Springer, Heidelberg (2006)
17. Mirkovic, J., Reiher, P.: A taxonomy of DDoS attack and DDoS defense mechanisms. ACM SIGCOMM Computer Communication Review 34(2), 39–53 (2004)
18. Werla, M.: Możliwości wykorzystania katalogu centralnego w sieci rozproszonych bibliotek cyfrowych. In: Konferencja Rola Katalogu Centralnego NUKAT w Kształtowaniu Społeczeństwa Wiedzy w Polsce, Warszawa, January 23-25 (2008)
19. Apache Cassandra project website, http://cassandra.apache.org/
20. Mazurek, C., Sielski, K., Stroiński, M., Walkowska, J., Werla, M., Węglarz, J.: Transforming a Flat Metadata Schema to a Semantic Web Ontology: The Polish Digital Libraries Federation and CIDOC CRM Case Study. In: Bembenik, R., Skonieczny, L., Rybiński, H., Niezgodka, M. (eds.) Intelligent Tools for Building a Scient. Info. Plat. SCI, vol. 390, pp. 153–177. Springer, Heidelberg (2012)
21. Apach Solr project website, http://lucene.apache.org/solr/

Author Index

Printed in the United States
By Bookmasters